Green Energy and Technology

Climate change, environmental impact and the limited natural resources urge scientific research and novel technical solutions. The monograph series Green Energy and Technology serves as a publishing platform for scientific and technological approaches to “green”—i.e. environmentally friendly and sustainable—technologies. While a focus lies on energy and power supply, it also covers "green" solutions in industrial engineering and engineering design. Green Energy and Technology addresses researchers, advanced students, technical consultants as well as decision makers in industries and politics. Hence, the level of presentation spans from instructional to highly technical.

Indexed in Scopus.

Indexed in Ei Compendex.

M. Ziya Sogut • Murat Koray
Editors

Energy Rationality and Management for Decarbonization

Springer

Editors
M. Ziya Sogut
Maritime Faculty
Piri Reis University
Tuzla, Istanbul, Türkiye

Murat Koray
Maritime Faculty
Piri Reis University
Tuzla, Istanbul, Türkiye

ISSN 1865-3529 ISSN 1865-3537 (electronic)
Green Energy and Technology
ISBN 978-3-031-68496-8 ISBN 978-3-031-68497-5 (eBook)
https://doi.org/10.1007/978-3-031-68497-5

This Springer imprint is published by the registered company Springer Nature Switzerland AG
The registered company address is: Gewerbestrasse 11, 6330 Cham, Switzerland

Contents

1 **A Phenomenon of Social Transition: Behavioral Responsibility for Energy Efficiency** 1
M. Ziya Sogut

2 **Exergy Rationality in Sustainable Decarbonization** 19
Birol Kılkış

3 **Implementing Cold Ironing Methods in Georgian Ports** 77
Rati Janjghava, Avtandil Tsitskishvili, and Givi Tsitskishvili

4 **Study of Combustion Characteristics and Prediction of Cyclic Variability of a Diesel Engine Fueled by Blends of Hydrotreated Waste Cooking Oil and Diesel Using Artificial Neural Network** 83
Ankit Sonthalia, Naveen Kumar, Mukul Tomar, Edwin Geo Varuvel, and S. Thiyagarajan

5 **AI and Energy Management in Organizations: A Systematic Literature Review** 107
Abdülkadir Akturan and Aykut Arslan

6 **Exploring Energy Management Practices in BIST 50 Enterprises: Insights and Implications** 125
Aykut Arslan

7 **Design and Analysis of a New Double-Flash Geothermal Power Plant Integrated with Multistage Flash Desalination and Reverse Osmosis Desalination Units for Power, Fresh Water, and Hot Water Generation** 143
Fatih Yilmaz and Resat Selbas

8 **The Role of Innovative Ship Design in Decarbonization** 159
Ahmet Taşdemir and Mehmet Akman

9 **Post-COVID-19 Energy Transitions: Strategies for Global Recovery** ... 177
Islam Md Rizwanul Fattah, Ibham Veza, Selçuk Sarıkoç, Shurooq T. Al-Humairi, Oyetola Ogunkunle, and Hafiz Muhammad Ali

10 **Solar Photovoltaic Systems: Assessing Their Impact on the Advancement of Marine Decarbonization** ... 195
Gülşen Akın Evingür and Büşra Osma

11 **Carbon Management Based on Circular Chain in Decarbonization and Economic Sustainability** ... 207
M. Ziya Sogut and Oral Erdogan

12 **Investigation of Organic Rankine Cycle for Power Generation from Marine Diesel Generator Waste Heat** ... 225
Cüneyt Ezgi, Haydar Kepekci, and Mehmet Ziya Sogut

13 **Development and Performance Analysis of a Solar-Powered Thermoelectric Air Conditioning System for Space Conditioning** ... 249
Manuel Fernando Ariza Taba and Koray Ülgen

14 **An In-depth Analysis of the Energy and Exergy Performance of Radiant Heating Integrated with Ground Source Heat Pumps in Building Environments** ... 273
Ahmet Dogan, Nurullah Kayaci, and Hakan Demir

15 **Energy, Exergy, Exhaust, and Economic Attributes of a CI Engine Powered with Binary Blends of Watermelon Seed and Pine Oil, with Diethyl Ether as a Fuel Additive** ... 287
Manikandaraja Gurusamy, Balaji Subramaniyan, Malarmannan Subramaniyan, Olusegun D. Samuel, Christopher C. Enweremadu, and Selçuk Sarıkoç

16 **Management of Energy Efficiency and the Environment for Process Cooling Through the Use of Free Cooling** ... 311
M. Ziya Sogut

17 **Assessment of Thermo-Economic Performance with Sustainable Heat Management in Trigeneration Processes** ... 337
Özay Kas and M. Ziya Sogut

18 **Pneumatic Propulsion of Vehicles** ... 355
Abdul Hai Alami, Mohamad Ayoub, Adnan Alashkar, Fatima Alshamsi, Ayman Mdallal, Siren Khuri, and Selçuk Sarıkoç

19 **Navigating Toward a Greener Future: An Analysis of Emission Reduction in Shipping** ... 373
Igor Gritsuk, Andrii Golovan, Iryna Honcharuk, and Rima Mickiene

20 Modification of Low-Speed Marine Diesel Engine for Slow Steaming Operation 381
Dimitar Popov

21 Sustainability-Based Assessment of Aircraft Pre-conditioned Air Supplying Systems in Green Transition 393
Enver Yalcin and T. Yiğit Gencel

22 Can Green Energy Transition Be Achieved Despite Investment Disputes? 401
Ceyda Süral Efeçınar and Başak Başoğlu

23 Investigation of Energy Efficiency Practices in Buildings and Determination of Roadmap for Zero-Energy Buildings Target 413
Umit Unver

24 Energy Efficiency in the Road Transportation Sector 427
A. Murat Köseoğlu

25 Driving Sustainable Energy Efficiency in Industry: An Investigative Approach 445
M. Ziya Sogut

26 Walk-Through Audit Model for Enterprise's Decision Processes Regarding Energy Management 467
M. Ziya Sogut

27 Analyzing CO_2 Emission: A Case Study of Batumi State Maritime Academy 491
Guram Vadachkoria, Avtandil Tsitskishvili, and Givi Tsitskishvili

28 The Analysis of Climate Change Awareness and Environmental Attitudes at Batumi State Maritime Academy 497
Iakob Makharadze and Givi Tsitskishvili

29 Advancing Sustainable Aviation: Integration of Emerging Energy Technologies in Aviation Infrastructure 507
Omar Alharasees, Utku Kale, and Daniel Rohacs

Index 523

Chapter 1
A Phenomenon of Social Transition: Behavioral Responsibility for Energy Efficiency

M. Ziya Sogut

1.1 Introduction

The relationship between energy and environmental processes has had a direct impact on both processes and lives, serving as an important indicator of social life. The connection between human beings and energy resources has brought about technological change and interaction among all components and processes in the social structure (Kiatcharoenpol & Sirisawat, 2020). The development of urbanization culture has significantly impacted essential tools and transformed the relationship between humans and energy into an addiction. Societies have become increasingly dependent on energy due to changing vital needs and advancements in technology, resulting in a constant increase in demand. As sector players, they have constantly renewed the types of demand that have direct impact indicators in society due to the phenomenon of energy. This has triggered an intense waste in the use of energy-based vehicles (Zhang et al., 2023). Although the development of effective tools in social life was triggered by this change, which was mostly caused by oil, it also led to increased environmental pollution and the rapid extinction of life forms, with all its consequences. The effect described has resulted in a dual threat to both social development and ecological life. On the one hand, there is the disappearance of forms and pollution, while on the other hand, there is an increasing demand for urbanization, leading to uncontrollable destruction of agricultural areas and the destruction of natural habitats due to the inability to control migration mobility (Anderies et al., 2022).

It is the responsibility of human beings to protect the environment they live in, particularly by preventing the destruction caused by oil-based pollution. The

M. Z. Sogut (✉)
Marine Engineering, Piri Reis University, Tuzla, Istanbul, Türkiye
e-mail: mzsogut@pirireis.edu.tr

M. Z. Sogut, M. Koray (eds.), *Energy Rationality and Management for Decarbonization*, Green Energy and Technology,
https://doi.org/10.1007/978-3-031-68497-5_1

consumption processes and industrial mobility of today's world, which are developing due to increasing competition, should be directed toward this goal. The introduction of new frameworks has been brought to the fore due to the development of environmental awareness in technology, energy, raw materials, and production, which is considered a valuable aspect of social development. Today, the competitive and technological advancements that trigger change necessitate an economic shift from raw materials to all expenses in production. The development of smart materials by human beings has introduced new features that support this change, such as recyclability (Barr et al., 2021). However, existing technologies have irreversible burdens that negatively affect environmental sustainability. This shows that compatibility with the environment requires a responsibility in a manageable structure. However, the manageability of energy demand in all processes and life cycles from production to product produces an advantageous result. Comfort processes that have developed and changed since the Industrial Revolution have triggered an increase in the demand for energy in every direction from production to consumption. This can also be considered a direct threat to economic and environmental sustainability. However, the manageability of all these means of change also triggers an effective responsibility for human beings.

Toward the end of the twentieth century, the depletion of the ozone layer highlighted the longer-term issue of global warming and climate change. The role of human activity in environmental pollution, caused by uncontrolled growth, has unfortunately accelerated the onset of unavoidable and undesirable changes. The challenges presented by these threats have raised awareness among humans regarding the environment and the need for action. The United Nations Framework Convention on Climate Change, held in Rio de Janeiro in 1992, can be considered the first international response to this issue (Almedia et al., 2024). This protocol, which has unfortunately become ineffective, has not succeeded despite being signed by approximately 160 countries. The process reveals that responsibility and management are crucial not only for raising awareness but also for developing a sense of common consciousness and behavioral frameworks (Stewart et al., 2021).

Today, various tools have been developed to address this issue. Institutional measures have been implemented to promote energy efficiency. The concept of a "green call," which has international frameworks, has gained global recognition due to the effects of ozone depletion. Additionally, decarbonization, a current conceptual value, is primarily reflected in national goals. However, this concept has also led to effective changes for individuals at all levels, particularly in social life. Nevertheless, it can be argued that there is a conceptual gap in all processes. For instance, the development of change tools and effective control measures against obstacles to change have emerged in this context. However, it is imperative to acknowledge that human agency plays a pivotal role in effecting this transformation. In this respect, focal points or basic indicators for the realization of social change come to the fore. But above all, the focus is on people. Indeed, individual responsibility and the social change processes that will develop accordingly will directly affect both technology management and demand management. All approaches that target decarbonization, such as the 2050 and 2100 scenarios that will trigger social change today, show that there will be a continuous agenda in the future of human beings. However, it would be a realistic process to consider a gradual change in all these expectations and social change. Despite the tools that are still not sufficiently understood and technology that is changing rapidly, the emerging concept of green transition and energy efficiency can be seen as pioneering steps for this process (Wenyao et al., 2023).

Energy is one of the basic inputs in life process analysis, which is an industrial process in the world, from raw materials to products and from products to waste or garbage. In terms of the structure of technology, it can be seen that many control tools that trigger change are developed at each stage of these processes (Ronei de Souza Teixeira, 2024). However, individual awareness and responsibility are crucial components for supporting change and development in the sustainability and effectiveness of technology. It is evident that numerous initiatives and training programs have been developed to address this structural need. Awareness training, particularly in relation to energy efficiency, has also been conducted. Research and its results are often presented in sectoral and scientific studies. For example, Gaspar and Antunes (2011) examined consumer behavior in energy conservation and waste management and investigated directly related variables. Frederiks et al. (2015) examined the impact parameters in the vital areas of individuals. In the study, which was discussed together with the literature, individual responsibilities were evaluated, especially in household life processes. Another study is the development of focal point responsibility in households. In this respect (Abrahamse et al., 2005), it was considered a behavioral assessment of the use of thermostats, lighting control, and air conditioning.

Energy management and sustainability have become a process that needs to be evaluated together. In this respect, thermal comfort, which has developed as an indicator, has developed as a sought-after feature in the modern world. However, in modern societies, energy and thermal comfort have developed as vital needs as features. However, the debate on energy poverty that has emerged in recent years has also contributed to this. As a matter of fact, according to the definition of the European Union (EU), energy poverty is defined as a measure that will adversely affect the health and well-being of a household. He attributed this to three main reasons: households spending a significant proportion of their income on energy, low income that does not meet vital needs, and low energy performance of energy-consuming systems. As a matter of fact, according to the EU data, the rate of insufficient energy use, which was 8% in 2020, increased by 16.25% in 2022 (EU, 2023). (Wang et al., 2016, 2017). Indeed, inefficiencies in energy use behaviours also trigger this process. The development of an institutional framework is contingent upon a shift in consumer behaviour, which should be regarded as a primary and attainable objective. This study has been prepared primarily with a responsibility based on developing it. A framework for the development of a culture of energy efficiency behavior as a phenomenon in social development and change was presented (Ulrika et al., 2014). Then, the effect indicators of this were revealed by sampling.

1.2 Energy Efficiency Behavior Concept

The concept and relationships of behavior in energy efficiency directly affect the effective and sustainable management of energy, and this behavioral ability is a process of action that directly supports change within service and production activities. While improving the efficiency of all defined structural processes, it is also based on reducing costs and environmental pollution. In this respect, energy efficiency for corporate and individual processes can be expressed as reducing energy consumption and costs that support environmental sustainability without changing the quality of production and service in the directly defined process. Energy management in corporate structures and businesses is a necessity that has the potential to engender structural change in corporate behaviours. In addition to energy efficiency, expectations support the development of a management model that will enable this process to become a behavioral culture (Ulrika et al., 2014). First of all, this should be seen as a vital need in individuals and should form the framework for individuals to return to a desired behavior. This situation is the transformation of desired behaviors that highlight a sustainable change in social life into psychomotor behavior while directly developing awareness.

Adult and trained manpower, which is needed in social life and the business world, create a holistic environment that uses its direct needs in an effective and efficient structure and evaluates all tools for these action processes in an integrated manner. In this context, the concept of behaviour based on energy efficiency (EE) should be evaluated and developed with integrity within the formal structure of adult education to be given for institutional structures, and with effective action plans. The concept of behavior, which has four action processes, and its flow diagram are given in Fig. 1.1.

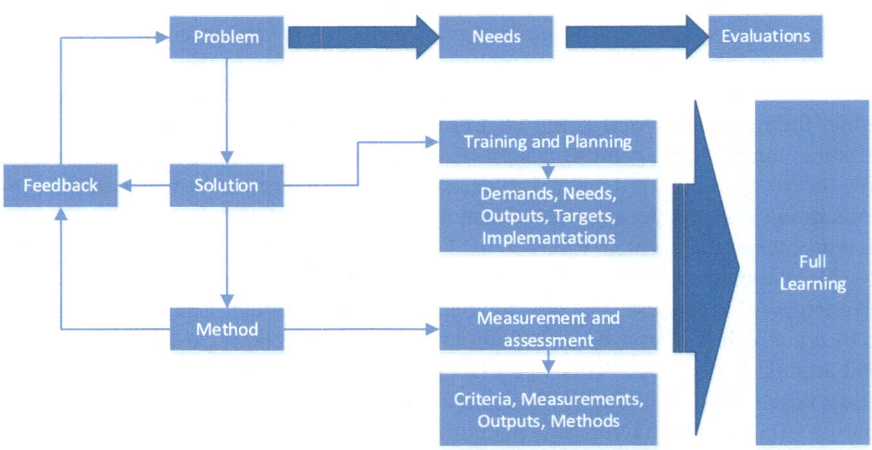

Fig. 1.1 Concept of behavior in adult education

Vocational education in behavioral science is a framework based on providing the individual with the desired behavior, especially the needs of the labor market. While this framework is based on gaining competence in business life, it considers the basic functions of providing the person with professional knowledge, skills, and work habits and developing psychomotor skills. In fact, this directly serves the purpose of raising desirable individuals in terms of national employment targets. For example, in the 2012–2014 National Employment Strategy Action Plan prepared by the Ministry of Labor in Turkey, the first policy axis is defined as "Strengthening the Education-Employment Relationship." The main purpose of this axis is is stated as "increasing the adequacy of education and training to meet labour market needs and making it accessible to all." A robust strategy has been developed to achieve this objective, encompassing policies for vocational education that are aligned with the fundamental principles outlined.

An accessible education system in the modern world begins with preschool education. This is the basic condition for gaining a qualification, and it is also related to the quality of vocational education. However, this educational process is shaped by many structures and disciplines, along with lifelong learning. Therefore, creating accessible educational environments and encouraging continuous and sustainable education are critical points. In modern education, there has been a growing emphasis on entrepreneurship. This has led to the development of new approaches that can be trained and improved in the active workforce. These approaches support the development of policies and managerial organizations that strengthen the labor market and highlight solutions that increase individual responsibility in the educational structure (Kenar, 2012).

Vocational education is actually shaped as a need of the industrial society and has the tools to support the development of the industry in terms of social roles. This training aims to ensure that the success of the training output is possible by defining the training objectives, goals, and behaviors of the technical personnel depending on the needs of each occupational group in the labor market. Professional qualifications in the labor market relate to the hierarchy of employment in the production flow and have job descriptions ranging from qualified personnel to engineers. In the professional sense, an individual's ability to work can be graded as the accumulation of knowledge from practice to theory, depending on the occupational hierarchy. The definition of this need is given in Fig. 1.2.

This flow chain is actually a result of the individual's level of education. These cover a process that starts from apprenticeship and continues to engineering education. In reality, the framework of education is related to levels and competencies. The training process shapes how the training will be carried out through basic structures such as resource use and operational energy efficiency, as well as the individual's process management and operational business competence. The basic principle of education is that educational programs are created according to the objective-content-learning and teaching-process-evaluation cycle, depending on the level. Considering the occupational group in vocational education and engineering education processes, this cycle should be the same for each individual. For this reason, there should be structural integrity in energy efficiency training. In this context, occupational qualifications based on occupational levels have been established in the labor market and are presented in Table 1.1.

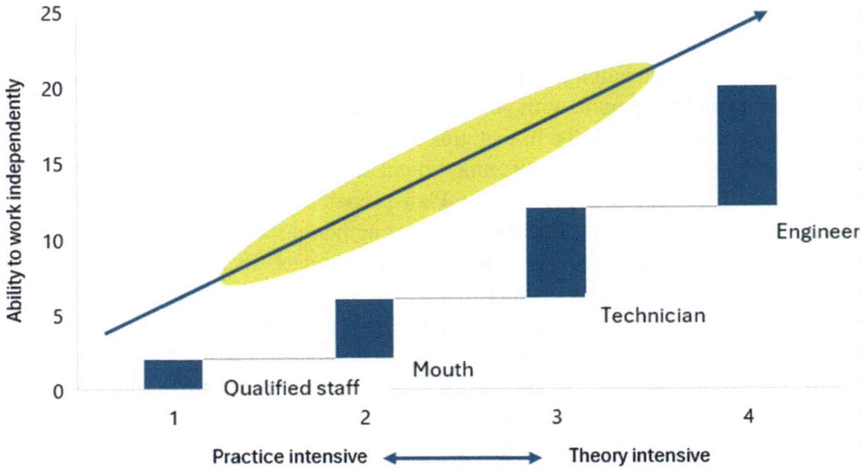

Fig. 1.2 Structure of qualified personnel required by the job in terms of competence in workplaces (Karcı, 2008)

Table 1.1 Qualifications related to professional levels (Sogut & Tasdoven, 2012)

Degree	Educational background	Staff	Qualifications
1	Primary	Apprentice Qualified staff	They are the employees who have the knowledge and ability to recognize simple operations in the labor market
2	Primary or vocational high school	Mouth Technician	They are the staff who can read and prepare documents related to the field they specialize in and the process they do and who can set up the system, make it operational, perform system fault detection and troubleshooting, produce simple system solutions, and read projects
3	Vocational school	Technician Foreman Maintenance manager	They are the employees who manage a group of qualified personnel or technicians; have coordination skills depending on the nature of the job; have time management skills; have programming, algorithm setup, and decision-making skills; and can solve problems at the cognitive level
4	Faculty	Engineer	They are the staff who have strong theoretical field knowledge, can design a system, and have the ability to develop a system, solve a problem and make analytical examinations, manage a production system and process, and perform and manage administrative planning, programming, and strategic analysis

In the labor market, training processes based on professional demands are questioned with qualifications linked to four levels. However, each degree has been handled in terms of target competencies, with programs developed and validated in the educational processes. For this purpose, while each class from nonformal education to formal education determines the level of behavior within itself, the desired behavior change creates an environment related to the processes created to provide them. However, adult education is also associated with a subject and process, and competencies are levels of education created for certain hierarchies. In educational sciences, this process is defined by a taxonomy linked to a matrix of needs. In the training process, target taxonomies were created depending on the competencies developed for the defined professional levels, the level at which they were defined, and the outputs in which the level of education to be taken as a basis for the outputs were defined and determined by Table 1.2.

In training programs, competencies are handled directly together with target taxonomy and behavioral competencies. In this context, simple action behaviors, knowledge of recognizing tasks, and an ability to do things refer to the level of knowledge at the cognitive level, the level of receiving at the affective level, and the level of arousal at the psychomotor level for qualified personnel. Adult education in vocational education, on the other hand, starts from the apprenticeship level and is

Table 1.2 Target taxonomies of degrees (Sogut & Tasdoven, 2012)

Cognitive Level	Degree				Affective Level	Degree				Psychomotor Level	Degree			
	1	2	3	4		1	2	3	4		1	2	3	4
Knowledge					Obtaining					Excitation				
Comprehend					Reacting					Do it in a grid control				
Implementation					Valuing					Skill				
Analysis					Organizing					Fitting to the Situation				
Feel					Personality					Creation				
Assessment														

shaped according to the needs together with the field expertise. In this context, a behavioral concept has been created in energy efficiency and energy management, which is shaped according to the needs and shaped by the field competencies based on the field competence, together with the field competence.

A concept has three stages: problem definition, needs, and solutions. For this systematic approach, it consists of the basic steps such as problem definition, solution proposal, method, application, and evaluation stages. In particular, the training processes defined in the process of creating the method usually consist of a discipline of action at the points created by the problem. The second stage of this concept is an effective process defined as training planning. The main point in training planning is the definition of needs and the determination of competencies, especially for the identified needs. Competencies are the educational outcomes expected from individuals and the results obtained for the problems to be created. In reality, formal education is shaped in two structures. These can be expressed as deduction or induction. However, adult education is a structure created to solve a problem or problems. For this reason, in such training models, it is more important to define the desired competencies first. This approach, which is generally shaped depending on the professional field, is handled directly within the professional competencies, and this is the result of a hierarchical order. For this purpose, a program framework that is not included in professional qualifications and that will make energy efficiency and especially energy management systems valuable for corporate structures has been evaluated. For this purpose, competencies for field energy efficiency and management have been developed, and job responsibilities related to the professional hierarchy have been examined (Table 1.3).

Table 1.3 Task taxonomy of energy efficiency behaviors

Degree	Cognitive level	Affective level	Psychomotor level
1	Recognizes energy systems	Understands the system and its functions	Assembles or operates as shown
2	Recognizes energy systems	Understands the system and its functions, comprehends its relationships	Understands the system and its functions, comprehends its relationships
3	Recognizes energy systems, knows the properties of systems	Understands the system and its functions, comprehends its relationships	Makes energy system capacity calculations and system assemblies, checks system settings, selects system elements, programs automatic control systems, makes and manages work programs, manages multidisciplinary processes
4	Recognizes energy systems, knows the properties of systems	Understands the system and its functions, comprehends its relationships	Makes capacity calculations of energy systems, modifies or develops systems, performs optimization analyses in elements, plans management and organization, makes cost analysis, plans or makes program algorithms, automation, and programming

The concept of behavior in the energy management system should develop a target-oriented standard for the desired behaviors expected from individuals in all energy-related processes. In this holistic energy management, the responsibilities of the individual are concrete behaviors that are directly related to the degrees described above. For this purpose, all relevant elements must be based on the competencies of all individuals defined in the labor market. It will be easy to define this concept through a defined program in enterprises with the ISO 50001 system. However, for businesses that do not depend on it, creating a concept is associated with a process. For the subject, where the program will be implemented depending on the concept, a summary of the training and a processing matrix should be created.

The team responsible for energy management must first address how to read the results linked to these evaluation criteria. These evaluation criteria and methods are indicators of the vision of the energy management team. There are some methods developed especially for management culture in the literature. Wising et al. (2015) developed such a method and considered this evaluation in eight criteria and five levels in their study. They meet the criteria in question: visibility, responsibility, collaboration, targeting, commitment, motivation, learning, and progress (Wising et al., 2014).

Wising et al.'s assessment of behavioral culture can be considered an individual criterion. However, for each individual in the labor market, production- and service-related metrics can also be developed. In this study, it was evaluated that the behavioral culture should contain concrete data related to production and service. The measurement and evaluation criteria to be made are discussed in this respect, and the criteria depend on the degree of labor force in the study: awareness, responsibility, efficiency, comprehension, management, and learning. However, it is necessary to examine the evaluation process in two ways. This concept of behavior has two basic elements. These are the acquisition and practice of behavior or the acquisition of psychomotor behavior. Energy management should make an assessment that takes these two aspects into account in such training, and at the end of each program, it should question the training process in terms of production, efficiency, time management, financial savings, and emission savings.

1.3 Results and Discussion

In this study, the concept of energy efficiency behavior is discussed in two ways. The first is a survey study of the defined taxonomy of behavior. The other is the assessment of the operational competence of energy efficiency behavioral competence. Sectoral samplings developed in this context provided an evaluation. First of all, the questionnaire evaluation is discussed. For this purpose, the behavioral criterion based on permanent behavior in individuals is discussed with six behavioral criteria. Individuals involved in energy management in the examined enterprise were evaluated depending on the developed concept. The third stage of the behavior concept includes the measurement and evaluation process.

Measurement tools for the level of knowledge in educational processes are readiness and performance evaluation tests. The criteria for psychomotor behavior are process-related practical tests. Both criteria should be done by scoring. However, a reference point must be set for all measurement processes. Below is a sectoral sampling of education according to the energy efficiency flow matrix. Full learning is aimed at the training to be given. In this context, the evaluation was handled with a scoring criterion of 5 and defined as 0: very poor, 1: poor, 2: inadequate, 3: medium, 4: sufficient, and 5: good. The same criteria will be used for practical tests. In this context, following the determination of a minimum of four as the success criteria, the evaluation of the business can be conducted through the utilisation of either a test or a survey. Evaluation criteria and success targets depending on the characteristics of the company are given in Fig. 1.3.

Evaluation criteria should be tailored to a business. In this context, readiness levels are defined at the test 1 level, especially for the personnel to be trained. However, for a basic level of personnel, businesses should also take into account evaluations or criteria based on energy efficiency in their personnel preferences. These criteria can also be taken into account during the adaptation process. As with any training program, energy education should include a measurement and evaluation system. In this study, a case study was developed for the structure of which the concept and matrix are described above. In this context, an energy efficiency training matrix for an enterprise was planned over 65 blue-collar first- and second-degree personnel, and energy efficiency training was implemented. As a result of the current situation analysis, the process reached in the theoretical training and the practical training before the training, tests with a total of 60 questions were applied. Depending on the readiness test, the process was implemented with an active learning method gradually with motivation and presentations including the objectives of the training, necessity, and needs assessment. On the second day, short-term training and practice were developed for the psychomotor behaviors of the training. At

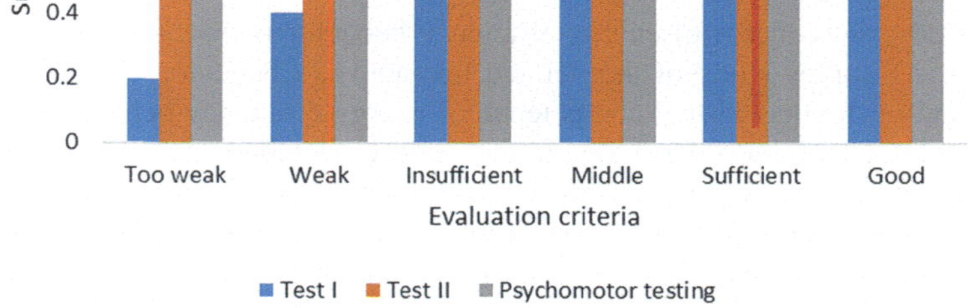

Fig. 1.3 Energy efficiency assessment criteria and achievement targets

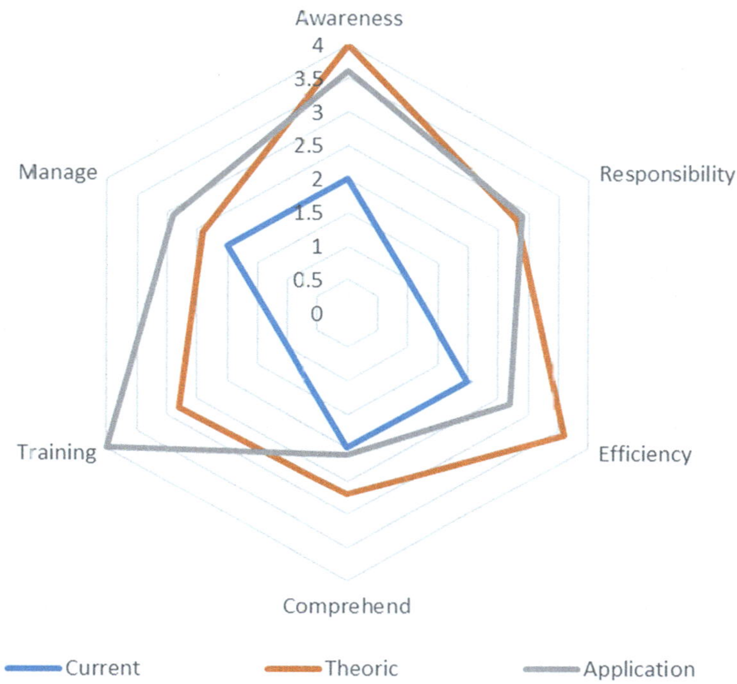

Fig. 1.4 Energy efficiency training program behavioral measures

Table 1.4 Exams and target deviation results

Behavioral criteria	Results		Deviations		Explanation
	Targeted	Application	Targeted	Application	
Awareness	4	3.6	0	0.1	Comprehend
Responsibility	2.8	2.9	0.3	0.275	Comprehend
Efficiency	3.6	2.7	0.1	0.325	Application
Comprehend	2.7	2.1	0.325	0.475	Comprehend
Training	3.7	4	0.075	0	Cognitive
Manage	2.4	2.9	0.4	0.275	Application

the end of the day, the process was completed with psychomotor tests to evaluate behaviors. During this process, behaviors were questioned separately for each individual. The training results developed in this context are given in Fig. 1.4.

Energy management is the permanent behavioral change expected in education processes. In this respect, the evaluation results of each training should be compatible with the goals and objectives of the training. The principal aim of the training programme is to achieve a minimum of four levels of proficiency at the cognitive and psychomotor levels. However, the results did not fully achieve the desired success in the program. First of all, deviations in behavior should be taken into account. According to these evaluations, the levels and deviation results provided for each criterion are given in Table 1.4.

According to the results of measurement and evaluation, although the desired goals in learning and awareness have been achieved, the desired goal has not been achieved at other points. In this case, it is necessary to reevaluate, especially theoretical and psychomotor training and the objectives related to this program. Depending on the goals and behaviors it defines, energy management should repeat such programs for all structures. However, all these structures should be defined in advance in the annual training program, and feedback should be given according to the results of measurement and evaluation during the training process.

Energy efficiency is an important gain in an energy management process. In this context, the individual responsibilities of energy efficiency include inquiries that will trigger change. For this reason, individuals who have gained energy efficiency behavior also improve their questioning skills in the operational processes of businesses. As a matter of fact, studies show that the right choices, especially in operational capabilities, provide up to 15% efficiency in systems. Energy efficiency behaviors aim to evaluate important energy users for many systems in operational processes. Below is a case study of a textile enterprise.

The enterprise examined in this study is a carpet factory with a total production of 10.56 million m^2 and an average total consumption value of 53.13 GWh, with an annual energy consumption of 36.09 GWh of electricity and 17.06 GWh of natural gas. The energy density of the operator was found to be 5.03 kWh/m^2. Sectoral predictions show that the unit energy density in such enterprises is 3 kWh on average. Although the carpet business shows different characteristics, it is seen that it has different consumption behaviors in energy consumption. In this context, first of all, the 3-year average energy intensities of the enterprise were discussed on a resource basis, the values related to consumption fluctuations were examined, and the distributions are given in Fig. 1.5.

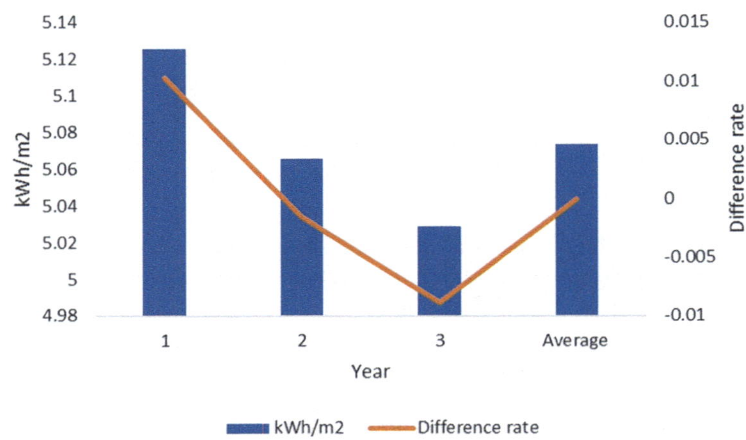

Fig. 1.5 Energy density and variations

Energy density, which is one of the reference values in the enterprise, can be evaluated by energy management. However, as seen in this business example, change does not correlate with consumption. This directly leads to a different direction of energy management for target operations. In this respect, the histogram is also examined as an option. In this respect, when the production–consumption histogram of the enterprise is considered together with the boundary conditions, the consumption change for electricity is 9.01%, while this value is 8.46% for natural gas, indicating an effective potential. According to the 3-year consumption behavior of the enterprise, the energy efficiency effect was found to be 13.89% according to the regression analysis. However, all these data show that the enterprise should be considered as an operational value over energy users. In the evaluations, a total of six important users of the enterprise, including important energy users for natural gas and Bio Combustion Fuel (BCF) machines, cooling, steam boiler, weaving machines, yarn covering machines, and fixing machines and important energy users for electricity, and four important energy users: yarn twisting machine, BCF compressed-air machine, weaving machines, and weaving pressure air process, stand out. For the effectiveness of these machines in the enterprise, operational parameters have been considered, and limit conditions have been developed for each parameter. The distributions for this are given in Fig. 1.6.

According to the management program implemented, a total of 37 critical operating parameters were defined for all major energy users. It has also been observed that the criteria defined in this process, in which a direct measurement system was not applied before the program, were direct impact parameters for each important energy user. In particular, these parameters, which can be measured in BCF, weaving, and boiler systems, are properties that can be set for defined nominal values. In addition, maintenance criteria in business management are also discussed in terms of energy efficiency. However, beforehand, on-the-job training of maintenance processes and maintenance personnel on energy-efficient maintenance procedures based on energy efficiency will make a greater contribution to this process. In this context, two actions have been developed on operational control management and energy-efficient maintenance training in the energy management process. The total annual energy-saving potential of the enterprise provided by the actions described below was approximately 0.28 GJh/year. The effect of these savings on the annual targeted savings corresponds to a rate of 9.4%.

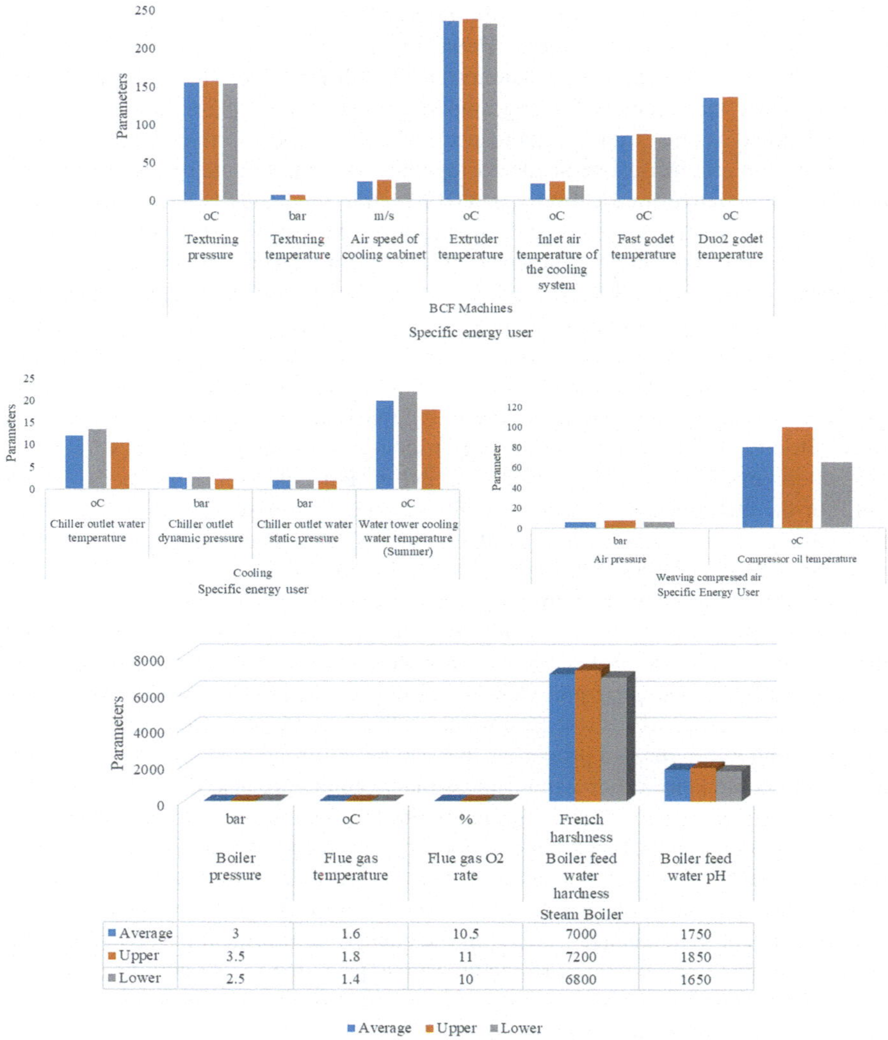

Fig. 1.6 Operational parameters

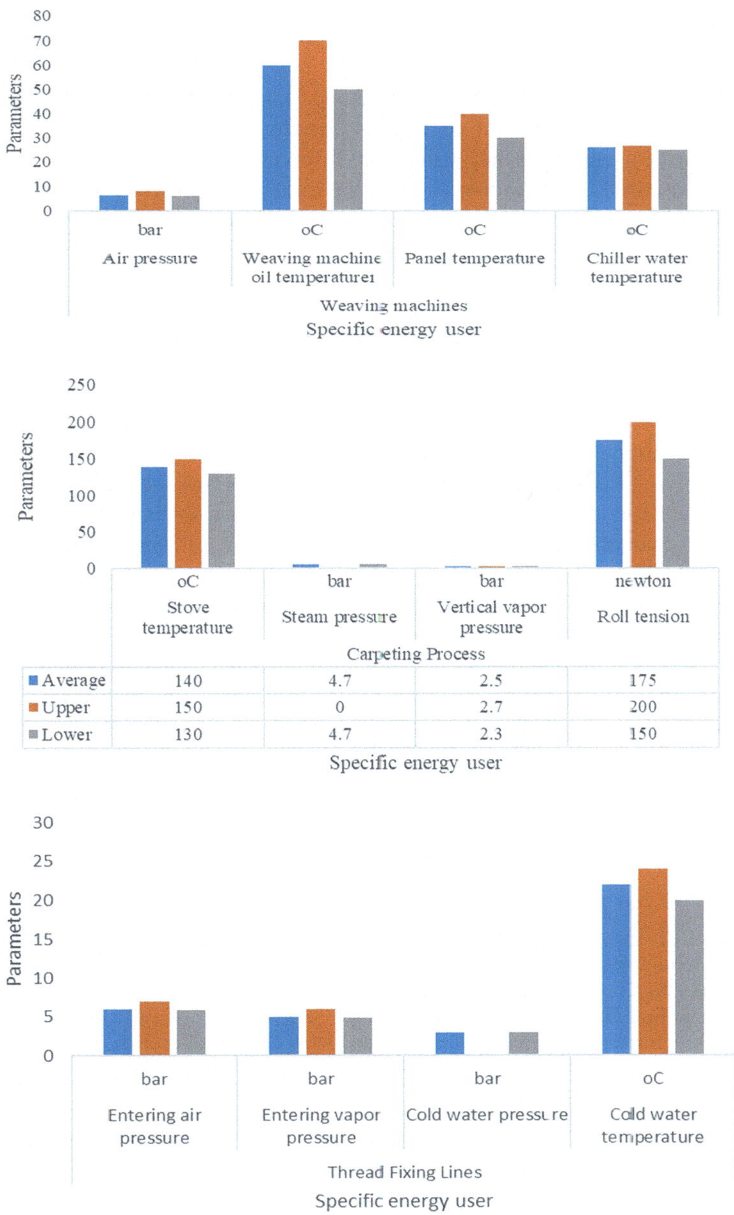

Fig. 1.6 (continued)

1.4 Conclusion

This study presents a behavioral approach to enhance energy and environmental sustainability. It addresses the impact of effective energy-efficient behaviors on operational processes for energy users, using two different methods. First, it presents a framework based on the awareness of a behavioral concept developed from the perspective of behavioral science. The study presents the results of a behavioral concept developed to improve energy efficiency and operational process efficiency in an industrial application.

A framework was developed, along with professional qualifications, and relevant criteria were defined separately. The concepts were evaluated separately, and the results are presented below. The text describes how to apply a concept using a business example.

- Based on the educational taxonomy criteria for behavior in education, the target was achieved in two criteria, while four criteria remained below the target.
- The success target for measured psychomotor training was not achieved. It is recommended to evaluate the training process and repeat the training by examining the goals and behaviors.
- The objective for energy efficiency should always be complete learning. It is also important to review the validity of the tests.
- While 37 critical operating parameters were evaluated in operational parameters, two action development potentials were identified for maintenance processes.
- The total efficiency potential of the business depending on operational parameters was found to be 9.4%.

References

Abrahamse, W., Steg, L., Vlek, C., & Rothengatter, T. (2005). A review of intervention studies aimed at household energy conservation. *Journal of Environmental Psychology, 25*(3), 273–291. https://doi.org/10.1016/j.jenvp.2005.08.002

Anderies, J. M., Cumming, G. S., Clements, H. S., Lade, S. J., Seppelt, R., Chawla, S., & Müller, B. (2022). A framework for conceptualizing and modeling social-ecological systems for conservation research. *Biological Conservation, 275*, 109769. https://doi.org/10.1016/j.biocon.2022.109769

Barr Stewart, Lampkin Sal, Dawkins Laura, Williamson Daniel (2021) Smart cities and behavioural change: (Un)sustainable mobilities in the neo-liberal city, Geoforum, Volume 125, Pages 140-149, ISSN 0016-7185, https://doi.org/10.1016/j.geoforum.2021.06.010.

EU. (2023). Energy poverty. Retrieved 29 November, 2023, from https://energy.ec.europa.eu/topics/markets-and-consumers/energy-consumer-rights/energy-poverty_en.

Frederiks, E. R., Stenner, K., & Hobman, E. V. (2015). Household energy use: Applying behavioural economics to understand consumer decision-making and behaviour. *Renewable and Sustainable Energy Reviews, 41*, 1385–1394.

Gaspar, R., & Antunes, D. (2011). Energy efficiency and appliance purchases in Europe: Consumer profiles and choice determinants. *Energy Policy, 39*(11), 7335–7346. https://doi.org/10.1016/j.enpol.2011.08.057

Karcı, H. (2008). Organization of Vocational Schools and European Expectations, IV. In *National Vocational Schools Directors Meeting, YÖK-Karadeniz Technical University, 05-08 November Türkiye*.
Kenar, N. (2012). *Possible effects of the National Employment Strategy on the role of the private sector in the vocational education system*. MESS Education Foundation.
Kiatcharoenpol, T., & Sirisawat, P. (2020). A selection of barrier factors affecting reverse logistics performance of thai electronic industry Int. *J. Intell. Eng. Syst, 13*. 117–126.
Ronei, D. A., & de Souza Teixeira, R. L. (2024). Impacts of the COVID-19/2020 pandemic on the waste sector of Rio de Janeiro municipality, Brazil: Assessment on solid waste production in 2018–2023. *Waste Management Bulletin, 2*(1). 162–171. https://doi.org/10.1016/j.wmb.2024.01.005
Sogut M.Z., Tasdoven M. (2012) Vocational and Technical Education and Bologna Process, 1st National Air Conditioning and Refrigeration Education Symposium, 13-15 September, Balıkesir (TR)
Stewart, B., Sal, L., Laura, D., & Daniel, W. (2021). Smart cities and behavioural change: (Un)sustainable mobilities in the neo-liberal city. *Geoforum, 125*, 140–149. https://doi.org/10.1016/j.geoforum.2021.06.010
Ulrika, W., Chirez, S., & Adams, B. (2014). Improving industrial energy efficiency by changing the energy culture. In *ECEEE 2014 industrial summer study—Retool for a competitive and sustainable industry*, pp. 659–666.
Ulrika, W., Liu, J., & Vetromile, J. (2015). Why is this facility different? Measuring energy culture to achieve continual improvement in energy efficiency. In *ACEEE summer study on energy efficiency in industry*, pp. 1–12.
Wang, Z., Wang, X., & Guo, D. (2016). Policy implications of the purchasing intentions towards energy-efficient appliances among China ' s urban residents?: Do subsidies work? *Energy Policy, 102*, 430–439. https://doi.org/10.1016/j.enpol.2016.12.049
Wang, Z., Wang, X., Guo, D., Zhang, B., Li, G., Alam, S. S., Hashim, N. H. N., et al. (2017). A moral extension of the theory of planned behavior to predict consumers' purchase intention for energy-efficient household appliances in Malaysia. *Energy Policy, 39*(2), 140–145. https://doi.org/10.1016/j.rser.2017.05.150
Wenyao, Z., Wei, Z., & Daim, T. U. (2023). The voluntary green behavior in green technology innovation: The dual effects of green human resource management system and leader green traits. *Journal of Business Research, 165*, 114049. https://doi.org/10.1016/j.jbusres.2023.114049
Wising Ulrika, Jingjing Liu, Julia Vetromile (2015) Why is This Facility Different? Measuring Energy Culture to Achieve Continual Improvement in Energy Efficiency, ACEEE Summer Study on Energy Efficiency in Industry, ss.1–12.
Wising Ulrika, Sophie Chirez, Bart Adams(2014) Improving Industrial Energy Efficiency by Changing the Energy Culture, ECEEE 2014 Industrial Summer Study – Retool for a Competitive and Sustainable Industry, ss.659–666.
Sogut, M.Z. & Tasdoven M.(2012), Vocational technical education and the Bologna process, 1. In *National air conditioning refrigeration education symposium, Balıkesir, 5-7 Eylül, Türkiye*.
Zhang Wenyao, Zhang Wei, Tugrul U Daim (2023) The voluntary green behavior in green technology innovation: The dual effects of green human resource management system and leader green traits, Journal of Business Research, Volume 165, 114049, ISSN 0148-2963, https://doi.org/10.1016/j.jbusres.2023.114049.

Chapter 2
Exergy Rationality in Sustainable Decarbonization

Birol Kılkış

NomenclatureSymbols

a	Constant of the linearized heat pump performance (COP), Eqs. (2.54) and (2.56)
b	Constant of the linearized heat pump performance (COP), Eqs. (2.54) and (2.56), K-1
A_L	Total land area occupied to generate 1 kWex power (installed capacity), m^2/kWex
ALT	Residence time in the atmosphere, year
c	Constant in Eq. (2.2), $Q \times T_{ref}/T_{sup}$
c_K	Unit CO_2 content of a fuel, kg CO_2/kW-h_{en}
$\overline{c}_K$	Average c_K of multiple fuel mix, kg CO_2/kW-h_{en}
CFI	Carbon footprint index
CO_2	Direct emission, kg CO_2/kW-h_{en}
CO_{2base}	The reference emission responsibility of the energy sector for both power and heat, 0.63 kg/kW-hen
COP	Coefficient of performance
COPEX	Exergy-based COP
CP	Specific heat, kW-hen/(kgK)
E	Electric power (load), or energy, kW_{en}, kW-h_{en}
ELC	Exergy-levelized cost, $
EPI	Environmental performance index
ESI	Environmental sustainability index
EUI	Energy utilization index
EUS	Energy utilization success rate
EXUI	Exergy-based energy utilization index

B. Kılkış (✉)
OSTIM Technical University, Ankara, Türkiye

M. Z. Sogut, M. Koray (eds.), *Energy Rationality and Management for Decarbonization*, Green Energy and Technology,
https://doi.org/10.1007/978-3-031-68497-5_2

E	Electric (load)
EJ	Final energy demand, kW-h
EM	Embodiment, \$/kg
EX	Exergy, kW-hex
EXH	Thermal exergy, kW-h_{ex}
EXS	Exergy star green building rating scale
Fc	Correction factor for COP, Eq. (2.61)
GWP	Global warming index
HTCF	Heat offset timing for minimum carbon emissions factor (Eq. 2.47), [kW-h_{ex} h]$^{-1}$
k	Emission responsibility factor, kg CO_2/(kW-h_{ex}/kW-h_{en}); k factor for multiple systems, kg CO_2/(kW-h_{ex}/kW-h_{en})
LCCA	Life-cycle carbon emission analysis
LCEXA	Life-cycle exergy analysis
Lmax	Maximum allowable one-way distance from the plant to the district grid, m, km
LUEEX	Exergy-based index for land-use effectiveness, m^{-2} mass flow rate, kg/h
MRT	Mean radiant temperature, K
$MtCO_2$	Megaton CO_2 emissions
ODI	Ozone depletion index
ODP	Ozone depletion potential
OF	Objective function
OT	Operative temperature, K
P	Power demand of a pump (including motor efficiency), kW-h
Pp	Pressure, Pa
PC	Market price of the panel, \$
PEF	Primary energy factor
PEX	Pumping exergy, kW-h_{ex}
PM	Prioritization metric
PMV	Predicted mean vote
Pop	Urban population, persons
PR	Radiant to total sensible heat transfer ratio in space heating or cooling
RR	Exergy recovery ratio
Q	Thermal energy (heat) or thermal power, kW-h_{en}, kW_{en}
RCO_2	$\Delta CO_2/CO_2$
REU	Remaining carbon emissions, kg CO_2
REX	Exergy-based annual mix of renewables in the energy stock
rex	Potential renewable penetration index
S	Emission savings (in terms of heat or electricity)
SEI	Sustainable environment index
T	Temperature, K or °C
Tf	Adiabatic flame temperature of fuel, K
TSI	Total solar irradiance, W/m^2
TSV	Thermal sensation vote
t	Time, s, h, year

Vair,	Air flow rate, m^3/h
w1, w2	Weighing function of the prioritization metric, PM (Eq. 2.18)
W	Humidity ratio (ASHRAE Standard 55-2023)
WBR	Waste heat utilization rate, kW-hex/kW-hen
Wp	Weight of the panel, kg
Wr	Waste heat ratio, recoverable down to a temperature of 308 K (sewer heat is the limit)
X	Mixing ratio
x	Distance in the HRV (or ERV) heat exchanger from the indoor side, m
Y	Demand amount from TES tank (see Fig. 2.31), or year
Z	Demand from the first tank (see Fig. 2.31)

Greek Symbols

ε	Unit exergy, $kW\text{-}h_{ex}/kW\text{-}h_{en}$
ψR	REMM efficiency; ψR of multiple systems
ψRref	Reference ψR value
ηI	First-law efficiency
ηPm	First-law efficiency of the pump and its motor
ΔCO_2	Nearly avoidable CO_2 emission responsibility, kg $CO_2/kW\text{-}h_{en}$
ΣCO_2	Total emissions, $\Delta CO_2 + CO_2$, kg CO_2/kWen
ΔT	Temperature difference, K

Subscripts

a	Atmospheric, ambient
B	Boiler
correct	Corrected
dem	Demand
des	Destroyed (exergy)
e	Exit, exhaust
E	Exit, electric
EX	Exergy
em, m	Embodied
eq	Equivalent
f	Source, fuel
g	Geothermal
H	Heat
h, c	Constants of Eq. (2.35) depending on district piping and flow properties
high	High (temperature)
inv	Power inverter

LH	Latent heat
low	Low (temperature)
net	Net
o	Outdoor
p	Peak, Eq. (2.54)
panel	Solar panel
peaked	Peaking (Temperature), Eq. (2.54)
pm	Pump motor
pp	Power plant
R	Rational, recovery (Fig. 2.43)
ref	Reference
s	Supply
solar	Sun-related
ST	Standard (COP), Eq. (2.59)
sup	Supply
TT	Transmission and transformer (losses in the power grid)
T	Transmission
w	Waste
wh	Waste heat
WT	Wind turbine

Superscripts

` Virtual, or modified; Average

Acronyms

ABS	Absorption cooling machine
ADS	Adsorption cooling machine
CHP	Combined heat and power
CMBRT	Cosmic microwave background radiation temperature (2.74 K)
COP27	Conference of Parties, 27th meeting
CP	Circulating pump
EAHE	Earth air heat exchanger
DHW	Domestic hot water
ERV	Energy recovery ventilation
EU	European Union
EU 20-20-20	Saving, renewable, efficiency goals by 20% each up to 2020
FC	Fuel cell
FPC	Flat-plate collector
GHG	Greenhouse gas (emission)

GSHP	Ground-source heat pump
HP	Heat pump
HRV	Heat recovery ventilation
IAQ	Indoor air quality
IEA	International Energy Agency
IPCC	Intergovernmental Panel for Climate Change
LoeEx	Low exergy (building or energy source)
NG	Natural gas
nZEXB	nearly zero exergy building
ORC	Organic Rankine cycle
PCM	Phase-change material
PP	Power plant
ppm	Parts per million
PV	Photovoltaic (Panel)
PVT	Photovoltaic–thermal (Panel)
REMM	Rational Exergy Management Model
SDEWES	Sustainable Development of Energy, Water, and Environment Systems
SER	Sufficiency, Efficiency, Renewables
TEG	Thermoelectric generator
TES	Thermal energy storage
UN	United Nations
WT	Wind turbine
ZEB	Zero energy building

2.1 The Second Law of Thermodynamics and Environmental Sustainability

2.1.1 The Second Law of Thermodynamics and Environmental Sustainability

The second law of thermodynamics (Britannica, 2024), when defining the direction of heat flow, shows that the flow will be from lower temperature to higher temperature in a cyclical process. Physicist Rudolf Clausius (Britannica, 2024), who first formulated this law of thermodynamics, defined it as "It is impossible to transfer heat from an object at a certain low temperature (T_{low}) to an object at a higher temperature (T_{high}) in a cyclic process." This law of thermodynamics is used in this action process. According to dead state conditions, the amount of useful work to be obtained from the amount of heat or power transferred is called exergy (E_x) and defines the amount of useful energy generated. According to Eq. (2.1) describing an ideal Carnot cycle, E_X is always lesser than Q. The basics of the second law are summarized in the *2025 ASHRAE Handbook Fundamentals*, Chapter 2.

$$E_X = \left(1 - \frac{T_{\text{low}}}{T_{\text{high}}}\right) \times Q \tag{2.1}$$

The importance of the second law in terms of global warming today goes beyond the fundamentals and needs to be expanded to directly address the global warming issue with the greenhouse gas emission responsibilities of exergy destructions. In the quest of decarbonization, however, sustainability is defined only in terms of the first law, i.e., primarily energy savings and the wider utilization of renewable energy sources. However, renewable energy systems also carry CO_2 emission responsibility in proportion to the exergy that they destroy between the supply exergy, ΔCO_2. For example, a solar photovoltaic (PV) panel generates high-exergy electric power at a modest first law efficiency but destroys the opportunity of generating thermal power, yet at a lower exergy, downstream. After equating the low temperature to the environmental equilibrium temperature, T_{ref}, and denoting demand and supply temperatures by T_{dem} and T_{sup}, respectively, the exergy difference between the supply and demand translates to irreversible exergy destruction, causing the so-called nearly avoidable CO_2 emissions, ΔCO_2:

$$\Delta CO_2 = k \times \Delta E_{X\ \text{destruction}} = \frac{Q \times T_{\text{ref}}}{T_{\text{sup}}} \left(\frac{T_{\text{sup}}}{T_{\text{dem}}} - 1\right) = c\left(\frac{T_{\text{sup}}}{T_{\text{dem}}} - 1\right) \tag{2.2}$$

According to Fig. 2.1, the exergy destruction–related ΔCO_2 increases with the gap between the supply and demand temperatures. This relationship is not recognizable by the first law, and therefore, the efforts of decarbonization will continue to fall short until the second law is widely recognized. The Environmental Sustainability Index, ESI, and the Environmental Performance Index, EPI (NASA, 2024), are not sufficient for sustainable decarbonization because they rely on the first law.

Figure 2.1 shows that the first and second laws must go hand in hand to cover the entire aspects of global warming and provide more sustainable and effective measures. Within the framework of 17 sustainable development goals (SDGs) developed by the United Nations (UN) (UN, 2024), The Environmental Performance Index, EPI, utilizes a proximity-to-target methodology focused on a set of environmental outcomes linked to policy goals. Policy goals are based on quasi-linear economic principles, and environmental goals are also based only on the first law (Yale, 2024). Environment involves nonlinearities as exemplified in Eq. (2.1). Similarly, according to the Environment Sustainability Index (ESI), which is a measure of the overall progress toward environmental sustainability, progress is evaluated with the first law (NASA, 2024; EUI, 2019). Therefore, both EPI and ESI do not factor in the ΔCO_2 term in their findings, projections, and policies overviewed. Today, the four pillars of sustainability, namely human, social, economic, and environmental, are considered to be linear.

This drawback is exhibited by the following quote from IEA Annex 64 (EBC, 2019): "The term exergy is important... because it takes into account of the energy

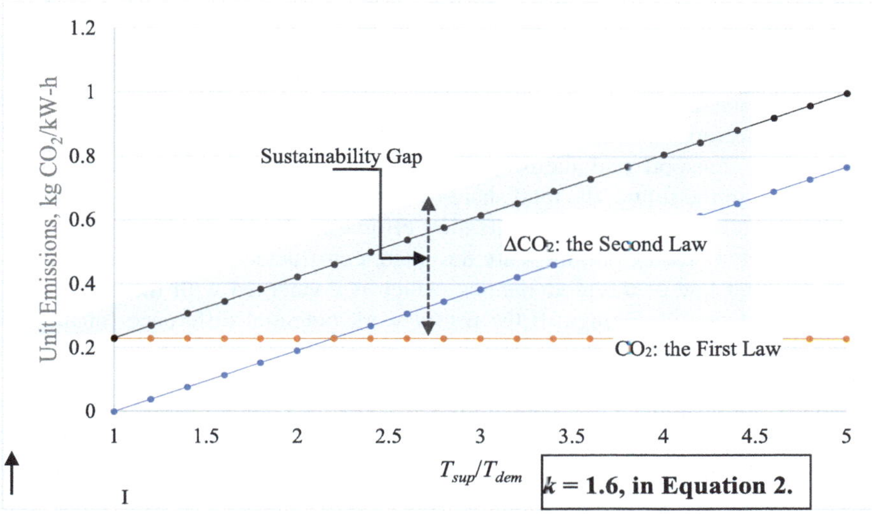

Fig. 2.1 Sustainability gap according to the exergy destructions due to mismatches. Sample case, $Q = 1$ kW-h, $T_{sup} = 2235$ K (T_f, adiabatic flame temperature of natural gas), $T_{ref} = 283$ K

quality and can help to ensure that the most appropriately used for a given application. For example, in space heating processes, natural gas, a high quality/high exergy source, is often burned to heat water to 70-80°C, which is in turn is used for a low exergy application of heating rooms in a home to 20°C. The same gas could be burned to deliver temperatures of 1000°C in industrial processes and the waste heat could be more appropriately used for low-exergy applications as space heating."

This chapter shows that the first law is necessary but not sufficient to cover all sustainable measures against global warming. Therefore, the second law must take over the mission at large along with the first law. American Society of Heating, Refrigerating, and Air Conditioning Engineers, Inc. Technical Committee 7.4 Exergy for Sustainable Buildings was established to serve the members and the global community by bringing in sustainable solutions against global warming, meet the Paris Agreement goals by closing the exergy mismatch, and reduce exergy destruction–led emissions, primarily in the built environment and the building sector.

2.1.2 Global Warming and Buildings

We pay for energy but use exergy

Peter Novak quoted at the ASHRAE Annual Meeting, 1996, in Quebec City.

2.1.2.1 Human Responsibility in Buildings

According to Fig. 2.2, the built environment is responsible for 42% (15.3 GT/year) of global annual direct and indirect CO_2 emissions for energy and process (36.3 GT/year) (Architecture2030, 2024). This emission responsibility is higher than in other sectors, like transport and industry. Furthermore, urbanization is growing fast enough that, in the future, the total share of buildings will reach 50% soon. This is a clear indication that the decarbonization efforts need to focus on buildings for decarbonization. Today, buildings are designed, constructed, and operated according to the first law of thermodynamics, which is concerned with the quantity of energy and does not distinguish the useful work potential difference (quality of energy: exergy) among several energy sources and demands (Kılkış & Kılkış, 2019). For example, electrical energy has a unit exergy of 0.95 kW-h_{ex}/kW-h_{en}, whereas, a hydronic floor heating supply temperature at 320 K has a unit exergy of 0.16 kW-h_{ex}/kW-h_{en} at a reference temperature of 283 K. If water is heated in an electric boiler, the unit exergy mismatch will be responsible to CO_2 emission due to the so-called irreversible exergy destruction (Kılkış, 2023). Unless it is recognized that energy sources and demands are not the same, the decarbonization efforts will fall short (SE, 2016). This amount is in addition to direct emissions due to the electric power generation at a power plant using fossil fuel. This chapter takes the second law of thermodynamics, which is concerned with exergy, one step further to relate it directly to global warming in terms of CO_2 emission by bringing a deeper insight into emission responsibilities and their root causes. Therefore, exergy, which measures energy and resource quality, should be integrated into policy, law, and everyday practice for the rational use of limited natural resources (SE, 2016).

It is also pointed out that when the second law comes clear in the picture of sustainability, a multivariable optimization is required to optimally satisfy all the goals for a sustainable environment (Wang et al., 2016; Kilkis, 2011, 2014).

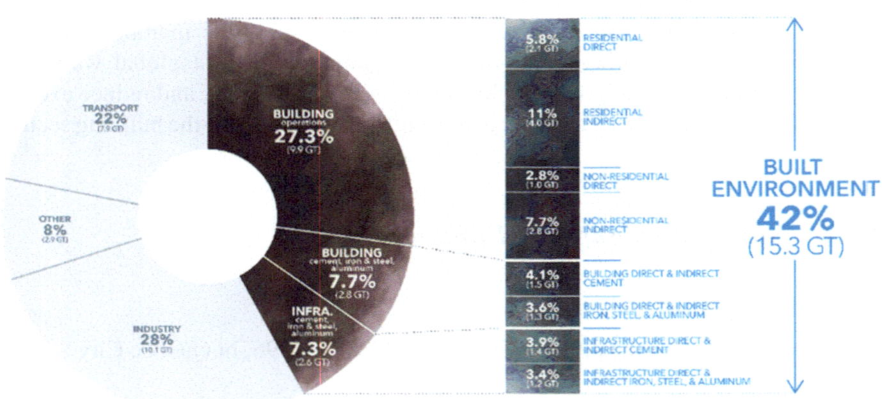

Fig. 2.2 The share of CO_2 emissions of the built environment (Architecture2030, 2024)

2.1.2.2 A Progressive Approach to the Paris Agreement in the Built Environment

Figure 2.3 shows the projections about the CO_2 concentration in the atmosphere compared to the Paris Agreement goals to be accomplished by the year 2053 such that the global temperature rise will be limited to 1.5 °C. In this respect, the European Union (EU) aims to be climate-neutral by 2050, an economy with net-zero greenhouse gas emissions. If the business as usual is continued, the CO_2 concentration will keep rising. A green growth, and even greener growth, with carbon capture and storage will only satisfy the 450 ppm stabilization target by the year 2030. The Paris Agreement goal envisions reducing this concentration to about 280 ppm, which means that some theories and practice in decarbonization are missing. The missing piece of the puzzle is the recognition of the second law (Kilkis, 2023). In 2030, the IPCC expects a peak, but the question of how remains unanswered without the second law. The solution may be visualized in Figs. 2.4 and 2.5, which denote that the ΔCO_2 term from the second law is the dominant factor in calculating the total emissions, and therefore, it may solve the puzzle, as indicated in Eq. (2.2).

The (k) value depends on whether the major exergy destruction takes place upstream or downstream of the useful work obtained in a process. Figure 2.4 relates to the major exergy destruction taking place downstream like in a solar PV panel or a power generator. The (k) value represents a series of offsets for the exergy destruction. In the PV case, the thermal exergy destroyed downstream needs to be offset by someone, somewhere, at some time, and it echoes down in a diminishing manner but practically never comes to zero. With today's global average exergy-based mix, R_{EX} of 0.18, the (k) value is about 1.1. For the second case, which exemplifies a solar flat-plate collector (FPC) or a boiler, the (k) value is about 2.1. The (k) factor

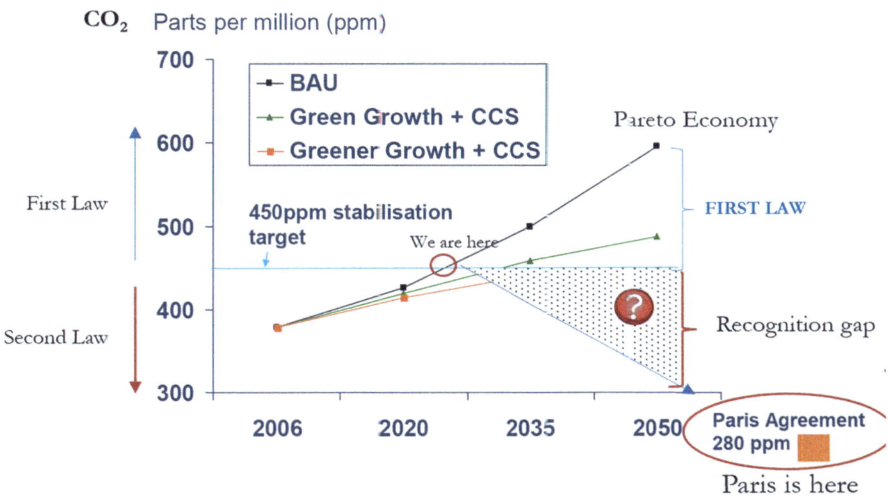

Fig. 2.3 The future scenarios for atmospheric CO_2 concentration (Kilkis, 2023)

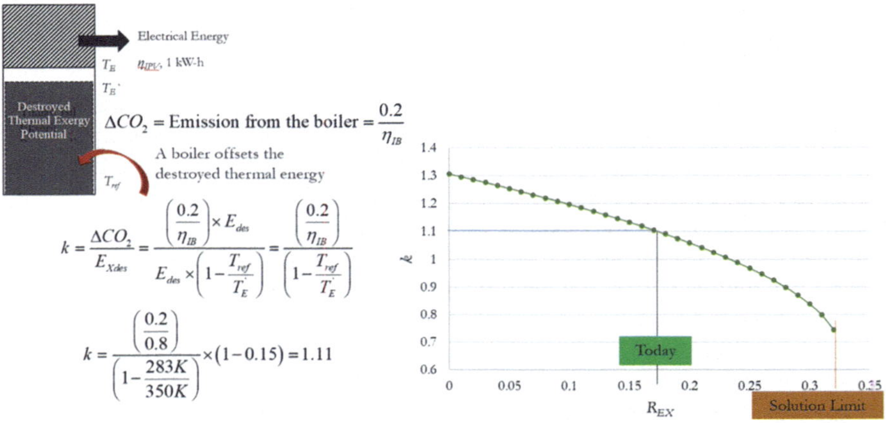

Fig. 2.4 Value for solar PV panel: major exergy is destroyed downstream

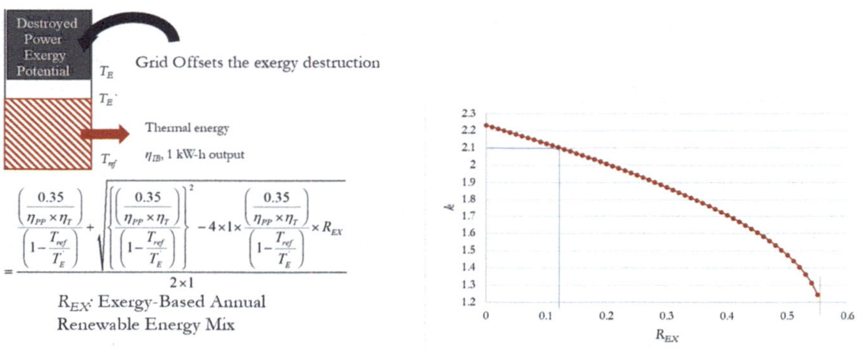

Fig. 2.5 The (k) value for solar flat-plate collector: major exergy is destroyed upstream

represents the offsetting echo of exergy destructions of PV and PVT pairs to offset, as shown in Fig. 2.6, in a diminishing format. The thermal exergy destruction of the first PV panel is offset by the adjacent flat-plate collector (FPC). This pair is repeated because the exergy destruction upstream of the thermal offset by the FPC needs to be offset by the PV panel in the second pair, and this continues to repeat itself. Figure 2.7 shows that during pairing, ΔCO_2 emission responsibility cumulates faster than the CO_2 mitigation cumulation with (i), which represents fossil fuel replacement in the energy sector.

$$\Sigma \Delta CO_{2\,\mathrm{responsibility}} = \sum_{n}^{i=1} \left(\frac{1}{\eta_{\mathrm{IPVi}}} - 1 \right) \left\{ \left(1 - \eta_{\mathrm{IFPCi}}\right) \times \mathrm{PEF} \times 0.2 \right\} \times \left(1 - \eta_{\mathrm{PVi}}\right)^{i-1} \tag{2.3}$$

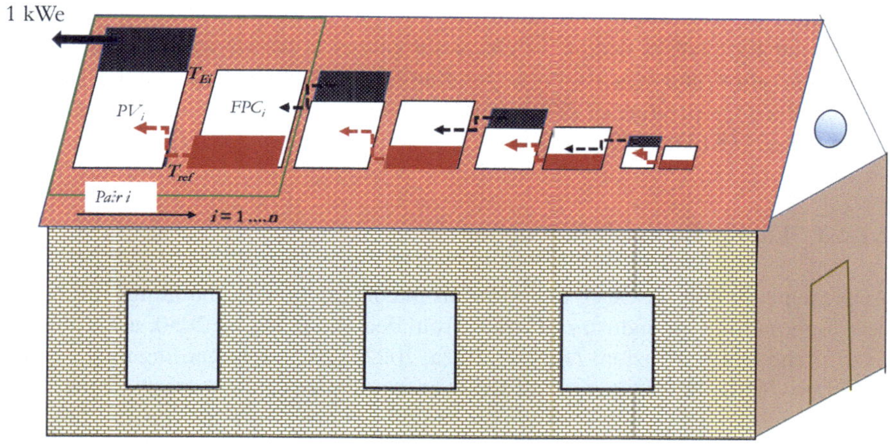

Fig. 2.6 Diminishing set of exergy destructions offset on a sustainable building roof

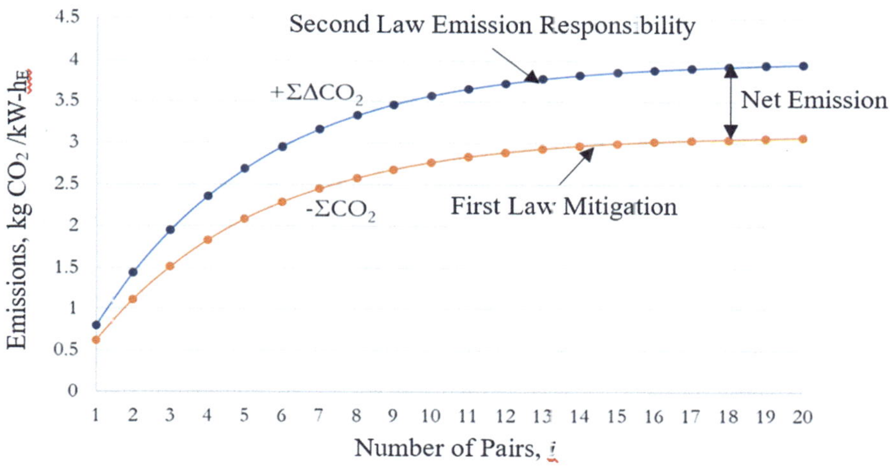

Fig. 2.7 Minimum net CO_2 emissions have no root for $i > 1$. Net CO_2 emissions increase with i

$$\Sigma CO_{2\text{mitigated}} = \sum_{n}^{i=1} 0.2 \times \left[\text{PEF} + \frac{(1-\eta_{\text{IPP}})}{\eta_{\text{IB}}} \right] \times (1-\eta_{\text{PVi}})^{i-1} \tag{2.4}$$

The term (k_i) may also be written in terms of the preceding temperature (T_{i-1}) of an application (i) in a chain of applications.

$$k_i(T_{\text{in}-i}) = 1.1 + 0.0024 \times (T_{\text{in}-i} - T_{\text{ref}}) \quad \{T_{\text{in}-i} \leq 700\,\text{K}\}, \quad \text{or :} \tag{2.5a}$$

$$k_i(\varepsilon_{\text{desi}}) = 2.1 \times \varepsilon_{\text{desi}} \tag{2.5b}$$

Example 1
For a solar flat-plate collector (FPC), major exergy is destroyed upstream of hot water generation, and then, the corresponding k_i value from Eqs. (2.11a) and (2.11b) at a reference temperature of 283 K will be 1.84. The following is a short list of the rating metrics (Kılkış & Kılkış, 2023).

2.1.2.3 Exergy Rationale in Buildings

As presented in Fig. 2.3, CO_2 emissions from construction and buildings reached a new high, putting the industry off track from decarbonizing by 2050, according to COP27 and UN conclusions (Kilkis, 2022a, 2022b, 2022c). Regarding their capacity for productive work (exergy), different energy sources are not created equal. For example, in net-zero energy buildings, building temperature and energy balance cannot be achieved. If Y kW-hen of thermal energy [kW-hen] is given for the energy demand of the building at 315 K on average, this has the same value at 340 K (Kilkis, 2022a, 2022b, 2022c). Unfortunately, the first law of thermodynamics is insufficient to describe this imbalance given below.

$$\text{Exergy Deficit} = Y \times \left(1 - \frac{315\ \text{K}}{340\ \text{K}}\right) = 0.0735 \times Y \quad [\text{kW} - \text{h}_{\text{ex}}] \tag{2.6}$$

The resulting temperature deficit in this context can be met by burning fossil fuels that emit CO_2 emissions in regional power plants or producers.

$$\Delta CO_2 = k_i \times \text{Exergy Deficit}_i \tag{2.7}$$

Figure 2.8 shows a prosumer building, which exchanges power, heat, and cold throughout the year. The building exchanges a lump-sum quantity of power (Q_1), heat (Q_2), and cold (Q_3). As long as the exchange of the lump sum is equal, this

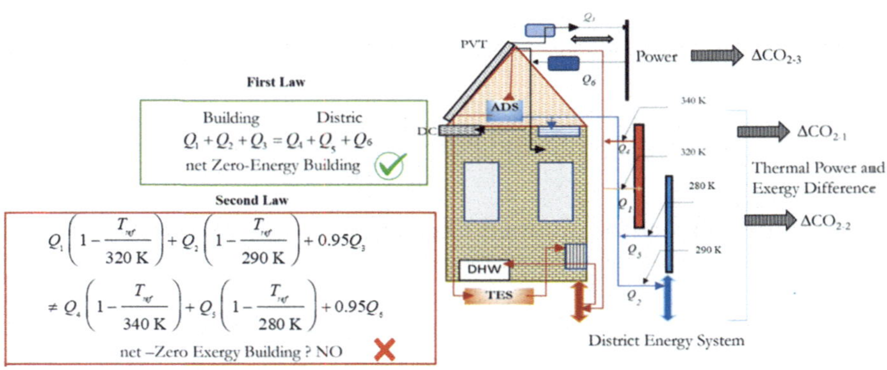

Fig. 2.8 A prosumer building, which is a net-zero energy but not net-zero exergy building

building is a net-zero energy building, according to the first law. This law does not recognize the different qualities or their shares of the power, heat, and cold in the lump sum.

What makes the difference is the ΔCO_2 term, which recognizes the breakdown of Q_1, Q_2, and Q_3 and their unit exergy. With the given sample data representing the average temperatures of the supply and demand temperatures of thermal exchanges, this building is not a net-zero exergy building and is responsible for several ΔCO_2 terms, which must be offset at the district plant.

$$\Delta CO_{2-1} = 2.1 \times \left(\frac{T_{\text{ref}} \times Q_1}{320\,\text{K}} - \frac{T_{\text{ref}} \times Q_4}{340\,\text{K}} \right); \Delta CO_{2-2} = 2.1 \times \left(\frac{T_{\text{ref}} \times Q_5}{280\,\text{K}} - \frac{T_{\text{ref}} \times Q_2}{290\,\text{K}} \right); \Delta CO_{2-3} = 1.1 \times (Q_6 - Q_3) \times 0.95$$

Research about the district energy system concerning the city of Östra Sala Backe in Uppsala, Sweden as a pilot, near-net-zero exergy district, with a planned population of 20,000 defines that a net-zero exergy district produces as much energy at the same grade or quality as consumed on an annual basis (Kilkis, 2014; Van der Leer et al., 2023).

2.2 Fundamentals of Exergy Rationale

2.2.1 REMM

The REMM combines the first and second laws to determine total CO_2 emission responsibility of a building. It leads the way for utilizing the quality of available energy as much as possible. In other words, the objective is to leave minimum part of the quality of energy behind. Consider an on-site boiler (B), which destroys the electric power generation potential upstream of satisfying the thermal load, Q. The destroyed exergy potential is offset at the grid connected to a thermal power plant (PP) with a load (Kilkis, 2011). In turn, the thermal power plant destroys thermal power generation downstream its electric power generation, and this series of destructions and offsets carries on. Equation (2.8) excludes the forward-looking and backward-looking echoes (backdrops) of offsets to be made for continuing series of exergy destructions (see Fig. 2.6). REMM efficiency, ψ_R, is a measure of how rational and exergy balanced the utilization of a given exergy source is. They are expressed by Eqs. (2.9) and (2.10). For application temperatures below 373 K, both equations may be used interchangeably.

$$\Sigma CO_2 = CO_2 + \Delta CO_2 = \overbrace{\left[\underbrace{\frac{c_{KB}}{\eta_{IB}}}_{\text{First law}} + \underbrace{\frac{k \times c_{KB}}{\eta_{IB}} \times (1 - \psi_{RB})}_{\text{Second law}} \right]}^{\text{Boiler}} \times Q + \overbrace{\left[\underbrace{PEF \times c_{KPP} + \frac{k}{\eta_{ITT}} \times (2 - \psi_{RPP})}_{\text{First and second law}} \right]}^{\text{Power plant}} \times E \quad (2.8)$$

$$\psi_R = \frac{\varepsilon_{dem}}{\varepsilon_{sup}} \{\text{When exergy destroyed upstream}\} \quad (2.9)$$

$$\psi_R = 1 - \frac{\varepsilon_{des}}{\varepsilon_{sup}} \{\text{When exergy destroyed downstream}\} \quad (2.10)$$

The exergy destruction in the universe is minimal; thus ψ_R is maximum. Consider that the average surface temperature of the sun in our solar system is 5778 K and the cosmic microwave background radiation temperature (CMBRT) is 2.74 K. Then, using Eq. (2.10) with an ideal supply exergy of 1 $kW\text{-}h_{ex}/kW\text{-}h_{en}$ from the sun and using the ideal Carnot cycle,

$$\varepsilon_{des} = \left(1 - \frac{2.74\,K}{5778\,K}\right) = 4.74 \times 10^{-4}\ kW - h_{ex} / kW - h_{en}$$

$$\psi_R = 1 - \frac{\varepsilon_{des}}{\varepsilon_{sup}} = \frac{\left(1 - \frac{2.74\,K}{5778\,K}\right)}{1} = 0.9995$$

Example 2

A natural-gas, noncondensing boiler generates thermal power, Q, but misses the opportunity of generating electric power, E, on-site, upstream, in contrast to a cogeneration system. Therefore, k_i = 2.1 (see Fig. 2.5 and Eq. (2.7)). The power plant destroys exergy downstream because it does not utilize the waste heat. Therefore, its k_i is 1.1 $\{i = 1\}$. See Fig. 2.6.

Problem Inputs:

c_{KB} = 0.2 kg $CO_2/kW\text{-}h_{en}$ (natural gas), η_{IB} = 0.85, ψ_{RB} = 0.25, Q = 15 kW; PEF = 2.5, c_{KPP} = 0.2 kg $CO_2/kW\text{-}h_{en}$ (natural gas), ψ_{Rpp} = 0.6 (combined cycle power plant), E = 10 kW, T_{ref} = 293 K, T_E = 600 K, T_{ref} = 283 K, η_{ITT} = 0.85. Using these inputs:

Solution:

$$\Sigma CO_2 = CO_2 + \Delta CO_2 = \left[\frac{0.2}{0.85} + \frac{2.1 \times 0.2}{0.85} \times (1-0.25)\right] \times 15 + \left[2.5 \times 0.2 + \frac{1.1}{0.85} \times (1-0.6)\right] \times 10 = 19.26\,\mathrm{kg\,CO_2/h}$$

The following is the breakdown between ΔCO_2 and CO_2:

$$\Delta CO_2 = \left[\frac{2.1 \times 0.2}{0.85} \times (1-0.25)\right] \times 15 + \left[\frac{1.1}{0.85} \times (1-0.6)\right] \times 10 = 10.73\,\mathrm{kg\,CO_2/h}\ \{\text{Emissions unseen on} - \text{site}\}$$

$$CO_2 = \left[\frac{0.2}{0.85}\right] \times 15 + [2.5 \times 0.2] \times 10 = 8.53\,\mathrm{kg\,CO_2/h}\ \{\text{Emissions seen and measured on} - \text{site}\}$$

Note that the unrecognized (unseen) emissions by the first law, ΔCO_2, are more than the recognized (seen) emissions, CO_2.

Exergy Destructions:

$$\left[\frac{0.2}{0.85} \times 2.1 \times (1-0.25)\right] \times 15 = 5.56\,\mathrm{kW-h_{ex}/kW-h_{en}}\ \{\text{Boiler}\}$$

$$\left[\frac{1.1}{0.85} \times (1-0.6)\right] \times 10 = 5.18\,\mathrm{kW-h_{ex}/kW-h_{en}}\ \{\text{Power plant}\}$$

2.2.2 Seen and Unseen Emissions

When the exergy destruction offset train is considered, Fig. 2.9 is obtained. In this case, the same on-site boiler is considered, which supplies heat at a supply temperature of 363 K. The return temperature from the building heating system is 343 K. The boiler uses natural gas with an adiabatic flame temperature of 2235 K. Because exergy is destroyed upstream, k is 2.1. The small exergy destruction downstream of the heating process is neglected. Out of unit exergy of 0.87 kW-h_{ex}/kW-h_{en}, 0.815 kW-h_{ex}/kW-h_{en} is destroyed, leading to a ΔCO_2 responsibility of 1.71 kg CO_2/kW-h_{en}. Direct emission from the boiler stack is 0.235 kg CO_2/kW-h_{en}. In this case, the $\Delta CO_2/CO_2$ ratio is 7.3.

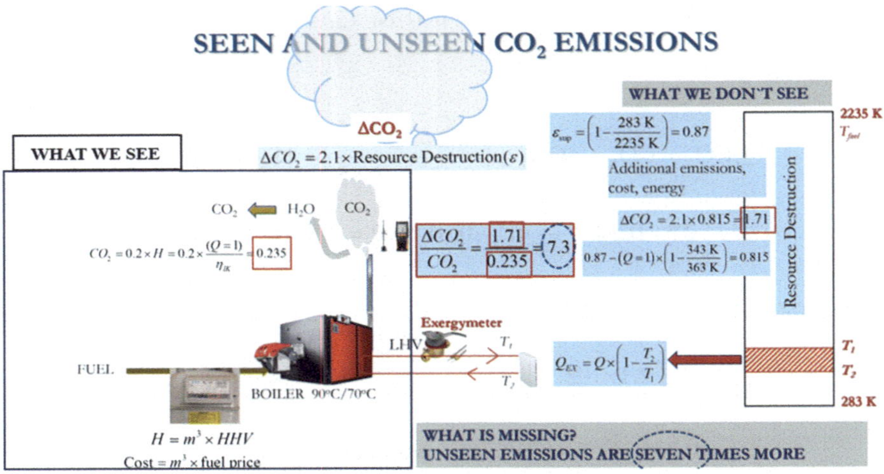

Fig. 2.9 The seen and unseen CO_2 emissions from a boiler in terms of the first and second laws

2.2.3 *The SDEWES Index for Sustainable Cities*

Based on environmental sustainability and development, the Sustainable Development of Energy, Water, and Environmental Systems (SDEWES) basically addresses cities in seven dimensions, 35 main indicators, and nearly 25 sub-indicators (SDEWES, 2024; Kılkış, 2018). While this index aims to provide a sustainability-based rating for cities, the general structure of the index is designed as a maximum of 500 points, and the performance of an average city is predicted to be 28,441 points.

2.2.4 *Energy Savings and the First Law*

The following illustration, as shown in Figs. 2.10 and 2.11, indicates that the $\Delta CO_2/CO_2$ ratio, namely RCO_2 does not change with the thermal load, Q, although direct emissions, CO_2, decrease with decreasing Q. The only way to decrease this ratio is to improve the term as shown in Eqs. (2.8), (2.9), and (2.10).

$$\begin{aligned}\mathrm{RCO_2} &= \frac{\Delta\mathrm{CO_2}}{\mathrm{CO_2}} = \left(2-\psi_R\right)+\mathrm{PEF}\times\left(1-\psi_R\right)\\ &= \left(2-0.3\right)+2.5\times\left(1-0.3\right)=5.55\end{aligned} \tag{2.11a}$$

$$\mathrm{CO_2} = \frac{c_K}{\eta_I}\times Q\left\{\text{First law}\right\} \tag{2.11b}$$

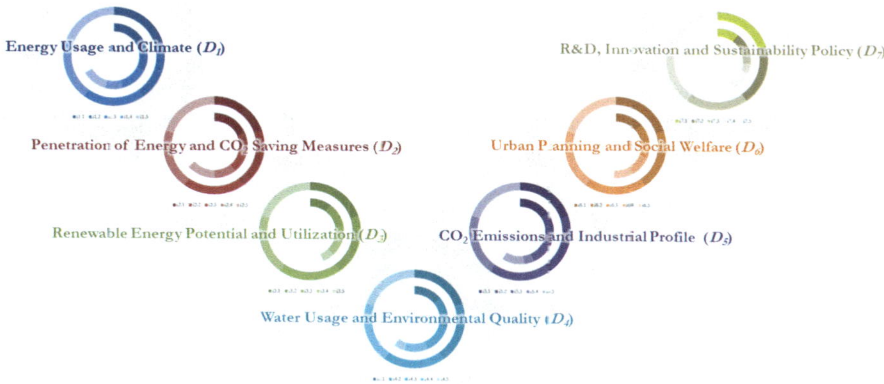

Fig. 2.10 The seven dimensions of the SDEWES index (SDEWES, 2024; Kılkış, 2018)

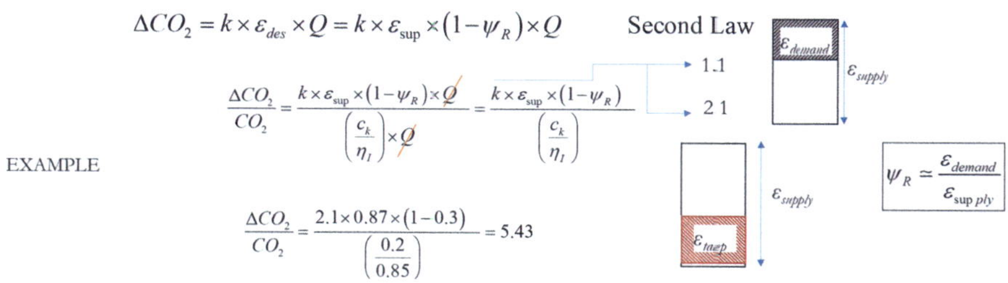

Fig. 2.11 Comparison of the first and the second laws in terms of emission responsibility

According to Fig. 2.11, current RCO_2 in the energy sector is high and independent of the amount of heat involved in a process, Q, although the individual reductions in ΔCO_2 and CO_2 may be reduced in the same order by the first law.

2.2.5 *Exergy-Based Roadmap for Decarbonization*

According to REMM, there is a strong dependence of global warming on exergy analysis in its most recent format, which extends the conventional exergy analysis by directly linking CO_2 emission responsibility of buildings and Heating, Ventilating, and Air Conditioning (HVAC) systems with exergy destructions. This is a departure from the conventional exergy analysis, which does not relate exergy destructions with emission responsibility. Yet this issue is not covered in IPCC reports (2022). In a similar format to Eq. (2.8):

$$\begin{aligned}\frac{\sum CO_2}{CO_2} &= (2-\psi_R) + 0.95 \times \text{PEF} \times (1-\psi_R) \\ &= \left[(2+0.95 \times \text{PEF}) - \psi_R \times (1+0.95 \times \text{PEF})\right]\end{aligned} \tag{2.12}$$

$$\frac{\Sigma CO_2}{CO_2} = [4.375 - 3.375 \times \psi_R] \quad \{PEF = 2.5, EU Average\} \tag{2.13}$$

Figure 2.12 shows the roadmap for decarbonization such that every year, about 2.7 times more effort and finance is required to reduce the emission responsibility compared to reducing direct emissions. Figure 2.13 also illustrates the additional

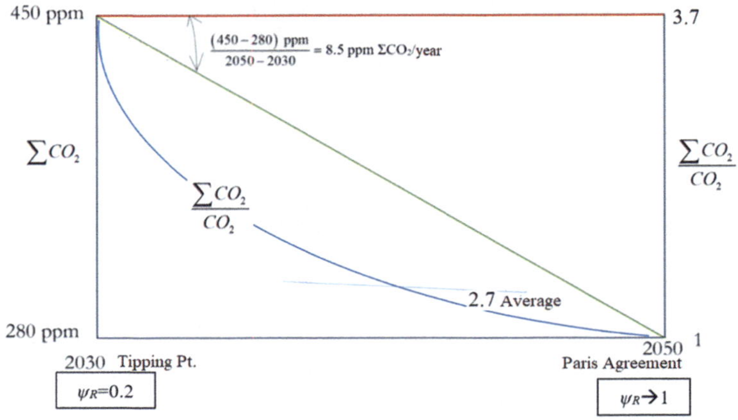

Fig. 2.12 The roadmap for reducing emissions with ΔCO_2 reduction until 2050

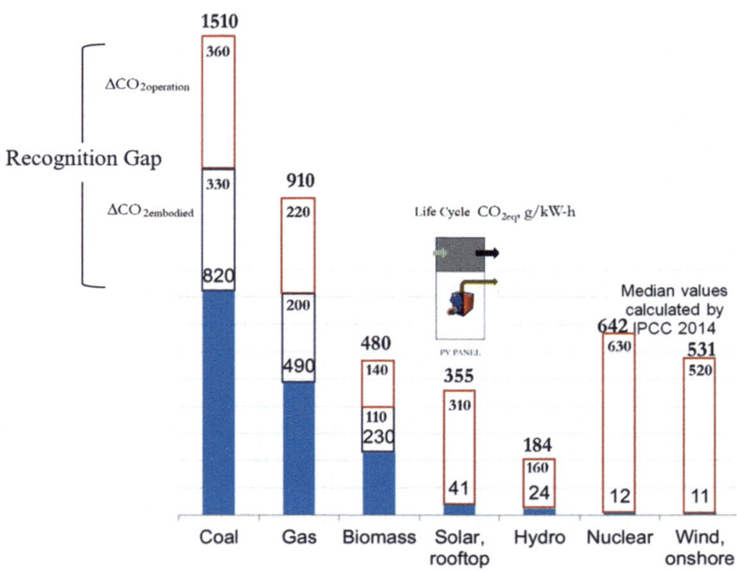

Fig. 2.13 ΔCO_2 emissions for embodiments and during operation. (Modified from IPCC et al., 2014; Solar, 2024)

decarbonization task about the predictions of several sectors related to the direct embodied emissions only (IPCC et al., 2014; Solar, 2024).

2.3 Human Responsibilities in Global Warming

Between 1990 and 2019, population growth was responsible for 28% of the increase in global emissions from residential buildings. The lack of sufficient policies accounted for 52% of the growth in floor area per capita. The increasing carbon intensity of the global energy mix accounted for 16% of emissions. Efficiency improvements contributed to a 49% reduction in global emissions, with no regions realizing their sufficiency potential over the period. Supply decarbonization remained untapped in developing countries and partially in developed Asia-Pacific. South Asia had the highest untapped sufficiency and supply decarbonization potential, leading to a 185% increase in emissions due to a lack of sufficiency measures and a 340% increase due to the high carbon intensity of the energy mix. Among developed countries, Asia-Pacific developed countries had the highest untapped sufficiency potential. The Middle East was the only region where no efficiency potential was realized.

2.3.1 Global Warming

Atmospheric CO_2 increase is a function of human policies at large, especially on the urban scale. Therefore, the global warming issue must be handled in terms of sustainability and carbon neutrality, not only on the building scale but at the same time on the city scale and beyond. The SDEWES index has predicted that the rate of challenging cities or cities affected by such difficulties among the world's cities is 0–25% among the cities where it is applied. On the other hand, cities seeking solutions were in the 25–50% band in the second quarter. While these cities, despite the solution, need strategic solutions, the third category, described as pioneer cities, are cities with strengths. These cities are expressed with strengths in the 25% performance, while transitional cities have strengths in the 50–75% band above the average value. All of these efforts and benchmarks can be viewed as efforts that directly support decarbonization (Kilkis, 2014). Altamirano-Avila and Manuel Martínez (2021) also concluded that very few cities in the world have a planning process based on a vision of sustainable development, even though 70% of the world's population lives in urban areas. Figure 2.14 shows global warming scenarios until 2022 (Reuter, 2019). Many scientific reports estimate that temperatures could rise by 3 °C in the next 30 years, which would cause sea levels to rise all over the world.

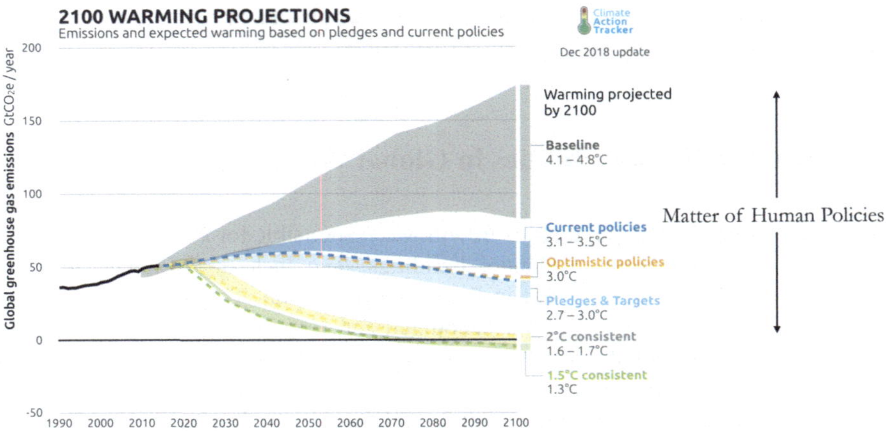

Fig. 2.14 Global warming scenarios until 2100 (Reuter, 2019)

2.3.2 IPCC Forecasts

Over a time frame of about 30 years between 1990 and 2019, CO_2 emissions from the building sector worldwide increased by 50%, and total greenhouse gas (GHG) emissions reached 12 $GtCO_2eq$ on an annual basis in 2019. Taking into account all sources of emissions across all sectors, including agriculture, forestry, and other land use, the building sector represented 21% of global GHG emissions and 31% of global CO_2 emissions. Of this global share, 57% of CO_2 emissions were attributed to off-site electricity and heat generation, and 24% came from direct CO_2 emissions on-site. 18% was based on CO_2 emissions from material production (IPCC, 2022a, 2022b). Using an index decomposition analysis with building-specific identities and representing the three pillars of the Sufficiency, Efficiency, Renewables (SER) framework, the building-specific drivers of GHG emissions in the four scenarios mentioned above are evaluated. A Kaya decomposition is used to analyze the population and GDP, two major sources of GHG emissions. Using an index decomposition analysis with building-specific identities and representing the three pillars of the Sufficiency, Efficiency, Renewables (SER) framework, the building-specific drivers of GHG emissions in the four scenarios mentioned above are evaluated (Lamb et al., 2020) for the impact of building-specific drivers on energy demand and not on CO_2 emissions (IPPC, 2015; Creutzig et al., 2018; IEA, 2023; ODYSSEE-Mure, 2020). For this assessment, the decomposition was conducted for energy-related CO_2 emissions for residential buildings only due to a lack of data for nonresidential buildings.

The attribution of changes in emissions in the use phase to changes in the drivers of population, sufficiency, efficiency, and carbon intensity of energy supply is calculated using additive log-mean Divisia index decomposition analysis (Ang & Zhang, 2000). The decomposition of emissions into four driving factors is shown in

Eq. (2.1), where m^2 refers to total floor area, EJ refers to final energy demand, and $MtCO_2$ refers to the sum of direct and indirect CO_2 emissions in the use phase. As an example, the following original equation from IPCC needs to be modified with ΔCO_2. Equation (2.14) exemplifies that those standards, codes, and metrics, which are all based on the first law, may be upgraded to the second law simply by adding appropriate terms concerning ψ_R, like in this case, $MtCO_2$ term is divided by $(1 - \psi_R)$ (Kilkis, 2019; TTMD, 2017).

$$CO2^k_{\text{total}} = \text{Pop} \times \frac{m^2}{\text{Pop}} \times \frac{\text{EJ}}{m^2} \times \frac{\text{Mt}_{CO2}}{\text{EJ}} / (1 - \psi_R) \tag{2.14}$$

2.3.3 *Effect of Human Activities on Exergy Destruction*

The current emissions of 15.3 $GtCO_{2eq}$/year × 1.14 × 10^{-3} K/Gt CO_2 = 0.0174 K/year of global temperature rise give a prediction of about 0.35 K in the next 25 years, which is 0.436 K just from buildings. But building construction will double in the next 25 years. Then, it may reach almost 1 K, which exceeds the 1.5 K target of the Paris Agreement until 2050.

Today, the globe is already at the 1 K level. IPCC can only predict 0.45 × 10^{-3} K/$GtonCO_{2eq}$, giving a lower prediction of 0.41 K instead of 1 K, as predicted by the second law. The difference is mainly due to the unrecognized and unaccounted ΔCO_2 term. See Fig. 2.15.

2.3.4 *Decoupling of Economic Growth and Emissions*

The first law does not recognize the emission responsibilities of exergy destructions in the built environment, and decarbonization efforts are all based on it with the Pareto economy, which is quasi-linear. Decoupling is possible by following the

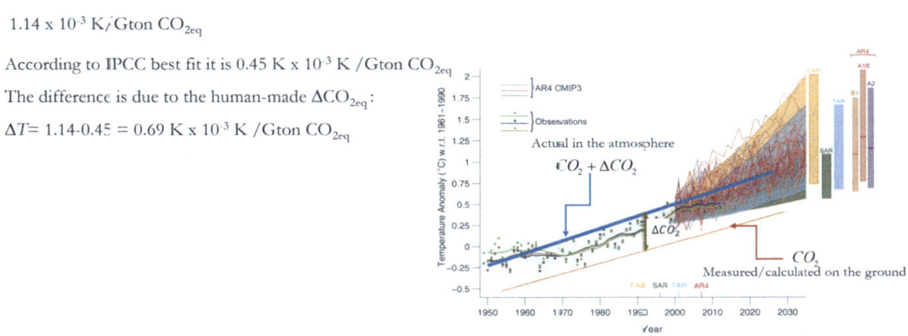

Fig. 2.15 The effect of introducing the ΔCO_2 term into atmospheric CO_2 concentration history

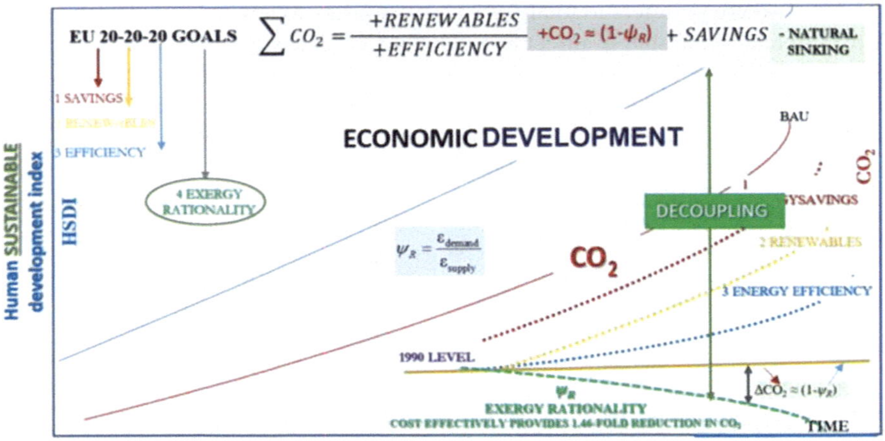

Fig. 2.16 The role of the second law on decoupling economic development and CO_2 emissions (Kilkis, 2022a)

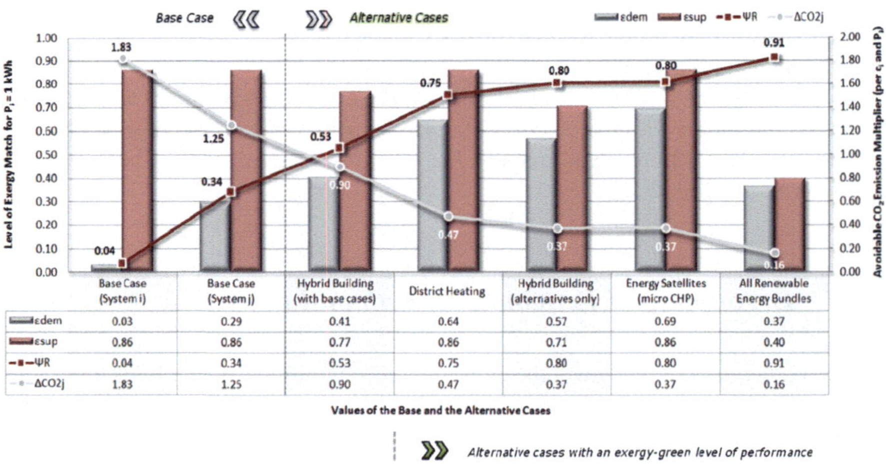

	Base Case (System i)	Base Case (System j)	Hybrid Building (with base cases)	District Heating	Hybrid Building (alternatives only)	Energy Satellites (micro CHP)	All Renewable Energy Bundles
εdem	0.03	0.29	0.41	0.64	0.57	0.69	0.37
εsup	0.86	0.86	0.77	0.86	0.71	0.86	0.40
ΨR	0.04	0.34	0.53	0.75	0.80	0.80	0.91
ΔCO2j	1.83	1.25	0.90	0.47	0.37	0.37	0.16

Fig. 2.17 The relationship between ΔCO_2, unit demand exergy, and unit supply exergy (Kilkis, 2011)

REMM Methodology (Kilkis, 2011). Furthermore, the linear economy needs to be adaptable to nonlinear environmental and atmospheric issues pertaining to global warming (Kilkis, 2022a). Figure 2.17 supplements Fig. 2.16 by showing the relationship between ΔCO_2 and unit supply and demand exergy for different scenarios.

2.4 Exergy-Based Design and Rating

2.4.1 Design Rules

Figures 2.18 and 2.19 summarize the priorities to be given for setting sustainable roadmaps for decarbonization in sustainable buildings and cities. Figure 2.18 shows

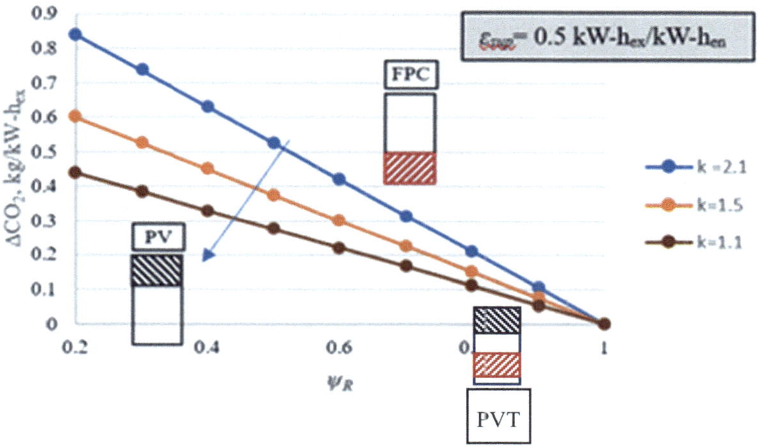

Fig. 2.18 The comparison of ΔCO_2 for different systems and applications

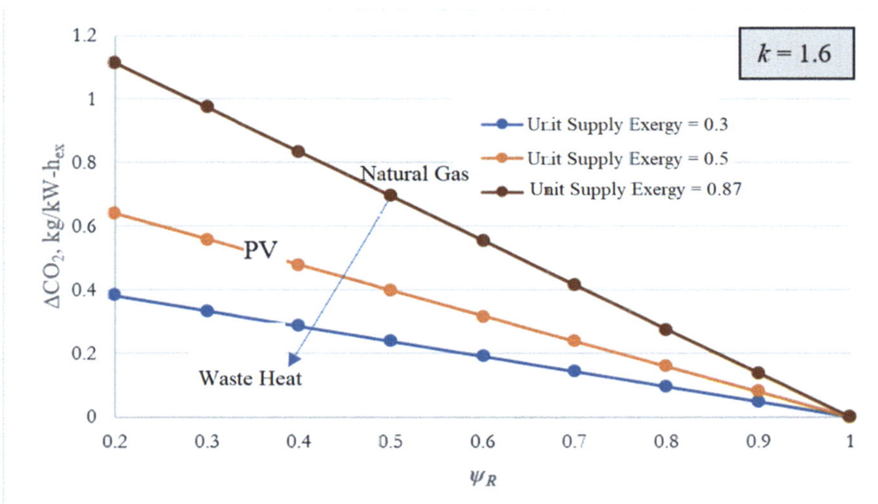

Fig. 2.19 Importance of the utilization of waste heat to replace fossil fuels at an average *k* value of 1.6

that ΔCO_2 simultaneously depends on the k value, which corresponds to the energy conversion system type and the exergy rationality, ψ_R. For example, as a sustainable policy for the rational utilization of solar energy, solar photovoltaic (PV) panels should be preferred over solar flat-plate collectors (FPCs) for a wide range of ψ_R, which varies with the type of use. For example, even if the electric power generated by the PV panel is used for space heating with an electric resistance heater with a ψ_R value even less than 0.20, ΔCO_2 will be less than the FPC panel with a higher ψ_R value, even if the generated heat is directly used as heat for the same purpose. Obviously, a solar photovoltaic–thermal (PVT) system would be much better. Figure 2.19 shows that the higher the unit supply exergy for sources like fossil fuels, the more ΔCO_2 increases.

2.4.2 *Impact of Renewable Energy Systems*

Renewable energy system, which is an important step in decarbonization processes, has also become the main criterion for the Paris Agreement targets. Table 2.1, developed for this purpose, can be considered as a criterion that will support all sectoral

Table 2.1 Comparison of exergy rationality and energy efficiency of renewable energy systems (Kilkis, 2022a, 2022b, 2022c)

System		ψ_R	$\bar{\eta}_I$	η_{II}	Exergy Flow Bar	ΔCO_2	r_{EX}
SOLAR, I_n = 750 W/m², ε_{so} = 0.58 W/W, T_f = 591 K	PV	0.62	0.20	0.327	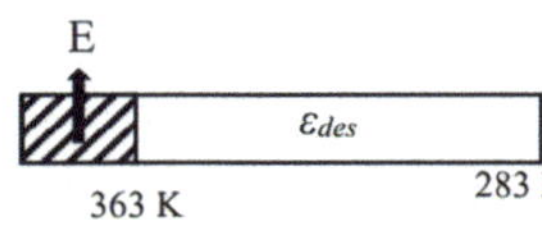	0.06	0.335
	PVT	0.74	0.80 (0.2+0.6)	0.414	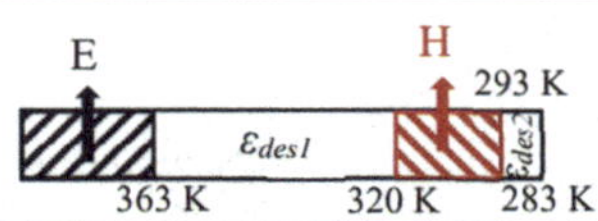	0.04	0.505
	FPC (Flat-Plate Collector)	0.217	0.75	0.164	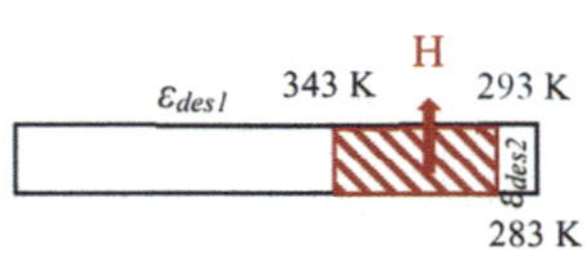	0.286	0.059
Geothermal ORC+Waste H. (120°C) ε_g= (1-283/393) =0.2798		0.713	0.70 (0.1+0.6)	0.489	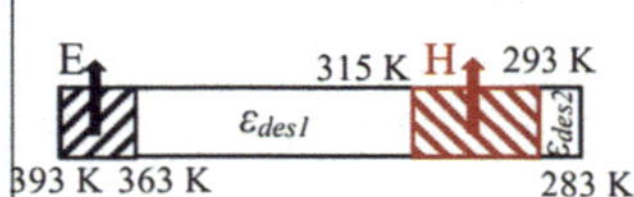	0.045	0.28
Wind Turbine		0.40	0.40	0.40	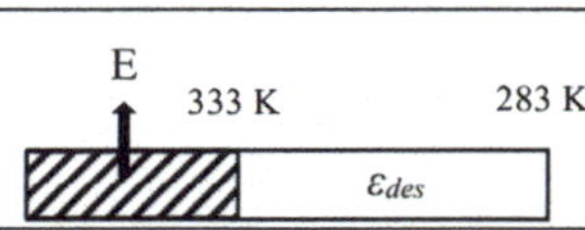	0.0405	1.0

stakeholders in their decision processes. In this context, r_{EX} (The new Potential Renewable Penetration Index) defines the decarbonization potential of a renewable energy system and refers to a natural gas ($\varepsilon_f = 0.87$ kW-h$_{ex}$/kW-h$_{en}$) power plant with 600 K and a reference temperature of 283. Therefore, ψ_{Rref} is equal to 1-(1–283 K/600 K)/0.87) = 0.40.

$$r_{EX} = \left(\frac{\varepsilon_{sup}}{\varepsilon_f}\right) \times \left(\frac{\psi_R}{\psi_{Rref}}\right) \times \eta_{II} \tag{2.15}$$

According to Table 2.1, r_{EX} is the highest penetration index related to wind turbines, and the next highest index belongs to PVT. A combination of a wind turbine and PVT with ancillaries is shown in Fig. 2.20. In this case, r_{ex} values are additive: 1 + 0.505 = 1.505. This shows that the hybridization of renewable energy systems and equipment gain priority. LUE$_{EX}$ is the new exergy-based index for land-use effectiveness given in Eq. (2.16).

$$\mathrm{LUE}_{EX} = \frac{\sum_{n}^{i=1} \eta_{I_i} \times \varepsilon_{sup_i}}{A_L} \tag{2.16}$$

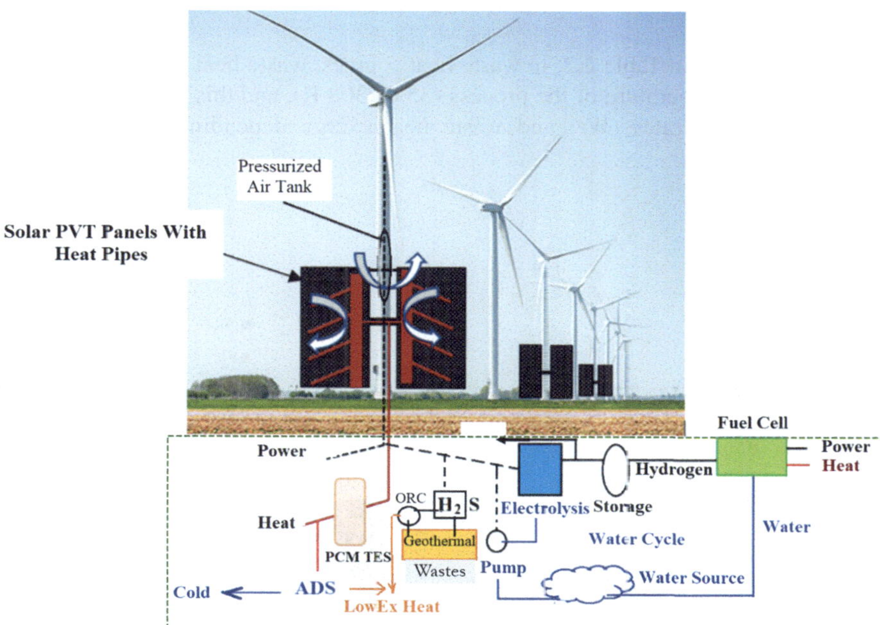

Fig. 2.20 A synectic combination of wind turbines and PVT panels

Here, (n) refers to the number of different exergy outputs. We can exemplify this for ORC in a geothermal process. As a matter of fact, n is two in ORC's electric power and waste heat production. AL refers to the total land area occupied for each kW_{ex} (UN, 2024; Kilkis, 2022a, 2022b, 2022c).

Example 3

Regarding the geothermal ORC and waste heat use system from Table 2.1, with A_L equal to 0.27 m^2/kW_{ex}, power generation efficiency of 0.12, and thermal output efficiency of 0.65:

$$LUE_{EX} = \frac{0.12 \times 0.95 \text{kW} \quad \text{h}_{ex} / \text{kW} \quad \text{h}_{en} + 0.65 \times \left(1 - \frac{293\text{K}}{315\text{K}}\right) \text{kW} \quad \text{h}_{ex} / \text{kW} \quad \text{h}_{en}}{0.27 \text{m}^2 / \text{kW}} = 0.59 \text{kW} / \text{m}^2$$

See also Fig. 2.21 for different renewable energy systems.

New criteria for prioritizing climate emergencies relative to environmental conditions are required, in terms of the waste heat utilization rate (WBR), in addition to r_{EX}:

$$\text{WBR} = \frac{W_r \times \varepsilon_w \times \eta_{Iw}}{\left(\frac{c_K}{0.2}\right)} \quad \{\text{Maximum}, c_K > 0\} \tag{2.17}$$

As can be seen in Table 2.2, in waste heat sources, waste heat can be recovered up to the limit temperature of the process (35 C, 308 K), and this potential has different waste heat rates (Wr) and waste heat exergy depending on the source

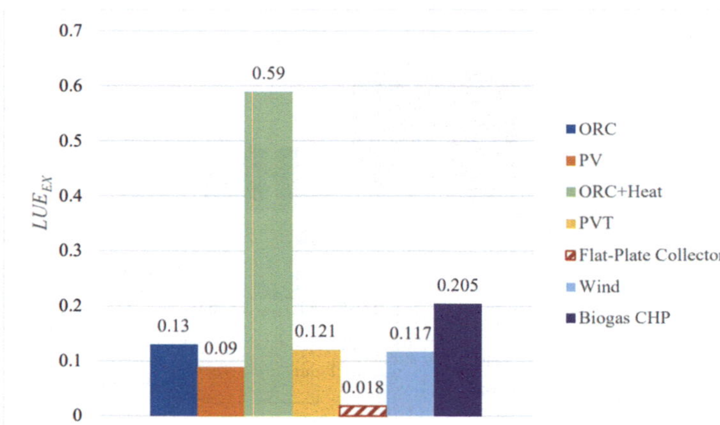

Fig. 2.21 LUE_{EX} of different renewable energy systems

Table 2.2 Waste heat potential of different sources, unit exergy, WBR, and PM. $T_{ref} = 283$ K

Source		W_r	ε_W	η_{IW}	c_K	WBR	ΔCO_2	PM[b]
Power plants	Natural gas, combined cycle	0.40	0.096[a]	0.40	0.20	0.0154	0.056	0.0116
	Solid fuel	0.60	0.402	0.70	0.65	0.0519	0.235	0.4518
Low-enthalpy geothermal		0.70	0.198	0.80	0.45	0.2494	0.115	0.1661
Buildings with:	Condensing B.	0.10	0.150	0.15	0.20	0.0022	0.087	0.0324
	Non-condensing B.	0.25	0.331	0.75	0.20	0.0620	0.194	0.1008
Roof type solar PV		0.60	0.220	0.70	0.27	0.028	0.129	0.0604
Solar PVT		0.15	0.095	0.25	0.27	0.0684	0.040	0.0486

[a] At the cooling tower inlet

[b] CO_{2base} is 0.63 kg CO_2/kW-h_{exergy} for both power and heat in equal proportions $w_1 = w_2 = 0.5$

temperature. In this context, the first-law efficiency of thermodynamics (η_{IW}) directly shows the efficiency of the heat recovery system. In the table, waste heat utilization rate (WBR) refers to c_K (0.2 kg CO_2/kW-h_{ex}) indexed to natural gas, which is the CO_2 content of the fuel source. As a matter of fact, in renewable energy sources, c_K is either 1.1 or 2.1 kg CO_2/kW-h_{ex}. This value depends on the upstream or downstream exergy destruction of the useful work potential that occurs during the flow of exergy in the process. The WBR value is directly affected by waste heat exergy, recovery potential, first-law efficiency, and source-dependent carbon factor (c_K). However, WBR is also prioritized to generate PM for carbon removal strategy directly linked to the maximum utilization potential of waste heat. However, w_1 and w_2 are the factors that will affect it as given in Eq. (2.18).

$$\mathrm{PM} = w_1 \times \mathrm{WBR} + w_2 \left(\frac{\Delta CO_2}{CO_{2\mathrm{base}}} \right) \quad \{\mathrm{Maximum}\} \tag{2.18}$$

CO_{2base} is the reference emission responsibility of the energy sector for both power and heat (0.63 kg/kW-h_{en}). Both WBR and PM must be maximum for the highest priority to penetrate the energy market due to their urgency of converting from waste to useful work (Kilkis, 2022a, 2022b, 2022c).

- Exergy-Levelized Cost, ELC

  The concept of standard cost is expanded in a levelized structure by the following two key regulatory measures that include supply exergy, first-law efficiency, and panel area. Direct costs are shaped by exergy potential. ELC primarily involves the rationality of exergy and environmental costs and is expressed directly in relation to energy and CO_2 regulations. WP defines panel weight here, and PC is the panel market price.

$$\mathrm{ELC} = \left(\frac{\mathrm{PC} + \mathrm{EM} \times W_P}{I_{\mathrm{test}}\, A_P \Psi_R \left[\varepsilon_{\sup E} \eta_E + \varepsilon_{\sup H} \eta_H \right]} \right) \quad \{ ¢ / \mathrm{kW_{EXpeak}} / \mathrm{m}^2 \} \tag{2.19}$$

- Exergy-Levelized Emissions, ELC_{CO2}

A similar equation may be derived for CO_2 embodiments levelized to exergy destructions during operation at a given renewable energy ratio, R_{EX}:

$$ELC_{CO_2} = \left(\frac{EM_{CO_2} \times W_P + 0.27(1 - R_{EX}) I_{test} A_P (1 - \Psi_R)}{I_{test} A_P \Psi_R \left[\varepsilon_{\sup E} \eta_E + \varepsilon_{\sup H} \eta_H \right]} \right) \tag{2.20}$$

$$\{kg\,CO_2 / kW_{EXpeak} / m^2\}$$

Energy Utilization Success Rate, EUS

If the result is negative or zero, then EUS is zero (no mitigation success). 0.63 kg CO_2/kW-hen is the threshold value.

$$EUS = 1 - \frac{\Sigma CO_2}{0.63} \tag{2.21}$$

Carbon Footprint Index, CFI

$$CFI = \frac{\Sigma \varepsilon_{des}}{\varepsilon_{sup}} \quad \{CFI \to 0\} \tag{2.22}$$

Sustainable Environment Index, SEI

$$SEI = \overline{\eta}_I \times \overline{\psi}_R (1 - ODI) \quad \text{or} \quad COP_{system} \times \overline{\psi}_R (1 - ODI) \tag{2.23}$$

$$ODI = \frac{wGWP^y}{(1 - ODP)} \times \left(\frac{ALT}{1} \right)^z \quad \{ODP < 1\} \tag{2.24}$$

Ideally, SEI should be close to one. Concerning Eq. (2.24), typically, $w = 0.1$, $y = 0.045$, and $z = 0.01$. The term (y) includes the water vapor effect. Any refrigerant with zero ODP is not free from the ozone depletion effect. Refrigerant leakages have ΔCO_2 emission responsibility.

$$\Delta CO_{2-\text{refrigerant}} \equiv 1/g \left[1 - \{wGWP^y ALT^z\} / ODI \right]^{1/x} \tag{2.25}$$

The ratio of ΔCO_2 to CO_2 is an indicator of how much CO_2 content measured in the atmosphere is traceable on the site of the emissions. The terms (g) and (x) depend on the heat pump type, size, and refrigerant. Usually, the cause and origin of the ΔCO_2 emissions are not recognized, and therefore, carbon mitigation efforts fall short of reaching the Paris Agreement. R_{CO2} must be close to zero. Yet in conventional systems like boilers, ΔCO_2 may reach seven (see Fig. 2.9), and mitigation efforts fall short by the same magnitude.

$$R_{CO_2} = \Delta CO_2 / CO_2 \quad \{\text{For } CO_2 > 0\} \tag{2.26}$$

Virtual Adiabatic Flame Temperature, T_f^*

To draw the temperature-based exergy flow bar for nonthermal processes and energy sources, an ideal Carnot cycle–based virtual source temperature must be defined. The so-called T_f^* is given by the following equations.

For Solar Energy

$$T^*_{f\text{-solar}} = \frac{T_{ref}}{1 - 0.95 \times \dfrac{I_n}{TSI}} \tag{2.27}$$

The nominal value of total solar irradiance (TSI) is 1.3675 ± 0.5% kW/m^2 (GSFC, 1999).

Example 4

If the instant solar irradiation normal to the receiver surface is 0.85 kW/m^2, and T_{ref} is 283 K:

$$T^*_{f\text{-solar}} = \frac{T_{ref}}{1 - 0.95 \times \dfrac{I_n}{TSI}} = 691\,\text{K}$$

The corresponding ε_{solar} is (1–283 K/691 K) = 0.590 kW-h$_{ex}$/kW-h$_{en}$, according to the ideal Carnot cycle.

For Wind Turbines

The virtual source temperature may be calculated as follows:

$$T^*_{f\text{-WT}} = \frac{T_{ref}}{(1 - 0.95 \times \eta_{IWT})} \tag{2.28}$$

Example 5

The first-law efficiency of a given wind turbine is 0.5.

$$T^*_{f\text{-WT}} = \frac{283\,\text{K}}{(1 - 0.95 \times 0.5)} = 539\,\text{K}$$

In this example, ε_{WT} is (1–283 K/539 K) = 0.475 kW-h$_{ex}$/kW-h$_{en}$, according to the ideal Carnot cycle.

For Geothermal Steam or Any Other Steam

$$T^*_{f\text{-steam}} = \frac{T_{ref}[\text{K}]}{\left[1 - \dfrac{E_{Xsteam}[\text{kW}-\text{h}/\text{kg}] \times \dot{m}_{steam}[\text{kg}/\text{h}]}{\dot{Q}_{steam}[\text{kW}]}\right]} \tag{2.29a}$$

Another simple expression is from Akbar and Kilkiş (2024):

$$\varepsilon_{\text{steam}} \approx 0.00198 \times \left(T_{\text{steam}} - 283\ \text{K}\right) \tag{2.29b}$$

$$T^{`}_{f-\text{steam}} = \frac{T_{\text{ref}}}{\left(1 - \varepsilon_{\text{steam}}\right)} \tag{2.29c}$$

For Carnot-Cycle-Equivalent Pressure

With a direct analogy between the mechanical energy potential of a pressurized volume and P_p, in reference to the standard atmospheric pressure, P_a, and T_{ref} taken equal to the ambient air temperature, T_a:

$$T^{`}_{f-\text{pressure}} = \frac{T_a}{1 - 0.95\left|1 - \frac{P_a}{P_p}\right|} \tag{2.30}$$

For Commodity Cost

For a given commodity, if CO_2 embodiments are recovered in Y years by CO_2 savings compared to 0.63 kg CO_2/kW-h, which is the base value, and the present value of CO_2 savings remains constant, then a Carnot-cycle equivalent may be calculated:

$$T^{`}_{f-\text{cost}} = \frac{T_{\text{carnot}}}{1 - \left[1 - \frac{\sum CO_{2\text{em}}}{0.63 - Y \times \sum CO_{2\text{sav}}}\right]} = \frac{383.3\,\text{K}}{1 - \left[1 - \frac{\sum CO_{2\text{em}}}{0.63 - Y \times \sum CO_{2\text{sav}}}\right]} \tag{2.31}$$

District Energy Systems

Figure 2.22 shows a district energy system that utilizes wind turbines to supplement the grid power, and the waste heat from the data centers is temperature peaked by the heat pump station to serve the district heating system. According to the first law, this system seems to be energy-efficient and environmentally useful. However, the second law indicates the following issues:

1. The district size, expressed by the distance between the heat pump station and the district, does not satisfy Constraint 1, L_{max}.
2. The electric power used by the district energy system pumps is not included in the analysis (Constraint 2).
3. The CO_2 equivalence of the ozone depletion index (ODI) is not considered.
4. The exergy destruction–related ΔCO_2 due to the exergy imbalance between the electric power and the district heat supplied is not considered.

Differentiating Eq. (2.7) and equating it to zero gives the optimum diameter:

$$\Sigma CO_{2-1} / \Sigma CO_{2-0} = a + \text{bID} + \text{cID}^{-\text{d}} \tag{2.32}$$

Here, a, b, c, and d, are factors depending on the specific pipe grid design

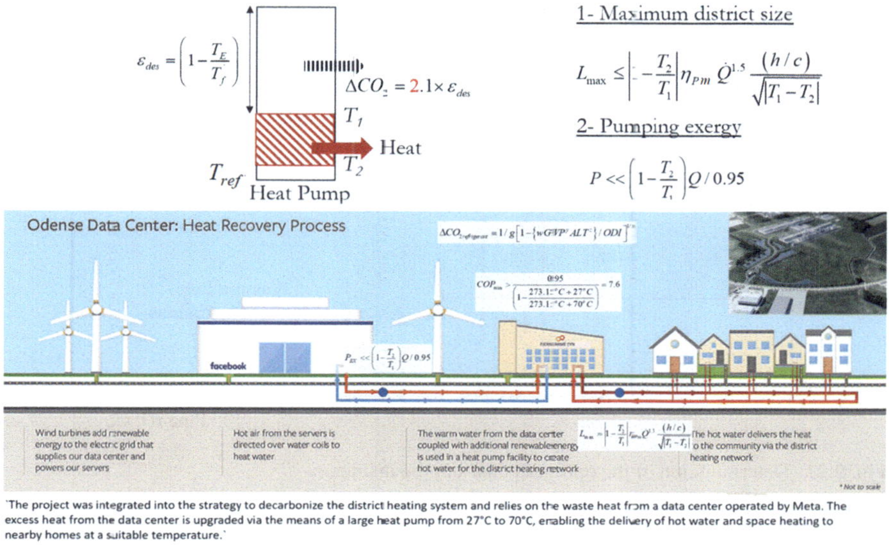

Fig. 2.22 Odense Data Center, how green is it? (META, 2023)

$$\frac{\delta \sum CO_2}{\delta ID} = b - dcID^{-(d+1)} = 0 \tag{2.33}$$

$$ID_{opt} = \sqrt[(d+1)]{\frac{b}{dc}} \tag{2.34}$$

Equation (2.35) shows that L_{max} is a direct function of Q exchanged in the district. To increase L_{max} to serve a wider district with minimum ΣCO_2, Eq. (2.32) must be minimized.

$$L_{max} = \left|1 - \frac{T_2}{T_1}\right| \eta_{IPM} \dot{Q}^{1.5} \frac{(h/c)}{\sqrt{|T_1 - T_2|}} \quad \{\text{Optimization Constraint 1}\} \tag{2.35}$$

Equation (2.36) is the major constraint for the pumping exergy, P_{EX}, in the optimization process:

$$P_{EX} \ll \left(1 - \frac{T_2}{T_1}\right) Q_E / 0.95 \quad \{\text{Optimization Constraint 2}\} \tag{2.36}$$

Figure 2.23 expresses the emission responsibility per meter in a conceptual structure in optimum regional systems. In this context, correlations show that the optimum pipe diameter can be defined to reduce emissions. As a matter of fact, in the flow form, there is an inverse correlation between pipe diameter and pressure loss, which directly reduces exergy demand (Akbar & Kilkiş, 2024). However, as energy demand increases in large pipe diameters, emission and exergy control is

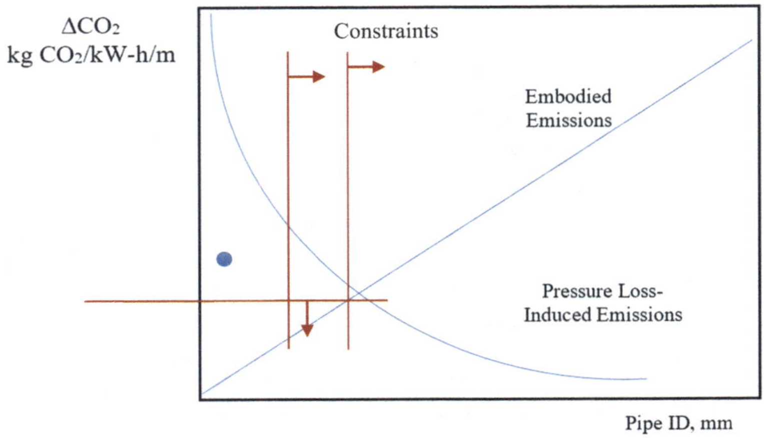

Fig. 2.23 Determination of the optimum district pipe diameter

critical. In this context, while the emission can be defined in a linear structure, the relationship of ΔCO_2 with ID can be estimated with the possible power curve based on the CO_2 and ΔCO_2 and energy input mixture associated with the pumps, as can be seen in Eq. (2.32).

2.4.3 How Close It Can Be to Net-Zero?

2.4.3.1 Waste Heat Utilization

Waste heat from many sources is widely available worldwide, but most of them are not mobilized for useful work because they are considered to have low enthalpy. Generally, if ever used, temperature-conditioning heat pumps are employed, which have nearly avoidable CO_2 emission responsibility. This renders the solution for many cases environmentally irrational, from the viewpoint of the second law of thermodynamics. However, the advent of the very-low- and ultralow-temperature district heating systems offers the ability to accommodate such low-enthalpy energy sources as rational to be used.

This means that the waste heat from a thermal power plant, which, otherwise, is released to the ambient with water consumption, CO_2, cost, and other environmental footprints, may become a viable solution in district energy systems by replacing fossil fuels thus CO_2 emissions, and nearly avoidable CO_2 emissions responsibilities even for renewable energy systems like solar systems. Waste heat sources (1) in Fig. 2.24, are not limited to data centers, supermarkets, cold warehouses, sewer heat, metro stations, industry, and thermal power plants using fossil fuels. On the demand side, passive buildings, prosumers, and nZEXB (2) in Fig. 2.24 are energy-efficient and decrease the thermal loads, even supplying thermal energy to the district energy system (negative-energy buildings). This saves CO_2 and ΔCO_2, but at

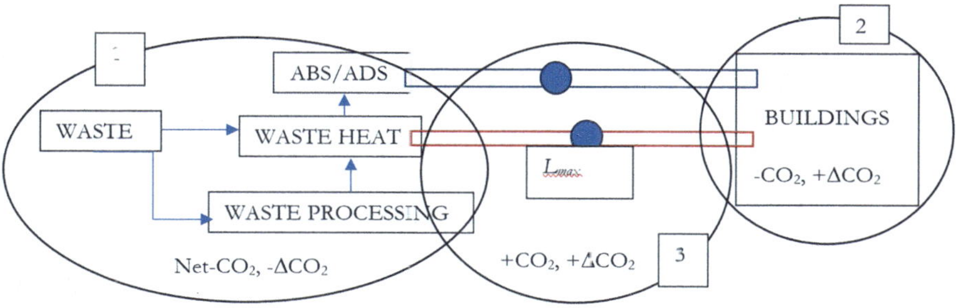

Fig. 2.24 Three prongs of utilizing waste heat versus zero energy buildings

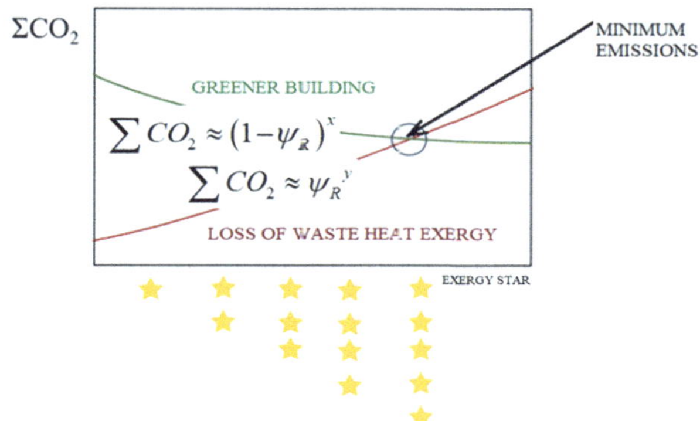

Fig. 2.25 Zero energy building versus waste heat recovery in terms of ΔCO_2 (Energy Star, 2024)

the same time, reduces the marketability and their environmental benefits. The link between the waste heat sources and the buildings in the district (3) by means of pipelines demand pumping electricity during operation and come with embodiments before operation. Thus, the issue becomes a three-pronged optimization problem for minimum CO_2 emission and responsibility in total. The prime question is how much the buildings can close to be nZEXB and/or passive buildings for energy efficiency and the level of prosuming to optimally utilize the waste heat. Linking prosumers in a district energy system also makes thermal sharing and peak-load shaving possible, all of which may contribute to decarbonization starting from the equipment and piping selection level. In this respect, passive houses may not be necessary to be linked because they do not produce energy on-site. Emission predictions must include all embodiments. The objective is to minimize the sum of all CO_2 and ΔCO_2 involved. Figs. 2.24, 2.25, and 2.26 show that if there is a nearby waste heat source, like a power plant or industry, the waste heat may be utilized instead of making the buildings net-zero energy and rejecting the waste heat to the ambient.

If the condition given in Eq. (2.35) is satisfied, then an optimum level of making buildings nearly zero energy and partly utilizing the waste heat has a solution. There is a total CO_2 emission–based break-even point between net-zero building emission savings and the amount of waste heat reclaimed, which, otherwise, would be rejected to the ambient, causing CO_2 emission responsibility.

2.4.3.2 Exergy Star Green Building Rating System

Existing green building metrics do not factor in the second law. They consider only energy efficiency and the utilization of renewable energy systems and mostly end up with metrics such as annual energy consumption per floor area and total CO_2 emissions with a limited concern about secondary emissions (Erten & Kılkış, 2022). The new exergy star rating system includes CO_2 and its root causes in terms of the REMM efficiency, ψ_R, which must be equal to or greater than 0.7 to earn exergy stars (Figs. 2.24 and 2.25). This rating system may be applied to buildings, rather than household equipment, as in the Energy Star program in the United States (Energy star, 2024).

$$\sum CO_2 = \varepsilon_{\text{sup}} \times \bar{k} \times (1-\bar{\psi}_R) + \frac{\bar{c}_K \times \varepsilon_{\text{sup}}}{\eta_{\text{IB}}\left(1-\frac{T_{\text{ref}}}{T_{\text{in}}}\right)} = \varepsilon_{\text{sup}}\left[\bar{k} \times (1-\bar{\psi}_R) + \frac{\bar{c}_K}{\eta_{\text{IB}}\left(1-\frac{T_{\text{ref}}}{T_{\text{in}}}\right)}\right] \tag{2.37}$$

For minimum green application conditions of $\psi_R = 0.70$, Eq. (2.31) becomes Eq. (2.32), which provides an optimization constraint for green and sustainable buildings to be eligible for exergy stars. The optimization condition is given by Eq. (2.38).

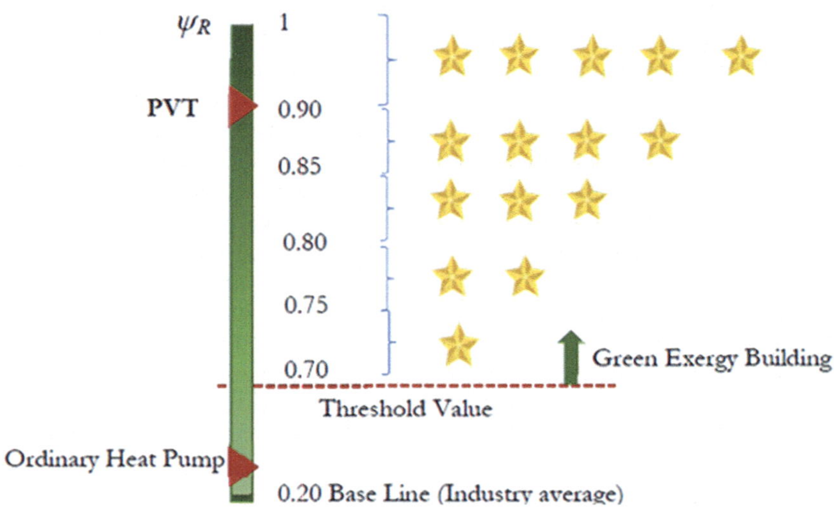

Fig. 2.26 Exergy star rating scale according to ψ_R (Energy Star, 2024)

$$\Sigma CO_2 \leq \varepsilon_{\text{sup}} \left[\bar{k} \times (0.30) + \frac{\bar{c}_K}{\eta_{\text{IB}} \left(1 - \frac{T_{\text{ref}}}{T_{\text{in}}}\right)} \right] \{\text{To be green}\} \tag{2.38}$$

$$\left[Q_{\text{WH}} \left(1 - \frac{T_1}{T_2}\right) + Q_c \left(1 - \frac{T_3}{T_4}\right) \right] - 0.95\ \text{P} \equiv Q_{\text{nZEB}} \left(1 - \frac{T_4}{T_5}\right) \{\text{Exergy Balance}\} \tag{2.39}$$

2.4.4 Importance of Time and Exergy Match in Thermal Energy Storage

Thermal energy storage is an important issue for shaving the peak thermal loads of buildings, reducing the installed capacity of the equipment and their embodiments, letting them run at peak or near-peak efficiency most of the time, and following the various power and heat tariffs for reduced operational costs. See Chapter S50 in *ASHRAE HB HVAC Systems and Equipment*, for information based on the first law (energy quantity). However, thermal energy storage and demand management need to have a careful exergy balance, which especially holds true with the advent of utilizing renewable and waste heat sources. Figure 2.27 shows that exergy coincidence beyond the coincidence of the amounts of stored energy and demand energy is important, which, otherwise, will result in ΔCO_2 emission responsibilities at an

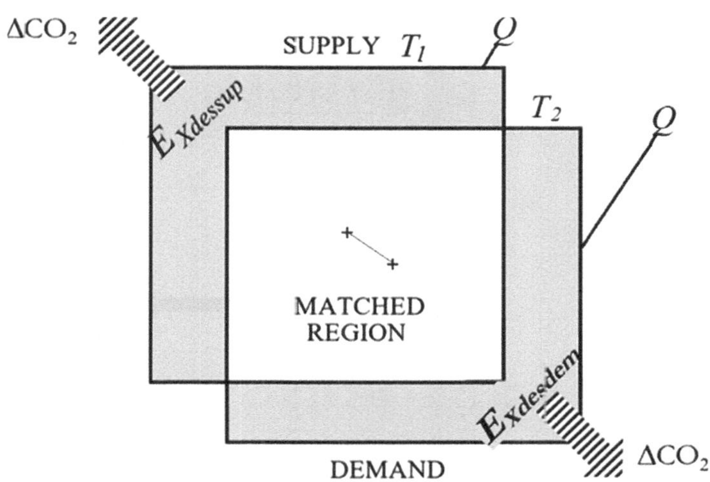

Fig. 2.27 Exergy must be time- and quantity-matched (Kilkis, 2023a)

amount proportional to the exergy mismatch. Consider Fig. 2.27 where the supply energy available at the thermal energy storage (TES) system is sufficient to satisfy the demand energy, as shown by equal square areas. However, the supply exergy does not match the demand exergy completely (the white area) (Kilkis, 2022b). The remaining gray areas representing the exergy mismatch will cause ΔCO_2. This usually happens when a thermal energy storage tank is continually supplied with adequate waste heat at a nearly constant temperature, T_1, where the demand temperature varies by time, $T_2(t)$. If heat, Q, is supplied to satisfy the same quantity of demand, then, the time integral of the exergy differences will be responsible for ΔCO_2, over a period of time (t):

$$\sum \Delta CO_2 = k \times \Delta E_X = k \times Q \times \int_t^0 \left[\left(1 - \frac{T_{\text{ref}}}{T_1} \right) - \left(1 - \frac{T_{\text{ref}}}{T_2(t)} \right) \right] dt \tag{2.40}$$

A similar problem arises when there is a time mismatch. For example, the waste heat supply temperature, T_1, is controllable to follow the time variations of T_2, but there is a response time lag, Δt. Then:

$$\sum \Delta CO_2 = \int_{\Delta T}^0 \left(1 - \frac{T_1(t)}{T_2(t)} \right) dt \tag{2.41}$$

Equation (2.35) assumes that any surplus supply exergy is not wasted and is conserved in the TES tank.

Thermal energy at different temperatures may not be arbitrarily mixed without referring to their unit exergy. There may be three distinct cases. In the first case, there is a single TES tank, which is supplied heat by different sources at different temperatures. Figure 2.28 shows a case with two different supplies serving the same TES tank at Q_1, T_1 and Q_2, T_2, respectively.

$$T_{\text{mix}} = \frac{Q_1 \times T_1 + Q_2 \times T_2}{Q +_1 Q_2} \tag{2.42}$$

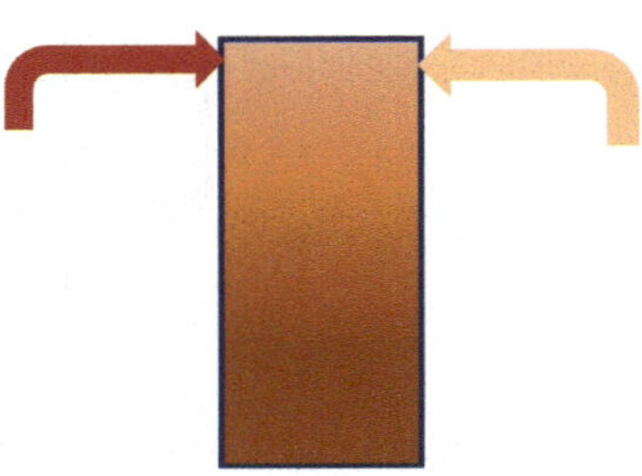

Fig. 2.28 Two energy sources serve a single TES tank

$$\left(Q_1+Q_2\right)\times\left(1-\frac{T_{\text{ref}}}{T_{\text{mix}}}\right)?Q_1\times\left(1-\frac{T_{\text{ref}}}{T_1}\right)+Q_2\times\left(1-\frac{T_{\text{ref}}}{T_2}\right)\{\text{Case 1}\} \tag{2.43}$$

Example 6

Let $Q_1 = Q_2 =$ one unit each, $T_1 = 360$ K, $T_2 = 300$ K, $T_{\text{ref}} = 283$ K. Then, $T_{\text{mix}} = 330$ K.

$$(2)\times\left(1-\frac{283\,\text{K}}{330\,\text{K}}\right)?1\times\left(1-\frac{283\,\text{K}}{360\,\text{K}}\right)+1\times\left(1-\frac{283\,\text{K}}{300\,\text{K}}\right);0.285>0.270.$$

In this case, mixing is admissible. However, when Q is 1.5 units and Q_2 is 0.5 units at the same given temperatures:

$$(2)\times\left(1-\frac{283\,\text{K}}{330\,\text{K}}\right)<1.5\times\left(1-\frac{283\,\text{K}}{360\,\text{K}}\right)+0.5\times\left(1-\frac{283\,\text{K}}{300\,\text{K}}\right);0.285<0.349$$

Accordingly, mixing is not admissible. This example shows that whether or not to mix different sources at different temperatures may have different solutions either at the design or operation stage due to instantaneous conditions of supply and demand. Therefore, keeping each energy source at different, dedicated tanks must be preferred. However, the following two cases show that this option has restrictions.

In Case 2, there are two separate tanks, namely hot TES and warm TES tanks, containing the same volume of fluid. Fluid is transferred from the hot TES tank to the second one, which has double the volumetric capacity to accommodate the condition where the hot tank is completely transferred to the warm tank ($X = 1$). Figure 2.28, using Eq. (2.34), compares two different warm TES temperatures, namely 320 K and 300 K. In this case, the mixing process uses a pump, which demands electrical energy.

$$\underbrace{\left(1-\frac{T_{\text{ref}}}{T_{\text{hot}}}\right)\times(1-X)+\left(1-\frac{T_{\text{ref}}}{\left[T_{\text{hot}}\times X+T_{\text{warm}}\right]/\left[1+X\right]}\right)}_{\text{With mixing}, X>0}\times(1+X)<?> \\ \underbrace{\left(1-\frac{T_{\text{ref}}}{T_{\text{hot}}}\right)\times 1+\left(1-\frac{T_{\text{ref}}}{T_{\text{warm}}}\right)\times 1}_{\text{Without mixing}, X>0}\ \{\text{Case 2}\} \tag{2.44}$$

Figure 2.29 shows that whether mixing is exergy-rational or not largely depends on the hot and warm temperatures. If T_{warm} is 300 K, then, there is a comparably large exergy destruction during the mixing process, resulting in more ΔCO_2. Therefore, Eq. (2.44) must be checked every time two or more different thermal energy sources at different temperatures may be mixed, and mixing may be controlled or cut off on a time scale, depending on the current demand temperature(s) and the energy demand. Figure 2.30 shows the ΔCO_2 responsibility when T_{warm} is 300 K. This case shows that mixing is not exergy-rational.

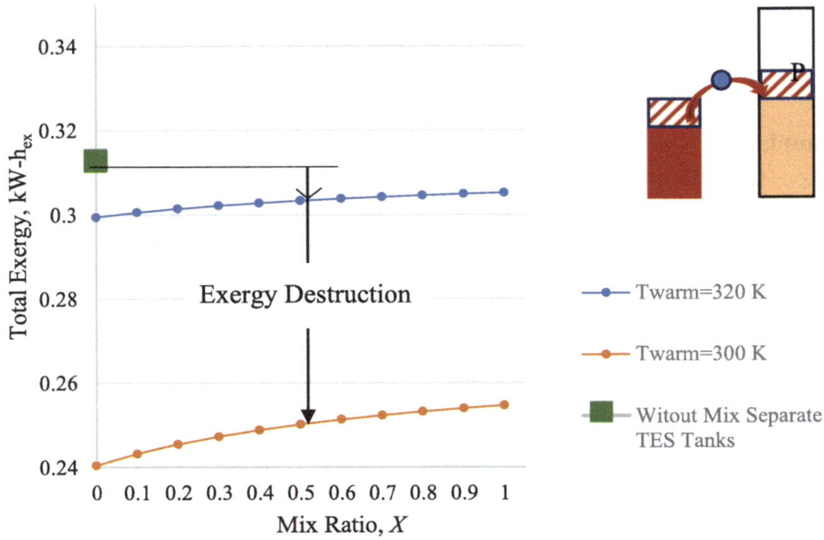

Fig. 2.29 Variation of total exergy with the mix ratio, X. T_{hot} is 360 K, and T_{ref} is 283 K

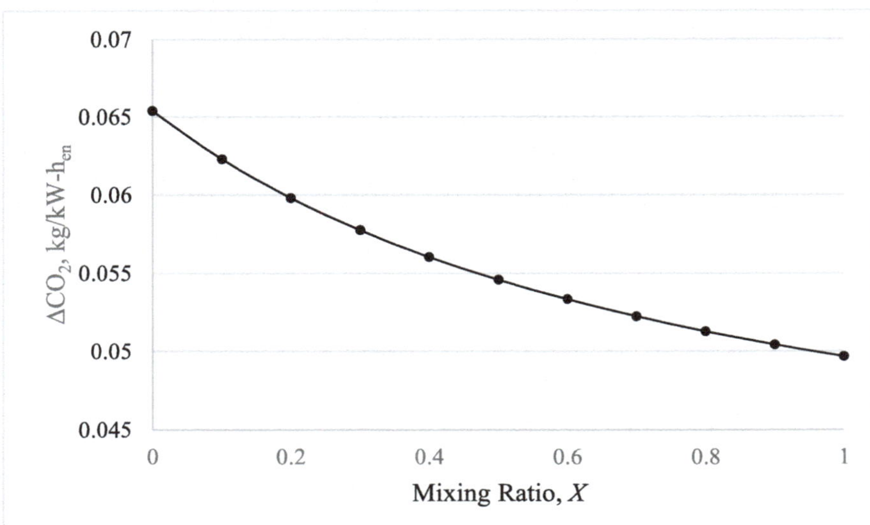

Fig. 2.30 ΔCO_2 responsibility of mixing, Case 2: mixing TES tanks

In the third case, TES tanks are not mixed, but the supply fluids downstream of the TES tanks are mixed on the demand side depending upon the demand temperature and amount of energy demand. Then, Eq. (2.45) and Fig. 2.31 apply.

- Total exergy without mix: 0.329 kW-h$_{ex}$

$$\left(1-\frac{T_{ref}}{T_{mix}}\right)\times(Y+Z)?\left(1-\frac{T_{ref}}{T_{hot}}\right)\times Z+\left(1-\frac{T_{ref}}{T_{warm}}\right)\times Y\,\{\text{Case 3}\} \tag{2.45}$$

$$T_{mix}=\frac{Z\times T_{hot}+Y\times T_{warm}}{(Z+Y)}=\frac{T_{hot}+\left(\frac{Y}{Z}\right)\times T_{warm}}{\left(1+\left(\frac{Y}{Z}\right)\right)},\quad Z>0\{\text{Case 3}\} \tag{2.46}$$

Example 7

$$T_{hot}=360\,\text{K}, T_{warm}=300\,\text{K}, T_{ref}=283\,\text{K}, Y/Z=1.$$

$$T_{mix}=\frac{360\,\text{K}+1\times 300\,\text{K}}{2}=330\,\text{K}\text{, from Eq. (2.35):}$$

$$\left(1-\frac{283\,\text{K}}{330\,\text{K}}\right)\times(2)?\left(1-\frac{283\,\text{K}}{360\,\text{K}}\right)\times 1+\left(1-\frac{283\,\text{K}}{300\,\text{K}}\right)\times 1$$

$$0.142<\left(1-\frac{283\,\text{K}}{360\,\text{K}}\right)\times 1+\left(1-\frac{283\,\text{K}}{300\,\text{K}}\right)\times 1=0.270$$

This example shows that mixing the two TES tanks downstream for a given demand is not exergy-rational. Figure 2.31 shows this case for ΔCO_2 emission responsibility at different *Y/Z* ratios. Therefore, instead of mixing the fluid downstream from multiple TES tanks, each tank should serve more exergy-compatible

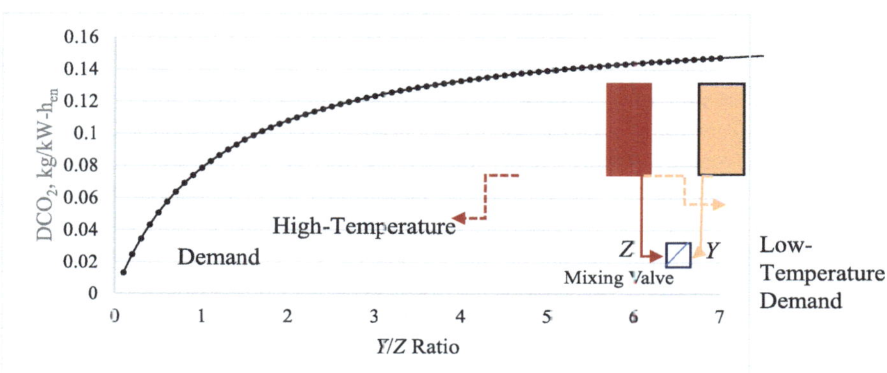

Fig. 2.31 ΔCO_2 responsibility with (*Y/Z*) ratio of TES mix downstream on the demand side

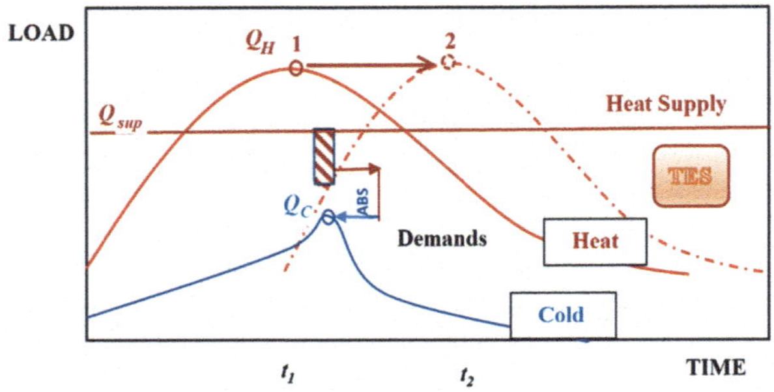

Fig. 2.32 Elimination of the time gap between peak heat supply and cooling loads by TES

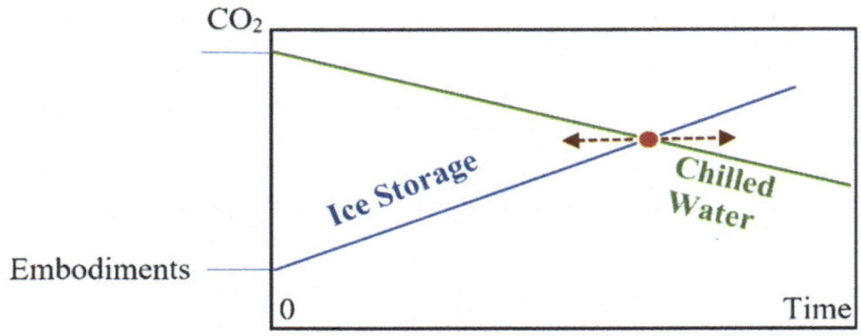

Fig. 2.33 Optimum system selection for cold storage

demands in a multitude of demands at different temperatures, diversified during the design.

Cold supply, which has a lower ψ_R compared to heat, must be supported by thermal energy storage (TES) systems. Peak heat load is shifted from time t_1 to t_2 to allow surplus heat during peak cold loads, as shown in Fig. 2.32. Then, the stored heat is utilized by absorption and or adsorption machines to produce cold to cover the peak cooling load. This method reduces the problem of the lower ψ_R values for cooling. The time gap between peak heat supply and peak cooling loads is shown in Fig. 2.32.

$$\text{HTCF} = \left(Q_H - Q_C\right)\left(\frac{\varepsilon_H - \varepsilon_C}{t_2 - t_1}\right) \tag{2.47}$$

HTCF defines the exergy-minimum TES requirement by load management and deep building renovations (EPEE, 2020). Ice storage has less embodied CO_2 (smaller tank), but ice-making requires deep chillers with lower COP, which means more CO_2 emissions during operation. A GSHP has a higher COP for chilled-water preparation; thus, less electricity is spent during operation but more embodied emissions due to the manufacturing operation for increased tank size. Therefore, an optimum solution must be sought regarding the life cycle, COP values of the deep chiller and the heat pump, and their CO_2 embodiments (Fig. 2.33).

2.5 Illustrative Calculations

2.5.1 *Heat Pumps*

2.5.1.1 Heat Pumps and Decarbonization

Heat pump is increasingly recognized as one of the critical technologies for thermal decarbonization, receiving focused policy support in several countries (IEA, 2024). In the same token, heat pumps are considered to be the best options for decarbonization of thermal systems with the roadmaps of near-total electrification with renewable energy sources (EU, 2023a, 2023b). The *ASHRAE Handbook* chapters A9, A35, and S48 give information about heat pumps in terms of the first law. Heat pumps are widely used for temperature-peaking (in heating) and temperature-decreasing (in cooling) in a better and wider recovery and utilization of the waste heat and refrigeration heat (ASHRAE Handbook Chapters A9). Despite several advantages of heat pumps according to the first law, the second law recognizes that heat pumps use electric (or mechanical) power to generate thermal power (power-to-heat). Electricity has high unit exergy, where the unit thermal energy is much lower. Therefore, heat pumps may destroy exergy leading to nearly avoidable CO_2 emissions, ΔCO_2, if their COPs are not high enough.

In particular, heat pumps are favored in nearly zero (nZEBs) and zero energy buildings (ZEBs), especially in schools, coupled with solar PV panels. Torcellini and Drury (2006) give information about zero energy buildings and provide a case study regarding the Lewis Center at Oberlin College in Oberlin, Ohio. The building features daylighting, radiant heating, natural ventilation, evaporative cooling, ground-source heat pumps, solar PV panels, and passive solar strategies. Based on the first law only, they found that the buildings' energy consumption was 25–70% lower than the ASHRAE 90-1 code. They rated four different scenarios, based on the energy use intensity, EUI, which deals only with the first law.

The type and unit exergy of different energy uses are neglected such that EUI definition should be upgraded according to the second law (EXUI).

Another case study is about the new Music and Science Building at the Hood River Middle School located in Portland Oregon (Kalkan et al., 2011). The building features a ground-source heat pump, radiant panel heating and cooling concrete floors, outside air preheating, solar duct, heat recovering ventilation, and roof-mounted PV panels. Their calculations were based on the first law, and the average COP values for the heat pumps for heating and cooling were not mentioned.

A further sample application is about the PV-assisted thermal load satisfaction of the Highfield Campus buildings at South Hampton University by air-source heat pumps (Brown & Frichtl, 2013). They reported a seasonal COP value of 3.7. The authors argue that if the COP is three, then, the heat pump is "300% efficient," meaning that for every kW-h of electricity, 3 kW-h of heat is obtained. Obviously, this is not a correct statement because the first-law efficiency may not exceed 100%, and that is why the term COP is defined, which excludes the ambient energy input.

Their research does not recognize the unit exergy difference between electric and thermal powers involved.

2.5.1.2 Heat Pumps and Exergy Destructions

Although the heat pump industry has a background of over a century, the knowledge base is limited by the first law of thermodynamics to define their performance and benefits. More importantly, the related terminology about the most fundamental concepts and units is not properly used. For example, the European Heat Pump Association (EHPA) claims that for each "kW" of electricity consumed by a heat pump, about 4 kW of "thermal energy" is generated (Cu, 2018). They further claim that:

this corresponds to a 300% efficiency.

It is ignored that electric power is used for thermal power, both of which have different useful work potentials (quality of energy: exergy). Scientifically sound unit exergy of electricity, $\varepsilon_{E,}$ may be defined in the following manner:

$$\varepsilon_E = \left(1 - \frac{288.45\,\mathrm{K}}{5500\,\mathrm{K}}\right) = 0.94755\,\mathrm{kW - h_{ex}\,/\,kW - h_{en}} \approx 0.95\,\mathrm{kW - h_{ex}\,/\,kW - h_{en}}$$

Here, 5500 K is the average surface temperature of the sun (NASA, 2024). The ideal Carnot cycle in the solar system is closed by assuming that an ideal photovoltaic cell located on Earth having an average surface temperature of 288.45 K (in 2023) (EU, 2023a, 2023b) is directly facing the sun. This is the maximum unit of exergy that may be obtained on earth. In contrast to this ideal condition, sometimes ε_E is rounded to one for practical purposes (Kilkis, 2021). However, this is not admissible unless either the reference temperature is absolute zero (0 K) or the source temperature is infinite, both of which are not available in the solar system. On the other hand, if a heat pump generates thermal energy between a supply and return temperature of 325 K and 315 K, respectively, the quality of thermal power will be:

$$\varepsilon_H = \left(1 - \frac{315\ \mathrm{K}}{325\ \mathrm{K}}\right) = 0.031\,\mathrm{kW - h_{ex}\,/\,kW - h_{en}}$$

Although a heat pump receives heat (in heating) or rejects heat to the ambient (in cooling) and the heat pump supplies a total amount of heat at a coefficient of performance (COP), which is greater than one, there is an exergy imbalance between exergy supply and exergy provided. This is irreversible exergy destruction:

$$\text{Exergy destruction} = (0.031 \times \mathrm{COP} - 0.095)\,\mathrm{kw - h_{ex}\,/\,kW - h_{en}}$$

This imbalance (if negative) will lead to avoidable CO_2 emissions, ΔCO_2, because someone, somewhere, and some time will need to offset the irreversibly destroyed useful work opportunities, most possibly by using some fossil fuels.

$$R_{\mathrm{EX}} = \frac{\sum_{n}^{i=1} Q_{ri} \times \varepsilon_{ri}}{\sum_{n}^{i=1} Q_{ri} \times \varepsilon_{i} + \sum_{m}^{j=1} Q_{ji} \times \varepsilon_{fj}} \quad \{Q'\text{s are on annual basis, kW} - \mathrm{h_{ex}} / \text{year}\} \quad (2.48)$$

2.5.1.3 Sample Case (Fig. 2.34)

Each source may have multiple supplies like power, heat, and cold. Although for a unit input electrical power of 1 kW_{en}, thermal power is generated by a factor of COP, which in this case is four, there remains an exergy deficit according to the second law:

$$\text{Electrical exergy input} = 1\,\mathrm{kW_{en}} \times 0.95\,\mathrm{kW-h_{ex}} / \mathrm{kW-h_{en}} = 0.95\,\mathrm{kW_{ex}},$$

$$\text{Thermal exergy output} = 4 \times 1\,\mathrm{kW_{en}} \times (1 - 300\,\mathrm{K} / 320\,\mathrm{K}) = 0.25\,\mathrm{kW_{ex}}.$$

Example 8

Here, 300 K is a typical input (return) temperature to the heat pump, T_1, and 320 K is the supply temperature, T_2, for example, in a low-temperature heating schema. The presence of a different k value (1.1) for the small exergy destruction downstream is neglected ($\varepsilon_{\text{des-2}} = (1–283\ \text{K}/300\ \text{K}) = 0.057\ \text{kW-h}_{\text{ex}}/\text{kW-h}_{\text{en}}$).

The threshold COP of this application for the balance between input and output exergy is about:

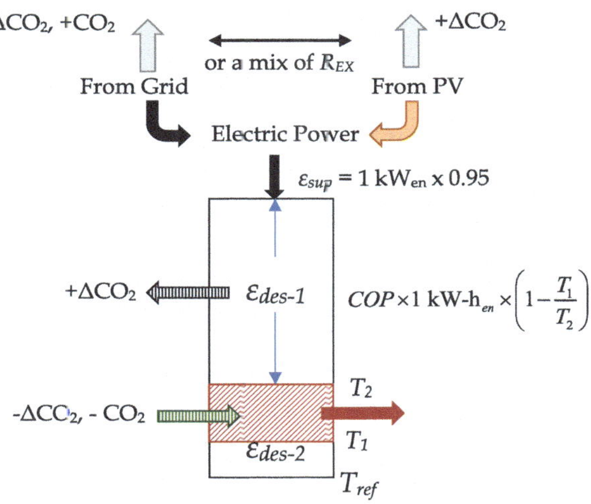

Fig. 2.34 Sample heat pump operation

$$\mathrm{COP}_{\min} \geq \frac{0.95}{\left(1-\dfrac{300\,\mathrm{K}}{320\,\mathrm{K}}\right)} = 15.2 \tag{2.49}$$

Such a high COP value is not available in the industry today, even if the heat pumps are cascaded (not more than two: see (Kilkis, 2021). Instead, one should generate more useful work output by filling the gap between the electrical power exergy and total thermal exergy output, thus reducing $\varepsilon_{des\text{-}1}$. The present literature about heat pumps does not relate their COP and exergy destructions concerning emissions, namely nearly avoidable ΔCO_2 emission responsibility and direct CO_2 emissions at the source of electric power generation. In his Ph.D. thesis, Velazquez (2008) compiled 104 papers, starting in 1995, about ground-source heat pumps and identified five categories of research, namely, thermal characterization, operational and feasibility analysis, computer-aided simulations, experimental models, and environmental variable characterization [55]. Only 16 papers regarding "environment" were listed, which were concerned about contamination, ground conductivity, moisture effects, boreholes, underground water flow, etc. None of the papers addressed CO_2 emissions due to their power demand, exergy destructions, and associated nearly avoidable ΔCO_2 emission responsibility, all attributable to heat pumps. Only Zogou and Stamatelos (2007) considered the connection of the thermal performance of a building with a ground-source heat pump system. As a result, COP must be modified with the second law to recognize the actual carbon footprint of heat pumps. In this example, the exergy-based COP (equal to four), namely COPEX, is less than one:

$$\mathrm{COPEX} = \mathrm{COP} \times \frac{\left(1-\dfrac{T_1}{T_2}\right)}{0.95} = 4 \times \frac{\left(1-\dfrac{300\,\mathrm{K}}{320\,\mathrm{K}}\right)}{0.95} = 0.263 \lll 1 \tag{2.50}$$

$$\Delta CO_{2\ HP} = +2.1 \times \varepsilon_{des} = +2.1 \times \left(\varepsilon_{sup} - \varepsilon_{dem}\right) = +2.1 \times \left(0.95 - 4 \times \left(1 - \frac{300\,\mathrm{K}}{320\,\mathrm{K}}\right)\right)$$
$$= +1.47\,\mathrm{kg\,CO_2 / kW\ h_{ex}}$$

The same result may be obtained by the following formula for a heat pump:

$$\begin{aligned}\Delta CO_2 &= 2.1 \times 0.95 \times (1 - \mathrm{COPEX}) = 1.995 \times (1 - 0.263) \\ &= 1.47\,\mathrm{kg\,CO_2 / h}\,\{\text{Second law}\}\end{aligned} \tag{2.51}$$

For other systems with multiple exergy destructions, the following approximate formula may be used:

$$\Delta CO_2 = 1.995 \times (1 - w \times \mathrm{COPEX})\{0.85 \leq w \leq 1\} \tag{2.52}$$

2.5.1.4 If Electricity Is Supplied from Central Power Plants

If electricity of 1 kW-h_{EL} to run the heat pump directly comes from thermal power plants through the grid (with an approximate mix of natural gas and anthracite coal power plants, $c_K = 0.3$ kg CO_2/kW-h_{FUEL}) and thermal power plants do not supply heat to the district, then, the CO_2 responsibility of the heat pump will be:

$$\begin{aligned} CO_{2\ PP} &= +0.3 \times \text{PEF} \times 1\,\text{kW}_{EL} \\ &= +0.75\,\text{kg}\,CO_2\,/\,\text{h}\{\text{First law,PEF} = 2.5,\text{at the heat pump plug}\} \end{aligned}$$

$$CO_{2-PPS} = +0.3 \times 1\,\text{kW}_{EL} = +0.3\,\text{kg}\,CO_2\,/\,\text{h}\{\text{At the power plant stack}\}$$

$$\begin{aligned} \Delta CO_{2\ PP} &= +\frac{1.1 \times \left(1 - \dfrac{T_{\text{ref}}}{T_E}\right) \times 1\,\text{kW}_{\text{en}}}{(\eta_{\text{ITT}})} \\ &= +0.684\,\text{kg}\,CO_2\,/\,\text{h}\{\text{Second law}, T_E = 600\,\text{K}, \eta_{\text{ITT}} = 0.85, k = 1.1\} \end{aligned}$$

$$\Sigma CO_{2-PP} = (+0.75 + 0.684)\,\text{kg}\,CO_2\,/\,\text{h} = +1.434\,\text{kg}\,CO_2\,/\,\text{h}.$$

$$\begin{aligned} &\text{TOTAL emission responsibility of the heat pump and the power plant} \\ &= 1.47 + 1.434 = +\mathbf{2.904}\,\text{kg}\,CO_2\,/\,\text{h}. \end{aligned}$$

Note that the overall second-law emission responsibility, ΔCO_2, including the heat pump and the power plant [(1.47 + 0.684)/0.75], is almost three times more than what the first law predicts for $CO_{2\text{-pp}}$. Therefore, designers and decision-makers for heat pumps for decarbonization need to consider the second law as well.

2.5.1.5 Exergy Rationality, ψ_R.

At the power plant, $\psi_{R\text{-PP}}$ is:

$$\begin{aligned} \psi_{R\ PP} &= 1 - \frac{\varepsilon_{\text{des}}}{\bar{\varepsilon}_{\text{sup}}} \times (1 - \eta_{\text{IPP}}) = 1 - \frac{(\Delta CO_2 \times \eta_{\text{ITT}})/k}{0.87 \times Y + (1 - Y) \times 0.89} \times (1 - \eta_{\text{IPP}}) \\ &= 0.67\{Y = 0.5\} \end{aligned} \tag{2.53}$$

At the heat pump, $\psi_{R\text{-HP}}$ is:

$$\psi_{R\ \text{HP}} = 1 - \frac{\varepsilon_{\text{des}}}{\varepsilon_{EL}} \times (1-0) = \frac{\varepsilon_{\text{dem}}}{0.95} = \frac{\left(1 - \frac{300\,\text{K}}{320\,\text{K}}\right)}{0.95}$$
$$= 0.066 \{\text{No electricity generated but consumed}, \eta_{\text{I-PP}} = 0\}$$

Taking an exergy-weighted average:

$$\bar{\psi}_R = \frac{\psi_{R\text{-pp}} \times \varepsilon_{\text{EL}} \times \eta_{\text{ITT}} + \psi_{R\text{-HP}} \times \text{COP} \times \left(1 - \frac{300\,\text{K}}{320\,\text{K}}\right)}{\varepsilon_{\text{EL}} \times \eta_{\text{ITT}} + \text{COP} \times \left(1 - \frac{300\,\text{K}}{320\,\text{K}}\right)} = 0.527$$

The overall ψ_R (0.527) is less than 0.7. So this application is not green.

2.5.1.6 If the Heat Pump Is Driven by On-Site PV Panels

Because the solar PV system replaces the power plant demand for electric power supply:

$$CO_{2\ \text{PV}} = 0 - (CO_{2\text{-pp}} + \Delta CO_{2\text{pp}})$$
$$= -1.434\,\text{kg}\,CO_2/\text{h}\{\text{If PV power is utilized on} - \text{site}\}$$

$$CO_{2\ \text{PV}} = 0 - \left(CO_{2\text{-pp}} - \Delta CO_{2\text{pp}}\right) \times \left(\eta_{\text{IINV}} \times \eta_{\text{ITT}}^{\ 2}\right)$$
$$= -0.932\,\text{kg}\,CO_2/\text{h}\{\text{If PV power is fed to the grid}\}$$

$$\Delta CO_{2\ \text{PV}} = + \frac{1.1 \times \left(1 - \frac{T_{\text{ref}}}{T_E}\right) \times 1\,\text{kW}_{\text{E}}}{\left(\eta_{\text{IINV}} \times \eta_{\text{IPV}}\right)}$$
$$= +1.347\,\text{kg}\,CO_2/\text{h}\{T_E = 363\,\text{K}, \eta_{\text{IINV}} = 0.9, \eta_{\text{IPV}} = 0.2\}$$

$$\Sigma CO_{2\text{-PV}} = (1.347 - 1.434)\,\text{kg}\,CO_2/\text{h} = -0.087\,\text{kg}\,CO_2/\text{h}$$
$$\{\text{If PV power is utilized on} - \text{site}\}[\text{Emissions savings are negligible}]$$

$$\Sigma CO_{2\text{-PV}} = (1.347 - 0.932)\,\text{kg}\,CO_2/\text{h} = +0.415\,\text{kg}\,CO_2/\text{h} \sim 0$$
$$\{\text{If PV power is fed to the grid}\}[\text{There are no emission savings}]$$

$$\text{TOTAL emission responsibility} = 1.47 - 0.087 = +\mathbf{1.383}\,\text{kg}\,CO_2/\text{h}$$
$$\{\text{With HP responsibility, PV power on} - \text{site}\}$$

$$\text{TOTAL emission responsibility} = 1.47 + 0.415 = +1.885\,\text{kg}\,CO_2\,/\,\text{h}$$
$$\{\text{With HP responsibility, PV power on grid}\}$$

In this example, the PV + HP combination does not mitigate CO_2 in all cases, but this conclusion must be corrected because the heat pump replaces a boiler to supply the same heat.

2.5.1.7 Replacement of the Boiler by the Heat Pump

If this heat pump is replacing a natural gas–fired ($\varepsilon_{NG} = 0.87$ kW-h$_{ex}$/kW-h$_{en}$), on-site condensing boiler with $\eta_{IB} = 0.85$ (average) and it saves the following emissions for four kW$_{en}$ of heat at the same supply temperature (320 K) generated by the boiler, the emission mitigation due to this replacement:

$$CO_{2-B} = -4\,\text{kW}_{en} / 0.85 \times 0.2 = -0.94\,\text{kg}\,CO_2\,/\,\text{h}$$

$$\Delta CO_{2-B} = -2.1 \times 4\,\text{kW}_{en} \times \left[0.87/0.85 - (1 - 283\,\text{K}/320\,\text{K})\right]$$
$$= -7.626\,\text{kg}\,CO_2\,/\,\text{h}$$

$$\Sigma CO_{2B} = -8.566\,\text{kg}\,CO_2\,/\,\text{kW} - \text{h}_{ex}$$

Comparing this mitigation potential with the total CO_2 responsibility of the heat pump driven by the grid:

$$+\mathbf{2.904}\,\text{kg}\,CO_2\,/\,\text{h} - 8.566\,\text{kg}\,CO_2\,/\,\text{h} = -5.662\,\text{kg}\,CO_2\,/\,\text{h}$$

If connected to on-site PV panels:

$$-0.087\,\text{kg}\,CO_2\,/\,\text{h} - 8.566\,\text{kg}\,CO_2\,/\,\text{h} = -8.652\,\text{kg}\,CO_2\,/\,\text{h}$$

If connected to the grid first with on-site PV panels: +0.415 kg CO_2/h −8.566 kg CO_2/h = −8.151 kg CO_2/h.

2.5.1.8 Replacement of the Boiler with the Flat-Plate Collector (FPC) (Fig. 2.35)

With the advent of an increase in the renewable mix in power generation stock, the share of solar flat-plate collectors (FPCs) may also increase in the same order. In this case, the replacement of solar collectors instead of on-site boilers may affect the final decision. This is a dilemma, showing that on the one hand, if PV panels increase in stock, CO_2 mitigation potential increases, but on the other hand, the impact of PV

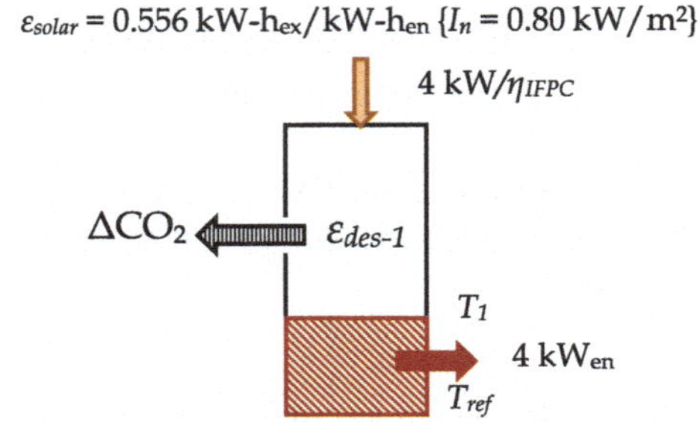

Fig. 2.35 Exergy destruction in a boiler or FPC

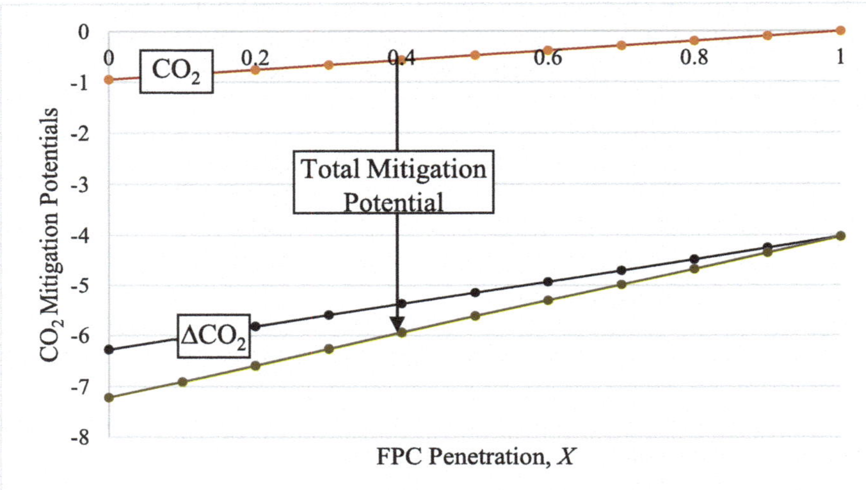

Fig. 2.36 Mitigation potential of the heat pump driven by on-site PV power

panels on CO_2 mitigation potential diminishes with the increase of FPC share in the energy stock, replacing boilers, which also seem to be banned in several countries soon (Home Building and Renovating (2024). The FPC panel does not replace any other system and misses solar power generation opportunity, and therefore, it is responsible for a net ΔCO_2:

$$\Delta CO_2 = +2.1 \times 4\ kW_{en}\left[\frac{0.556}{\eta_{IFPC}} - \left(1 - \frac{283\,K}{335\,K}\right)\right] = +5.37\,kg\,CO_2\,/\,h$$
$$\{\eta_{IFPC} = 0.70, T_1 = 335\,K, Tre_2 = 283\,K\}$$

$$CO_{2FPC} = 0\{\text{Neglecting embodiments}\}$$

The only advantage over boilers is that FPC does not emit direct emissions (except embodiments). Figure 2.36 shows the ΔCO_2, CO_2, and the total mitigation

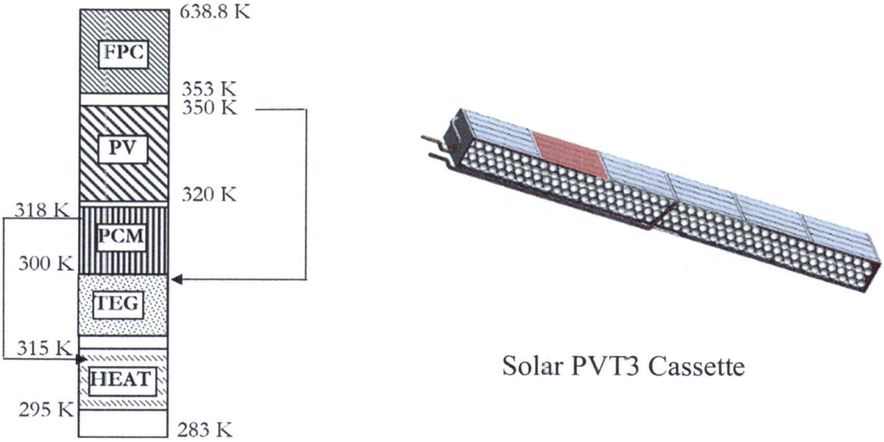

Fig. 2.37 Exergy flow bar for the cassette-type solar PVT3 (Kılkış, 2020) (Patent: TPE 2017/10622)

impact of the penetration of solar FPC panels, X, to replace boilers in the energy sector.

Figures 2.37 and 2.38 show an advanced PVT system, PVT3, which comprises several layers sandwiched together. The main layers are a flat-plate collector (FPC) at the top, a PV layer, a phase-change material (PCM) layer, a thermoelectric generator (TEG), and a heat collecting layer (HEAT). Heat is collected by tiltable heat pipes. During this process, a minuscule amount of TEG power in the opposite polarity is also possible (Fig. 2.39) (Kılkış, 2020). Exergy destructions are minimal, and the REMM efficiency is maximum. When part of the thermal power is temperature-peaked, on-demand, ψ_R decreases slightly to 0.85 on average, depending upon how much PVT3 thermal output is directed to the peaking PVT, downstream, which is not as exergy-efficient as PVT3. Figure 2.40 shows the exergy-feasible domain of PVT3 and PVT in Europe and the Mediterranean countries.

Solar PVT3 and conventional solar PVT for temperature-peaking in tandem arrangement may eliminate the use of heat pumps. According to the second law, such a combination mitigates total CO_2 because of three reasons:

1. ΔCO_2 of solar PVT3 and PVT are small.
2. They do not bear any direct emissions, CO_2.
3. If heat pumps are used for additional temperature-peaking, COP of the heat pump increases; thus, ΔCO_2 and CO_2 at the grid-feeding power plants decrease in the inverse proportion of the increase in COP.

If a temperature-peaking heat pump is required, the tandem arrangement reduces the temperature-peaking demand assigned to the heat pump, ΔT_p, which is the difference between the source temperature T_{source} and the peaked temperature T_{peaked}. This reduction increases the COP of the heat pump:

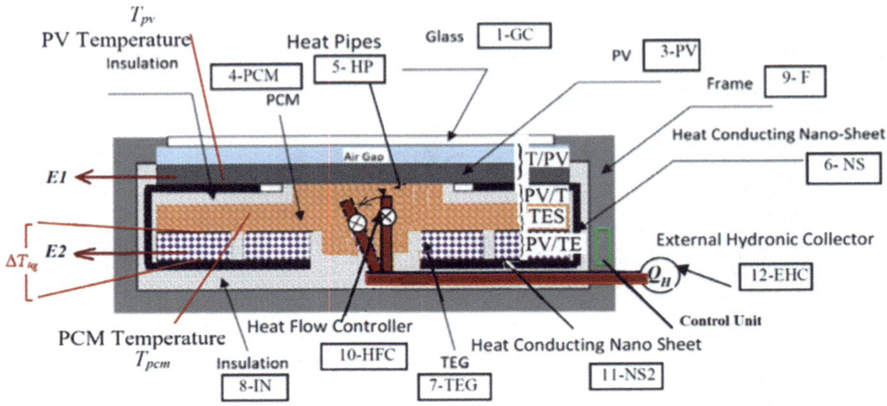

Fig. 2.38 Cross-sectional view of solar PVT3 (Kılkış, 2020)

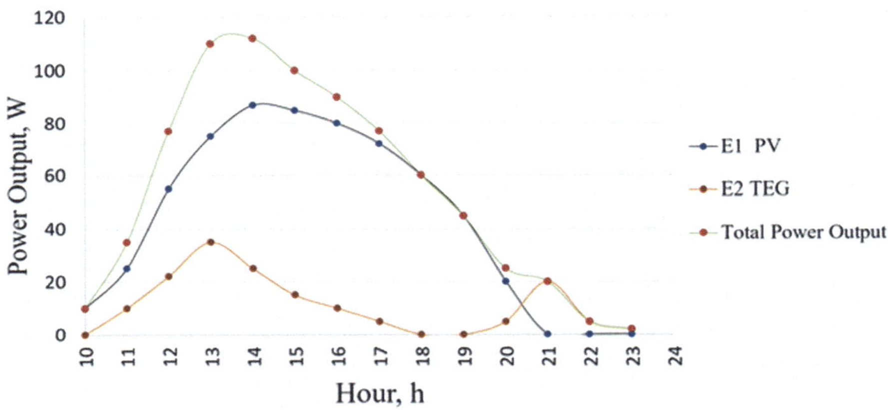

Fig. 2.39 Daily performance of solar PVT3 (Kılkış, 2020)

$$\mathrm{COP} = a - b\Delta T_p = a - b\left(T_{\text{peaked}} - T_{\text{source}}\right) = a - b\left(\Delta T_p\right) \tag{2.54}$$

Here, a and b are the linearization coefficients of the COP curve of a given heat pump. The following criteria have to be considered for 1 kW of electrical power input, E_{XE}, to supply thermal power, E_{XH}:

$$E_{\mathrm{XH}} > E_{\mathrm{XE}} \tag{2.55}$$

$$\mathrm{COP} = \left[a - b\Delta T_p\right] \times 1 > \frac{0.95}{\left(\dfrac{\Delta T_p}{T_{\text{peaked}}}\right)} \times 1 \tag{2.56}$$

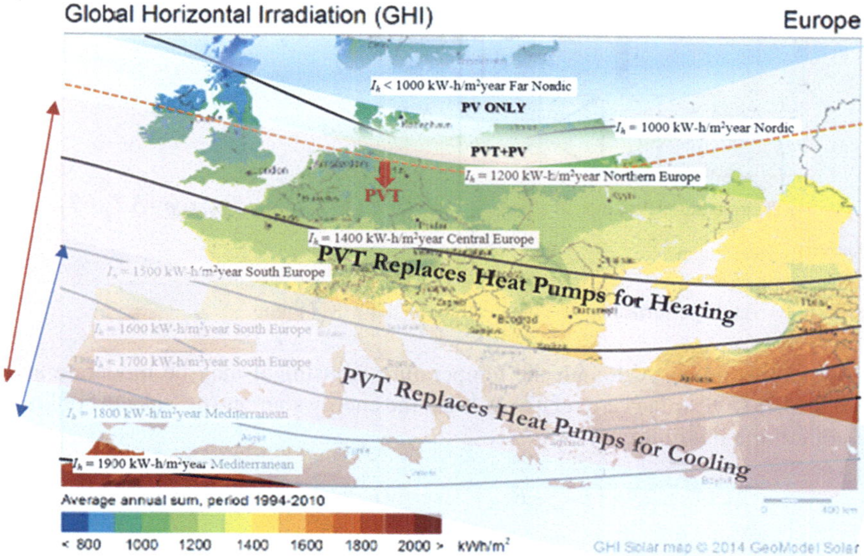

Fig. 2.40 Solar PVT feasibility map for Europe and the Mediterranean countries from the original GIS map

Example 9
If $b = 0.05\ \text{K}^{-1}$, $T_p = 320$ K, $\Delta T_P = 30$ K, $a = 4$, then:

$$E_{XH} = 2.5\,\text{kW}_{ex}, E_{XE} = 0.95\,\text{kW}_{ex}$$

Therefore, the necessary condition given in Eq. (2.25) is not satisfied, indicating that proportional to the difference between E_{XE} and E_{XH}, this power-to-heat system will always be responsible for ΔCO_2, unless COP is equal to or greater than 10.1 = 0.95/(30/320 K). The heat pump will also be responsible for CO_2 if grid power is used. As a result, instead of a temperature-peaking heat pump, a PVT alternative is preferable. In this case, the ΔCO_2 term will be in the range of 0.1 kg CO_2/kW-h of electricity generated, and ψ_R will be around 0.4 (compared to 0.85 for PVT3).

2.5.1.9 Corrections for COP

Conventional COP definition assumes that the ambient energy source like air, sea, or ground is practically infinite and the heat pump operation does not alter the thermodynamics of the ambient. However, today, heat pumps are widely used in the built environment and the industry to adjust (peak in heating) the supply temperature for the demand with limited waste heat or sewer heat in cities. There are large heat pumps used for fifth-generation district energy systems with low and ultralow

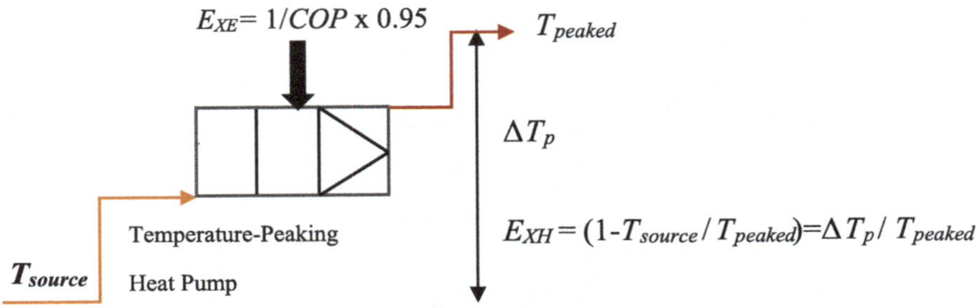

Fig. 2.41 Exergetic performance of a temperature-peaking heat pump

supply temperatures. Although the temperature adjustment may be moderate and therefore the COP values are relatively high, the energy sources (or sinks in cooling) are not infinite but finite.

$$\Delta CO_2 = 2.1 \times (E_{XE} - E_{XH}) \{\text{Exergy is destroyed upstream}\}$$

Consider a heat pump used for peaking the temperature of a low-exergy thermal source like municipal sewage water or waste heat from an industrial plant with limited supply to heat or cool indoors. These thermal sources are not infinitely available and not cost- and energy-free (Fig. 2.41). Therefore, as a first step, the COP equations should be modified. Only then, we may speak about the first-law efficiency, η_I, while the correct definitions will be less than one. Cooling, in fact, means heating the ambient in a reverse thermal flow compared to heating:

$$\text{COP}_{\text{cooling}} = \frac{Q_{ac}}{W + Q_c} = \eta_I < 1 \{\text{Heating the ambient}\} \tag{2.57}$$

$$\text{COP}_{\text{heating}} = \frac{Q_h}{W + Q_{ah}} = \eta_{Ih} < 1 \{\text{Heating indoors}\} \tag{2.58}$$

Based on the second law and the ideal Carnot cycle, this research has derived the following correction factor for such cases of limited energy sources (sinks):

$$\text{COPEX}_{\text{correct}} = \frac{\text{COP}_{\text{ST}} \times W_{\text{input}} \times \left(1 - \frac{T_2}{T_1}\right)}{W_{\text{input}} \times 0.95 + W_{\text{input}} \times (\text{COP}_{\text{ST}} - 1) \times \left(1 - \frac{T_{\text{ref}}}{T_R}\right)}$$
$$= \frac{\text{COP}_{\text{ST}} \times \left(1 - \frac{T_2}{T_1}\right)}{0.95 + (\text{COP}_{\text{ST}} - 1) \times \left(1 - \frac{T_{\text{ref}}}{T_R}\right)} \tag{2.59}$$

Then, for a given heat pump COP at given operating conditions and where the COPEX for a standard COP definition from the industry, COP_{ST} is:

$$COPEX_{ST} = \frac{Q_{supply} \times \left(1 - \frac{T_2}{T_1}\right)}{W_{input} \times 0.95} = COP_{ST} \times \frac{\left(1 - \frac{T_2}{T_1}\right)}{0.95} \tag{2.60}$$

Comparing Eqs. (2.5a), (2.5b), and (2.6) for deriving a corrected COP, the following correction factor, F_c, is used:

$$F_c = \frac{\dfrac{COP_{ST} \times \left(1 - \frac{T_2}{T_1}\right)}{0.95 + (COP_{ST} - 1) \times \left(1 - \frac{T_{ref}}{T_R}\right)}}{COP_{ST} \times \dfrac{\left(1 - \frac{T_2}{T_1}\right)}{0.95}} \{\text{Limited ambient resource}\} \tag{2.61}$$

$$COP_{correct} = F_c \times COP_{ST} \; \{F_c \leq 1\} \tag{2.62}$$

If the ambient energy source (sink) is assumed to be ideally infinite, then F_c is equal to one.

Example 10
When COP_{ST} equals four and with the given operational temperatures:

$$F_c = \frac{\dfrac{4 \times \left(1 - \frac{320\,\text{K}}{340\,\text{K}}\right)}{0.95 + (4-1) \times \left(1 - \frac{283\,\text{K}}{310\,\text{K}}\right)}}{4 \times \dfrac{\left(1 - \frac{320\,\text{K}}{340\,\text{K}}\right)}{0.95}} = \frac{0.19425}{0.2477} = 0.784$$

Then, from Eq. (2.55), the corrected COP value will be 4 × 0.784 = 3.14 (with limited ambient resources). Repeating the calculations given above, the CO_2 responsibility, if the ambient resource is limited, like in a municipal sewer heat in an urban area, will definitely be high: +0.619 kg CO_2/kW-h_{en}. The same application changes from (−0.21 – +0.619) kg CO_2/kW-h_{en}, which makes an absolute difference close to one kg CO_2/kW-h_{en} (0.83 kg CO_2/kW-h_{en}). Therefore, the design, rating, or a yes-or-no type of decision-making of using heat pumps also depends largely on whether the resources are limited or not. For example, any finite heat source at 40 °C might be utilized by the low-temperature heating equipment with heat pipe technology without a heat pump. Furthermore, a given temperature-peaking COP defined just in terms of the amount of peaking, ΔT, can be satisfied by many resource and heat pump supply temperature combinations. Yet the actual performance depends on individual temperatures too. Therefore, a better definition may be given in the following format:

$$\mathrm{COP} = a - b\left(T_{\mathrm{out}} - T_{\mathrm{in}}\right) \tag{2.63}$$

or in terms of input and output unit exergy values in temperature-peaking:

$$\mathrm{COP}' = c - d \times T_{\mathrm{ref}}\left(\frac{1}{T_{\mathrm{in}}} - \frac{1}{T_{\mathrm{out}}}\right). \tag{2.64}$$

References

Akbar, D., & Kilkiş, B. (2024). Exergy-based optimization constraints for solar PVT panels and district energy systems with onboard green hydrogen production by solar prosumers. *Energy Efficiency, 17*, 6. https://doi.org/10.1007/s12053-023-10184-8

Altamirano-Avila, A., & Manuel Martínez, M. (2021). Urban sustainability assessment of five Latin American cities by using SDEWES index. *Journal of Cleaner Production, 287*(10), 125495. https://www.sciencedirect.com/science/article/abs/pii/S0959652620355414.

Ang, B. W., & Zhang, F. Q. (2000). A survey of index decomposition analysis in energy and environmental studies. *Energy, 25*(12), 1149–1176.

Architecture2030. (2024). *Why the Built Environment?* Retrieved March, 2024, from Why The Built Environment—Architecture 2030.

Britannica. (2024, February 20). Second law of thermodynamics. The Editors of Encyclopaedia. Encyclopedia Britannica. Retrieved 28 March, 2024, from https://www.britannica.com/science/second-law-of-thermodynamics.

Brown, C., & Frichtl, A. (2013). *A building that teaches, high performance buildings, winter 2013* (pp. 35–46). ASHRAE.

Creutzig, F., Roy, J., Lamb, W. F., et al. (2018). Towards demand-side solutions for mitigating climate change. *Nature Clim Change, 8*, 260–263. https://doi.org/10.1038/s41558-018-0121-1

Cu. (2018). *Heat pumps integrating technologies to decarbonize heating and cooling*. European Copper Institute, Autmn. Retrieved from White_Paper_Heat_pumps.pdf (esc-production-2021.s3.eu-west-2.amazonaws.com).

EBC. (2019). *Annex 64-LowEx Communities, Fraunhofer IBP*. Retrieved from https://www.ibp.fraunhofer.de/en/projects-references/annex-64.html.

Energy Star. (2024). *The simple choice for saving energy*. Retrieved March, 2024, from https://www.energystar.gov/.

EPEE. (2020). *Statement on the renovation wave: Cooling and heating systems key for healthy, low-carbon buildings*. Retrieved 15 October, 2020. from https://www.epeeglobal.org/epee-statement-on-the-renovation-wave-cooling-and-heating-systems-key-for-healthy-low-carbon-buildings/.

Erten, D., & Kılkış, B. (2022). How can green Building certification systems cope with the era of climate emergency and pandemics? *Energy and Buildings, 256*, 111750. https://doi.org/10.1016/j.enbuild.2021.111750

EU. (2023a). *Energy-heat pumps*. EU Heat Pump Action Plan. Retrieved March, 2023, from Heat pumps (europa.eu).

EU. (2023b). *Copernicus: October 2023—Exceptional temperature anomalies; 2023 virtually certain to be warmest year on record*. Retrieved 8 November, 2023, from https://climate.copernicus.eu/copernicus-october-2023-exceptional-temperature-anomalies-2023-virtually-certain-be-warmest-year.

EUI. (2019). *Euro monitor international, The environmental sustainability index*. Retrieved March, 2024, from The Environmental Sustainability Index—Euromonitor.com.

GSFC. (1999). *Solar constant*. NASA Lecture No. 693. Retrieved from https://llis.nasa.gov/lesson/693.

Home Building & Renovating. (2024). *The gas boiler ban has been extended until 2035-What it means for your home*. Retrieved from https://www.homebuilding.co.uk/news/gas-boiler-ban.

IEA. (2023). *World Energy Outlook 2023*. Retrieved from https://www.iea.org/reports/world-energy-outlook-2023.

IEA. (2024). *Heat Pumps*. Retrieved from https://www.iea.org/energy-system/buildings/heat-pumps.

IPCC. (2022a). Sixth assessment report, working group III: Mitigation of climate change, climate change: Mitigation of climate change. Retrieved from https://www.ipcc.ch/report/ar6/wg3/.

IPCC. (2022b). *Sixth assessment report working group III: Mitigation of climate change, chapter 9: Buildings*. Cambridge University Press. https://www.ipcc.ch/report/ar6/wg3/chapter/chapter-9/

IPCC, Schlömer, S., Bruckner, T., Fulton, L., Hertwich, E., McKinnon, A., Perczyk, D., Roy, J., Schaeffer, R., Sims, R., Smith, P., & Wiser, R. (2014). Annex III: Technology-specific cost and performance parameters. In O. Edenhofer, R. Pichs-Madruga, Y. Sokona, E. Farahani, S. Kadner, K. Seyboth, A. Adler, I. Baum, S. Brunner, P. Eickemeier, B. Kriemann, J. Savolainen, S. Schlömer, C. von Stechow, T. Zwickel, & J. C. Minx (Eds.), *Climate change 2014: Mitigation of climate change. Contribution of working group III to the fifth assessment report of the intergovernmental panel on climate change*. Cambridge University Press.

IPPC. (2015) Buildings. In: Climate change 2014: Mitigation of climate change: Working group III contribution to the IPCC fifth assessment report. Cambridge University Press; 671–738.

Kalkan, N., Bercin, K. M., Cangul, O., Morales, M. G., Saleem, M. M. K. M., Marji, I., Metaxa, A., & Tsagkogianni, E. (2011). A renewable energy solution for Highfield campus of University of Southampton. *Renewable & Sustainable Energy Reviews, 15*(2011), 2940–2959.

Kilkis, S. (2011). *A rational exergy management model to curb CO2 emissions in the exergy-aware built environments of the future*. dissertation, KTH-The Royal Institute of Technology, Stockholm, Sweden. Microsoft Word—Siir KILKIS_KTH Doctoral Thesis_2011_Print Version_A4_Border_2 (diva-portal.org). Retrieved from https://www.researchgate.net/publication/266000386_A_Rational_Exergy_Management_Model_to_Curb_CO_2_Emissions_in_the_Exergy-Aware_Built_Environments_of_the_Future.

Kilkis, S. (2014). Energy system analysis of a pilot net-zero Exergy District. *Energy Conversion and Management, 87*(2014), 1077–1092. https://doi.org/10.1016/j.enconman.2014.05.014

Kılkış, Ş. (2018). Sustainable development of energy, water and environment systems (SDEWES) index for policy learning in cities. *International Journal of Innovation and Sustainable Development., 12*, 87. https://doi.org/10.1504/ijisd.2018.10009938

Kilkis, B. (2019). Decarbonization: Exergy to the rescue. *Rehva Journal, 3*, 24–30. https://www.rehva.eu/rehva-journal/chapter/decarbonization-exergy-to-the-rescue

Kılkış, B. (2020). Development of a composite PVT panel with PCM embodiment, TEG modules, flat-plate solar collector, and thermally pulsing heat pipes. *Solar Energy, 200*, 89–107. https://doi.org/10.1016/j.solener.2019.10.075

Kilkis, B. (2021). An exergy-based minimum carbon footprint model for optimum equipment oversizing and temperature peaking in low-Temperature District heating systems. *Energy, 236*, 121339. https://doi.org/10.1016/j.energy.2021.121339

Kilkis, B. (2022a). Net-zero buildings, what are they and what they should be. *Energy, 256*, 124442.

Kilkis, B. (2022b) Can a linear economy solve the climate crisis? A geothermal case study. In *Proceedings of the 14th international green energy conference IGEC 2022 July 4–8, 2022.* Retrieved from https://enover.com.tr/wp-content/uploads/2022/06/Kilkis-B.-2022.-Can-A-Linear-Economy-Solve-the-Climate-Crisis-A-Geothermal-Case-Study-IGEC-Proceedings-of-the-14th-International-Green-Energy-Conference-IGEC-2022.pdf.

Kilkis, B. (2022c). Exergy rationality of utilizing wastewater heat in districts and buildings. In *5th SEE conference, SDEWES, Vlore Albania, 22-26 May, 2022*, Paper No: SE2022.003.

Kilkis, B. (2023a). Exergy-optimum coupling of radiant panels with heat pumps for minimum CO_2 emission responsibility. *Energy Conversion and Management: X, 20*, 100439.

Kilkis, B. (2023b). *Performance metrics for the solar PVT and Enover radiator couple in winter*. Enover.

Kılkış, Ş., & Kılkış, B. (2019). *An urbanization algorithm for districts with minimized emissions based on urban planning and embodied energy towards net-zero exergy targets.*

Kılkış, Ş., & Kılkış, B. (2023). A rational exergy management approach for minimum CO2 emissions responsibility at the district scale in Glasgow. In *Proceedings of the 15th international green energy conference, IGEC2023–105, July 10–13, 2023, Glasgow, UK.*

Lamb W., Giulio M., Sebastian L., Timmons R., Stuart C., Felix C., Jan C. Minx, Finn Müller-Hansen, Trevor Culhane and Julia K. Steinberger, (2020) Discourses of climate delay. *Global Sustainability, 3*. https://doi.org/10.1017/sus.2020.13.

META. (2023). *Sustainability report*. Retrieved from https://sustainability.fb.com/wp-content/uploads/2023/07/Meta-2023-Sustainability-Report-1.pdf. https://sustainability.fb.com/wp-content/uploads/2020/12/FB_Denmark-Data-Center-to-Warm-Local-Community.pdf.

NASA. (2024). *Environmental sustainability index (ESI)*. Socioeconomic Data and Applications Center (SEDAC). Retrieved March, 2024, Retrieved from https://sedac.ciesin.columbia.edu/data/collection/esi/#:~:text=Introduction,indicators%20derived%20from%20underlying%20datasets.

ODYSSEE-Mure. (2020). *ODYSSEE data base*. Enerdata. https://www.odyssee-mure.eu/project.html

Reuter, T. A. (2019). Edition culture and foreign policy. In *German cultural diplomacy in Indonesia: Building Cooperation in a Changing World*. Retrieved from https://www.researchgate.net/publication/346098995_ifa_Edition_Culture_and_Foreign_Policy_German_Cultural_Diplomacy_in_Indonesia_Building_Cooperation_in_a_Changing_World/figures?lo=1.

SDEWES. (2024). *SDEWES index*. Retrieved 11 January, 2024, from https://www.sdewes.org/sdewes_index.php.

SE. (2016). Science europe. In *A resource-constrained world: Think exergy, not energy*. Retrieved from https://www.scienceeurope.org/media/0vxhcyhu/se_exergy_brochure.pdf.

Solar. (2024). *What is the carbon footprint of solar panels*? Retrieved March, 2024, from What is the Carbon Footprint of Solar Panels? | Solar.com

Torcellini, P. A., & Drury, B. C. (2006). Understanding Zero-Energy Buildings. *ASHRAE Journal*, 62–67. https://acrobat.adobe.com/id/urn:aaid:sc:AP:5f340f96-37ec-4825-aa3f-ffcd4b0d3a48

TTMD. (2017). Sustainability and decarbonization efforts of the EU: Potential benefits of joining energy quality (exergy) and energy quantity (energy) In *EU directives, a state-of-the-art survey and recommendations—Exclusive EU position report ©2017, TTMD 2017–1, Ankara Turkey.*

UN. (2024). *Sustainable development*. Seventeen Goals, SDG. Retrieved March, 2024, from https://sdgs.un.org/goals.

Van der Leer, J., Calvén, A., Glad, W., et al. (2023). Energy systems in sustainability-profiled districts in Sweden: A literature review and a socio-technical ecology approach for future research. *Energy Research and Social Science, 101*, 1–15. https://doi.org/10.1016/j.erss.2023.103118

Velazquez, A. D. V. (2008). *Characterization of a domestic horizontal ground sink heat pump system*. Ph. D. thesis, 211 pages, Tecnologico de Monterrey. Characterization of a domestic horizontal ground sink heat pump system | angel Valerio—Academia.Edu.

Wang, C., Kilkis, S., Tjernström, J., Nyblom, J., & Martinac I. (2016). Multi-objective optimization and parametric analysis of energy system designs for the Albano university campus in Stockholm. In *international high-performance built environment conference (iHBE), 17–18 November 2016, Sydney, Australia*.

Yale U. (2024). *EPI, Environmental Performance Index*. Retrieved March, 2024, from https://epi.yale.edu/.

Zogou, O., & Stamatelos, T. (2007). Optimization of thermal performance of a Building with ground source heat pump system. *Energy Conversion and Management, 48*(11), 2853–2863.

Chapter 3
Implementing Cold Ironing Methods in Georgian Ports

Rati Janjghava, Avtandil Tsitskishvili, and Givi Tsitskishvili

Nomenclature

BSMA	Batumi State Maritime Academy
CO	Carbon monoxide
CO_2	Carbon dioxide
CO_{2e}	Carbon dioxide equivalents
UN	United Nations
UNCTAD	United Nations Conference on Trade and Development

3.1 Introduction

In the twenty-first century, the world and the countries within it have faced challenges that need to be solved rapidly. One of the most concerning and important of these is climate change, which has been worrying scientists around the world for centuries. According to NASA's Global Climate Change website (NASA, 2019), the Earth's average global temperature has increased by about 1.0 °C since the late nineteenth century, and if we do not take action, it is unlikely to stop growing. Rising levels of global temperature can have a devastating impact on Earth, as they can cause various natural disasters like hurricanes and rising sea levels. The maritime industry has also been a massive contributor to climate change, as bunkers, which are used to power ships, and especially high-sulfur bunkers, emit dramatic amounts of carbon dioxide, which, on their part, creates a natural greenhouse effect, leading to a warming planet and causing climate change. In fact, the maritime industry is estimated to emit 2.5% of worldwide greenhouse gases, or, as suggested by the United Nations Conference on Trade and Development (UNCTAD)

R. Janjghava (✉) · A. Tsitskishvili · G. Tsitskishvili
Batumi State Maritime Academy, Batumi, Georgia

M. Z. Sogut, M. Koray (eds.), *Energy Rationality and Management for Decarbonization*, Green Energy and Technology,
https://doi.org/10.1007/978-3-031-68497-5_3

(UNCTAD, 2022), approximately 1.2 gigatons of carbon dioxide equivalents (CO_{2e}) in 2020. This problem exists in Georgia today, a country with several major ports that host hundreds of vessels annually. It should be noted that as a UN member, Georgia is required to take all reasonable steps to halt or slow climate change. According to the United Nations Framework Convention on Climate Change, which is the specialized body within the UN system that is focused specifically on addressing climate change, all countries that have signed the Paris Agreement are required to do everything in their power to achieve the goal of limiting global warming to well below 2 °C above preindustrial levels and pursue efforts to limit the temperature increase to 1.5 °C above preindustrial levels. Speaking for the maritime industry, there are not many ways to decrease pollution, especially when ships are free-sailing, without increasing the cost for the companies, but there is a solution for the ships that are docked. Cold ironing, also known as shore-to-ship power supply, stops diesel burning at the port and powers ships using electricity from shore. This method is not only eco-friendly but also costs less, as in Georgia's example. It has to be mentioned that Georgian port holders are also interested in establishing the cold ironing method in Georgia; for example, APM Terminals, which has its branch in Georgia's biggest port, Poti Sea Port, is interested in establishing the cold ironing method in their branch.

3.2 Materials and Methods

In order to fully and thoroughly understand how much better cold ironing is for the environment compared to the traditional way of getting energy through burning diesel, we used (Tokuslu, 2021) the research that Turkish researcher Aydin Tokuslu has conducted about emissions caused by vessels in Georgian ports. After getting valuable information that we used to find out how much CO_2 diesel-powered vessels emit annually, we compared it to the shore-to-ship power method, which mostly uses electricity from hydroelectric sources (the main source of electricity in Georgia). Additionally, we determined the cost of producing 1 kWh of energy using diesel and electricity. The price of low-sulfur diesel was taken from *Ship & Bunker* (Shipandbunker, 2023). Electricity's cost was obtained from Energo pro Georgia (2023).

3.3 Results and Discussion

For calculating the emissions, we used the following formula:

$$E_{\text{Hoteling}} = T \times \text{AE} \times \text{LF}_{\text{AE}} \times \text{EF}_{\text{hoteling}} \tag{3.1}$$

E_{Hoteling} = is the emission of the pollutant CO_2 while hoteling (units = ton)

Table 3.1 Time consumed in the operational mode of hoteling

Tanker	38 h
General cargo	39 h
Container vessel	14 h

Table 3.2 The power of the auxiliary engine by ship type

Average-sized tanker	232.5 KW
Average-sized general cargo	237.5 KW
Average-sized container vessel	330 KW

T = is the time consumed in the operational mode of hoteling (units = hours)
AE = is the power of the auxiliary engine (units = KW)
LF_{AE} = is the auxiliary engine load factor (units = %)
$EF_{Hoteling}$ = is the emission factor of hoteling operations. (units = g/kWh)

Subsequently, Aydin Tokuslu's research was utilized to determine how long each type of vessel stays hoteling in Georgian ports, the power of auxiliary engines in various ships, load factors depending on ship type, and the emission factor of hoteling operations by different ship types (Tables 3.1 and 3.2).

3.3.1 The Auxiliary Engine Load Factor

The auxiliary engine load factor:

- 60% in hoteling mode (for tankers)
- 40% in hoteling mode (excluding tankers) (Table 3.3)

3.3.2 Examining the Annual CO_2 Emissions from Vessel Hoteling in Georgia's Ports

The pollution caused by three of the ships that dock in Georgian ports most frequently in one visit will be assessed.

- Average-sized tanker = 3.86 tons (38 × 232.5 × 60/100 × 730/1,000,000)
- Average-sized general cargo = 2.49 tons (38 × 237.5 × 40/100 × 691/1,000,000)
- Average-sized container vessel = 1.27 tons (14 × 330 × 40/100 × 690/1,000,000)
- Average-sized tanker = 1805 tons ((568 × 5% + 584 × 44% + 130 × 65% + 98 × 100%) × 3.86)
- Average-sized general cargo = 915 tons ((568 × 39% + 584 × 25% + 130 × 0% + 98 × 0%) × 2.49)
- Average-sized container vessel = 384 tons ((568 × 44% + 584 × 9% + 130 × 0 % + 98 × 0%) × 1.27) (Table 3.4)

Table 3.3 The emission factor of hoteling operations

Tanker	730 g/kWh
General cargo	691 g/kWh
Container	690 g/kWh

Table 3.4 How much pollution is caused by three of the ships that dock in Georgian ports the most frequently annually (2018)

Georgian ports	Ship calls	Tanker (%)	General cargo (%)	Container (%)
The Poti Sea port	568	5	39	44
The Batumi port	584	44	25	9
The port of Kulevi	130	65	0	0
The port of Supsa	98	100	0	0

3.3.3 *Theoretical Reduction of CO_2 Emissions by Providing Shore Power to Vessels During Hoteling*

Since hydroelectric power stations are the primary sources of energy in Georgia, as suggested by Giorgi Muxigulishvili, the researcher of climate change and energetics in World Experience for Georgia (WEG), 102 g of CO_2 are emitted for every 1 kWh of energy. Formula:

$$\text{Emissions} = T \times \text{AE} \times 102 \tag{3.2}$$

T = is the time consumed in the operational mode of hoteling (units = h)
AE = is the power of the auxiliary engine (units = KW)
Shore power emission by types of vessels:

- Average-sized tanker = 0.9 tons (38 × 232.5 × 102/1,000,000)
- Average-sized general cargo = 0.92 tons (38 × 237.5 × 102/1,000,000)
- Average-sized container vessel = 0.47 tons (14 × 330 × 102/1,000,000)

Shore power emission by types of vessels annually:

- Average-sized tanker = 421 tons ((568 × 5% + 584 × 44% + 130 × 65% + 98 × 100%) × 0.9)
- Average-sized general cargo = 339 tons ((568 × 39% + 584 × 25% + 130 × 0% + 98 × 0%) × 0.92)
- Average-sized container vessel = 142 tons ((568 × 44% + 584 × 9% + 130 × 0 % + 98 × 0%) × 0.47)

Figure 3.1 shows just how much less CO_2 in tons is emitted when we use the cold ironing method

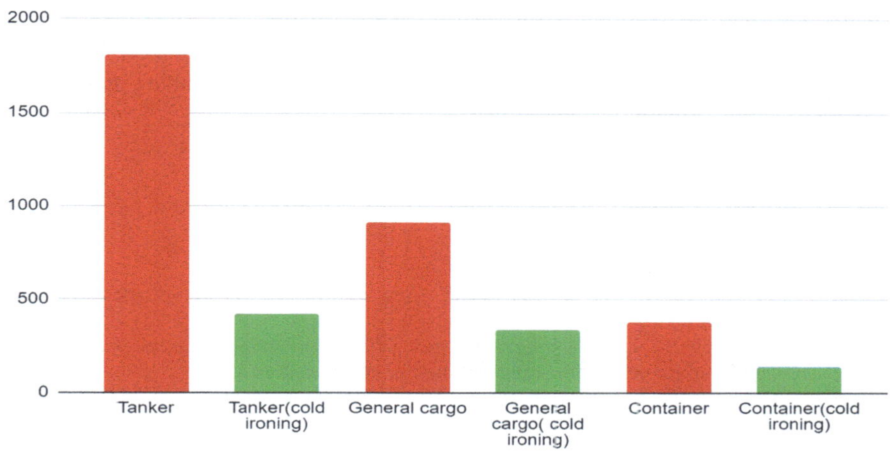

Fig. 3.1 CO_2 emission reduction through cold ironing

3.3.4 Comparing the Costs Associated with Using Both Energy-Generation Technologies

The price of low-sulfur diesel for March 29, 2023, in the ports of Istanbul was 728.2 dollars for 1 ton of low-sulfur diesel. For generating 1 kWh of energy, 0.4 L, or 0.336 kg, of diesel are burned on average (including efficiency). Thus, 0.24 dollars are required to generate 1 kWh if we power the ship the traditional way. According to "Energo Pro Georgia," the cost of 1 kWh of energy is 0.12 dollars for the country's ports. This clarifies that by using the cold ironing method, we save 0.12 dollars for every kilowatt-hour of energy generated. Therefore, ports can save up to 495,113 dollars annually on tanker vessels, 399,061 dollars on general cargo vessels, and 167,499 US dollars on container vessels.

Based on these calculations, it is clear that using hydroelectricity as a source is very beneficial for the planet, and the founding of the cold ironing method in Georgia is the best choice to make to avoid environmental issues today and in the long term. The results show that when using the shore-to-ship power supply method, 3.1 times less CO_2 is emitted, and in Georgia's example, we save two times more using the named method as well.

3.4 Conclusion

In conclusion, it can be asserted that the maritime sector holds the potential to contribute toward addressing the challenging issue of reducing global warming. Georgia, with its advantageous position, has the opportunity to safeguard the environment and secure a future where we can breathe safely and deeply. The calculations performed provide confidence that research on cold ironing in Georgia should persist, with a focus on acquiring additional essential information. By doing so, we can pave the way for finding necessary solutions and making substantial progress in combating global warming.

References

Energo pro Georgia. (2023). Retrieved from https://www.energo-pro.ge/ka/service/eleqtroenergiis-tarifi/23.

NASA. (2019, March 26). *Climate change evidence: How do we know? Climate change: Vital signs of the planet*. Retrieved from https://climate.nasa.gov/evidence.

Shipandbunker. (2023). Retrieved from https://shipandbunker.com/prices/emea/medabs/tr-ist-istanbul.

Tokuslu, A. (2021). Estimating greenhouse gas emissions from ships on four ports of Georgia from 2010 to 2018. *Environmental Monitoring and Assessment, 193*(7), 385. https://doi.org/10.1007/s10661-021-09169-w

UNCTAD. (2022, June 8). *Decarbonizing the maritime sector: Mobilizing coordinated action in the industry using an ecosystems approach*. Retrieved from https://unctad.org/news/decarbonizing-maritime-sector-mobilizing-coordinated-action-industry-using-ecosystems-approach.

Chapter 4
Study of Combustion Characteristics and Prediction of Cyclic Variability of a Diesel Engine Fueled by Blends of Hydrotreated Waste Cooking Oil and Diesel Using Artificial Neural Network

Ankit Sonthalia, Naveen Kumar, Mukul Tomar, Edwin Geo Varuvel, and S. Thiyagarajan

Nomenclature

ANN	Artificial neural network
CA	Crank angle in degrees
CA50	Crank angle at which 50% of the heat is released
CI	Compression ignition engines
CO	Carbon monoxide emission

A. Sonthalia (✉)
Center for Advanced Studies and Research in Automotive Engineering, Delhi Technological University, Delhi, India

Department of Automobile Engineering, Faculty of Engineering and Technology, SRM Institute of Science and Technology, Delhi - NCR Campus, Ghaziabad, Uttar Pradesh, India

N. Kumar
Center for Advanced Studies and Research in Automotive Engineering, Delhi Technological University, Delhi, India

M. Tomar
Research and Engineering, GTI Energy, Des Plaines, Illinois, USA

E. G. Varuvel
Department of Mechanical Engineering, Faculty of Engineering and Natural Sciences, Istinye University, Istanbul, Turkiye

Mechanical Science Institute, Vilnius Gediminas Technical University, Vilnius, Lithuania

S. Thiyagarajan
Department of Mechanical Engineering, Saveetha Engineering College, Thandalam, Chennai, India

M. Z. Sogut, M. Koray (eds.), *Energy Rationality and Management for Decarbonization*, Green Energy and Technology,
https://doi.org/10.1007/978-3-031-68497-5_4

COV	Coefficient of variance
D100	Test sample containing only diesel
FCC	Fluid catalytic cracking
FFA	Free fatty acid
H100	Test sample containing only hydrotreated vegetable oil
H10D90	Test sample containing 10% hydrotreated vegetable oil and 90% diesel on a volume basis
H20D80	Test sample containing 20% hydrotreated vegetable oil and 80% diesel on a volume basis
H30D70	Test sample containing 30% hydrotreated vegetable oil and 70% diesel on a volume basis
H40D60	Test sample containing 40% hydrotreated vegetable oil and 60% diesel on a volume basis
H50D50	Test sample containing 50% hydrotreated vegetable oil and 50% diesel on a volume basis
HC	Unburned hydrocarbon emission
HCR	Hydrocracking
HDO	Hydro-deoxygenation
HDM	Hydro-demetallation
HDN	Hydro-denitrogenation
HDS	Hydrodesulfurization
HTlc	Hydrotalcite-like
HVO	Hydrotreated vegetable oil
IMEP	Indicated mean effective pressure
LM	Levenberg–Marquardt
MAPE	Mean absolute percentage error
MMT	Million metric tons
MTOE	Million tons of oil equivalent
NOx	Oxides of nitrogen emission
RMSE	Root mean square error
SCG	Scaled conjugate gradient
TPES	Total primary energy supply
WCO	Waste cooking oil

4.1 Introduction

The economic and social growth of any country is dependent upon energy. The consumption of fossil fuels has increased exponentially. Figure 4.1 compares the total primary energy supply (TPES) of the world in 1973 and 2020. The world consumed 6101 MTOE (million tons of oil equivalent) energy in 1973, which increased to 13,294 MTOE in 2020 (IEA, 2017; BP, 2021), an increase of 117%. The major energy sources for 83.13% of the total energy consumption comprises oil, coal, and natural gas.

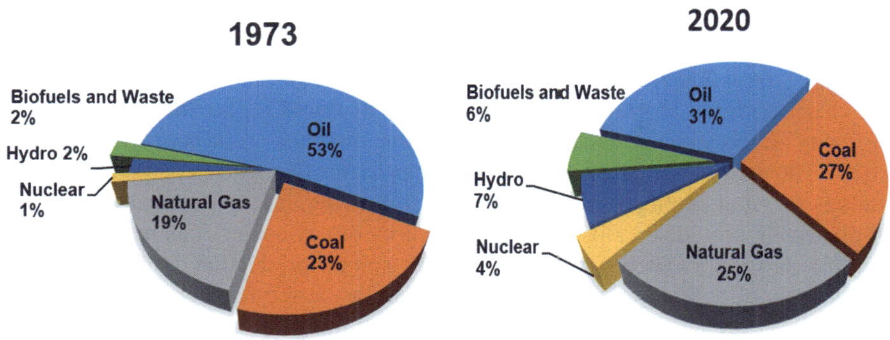

Fig. 4.1 Comparison of total primary energy supply in 1973 and 2020 (BP, 2021)

India is among the fastest-growing economies in the world, and energy is the wheel to its growth. However, the oil reserves in India are minuscule in comparison to its requirements. In 1947, the oil available was 0.25 million tons, which increased to about 38.09 million metric tons (MMT) in 2011–2012. India is highly dependent upon imports of crude oil to cater to its needs for agriculture, automobiles, and industrial applications. For the year 2019–2020, India produced nearly 32.17 MMT, whereas India imported nearly 226.95 MMT at an average cost of US\$60.47 per barrel. The cost to the exchequer for importing oil was INR 7,17,001 crore. 95% of the energy requirement of the transportation and agricultural sectors is fulfilled by imported oil. The diesel requirement in India was about 53 MMT in 2007–2008, which has grown to 82.6 MMT in 2019–2020 (IPNGS, 2020). Thus, India needs to take steps to reduce its dependence on oil with the growing need for energy.

Compression ignition (CI) engines are the backbone of the transportation and agricultural sectors. They are also used in off-road vehicles and for power generation. The main attractions to their use are the ruggedness in construction, usage simplicity, and ease of maintenance. Other features are high thermal efficiency, high torque, and low fuel consumption. Low pumping losses, lean air–fuel mixtures, and high compression ratio result in the high thermal efficiency of these engines. However, these engines emit gases that not only pollute the environment but are harmful to humans as well (Sonthalia & Kumar, 2021). The concentration of particulate matter emitted by diesel engines is generally above ten million particulates per cm^3. The primary carbon particles have a diameter varying from 0.01 to 0.08 microns, whereas the agglomerated particles have diameters varying from 0.08 to 0.25 microns, and the diameter of 90% of the particles is less than 1 ηm (Sonthalia et al., 2021). The particulate matter has been classified by several government agencies as "human carcinogen" or "probable human carcinogen." The risk of respiratory diseases and heart diseases is increased severalfold. Particulate matter containing hydrocarbons with two or more benzene rings are known human carcinogens. Four or five rings present in the soluble organic fraction of the particulate matter are the most harmful ones (Xiao et al., 2000). A long-lasting solution to the above-mentioned problems can be the use of renewable alternative fuels in the CI engines. To operate the diesel engine with an alternate fuel, the fuel should have properties close to diesel. Vegetable oils derived from plant seeds are among the few alternatives that match conventional diesel. Their cetane number and energy density are close to diesel. However, their viscosity is manyfold higher than diesel. High viscosity affects spray atomization, vaporization, and fuel–air mixing (Tomar et al., 2020). The use of vegetable oil offers significant cost-effective and perpetual opportunities for rural employment, particularly for growing, harvesting, and processing of plants that provide the oil. The energy security of any country can be attained by multifaceted actions such as a purposeful policy on the production of oil and natural gas, the use of hybrid and electric vehicles, and the search for low-cost alternative fuels, which can reduce dependence on oil imports. India, having 17% of the world's population, requires to use both renewable and exhaustible energy resources. The decentralized agro-based form of energy production can provide a sound base for a village-centered economy and also attain a high level of energy security (Kumar et al., 2020a, 2020b).

Enormous amounts of waste cooking oil (WCO) are generated from fast-food centers, restaurants, households, and the food processing industry. It is estimated that nearly 1,000,000–3,000,000 tons/year of waste cooking oil is generated in the European Union. In the United States, 5.2–7 billion gallons of used cooking oil are generated yearly, whereas Japan alone generates 600,000–1,000,000 tons of WCO annually (Hribernik & Kegl, 2009). In India, nearly 6.5 billion gallons of used cooking oil were generated in the year 2018 (FSSAI, 2018). The disposal and storage of WCO are both costly and troublesome. Improper disposal of used cooking oil results in the pollution of lakes, rivers, seas, and underground water, which leads to human health and environmental problems. Reusing the oil will reduce the burden on the waste treatment system and diminish the effect on the environment. WCO can be used as a fuel as it is readily available and is cheaper than diesel. Also, its physical and chemical properties are close to diesel. Table 4.1 shows the properties of waste cooking oil. Using waste cooking oil as a fuel can lead to considerable economic benefits since the industries and transportation sector can meet their

Table 4.1 Properties of waste cooking oil (Lapuerta et al., 2008)

Properties	Waste cooking oil
Density at 20 °C (kg/m^3)	880
Lower heating value (kJ/kg)	39,600
Viscosity (cSt) @ 30 °C	45
Flash point (°C)	218
Cetane number	40–45
Auto ignition temperature (°C)	320
Flame velocity (cm/s)	3–6
Flammability limits (vol.% in air)	0.1–0.3
Boiling point (K)	657
Molecular weight (g mol)	875.75
Carbon residue (%)	0.54
Sulfur content (% wt.)	0

energy requirements. Many researchers have converted used cooking oil into biodiesel using the transesterification process and tested it on different types of diesel engines (Kathirvel et al., 2016).

Vegetable oil can be upgraded (biodiesel) by the transesterification process wherein the triglycerides are converted into mono-alkyl esters using either methanol or ethanol. The catalyst amount used in the process is changed in accordance with the free fatty acid (FFA) content of the vegetable oil. However, the process is applicable when the FFA content is less than 2.5 wt.%. For oils having higher than 2.5 wt.% of FFA content, a pretreatment is required for reducing their FFA content so that transesterification can be carried out. The biodiesel produced from transesterification has low oxidation stability and cetane number because of the high degree of unsaturation and the presence of long chains (Knothe, 2010). Long-term biodiesel use in diesel engines causes corrosion and deposit problems in fuel injection systems (Kuronen et al., 2007). Other technologies for upgrading vegetable oil are gasification by the Fischer–Tropsch process, high pressure–temperature treatment, and fluid catalytic cracking (FCC). In high pressure–temperature treatment, no catalyst is used, and temperature and pressure greater than 250 °C and 250 bar, respectively, are utilized for upgrading vegetable oils. In the FCC method, any kind of heavy and complex hydrocarbon can be upgraded to diesel and gasoline. In the presence of transition metals, the syngas is converted to hydrocarbons. However, a significant amount of gaseous component is formed in the Fischer–Tropsch process, whereas the other methods suffer from coke formation (Patel & Kumar, 2016).

Hydroprocessing is a general term used to describe the reactions, such as hydrogenation, hydro-deoxygenation (HDO), hydro-denitrogenation (HDN), hydrodesulfurization (HDS), hydrocracking (HCR), and hydro-demetallation (HDM). To remove heteroatoms and for the cracking and saturation of aromatics and olefins, the petroleum refining industry uses the well-known cracking and hydroprocessing technique (Gary & Handwerk, 2001). The process can also be used for converting bio-based feedstock into straight-chain alkanes having C_{12}–C_{18} carbon atoms. The fuel has a high cetane number (55–65), and its properties are better than diesel produced via the transesterification process (Huber & Corma, 2007). As the fuel is similar to diesel, the existing infrastructure of diesel for storage, transportation, and retail can be utilized; the existing engines can be used without any modifications; and the process is flexible with the use of feedstock (Stumborg et al., 1996).

Many researchers have used different catalysts for producing renewable diesel. Some of the catalysts include sulfided nickel–molybdenum, cobalt–molybdenum, and several noble metals such as ruthenium (Ru), platinum (Pt), and palladium (Pd) on different supports (Kumar et al., 2020a, 2020b). Kikhtyanin et al. (2010) hydroprocessed sunflower oil using palladium catalyst. The authors found an increase in the C_{17}/C_{18} ratio with an increase in reaction temperature. Also, the cloud point and pour point of the fuel were lower than −50 °C. Jatropha oil was hydrotreated by Murata et al. (2010) using H-form Zeolite Socony Mobil-5 (HZSM-5)-supported Pt–Re catalyst. The authors found an optimum yield of 67%

of the C_{15}–C_{18} alkane mixture. During the reaction, the double bonds were first broken followed by hydro-deoxygenation, hydro-decarbonylation, and hydro-decarboxylation. Waste cooking oil was hydroprocessed with Ru/Al13-montmorillonite catalyst by Liu et al. (2012). The oil having a high acid value was easily deoxygenated. The authors found that the pour point of the oil with montmorillonite support was −15 °C, whereas with SiO_2 support, the pour point was 20 °C. Guo et al. (2015) used different supports with ruthenium for hydrotreating a blend of jatropha oil and stearic acid under mild conditions. The authors achieved 100% conversion efficiency of jatropha oil with $La(OH)_3$ support. The reaction was carried out at 200 °C and 4 MPa, which also resulted in 98% long-chain selectivity. It was also seen that the recycling of the catalyst up to four times resulted in 70% alkane selectivity under the same operating conditions. Patil and Vaidya (2017a, 2017b) added cerium to $La(OH)_3$-supported platinum and ruthenium catalysts. Using the catalyst, jatropha oil was deoxygenated in a fixed-bed reactor. The authors found the oil conversion and hydrocarbon yield to be 91.8 and 75.3% and 97.3 and 81% for Ru–Ce and Pd–Ce, respectively, at optimized conditions. Also, the catalyst conversion efficiency was not affected for 40 h of onstream time. The same research group used hydrotalcite-like (HTlc) supports for ruthenium and palladium catalysts in another study (Patil & Vaidya, 2017a, 2017b). They found that the hydrocarbon yield at optimum reaction conditions with ruthenium is 85.6%, and with palladium, it increases to 91.2%. The catalyst conversion efficiency was not affected for 30 h of onstream time.

There are many studies available that have used hydrotreated vegetable oil (HVO) as a fuel. Aatola et al. (2008) compared a 30% blend of HVO in diesel and 100% HVO with diesel for fuel consumption and NOx emissions. The authors observed a significant reduction in NOx emissions. Erkkilä et al. (2011) used partially isomerized HVO and fossil diesel and found 30% PM and 10% NOx reduction for iso-HVO in comparison to diesel. The properties of partially isomerized HVO were studied, and the comparisons of NOx and PM emissions of diesel, and HVO were made (Kuronen et al., 2007). Significant reductions in NOx, HC, and CO emissions were observed with HVO. In another study, PM emissions from petroleum diesel, HVO, and biodiesel engine operation were compared (Lapuerta et al., 2011). PM emissions were lower with HVO due to no aromatics and olefins, which are the main causes of soot formation. Buses representing various EURO emission certifications were field-tested for fuel consumption by Makinen et al. (2011). The energy consumption with neat HVO reduced by nearly 0.5% as compared to winter- and summer-grade diesel, whereas the volumetric consumption of HVO increased by 5.2 and 3.5%, in comparison to winter- and summer-grade diesel, respectively. A pickup was operated with 5% biodiesel and varying blends of iso-HVO in diesel (Caprotti et al., 2011). The fuel consumption was found to decrease when the HVO percentage was increased. The cold-start ability and drivability of the pickup with HVO blending was found to be similar to diesel.

It is generally known that the shelf life of biodiesel is very less as the fuel oxidizes over a period of time. Hence, to check the long-term effects of hydrotreated oil, a study was conducted over a period of 12 months, and its properties, namely density, kinematic viscosity, and calorific value were checked every month. The feasibility of the hydrotreated oil as a fuel was tested on a diesel engine. The engine was used in agricultural work as well as a power backup. The combustion characteristics of the engine were measured, and a neural network model was optimized for predicting the coefficient of variance (COV) of indicated mean effective pressure.

4.2 Preparation of Fuel

The oil used in the present study was obtained from the hostel mess of Delhi Technological University. Impurities were first removed from the oil through a filtration process. The oil was then heated to 100 °C such that any water content could be boiled off. The oil was hydrotreated in hydrogen in a batch reactor at 330 °C and 4 MPa. The catalyst used for the study was 2% ruthenium, cobalt, and manganese supported on alumina. After the reaction, the obtained bio-crude oil was distilled. The oil obtained in the range of 180–370 °C boiling temperature was considered as the hydrotreated fuel. The detailed procedure for obtaining hydrotreated oil can be studied in the previous work (Kumar et al., 2020a, 2020b). The properties of the fuel were obtained using American Society for Testing and Materials (ASTM) standards. Table 4.2 tabulates the properties of hydrotreated fuel and diesel. For the present study, diesel and hydrotreated oil were blended in various proportions on a volume basis. The blends contained 10, 20, 30, 40, and 50% of hydrotreated oil. The sample containing neat hydrotreated oil was designated as H100. The blend containing 90% diesel and 10% hydrotreated oil was designated as H10D90. The other blends were similarly designated as H20D80, H30D70, H40D60, and H50D50.

Table 4.2 Properties of diesel and hydrotreated waste cooking oil

Properties	Standard	Diesel	Hydrotreated waste cooking oil
Density @15 °C (kg/m^3)	ASTM D1298	835.07	817.7
Kinematic viscosity @40 °C (cST)	ASTM D445	2.51	2.9
Flash point (°C)	ASTM D93	58	68
Calorific value (MJ/kg)	ASTM D240	42.57	41.79
Cetane index	ASTM D4737	48.73	70.2

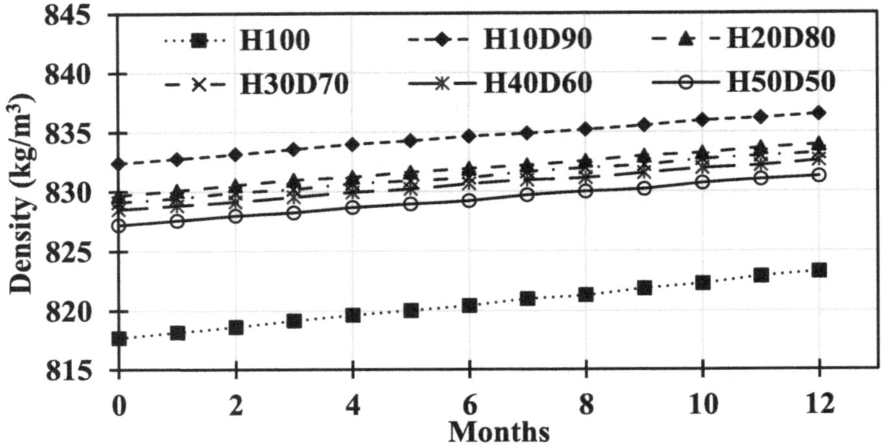

Fig. 4.2 Variation in density for blends of hydrotreated oil stored for 12 months

4.3 Effect of Storage on Fuel Properties

The blends of hydrotreated oil and diesel were stored in an air-tight glass container in a dark room where no direct sunlight enters. Three key properties namely, density, kinematic viscosity, and calorific value, were evaluated every month to understand the change in the fuel stored for 1 year. Figure 4.2 shows that the density of the test fuels changed slightly over a period of 12 months.

The rate of increase of density of H100 was higher than the blends. Since a small amount of oxygen present in the fuel molecules may have resulted in the formation of oxygenated products over a period of time, thereby increasing the density. As the blends contain less amount of hydrotreated oil, the oxygen content is less; therefore, the change in properties of the blends is less. As compared to freshly prepared fuel, the density of the stored fuel with H100, H50D50, H40D60, H30D70, H20D80, and H10D90 increased by 0.68%, 0.48%, 0.49%, 0.49%, 0.51%, and 0.48%, respectively. The change in kinematic viscosity of the test fuels stored for 12 months is shown in Fig. 4.3. The fuel injection system of a diesel engine is greatly affected by the viscosity of the fuel used.

The fuel injected by the injector will not be atomized properly if its viscosity is high, resulting in poor engine performance characterized by high fuel consumption and emissions (Demirbas, 2006). The high viscosity of the fuel can plug the fuel filters as well. With an increase in storage period, an increasing trend in the viscosity was observed. The viscosity of freshly prepared H100 was 2.9cSt, which increased to 3.33cSt in 1 year of storage. It was also seen that the kinematic viscosity of the blends also increased, but they were lower than H100. The variation of calorific value of H100, H10D90, H20D80, H30D70, H40D60, and H50D50 stored over 12 months is indicated in Fig. 4.4.

As the storage period increases, the calorific value decreases for all the test samples. The rate of decrease of calorific value for H100 was higher than the other fuel

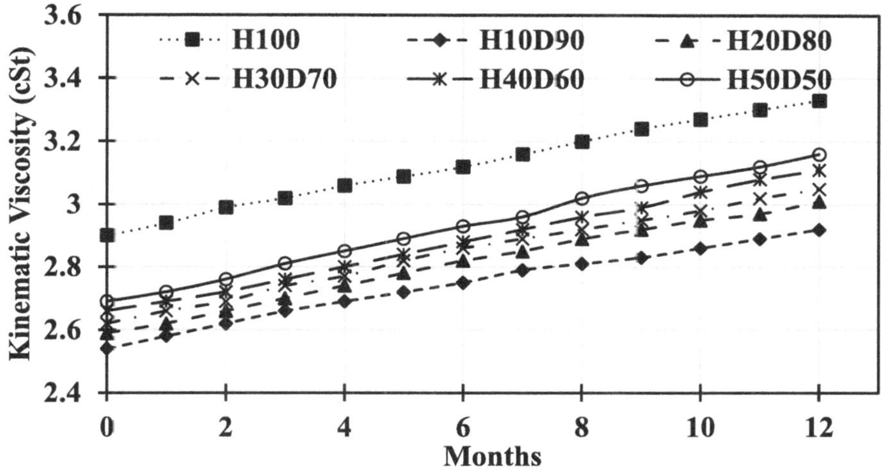

Fig. 4.3 Variation in kinematic viscosity for blends of hydrotreated oil stored for 12 months

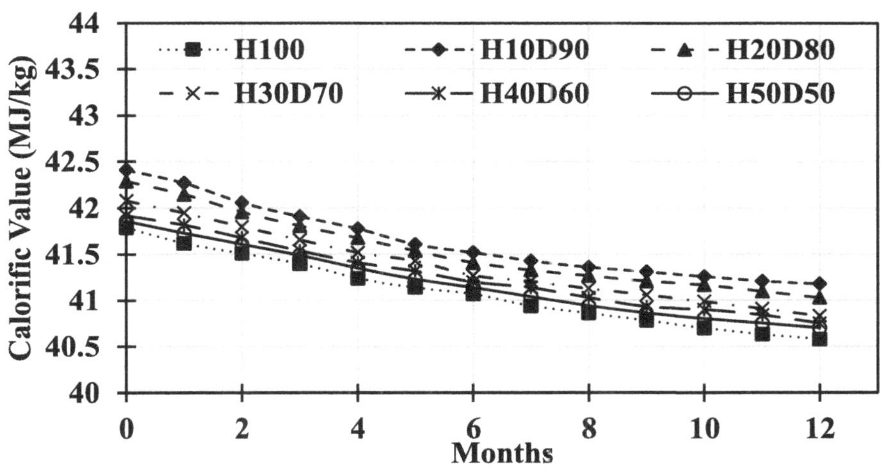

Fig. 4.4 Variation in calorific value for blends of hydrotreated oil stored for 12 months

samples. The calorific value of freshly prepared H100 was 41.79 MJ/kg, which decreased to 40.58 MJ/kg after 12 months, a decrease of 0.51%.

4.4 Experimental Setup for Engine Testing

A single-cylinder, direct-injection, diesel engine was used in the present study to characterize the combustion parameters. Table 4.3 tabulates the test engine specifications. The setup was equipped with a data acquisition system. Sensors were

Table 4.3 Test engine specification

Specification	Value
Rated power	3.5 kW
Rated speed	1500 rpm
No. of cylinders	1
Stroke	4
Displacement	661 cc
Stroke length	110 mm
Bore diameter	87.5 mm
Cooling system	Water-cooled
Injector	Solenoid, six-holed
Compression ratio	12–18

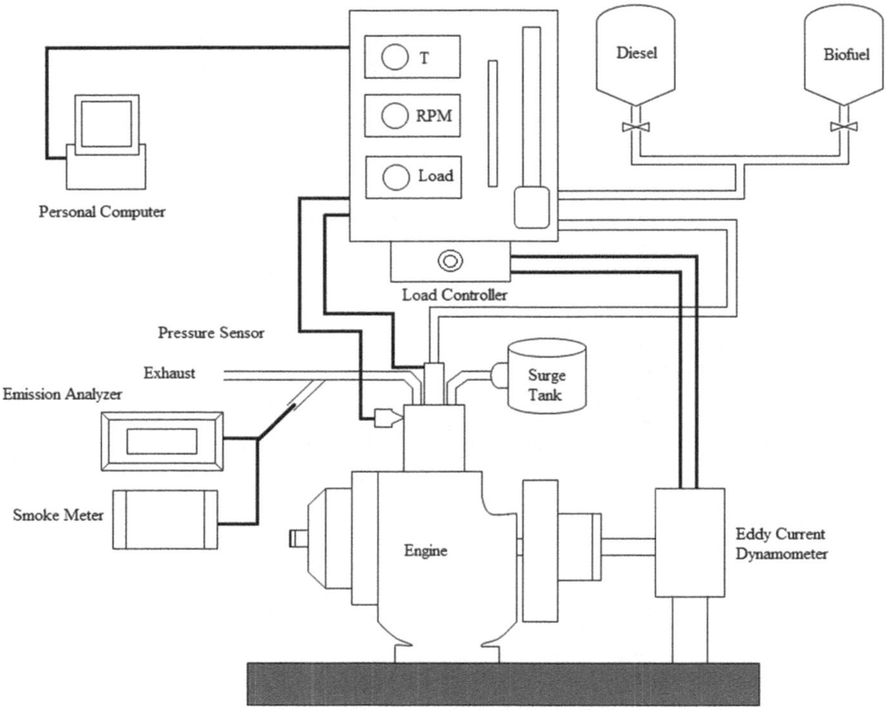

Fig. 4.5 Schematic of the experimental setup

used for the measurement of load and in-cylinder pressure. These sensors were sending signals to the data acquisition system, which converted them to machine-readable form. "Engine Soft" software installed on a personal computer processed the signals received from data acquisition system. The experimental setup is shown in Fig. 4.5.

The engine was loaded using an eddy current–type dynamometer, which was cooled by water. Using a piezoelectric pressure sensor (PCB USA), the in-cylinder pressure was measured. The sensor was fixed in the head of the engine, and it was air-cooled. On the other side of the dynamometer, a crank angle sensor (Kubler make) was placed to measure the crank angle. The in-cylinder pressure was calculated by averaging 50 consecutive cycles. The peak pressure was considered when the in-cylinder pressure was the highest. The ignition delay was measured from the point where the injection started to the point and where heat release starts. The combustion duration was considered to be starting from 5 to 90% of heat release. CA50 signifies the crank angle at which 50% of the heat was released. The engine tests were carried out at different loading conditions at a rated speed of 1500 rpm. The readings were recorded at 20, 40, 60, and 80% of full load and at full-load conditions.

4.5 Combustion Characteristics of the Engine

The variation in peak pressure at different loads for all the test samples is depicted in Fig. 4.6. As the load on the engine increases, the peak pressure increases.

Since the amount of fuel injected increases with the rise in load, the heat released is also increased, resulting in an increase in in-cylinder pressure (Karthickeyan et al., 2020). Moreover, the peak pressures at all the loads were lower for the test fuels in comparison to diesel. The lowest peak pressure was observed with neat hydrotreated oil, and as the percentage of the oil increases in the blend, the in-cylinder pressure decreases. The cetane index influences the combustion, causing

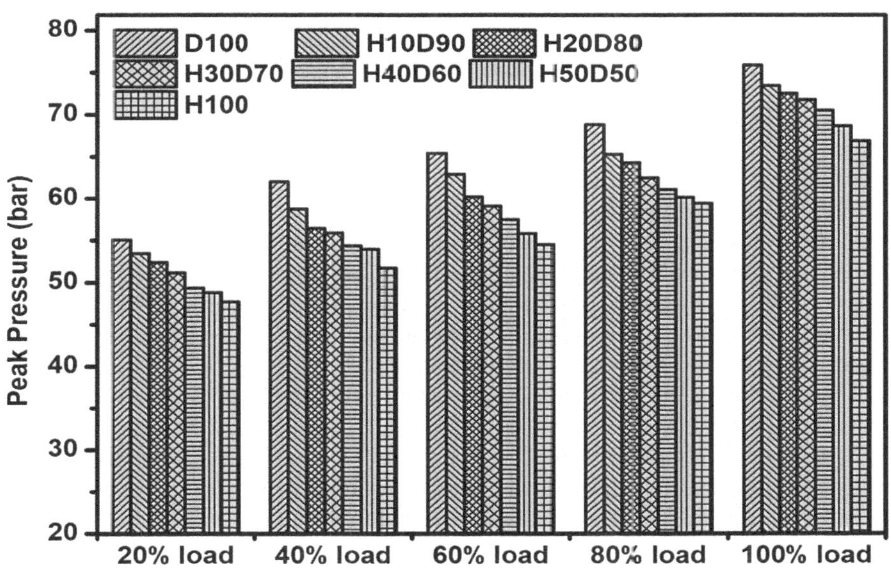

Fig. 4.6 Variation in peak pressure at various loads

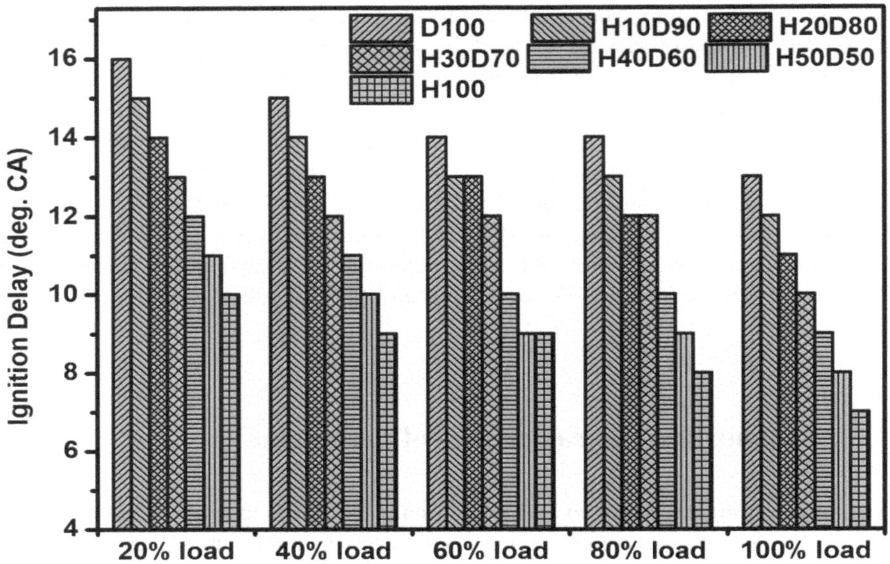

Fig. 4.7 Variation in ignition delay at various loads

the fuel to ignite earlier than diesel, and as less fuel is prepared to burn, the heat released in premixed combustion is less, thereby decreasing the in-cylinder pressure.

Figure 4.7 presents the modulation of ignition delay for all the fuel samples. For a diesel engine, the ignition delay is defined as the difference (in terms of degree crank angle) between when the fuel is injected and the time when the combustion begins (Geo et al., 2019). The onset of combustion was considered when the heat release became positive. For all the blends, the ignition delay reduces as the load on the engine increases. With the increase in load, the in-cylinder temperature along with pressure increases, which provides the necessary heat for the reaction to start earlier, thereby decreasing the ignition delay. It is seen that operating the engine with diesel resulted in a higher ignition delay than the hydrotreated oil–blended samples at all the loads. Due to the high cetane index of the oil, the ignition delay is shorter. These results also conform to the ignition probability test, presented in the previous study (Kumar et al., 2021), which shows that hydrotreated oil and its blends had a higher ignition probability than diesel.

Figure 4.8 indicates the combustion duration of the test fuels at 20%, 40%, 60%, 80%, and full load. The combustion duration with diesel and H100 was 57°CA and 45°CA, respectively, at full load. The trend shows that an increase in power output leads to an increase in combustion duration for all the test fuels. As the fuel requirement of the engine increases, the time required to burn the fuel increases thereby

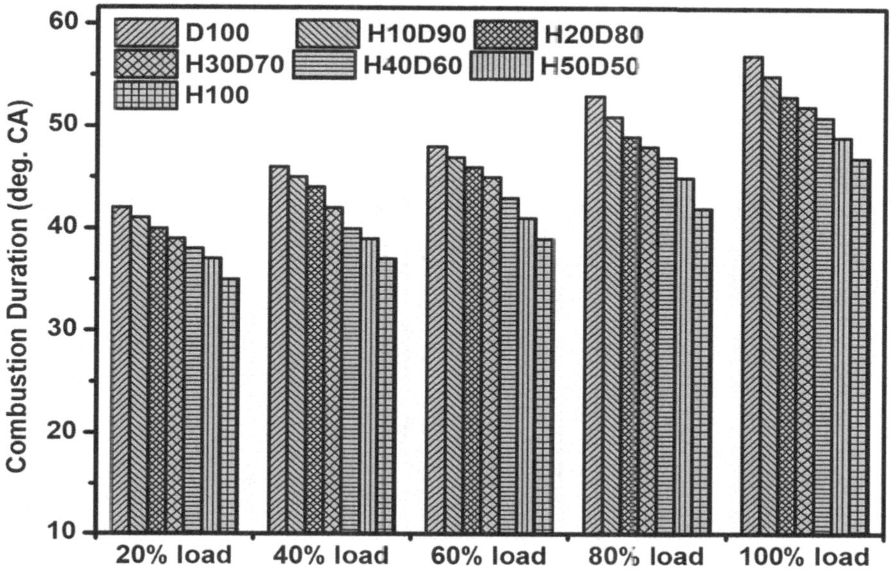

Fig. 4.8 Variation in combustion duration at various loads

increasing the combustion duration (Boopathi et al., 2018). Moreover, the combustion duration of the fuel samples containing hydrotreated oil was low as compared to diesel. Also, the combustion duration decreases as the percentage of the oil in the blend increases. As seen in our previous study (Kumar et al., 2021), rapid heat is released in the diffusion combustion phase for the blends. In the second combustion phase, the accumulated fuel burns earlier because of its higher cetane number. Due to this rapid release of heat, the time required in terms of crank angle for burning the fuel is less, thereby reducing the combustion duration. Similar trends were also observed by Subramanian et al. (2018) while operating a similar engine with a blend of karanja oil methyl ester and orange peel oil.

Figure 4.9 shows CA50, i.e., the crank angle at which the heat released is 50% for all the test fuels at all the loads. The trends reveal that CA50 increases with an increase in engine load. Zhang et al. (2019), while running an engine with hydrotreated oil and gasoline blends, observed similar results. Moreover, CA50 is the highest for diesel, and as the ratio of oil in the blend increases, CA50 decreases for all the loads. The decrease in CA50 for all the blends is due to the reduced ignition delay. The decrease in crank angle for CA50 leads to a uniform heat release as shown in our previous study (Kumar et al., 2021). This gives higher combustion stability, specifically at lower loads; the trends shown in Fig. 4.11 depict the same, i.e., the variation coefficient of IMEP for the blended samples is lower than diesel.

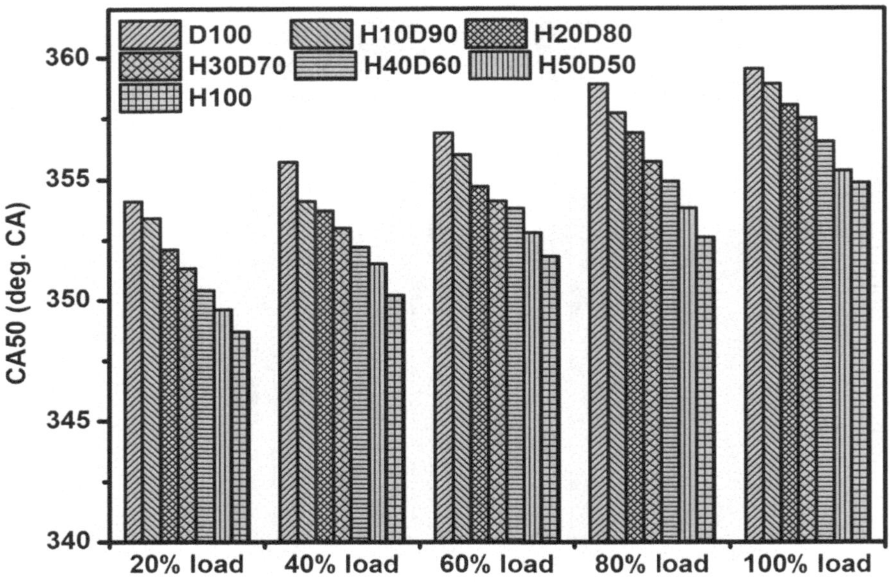

Fig. 4.9 Variation in CA50 at various loads

4.6 Predicting the Cyclic Variation Using Neural Network

It is well known that there is a variation in the combustion process for every cycle which has made it a topic of research as it adversely affects the engine performance that may lead to lower efficiency and higher emissions (Jang et al., 2012). The cyclic variation can be caused by air motion in the cylinder due to variations in swirl and turbulence. Also, the amount of fuel and air entering the cylinder and their mixing before the start of combustion can also invariably cause variations in the combustion process in every cycle (Bittle et al., 2010; Wirasingha et al., 2008). The cyclic variability is evaluated by indicated mean effective pressure (IMEP) as it reflects extensive pressure information of a cycle (Gürgen et al., 2018). The variation in IMEP can be understood by using the proper orthogonal decomposition method (Bonges & Lusk, 2016), the wavelet transform method (Lapuerta et al., 2010), and the coefficient of variance (COV) (Selim, 2005) method. In this study, the variance of IMEP was used to investigate the cyclic variability as given in Eq. (4.1).

$$\mathrm{COV}_{\mathrm{imep}} = \frac{\sigma_{\mathrm{imep}}}{\overline{\mathrm{imep}}} * 100 \tag{4.1}$$

where σ_{imep} (Eq. 4.2) is the standard deviation and $\overline{\mathrm{imep}}$ (Eq. 4.3) is the mean of IMEP.

$$\sigma_{\text{imep}} = \sqrt{\frac{\sum_{i=1}^{n}\left(\overline{\text{imep}} - \text{imep}_i\right)^2}{n-1}} \tag{4.2}$$

$$\overline{\text{imep}} = \frac{1}{n}\sum_{i=1}^{n}\text{imep}_i \tag{4.3}$$

Controlling the cyclic variations is important, as it will improve the engine efficiency and reduce emissions. To achieve low cyclic variations, a strategy needs to be developed. For developing the strategy, a model is required that can predict the cyclic variability for every load. Artificial neural network (ANN) is one such tool that is useful in modeling cyclic variability. ANN is a nonlinear tool that can work on multiple inputs and predict multiple outputs. It can be modeled without any prior knowledge of the processes. A well-trained neural network can predict faster than conventional models since no lengthy iterative calculations are required. However, selecting the network topology in terms of model simplicity and accuracy is important (Ghobadian et al., 2009; Oğuz et al., 2010).

The neural network contains neurons that are highly interconnected processing elements. The network contains an input, a hidden, and an output layer. Input is provided by the source to these neurons in the hidden layer, which perform nonlinear operations by using bias, weights, and activation functions and then provide the output. There can be more than one hidden layer in the network and one output layer. The external world then receives the data from the output layer. The network can approximate a function of any complexity with enough hidden neurons. However, there is a limit to the number of neurons as the network may tend to overfit during training, resulting in the introduction of random noise (Shojaeefard et al., 2012).

4.6.1 *Network Parameters*

The neural networks are nonlinear in nature, which makes them stronger than linear functions. For both the hidden and the output layers, a continuous differential log-sigmoid activation function was used. A form of automatic gain control is produced by the function, and it produces values between 0 and 1 (Roy et al., 2014). Equation (4.4) gives the activation function.

$$F_i = \frac{1}{1+e^{-x}} \tag{4.4}$$

where "x" is the weighted sum of inputs.

4.6.2 Normalization of Input and Output Data

The range of log-sigmoid function lies between 0.1 and 0.9; hence, it is necessary to normalize the input and output data for which Eq. (4.5) was used (Çay et al., 2012). The network's performance drastically improves by scaling the input and output parameters.

$$nv_i = 0.8 * \left(\frac{v_{\min} - v_i}{v_{\min} - v_{\max}} \right) + 0.1 \tag{4.5}$$

4.6.3 Learning Algorithm

An appropriate learning algorithm is required by the network, which adjusts the layer's weight such that the desired response can be achieved. Many algorithms exist for training the network; some of them are gradient descent with momentum and adaptive learning rule, gradient descent with adaptive learning rule (Davim et al., 2008), Levenberg–Marquardt (LM) (Shojaeefard et al., 2012), and scaled conjugate gradient (SCG) (Çay et al., 2012). In the present study, both LM and SCG were used to train the network, and their outputs were compared.

4.6.4 Training and Testing Data

The network was trained using two input parameters and one output parameter. Engine load and blends of hydrotreated oil and diesel were used as inputs, whereas the coefficient of variance of IMEP was used as the output parameter. In total, 35 tests were conducted. The test data was randomly divided into three datasets: training, validation, and test set. The network was trained using the training data set and the over-fitting of the training set was prevented by the validation set, whereas the performance of the model was determined using the test set. The training of the network was carried out using 70% of the experimental data, 15% of the data was used for validation, and the rest of the 15% data was used for testing the network.

4.6.5 Network Structure

A single hidden layer was used, and the number of neurons in the hidden layer were varied from 3 to 20. The iterations for each network topology were performed according to a loss criteria function, which should be minimized. Mean square error was used as the loss criteria function due to differentiability, symmetry, and convexity of the function. To prevent overtraining of the network, the training was stopped when the validation error started increasing. Figure 4.10

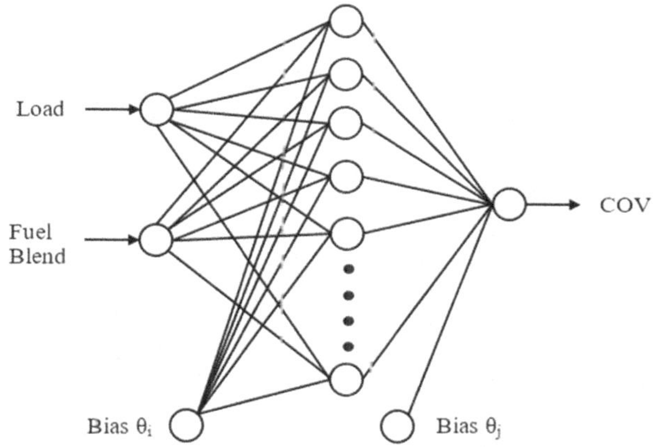

Fig. 4.10 ANN structure with one hidden layer

shows the neural network structure. MATLAB was used for the development of the model.

4.6.6 *Output Evaluation*

The statistical methods used for evaluating the prediction performance of the network are coefficient of determination (R^2) (Eq. 4.6), root mean square error (RMSE) (Eq. 4.7), and mean absolute percentage error (MAPE) (Eq. 4.8) (Javed et al., 2015; Parlak et al., 2006; Yusaf et al., 2010).

$$R^2 = 1 - \left(\frac{\sum_{i=1}^{n} (t_i - o_i)^2}{\sum_{i=1}^{n} (o_i)^2} \right) \tag{4.6}$$

$$\text{RMSE} = \sqrt{\frac{1}{n} \sum_{i=1}^{n} (t_i - o_i)^2} \tag{4.7}$$

$$\text{MAPE} = \frac{1}{n} \sum_{i=1}^{n} \left(\left| \frac{t_i - o_i}{t_i} \right| \right) * 100 \tag{4.8}$$

4.7 Results and Discussion

The coefficient of variance (COV) shows the extent to which the data in a sample varies in relation to the mean of the sample population (Geo et al., 2018). The coefficients of variance for all the test fuel samples at all the loads were calculated and are shown in Fig. 4.11.

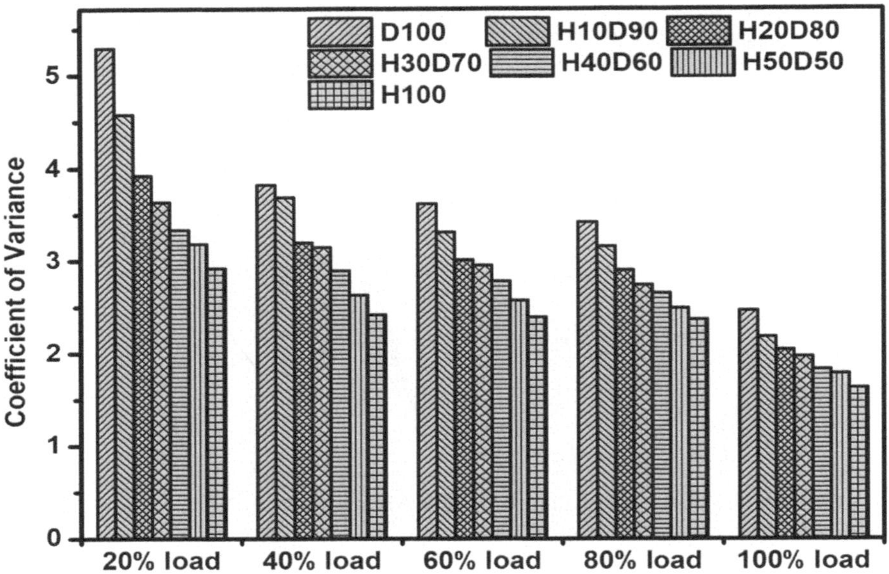

Fig. 4.11 Variation in coefficients of variance at various loads

The cyclic variation in the engine when operated with diesel was the highest, and as hydrotreated oil replaces diesel, the cyclic variation reduces, and it was observed to be the lowest with neat hydrotreated oil engine operation. At full load, the COVs for diesel, H10D90, H20D80, H30D70, H40D60, H50D50, and H100 were 2.46, 2.18, 2.04, 1.97, 1.83, 1.78, and 1.63, respectively. The decrease in variability with the increase of hydrotreated oil in the blend was caused by the high cetane number of the blend that reduces ignition delay and improves the combustion process. As the load increases, the air–fuel mixing process improves, which, in turn, improves the combustion process, thereby reducing the cyclic variability. It was observed during engine operation with the blends that the engine noise was less and the vibrations were also less as compared to diesel fuel operation. For developing the neural network, the input parameters were engine load and fuel blends, and the output parameter was cyclic variability. Both SCG and LM were used for training the network, and the neurons in the hidden layer were varied from 3 to 20. Table 4.4 shows R^2, RMSE, and MAPE for the various neurons.

It is seen that 16 neurons and the LM training algorithm resulted in the neural network model with the highest R^2 value of 0.9996 and the lowest RMSE and MAPE values of 0.0699 and 1.4245, respectively. Figure 4.12 shows the regression graphs of the neural network having 16 neurons.

Table 4.4 Statistical data and training algorithm

No. of neurons	Training algorithm	R^2	RMSE	MAPE
3	trainlm	0.99903	0.1097	2.5729
	trainscg	0.99597	0.2225	6.3126
4	trainlm	0.99909	0.1059	2.4366
	trainscg	0.99236	0.3111	9.0507
5	trainlm	0.99932	0.0922	2.1827
	trainscg	0.99685	0.1983	6.0480
6	trainlm	0.99935	0.0897	1.7840
	trainscg	0.99653	0.2083	6.7976
7	trainlm	0.99940	0.0863	2.0569
	trainscg	0.99675	0.2011	6.2348
8	trainlm	0.99920	0.0994	2.1464
	trainscg	0.99301	0.2970	8.7171
9	trainlm	0.99871	0.1268	2.8719
	trainscg	0.99277	0.2999	8.5528
10	trainlm	0.99932	0.0919	2.0271
	trainscg	0.99560	0.2349	6.7508
11	trainlm	0.99944	0.0838	2.0738
	trainscg	0.99147	0.3263	9.7434
12	trainlm	0.99834	0.1429	2.4583
	trainscg	0.99254	0.3042	8.9287
13	trainlm	0.99944	0.0830	1.8641
	tainscg	0.99188	0.3150	8.7220
14	trainlm	0.99944	0.0834	1.6558
	trainscg	0.99267	0.3011	8.8080
15	trainlm	0.99882	0.1219	1.8634
	trainscg	0.99208	0.3093	8.2816
16	**trainlm**	**0.99960**	**0.0699**	**1.4245**
	trainscg	0.99175	0.3196	9.2935
17	trainlm	0.99896	0.1133	2.5551
	trainscg	0.99172	0.3206	7.9569
18	trainlm	0.99903	0.1090	2.2621
	trainscg	0.99095	0.3328	9.8846
19	trainlm	0.99927	0.0947	2.4856
	trainscg	0.99241	0.3095	8.5175
20	trainlm	0.99831	0.0875	2.1960
	trainscg	0.99178	0.3195	9.1614

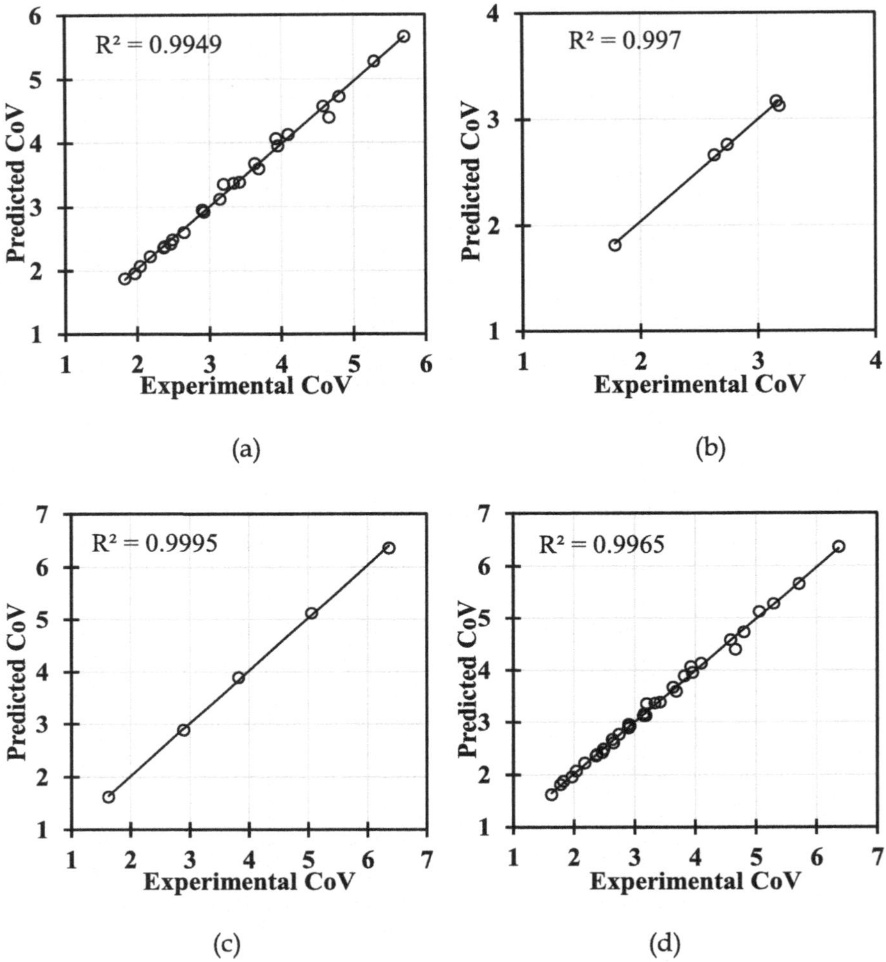

Fig. 4.12 Regression graphs: (**a**) training dataset, (**b**) validation dataset, (**c**) testing dataset, and (**d**) all datasets

4.8 Conclusion

The present research aims to utilize the hydrotreating process as a method for producing diesel-like fuel from used cooking oil. To understand the shelf life of hydrotreated oil, its blends with diesel were separately kept in a container for a period of 12 months. The properties of blends are affected when stored for 12 months, but the variation in properties is not significant, and the values are still lower than that specified in the standards of biodiesel. The peak pressure of the engine was the highest with diesel, and as the percentage of HVO increases in the blend, the peak pressure decreases. Due to the higher cetane index of hydrotreated oil, the fuel ignited earlier than diesel, resulting in lower premixed combustion, thereby decreasing the peak pressure. It was seen that the combustion duration was also shorter for the blends as compared to diesel. Since the heat released in the diffusion combustion phase is rapid, less time is required for the accumulated fuel to burn. The cycle-to-cycle variations were also studied by calculating the coefficient of variance of indicated mean effective pressure. The cyclic variability of the engine was less when hydrotreated oil was blended with diesel. The least cyclic variations were observed with neat hydrotreated oil. Due to the higher cetane index of the fuel, the rate of pressure rise was low, which, in turn, made the engine run smoothly with fewer vibrations. A neural network with neurons in the hidden layer varying from 3 to 20 was trained using Levenberg–Marquardt and scaled conjugate gradient algorithm for predicting the COV. The neural network trained using the LM algorithm containing 16 neurons resulted in the highest R^2 value of 0.9996 and the lowest RMSE and MAPE values of 0.0699 and 1.4245, respectively. The study shows that this method is a vital approach in tackling the problem of ever-changing crude oil prices, environmental pollution, and shelf life.

Appendix

Equation for predicting the CoV (Table 4.5).

Table 4.5 Weights between input layer and hidden layer

$$F_i = \frac{1}{1+e^{-(w_{i1}*\text{load}+w_{i2}*\text{blend ratio}+\theta_i)}}$$

i	w_{i1}	w_{i2}	θ_i
1	0.960	2.092	0.927
2	−3.272	−1.424	2.099
3	1.121	1.993	1.068
4	0.954	2.095	0.919
5	1.763	0.994	0.378
6	5.244	4.653	−4.278
7	6.482	14.598	−0.628
8	1.723	1.062	0.534
9	0.980	2.081	0.954
10	6.849	−0.648	−2.687
11	3.382	5.856	2.937
12	13.582	1.058	−14.214
13	9.232	5.473	−3.075
14	5.836	−2.647	−1.045
15	0.735	2.220	0.533
16	5.157	−4.706	3.280

$$X = -0.141 * F_1 - 6.585 * F_2 - 0.022 * F_3 - 0.144 * F_4 + 1.332 * F_5 - 3.27 * F_6 - 6.175 * F_7 + 1.156 * F_8 - 0.128 * F_9 - 3.388 * F_{10} - 2.561 * F_{11} - 5.921 * F_{12} - 2.799 * F_{13} + 2.52 * F_{14} - 0.231 * F_{15} - 2.747 * F_{16} \tag{4.9}$$

$$\text{COV} = \frac{1}{1 + e^{-(X+16.799)}} \tag{4.10}$$

References

Aatola, H., Larmi, M., Sarjovaara, T., & Mikkonen, S. (2008). Hydrotreated vegetable oil (HVO) as a renewable diesel fuel: Trade-off between NOx, particulate emission, and fuel consumption of a heavy duty engine. *SAE International Journal of Engines, 1*(1), 1251–1262.

Bittle, J., Knight, B., & Jacobs, T. (2010). *Biodiesel effects on cycle-to-cycle variability of combustion characteristics in a common-rail medium-duty diesel engine*. SAE Technical Paper 2010-01-0867.

Bonges, H. A., & Lusk, A. C. (2016). Addressing electric vehicle (EV) sales and range anxiety through parking layout, policy and regulation. *Transportation Research Part A Policy and Practice, 83*, 63–73.

Boopathi, D., Thiyagarajan, S., Edwin Geo, V., Madhankumar, S., & Gheith, R. (2018). Effect of geraniol on performance, emission and combustion characteristics of CI engine fuelled with gutter oil obtained from different sources. *Energy, 157*, 391–401.

BP. (2021). *Statistical review of world energy*. British Petroleum. Retrieved 5 April, 2024, from bp-stats-review-2021-full-report.pdf.

Caprotti, R., Tang, T., Ishibe, N., In-ochanon, R., et al. (2011). *Performance of diesel containing bio-hydrogenated component*. SAE Technical Paper 2011-01-1953.

Çay, Y., Çiçek, A., Kara, F., & Sağiroğlu, S. (2012). Prediction of engine performance for an alternative fuel using artificial neural network. *Applied Thermal Engineering, 37*, 217–225.

Davim, J. P., Gaitonde, V. N., & Karnik, S. R. (2008). Investigations into the effect of cutting conditions on surface roughness in turning of free machining steel by ANN models. *Journal of Materials Processing Technology, 205*(1), 16–23.

Demirbas, A. (2006). Biodiesel production via non-catalytic SCF method and biodiesel fuel characteristics. *Energy Conversion and Management, 47*(15–16), 2271–2282.

Erkkilä, K., Nylund, N., Hulkkonen, T., et al. (2011). *Emission performance of paraffinic HVO diesel fuel in heavy duty vehicles*. SAE Technical Paper 2011-01-1966.

FSSAI. (2018). *Repurpose used cooking oil*. Fssai's new law for reused cooking oil (Tpc/Tpm)—Agro & Food Processing. Retrieved from agronfoodprocessing.com.

Gary, J. H., & Handwerk, G. E. (2001). *Petroleum refining, technology and economics* (4th ed.). Marcel Dekker.

Geo, V. E., Sonthalia, A., Aloui, F., & Femilda Josephin, J. S. (2018). Study of engine performance, emission and combustion characteristics fueled with diesel-like fuel produced from waste engine oil and waste plastics. *Frontiers of Environmental Science & Engineering, 12*(4), 8–17.

Geo, V. E., Sonthalia, A., Nagarajan, G., Nagalingam, B., & Aloui, F. (2019). Effect of port premixed liquefied petroleum gas on the engine characteristics. *Journal of Energy Resources Technology, 141*(11), 1–8.

Ghobadian, B., Rahimi, H., Nikbakht, A. M., Najafi, G., & Yusaf, T. F. (2009). Diesel engine performance and exhaust emission analysis using waste cooking biodiesel fuel with an artificial neural network. *Renewable Energy, 34*(4), 976–982.

Guo, J. H., Xu, G. Y., Shen, F., Fu, Y., Zhang, Y., & Guo, Q. X. (2015). Catalytic conversion of Jatropha oil to alkanes under mild conditions with a Ru/La(OH)<inf>3</inf> catalyst. *Green Chemistry, 17*(5), 2888–2895.

Gürgen, S., Ünver, B., & Altın, İ. (2018). Prediction of cyclic variability in a diesel engine fueled with n-butanol and diesel fuel blends using artificial neural network. *Renewable Energy, 117*, 538–544.

Hribernik, A., & Kegl, B. (2009). Performance and exhaust emissions of an indirect-injection (IDI) diesel engine when using waste cooking oil as fuel. *Energy & Fuels., 23*(3), 1754–1758.

Huber, G. W., & Corma, A. (2007). Synergies between bio- and oil refineries for the production of fuels from biomass. *Angewandte Chemie International Edition, 46*, 7184–7220.

IEA. (2017). *Key energy world Statistics*. International Energy Agency.

IPNGS. (2020). *Indian petroleum & natural gas statistics*. Ministry of Petroleum and Natural Gas Economic and Statistics. PETROLEUM STATISTICS—Indian PNG Statistics | Ministry of Petroleum and Natural Gas | Government of India. Retrieved 5 April, 2024, from mopng.gov.in.

Jang, Y. J., Ko, Y. D., & Jeong, S. (2012). Optimal design of the wireless charging electric vehicle. In *IEEE international electric vehicle conference*. p. 1–5.

Javed, S., Satyanarayana Murthy, Y. V. V, Baig, R. U., & Prasada, R. D. (2015). Development of ANN model for prediction of performance and emission characteristics of hydrogen dual fueled diesel engine with Jatropha methyl Ester biodiesel blends. *Journal of Natural Gas Science and Engineering, 26*, 549–557.

Karthickeyan, V., Thiyagarajan, S., Ashok, B., Edwin Geo, V., & Azad, A. K. (2020). Experimental investigation of pomegranate oil methyl ester in ceramic coated engine at different operating condition in direct injection diesel engine with energy and exergy analysis. *Energy Conversion and Management, 205*, 112334.

Kathirvel, S., Layek, A., & Muthuraman, S. (2016). Exploration of waste cooking oil methyl esters (WCOME) as fuel in compression ignition engines: A critical review. *Engineering Science and Technology, an International Journal, 19*(2), 1018–1026.

Kikhtyanin, O. V., Rubanov, A. E., Ayupov, A. B., & Echevsky, G. V. (2010). Hydroconversion of sunflower oil on Pd/SAPO-31 catalyst. *Fuel, 89*(10), 3085–3092.

Knothe, G. (2010). Biodiesel and renewable diesel: A comparison. *Progress in Energy and Combustion Science, 36*(3), 364–373.

Kumar, N., Sonthalia, A., Pali, H. S., & Sidharth. (2020a). Next-generation biofuels—Opportunities and challenges. In A. K. Gupta, A. De, S. K. Aggarwal, A. Kushari, & A. Runchal (Eds.), *Innovations in sustainable energy and cleaner environment* (pp. 171–191). Springer Singapore.

Kumar, N., Sonthalia, A., & Koul, R. (2020b). Optimization of the process parameters for hydrotreating used cooking oil by the taguchi method and fuzzy logic. *Journal of Energy Resources Technology, 142*(12).

Kumar, N., Sonthalia, A., Tomar, M., & Koul, R. (2021). An experimental investigation on spray, performance and emission of hydrotreated waste cooking oil blends in an agricultural engine. *International Journal of Engine Research, 22*(7), 2305–2317.

Kuronen, M., Mikkonen, S., Aakko, P., & Murtonen, T. (2007). *Hydrotreated vegetable oil as fuel for heavy duty diesel engines*. SAE Technical Paper 2007-01-4031.

Lapuerta, M., Herreros, J. M., Lyons, L. L., García-Contreras, R., & Briceño, Y. (2008). Effect of the alcohol type used in the production of waste cooking oil biodiesel on diesel performance and emissions. *Fuel, 87*(15–16), 3161–3169.

Lapuerta, M., García-Contreras, R., Campos-Fernández, J., & Dorado, M. P. (2010). Stability, lubricity, viscosity, and cold-flow properties of alcohol−diesel blends. *Energy & Fuels, 24*(8), 4497–4502.

Lapuerta, M., Villajos, M., Agudelo, J. R., & Boehman, A. L. (2011). Key properties and blending strategies of hydrotreated vegetable oil as biofuel for diesel engines. *Fuel Processing Technology, 92*(12), 2406–2411.

Liu, Y., Sotelo-Boyás, R., Murata, K., Minowa, T., & Sakanishi, K. (2012). Production of bio-hydrogenated diesel by Hydrotreatment of high-acid-value waste cooking oil over ruthenium catalyst supported on Al-Polyoxocation-pillared montmorillonite. *Catalysts, 2*(4), 171–190.

Makinen, R., Nylund, N., Erkkilä Lng, K., Saikkonen, P., et al. (2011). *Bus fleet operation on renewable paraffinic diesel fuel*. SAE Technical Paper 2011-01-1965.

Murata, K., Liu, Y., Inaba, M., & Takahara, I. (2010). Production of synthetic diesel by Hydrotreatment of Jatropha oils using Pt–Re/H-ZSM-5 catalyst. *Energy & Fuels, 24*(4), 2404–2409.

Oğuz, H., Sarıtas, I., & Baydan, H. E. (2010). Prediction of diesel engine performance using biofuels with artificial neural network. *Expert Systems with Applications, 37*(9), 6579–6586.

Parlak, A., Islamoglu, Y., Yasar, H., & Egrisogut, A. (2006). Application of artificial neural network to predict specific fuel consumption and exhaust temperature for a diesel engine. *Applied Thermal Engineering, 26*(8), 824–828.

Patel, M., & Kumar, A. (2016). Production of renewable diesel through the hydroprocessing of lignocellulosic biomass-derived bio-oil: A review. *Renewable and Sustainable Energy Reviews, 58*, 1293–1307.

Patil, S. J., & Vaidya, P. D. (2017a). Catalytic Hydrotreatment of Jatropha oil over lanthanum hydroxide supported Noble metals: Effect of promotion with cerium. *ChemistrySelect, 2*(35), 11918–11925.

Patil, S. J., & Vaidya, P. D. (2017b). On the production of bio-hydrogenated diesel over hydrotalcite-like supported palladium and ruthenium catalysts. *Fuel Processing Technology, 2018*(169), 142–149.

Roy, S., Banerjee, R., & Bose, P. K. (2014). Performance and exhaust emissions prediction of a CRDI assisted single cylinder diesel engine coupled with EGR using artificial neural network. *Applied Energy, 119*, 330–340.

Selim, M. Y. E. (2005). Effect of engine parameters and gaseous fuel type on the cyclic variability of dual fuel engines. *Fuel, 84*(7), 961–971.

Shojaeefard, M. H., Etghani, M. M., Akbari, M., Khalkhali, A., & Ghobadian, B. (2012). Artificial neural networks based prediction of performance and exhaust emissions in direct injection engine using castor oil biodiesel-diesel blends. *Journal of Renewable and Sustainable Energy, 4*(6).

Sonthalia, A., & Kumar, N. (2021). Comparison of fuel characteristics of hydrotreated waste cooking oil with its biodiesel and fossil diesel. *Environmental Science and Pollution Research, 28*(10), 11824–11834.

Sonthalia, A., Garg, S., Sharma, R., Subramanian, T., & Kumar, N. (2021). Effect of electrostatic precipitator on exhaust emissions in biodiesel fuelled CI engine. *Environmental Science and Pollution Research, 28*(10), 11850–11859.

Stumborg, M., Wong, A., & Hogan, E. (1996). Hydroprocessed vegetable oils for diesel fuel improvement. *Bioresource Technology, 56*, 13–18.

Subramanian, T., Varuvel, E. G., Leenus, J. M., & Beddhannan, N. (2018). Effect of electrochemical conversion of biofuels using ionization system on CO2 emission mitigation in CI engine along with post-combustion system. *Fuel Processing Technology, 173*, 21–29.

Tomar, M., Sonthalia, A., Kumar, N., & Dewal, H. (2020). Waste glycerol derived bio-propanol as a potential extender fuel for compressed ignition engine. *Environmental Progress & Sustainable Energy, 40*.

Wirasingha, S. G., Schofield, N., & Emadi, A. (2008). Plug-in hybrid electric vehicle developments in the US: Trends, barriers, and economic feasibility. In: *IEEE vehicle power and propulsion conference*. p. 1–8.

Xiao, F., Liu, S., Hwang, J. W., et al. (2000). *Plasma treatment of diesel particulate for a minibus*. SAE Technical Papers Series 2000-01–3459.

Yusaf, T. F., Buttsworth, D. R., Saleh, K. H., & Yousif, B. F. (2010). CNG-diesel engine performance and exhaust emission analysis with the aid of artificial neural network. *Applied Energy, 87*(5), 1661–1669.

Zhang, Y., Li, Z., Tamilselvan, P., et al. (2019). Experimental study of combustion and emission characteristics of gasoline compression ignition (GCI) engines fueled by gasoline-hydrogenated catalytic biodiesel blends. *Energy, 187*, 115931.

Chapter 5
AI and Energy Management in Organizations: A Systematic Literature Review

Abdülkadir Akturan and Aykut Arslan

Nomenclature

AI	Artificial intelligence
CH_4	Methane
CO	Carbon monoxide
CO_2	Carbon dioxide
FT	Fischer–Tropsch

5.1 Introduction

In today's fast-changing business environment, organizations are continuously searching for creative ways to improve efficiency, cut costs, and lessen their environmental footprint. Artificial intelligence (AI) has emerged as a transformative tool for achieving these goals. While traditionally used for data analysis and automation, AI is now also reshaping energy management practices in organizations. This paper delves into the convergence of AI and energy management, highlighting how AI is changing the landscape by helping businesses streamline energy use, minimize waste, and advance toward sustainability.

Energy needs and electricity usage have always been a concern for home and business owners. As resources become scarcer and electricity costs continue to rise, it becomes more important for people, businesses, and countries to be aware of how

A. Akturan (✉) · A. Arslan
Piri Reis University, Tuzla, Istanbul, Turkiye
e-mail: a.akturan@pirireis.edu.tr; aarslan@pirireis.edu.tr

M. Z. Sogut, M. Koray (eds.), *Energy Rationality and Management for Decarbonization*, Green Energy and Technology,
https://doi.org/10.1007/978-3-031-68497-5_5

they use their energy and how to use it most efficiently. Nowadays, the need for energy management systems is increasing due to global climate change. The economical use and efficient production of energy have now become more prominent issues. Ensuring energy security and clean energy production is actively discussed by governments, companies, different international organizations, and research groups. The issue of using renewable energy resources to ensure energy security and clean energy production has started to increase in importance and has started to be discussed by almost every organization (Oyucu & Herdem, 2022).

AI has introduced a novel dimension to contemporary technology, particularly in the realm of energy design, leveraging sophisticated algorithms and artificial neural networks (Alassery et al., 2022). This technology enables the prediction of solar, wind, and water movements, among other energy sources, thereby mitigating the risk of loss and ensuring system reliability. Additionally, AI can effectively address various challenges that may arise in energy systems, offering swift and efficient solutions (Irshad et al., 2020).

Energy management systems are systems created to monitor and use energy resources effectively and efficiently (Sinoplu & Ceyhan, 2023). These systems are supported by concepts such as the Internet of Things and AI and are created with the support of sensors and actuators. Energy management with AI is a way to understand smart energy and how systems operate most efficiently. Smart energy management takes the methods currently used by homes and businesses to the next level, providing the highest level of energy savings. It is expected that the global renewable energy consumption rate will be 147% in 30 years, and compared to 2004, new global investments in clean energy increased almost tenfold in 2019 (Li et al., 2023). There is a growing demand for intelligent energy management through the utilization of AI to navigate these challenges effectively. This entails monitoring and coordinating the needs and capacities of all resources, suppliers, infrastructure operators, end users, energy transformers, and energy market stakeholders (Chai et al., 2014).

This chapter aims to summarize the development of artificial intelligence AI and its role in energy management within organizations. It first discusses the evolution of AI applications in energy management, highlighting the associated benefits and challenges. Building on this foundation, the chapter seeks to offer comprehensive insights through a systematic review of the business and management literature, providing a deeper understanding of how AI is shaping energy management practices in organizational contexts.

5.2 AI in Energy Management

AI aims to imbue computers or machines with human-like capabilities such as thinking, generalization, learning, and problem-solving, leveraging knowledge acquired through experience (Yılmaz & Soylu, 2022). In the realm of energy management, AI represents a cutting-edge technology that encompasses a variety of intelligent algorithms for tasks like energy planning, forecasting, demand response, and energy trading (Musa, 2023). AI enhances energy management by harnessing machine learning, advanced algorithms, and real-time data analytics. Like a living organism, AI can learn from user behavior, adjusting energy management and consumption accordingly, akin to human intelligence. Moreover, AI facilitates optimized energy utilization by enabling communication between different energy sources (Nutakki & Mandava, 2023). While human-centered innovation management is vital for organizations to reinvent themselves through exploratory initiatives, AI extends its assistance beyond human capabilities (Groves et al., 2013). The notion that AI can be applied in innovation environments is gaining traction with the rapid advancement of AI and machine learning, hinting at significant changes ahead (Haefner et al., 2021). AI technologies, including facial recognition, robots, autonomous vehicles, natural language processing, and various virtual applications, are extensively employed across various domains. Over half of businesses adopted some form of this technology wave in 2020, and its applications continue to proliferate (Berente et al., 2021).

Considering today's developments, the management of AI has become more important than ever. The role of managers is becoming more prominent in the developing and changing social process that includes AI and is greatly affected by it. Managers decide on the acquisition, development, and implementation of AI-based systems and constantly control this process. Managers use AI in decision-making processes, selection of target customer groups, and monitoring and evaluation of processes and results (Berente et al., 2021). Although AI offers opportunities to create value, it also brings challenges for managers that will require them to reconsider their duties and reevaluate their operating principles (Ballamudi, 2019). As cooperation between machines and humans increases, the division of labor and job descriptions will also change (Kolbjørnsrud et al., 2016). Technical issues will come to the fore among the challenges that managers will face, such as finding effective and new solutions for human interaction; tackling issues of trust, safety, and security; and being careful to avoid negative and unproductive consequences (Stone et al., 2022).

5.2.1 The History of AI in the Smart Energy Industry

The history of AI spans decades and is marked by significant milestones, breakthroughs, and paradigm shifts. The journey of AI began with the emergence of AI as a concept in the mid-twentieth century. This concept was led by Alan Turing and supported by his visionary ideas and ground-breaking research on computers and his intelligent machines (Turing, 2012). The 1956 Dartmouth Conference was a pivotal moment that formally coined the term "AI" and laid the foundation for AI as an interdisciplinary field (Russell & Norvig, 2010). During this period, the first applications of AI appeared. These applications primarily include logical theorems and chess games. The period from 1965 to 1970 can be called the Dark Ages when it comes to AI. During this period, there were not many developments in AI that could be tested (Maad, 2015). From 1970 to 1975, AI became popular again. Thanks to the success of AI systems, the foundations for today's AI were laid, for example, in the diagnosis of diseases (Delipetrev et al., 2020; Maad, 2015). Maad (2015) expressed the historical process of AI to date as follows. 1981: IBM builds the first personal computer. 1993: MIT begins production of the human-like robot Cog. 1997: Supercomputer Deep Blue defeats world-famous chess player Kasparov. 1998: Furby, the first artificially intelligent player, is released. 2000: A robot named Kismet is introduced that can communicate using gestures and facial expressions. 2005: ASIMO, the robot closest to AI and human skills and abilities, is introduced. 2010: Asimo was able to move using his head.

5.2.2 Predictive Analytics

As current insights into AI approach an all-time high, it is no longer a dream to see that algorithms applied in other industries can also be used in energy management services. AI algorithms can analyze historical energy usage patterns and forecast future demands. This enables organizations to proactively adjust their energy consumption, optimizing usage during periods of lower demand and avoiding costly peak rates. Google algorithms for automatic image classification can read CT scans of a patient with cancer or predict treatable blinding retinal diseases (Thompson et al., 2018; Kermany et al., 2018). AI methods used to predict the risk of loan default could be modified to predict the risk of sepsis or pneumonia (Caruana et al., 2015). AI derives all its adaptive capabilities from multiple intelligent algorithms working together rather than a single algorithm. There is not just one fixed formula on which the intelligent algorithm is based. Some of these algorithms make use of training and, unlike a specific mathematical formula, can update itself to gain a better performance, while others can variably adjust their operation based on data from inputs–outputs, making them more broadly applicable based on the type of activity (Li et al., 2023). These technologies enable organizations to create intelligent systems that continuously monitor and optimize energy consumption. AI gives organizations predictive analytics capabilities, allowing them to predict energy consumption patterns with remarkable accuracy. Considering historical data and understanding trends, AI algorithms can predict periods of peak demand and help organizations allocate resources more efficiently, avoid energy waste, and reduce costs (Lee et al., 2017).

5.2.3 Smart Building Systems

A study revealed that the energy consumption of buildings is higher than thought. International Energy Agency found that buildings consume 71% of the total electricity in urban areas and 39% of the final energy consumption in their global study. In addition, it has been revealed that buildings are responsible for nearly 40% of the total carbon dioxide emissions in urban areas (Farzaneh et al., 2021). AI has been shown to be extremely well-suited to flexibly manage urban energy systems using real-time data (Sharma & Saini, 2015). AI-powered smart building systems can autonomously control lighting, heating, ventilation, and air conditioning (HVAC) based on real-time occupancy data and external factors like weather conditions. This not only enhances comfort for occupants but also ensures that energy is used efficiently. AI-based forecasting models can minimize the operating cost of energy production by optimizing energy distribution using energy demand and supply forecasts. However, power plants meet the required demand as needed and operate within the operational constraints of the power network in constantly changing conditions (Farzaneh, 2019). The number of smart buildings is increasing day by day. AI-powered smart building management systems can optimize energy use by automating tasks such as lighting, heating, and cooling. Smart building management systems can reduce energy consumption and costs by monitoring occupancy patterns and adjusting these systems accordingly (Musa, 2023).

5.2.4 *Energy Consumption Optimization*

AI can optimize energy consumption by learning from real-time data, adjusting settings, and identifying opportunities for energy savings. This adaptability allows organizations to respond swiftly to changes in energy prices or operational requirements. AI can be used to analyze energy use patterns and identify areas where energy is wasted. By analyzing data from sensors and other sources using a variety of algorithms, AI can identify patterns that indicate inefficient energy use. This data can then be used effectively to reduce waste and optimize energy consumption (Ahmad et al., 2021; Musa, 2023). AI has the ability to monitor the conditions of different control devices, identify energy leakage issues, and understand energy consumption trends. Energy efficiency optimization depends on the availability of energy production and production assets. Energy optimization techniques and AI models with real-time adjustments are rapidly being used by utilities to increase return on investment and production efficiency (Ahmad et al., 2021).

5.2.5 *Demand Response*

Demand side management (DSM) is focused on the reduction of energy waste and the more efficient utilization of equipment (Faruqui et al., 2010). In the context of smart grid technology, effective management of demand from consumers (DSM) is crucial for ensuring the efficiency and reliability of electrical grids, particularly in optimizing peak load demand (Çakmak & Altas, 2020). Various demand response programs, such as critical peak pricing (CPP), are employed to enable the planning of home appliance usage in a manner that is both more efficient and effective

(Nutakki & Mandava, 2023). The demand response (DR) approach integrates both time-of-use (TOU) and emergency demand response program (EDRP) methods to identify the optimal incentives for a combined TOU and EDRP framework. Research suggests that TOU tariffs offer benefits for both electricity consumers and grid operators, indicating that pricing mechanisms can directly influence energy consumption patterns and grid operations (Aalami et al., 2008). Artificial intelligence (AI) plays a crucial role in enabling efficient demand response strategies by automating the reduction or adjustment of energy-intensive processes during peak demand periods. This not only leads to a decrease in energy costs but also contributes to the maintenance of grid stability.

5.2.6 Equipment Monitoring and Maintenance

AI can be used to predict the chances of equipment failure, allowing maintenance teams to take corrective action before a failure occurs. AI analyzes data from sensors and other sources, can identify potential equipment failures that may occur, and helps maintenance and repair teams plan repairs before failures occur (Musa, 2023). AI-enabled sensors and monitoring systems can detect anomalies in equipment performance, allowing organizations to identify and address potential issues before they lead to energy wastage or equipment failure (Abdalla et al., 2021). Predictive maintenance can prolong the lifespan of machinery and reduce downtime.

5.3 The Growing Importance of Energy Management

The increasing demand for renewable energy resources (RESs) and the rapid development and change in recent years require all organizations to take precautions and keep up with innovations (Antonopoulos et al., 2020). The transition to sustainable energy systems is one of the greatest challenges of our time. Most global emissions from burning fossil fuels are associated with energy sources and transportation, influencing climate change, which is largely an energy issue. In an era where environmental sustainability is a top priority, energy management has become a critical focus for organizations across industries. Efficient energy usage not only reduces operational costs but also aligns with corporate social responsibility goals, appealing to environmentally conscious consumers and investors. Traditional methods of energy management often involve static approaches, relying on historical data and predefined schedules. However, the dynamic nature of modern businesses demands more adaptive and responsive solutions, and this is where AI steps in. By leveraging the power of data analytics and smart systems, AI can help businesses and energy companies optimize energy use, reduce costs, and increase sustainability through the rational use of processes (Musa, 2023).

Digital technologies are also changing the competitive environment of the energy sector. Digitalization offers cost advantages that increase revenues while minimizing business risks. Businesses facing significant transformation opportunities can achieve success through digitalization and the use of AI instead of traditional methods (Göde et al., 2023). Ultramodern smart grid structures and similar systems can make it possible to use electricity grids more efficiently by adding the integration of renewable resources as these technologies become more complex and effective (Rinku, 2023). Smart grids are smart energy networks that use sensors, smart meters, and other advanced technologies to monitor, control, and manage energy consumption in real time (Alassery et al., 2022; Musa, 2023).

Load forecasting, supply, and management of energy are some of the core tasks of the utilities and energy industry (Saleem & Karmalkar, 2009). The estimation and prediction of loads are integral parts of a safe and efficient power system operation (Antonopoulos et al., 2020). Utilizing AI, electric load forecasting results can be integrated into electricity management systems to enhance production efficiency and minimize energy consumption. This is achieved by developing electricity generation plans in advance, ensuring optimal levels of electricity production, and avoiding inadequate or incomplete production (Hu et al., 2022). Therefore, it is very important to make accurate electrical load predictions in order to save money and use energy effectively and efficiently. AI allows us to more accurately predict the energy needed and optimize system efficiency (Jose et al., 2020). AI enhances efficiency in various processes, including system sharing, modular construction, waste material cascading, equipment refurbishment, and expert system learning. By minimizing human intervention, AI contributes to reducing or avoiding the greenhouse gas impact (Jose et al., 2020).

5.4 Benefits of AI in Energy Management

The integration of AI into energy management processes brings a multitude of benefits for organizations:

5.4.1 Cost Savings

Applications of AI in energy management lead to significant reductions in costs and improvements in operational efficiency. By intelligently managing energy consumption, organizations can streamline processes, reduce waste, and achieve significant savings in operational expenses (Lee et al., 2017). AI-driven energy management leads to significant cost savings by optimizing energy consumption, minimizing waste, and avoiding peak demand charges.

5.4.2 *Sustainability*

One of the most interesting benefits of AI in energy management is its contribution to environmental sustainability. AI-driven systems help organizations identify opportunities to reduce carbon emissions and adopt environmentally friendly practices in line with corporate social responsibility goals and regulatory requirements (Rong et al., 2019). Organizations can enhance their environmental sustainability by reducing their carbon footprint and demonstrating a commitment to eco-friendly practices.

AI applications in energy management lead to significant cost reductions and operational efficiency improvements. Through intelligent monitoring and analysis, organizations can identify and eliminate energy wastage, streamline processes, and achieve substantial savings in operational expenditures (Lee et al., 2017). AI streamlines energy management processes, automating routine tasks and allowing organizations to focus on core business activities.

5.4.3 *Regulatory Compliance*

One of the most important benefits of AI in energy management is its contribution to environmental sustainability. AI-driven systems help organizations identify opportunities to reduce carbon emissions and adopt environmentally friendly practices in line with corporate social responsibility goals and regulatory requirements (Rong et al., 2019). With AI's ability to monitor and report energy usage accurately, organizations can ensure compliance with increasingly stringent environmental regulations.

5.4.4 *Real-Time Monitoring and Adaptive Control*

Integrating AI into energy management systems facilitates real-time monitoring of energy consumption, allowing organizations to gain dynamic insights into their energy usage. Adaptive control systems driven by AI algorithms optimize energy use and minimize operating costs by rapidly responding to changes in demand (Mohsenian-Rad et al., 2010).

5.5 Challenges of AI in Energy Management

AI has become a buzzword in the energy sector, promising revolutionary changes in the way we produce, distribute, and consume energy. While the integration of AI into energy management has tremendous potential, it is crucial to critically examine the challenges and drawbacks associated with this transformative technology.

5.5.1 Integration Challenges with Existing Infrastructure

Many energy systems are built on legacy architectures, making integration with AI a challenging task. Reinforcing or upgrading existing infrastructure to accommodate AI technologies is not only technically challenging but can also incur significant costs. Striking a balance between modernization and continuity in energy management processes has become essential to prevent outages (Wang et al., 2018).

5.5.2 Data Security and Privacy Concerns

The effectiveness of AI in energy management largely relies on the analysis of large datasets covering user behavior, infrastructure details, and energy consumption patterns. This flow of sensitive data raises significant security and privacy concerns. Protecting against cyber threats and unauthorized access and ensuring compliance with privacy regulations such as General Data Protection Regulation (GDPR) are crucial to increasing public trust in AI-powered energy management systems (Acquisti & Fong, 2018; Musa, 2023).

5.5.3 Lack of Standardization and Interoperability

The lack of standardized protocols and interfaces is an obstacle to the seamless integration of AI into various energy management systems Without common standards, interoperability becomes difficult to achieve, and scalability and universal adoption of AI applications are hampered. Establishing industry-wide standards is essential to fostering a harmonious and collaborative AI ecosystem in energy management (Mell & Grance, 2011).

5.5.4 Energy Consumption of AI Models

Paradoxically, the energy efficiency gains that AI promises in managing energy systems are offset by the significant energy consumption of AI models. Training and running complex AI algorithms requires significant computing power, potentially offsetting the net positive impact on overall energy efficiency. Striking a balance between the benefits of optimization and the energy consumption of AI models remains a significant challenge (Strubell et al., 2019).

5.5.5 *Lack of Practical Moxie*

Many experts possess distinct areas of expertise. The task of identifying competent professionals capable of executing dependable processes that yield tangible and applicable advantages is a formidable undertaking. The utilization of advanced operational software in the realm of digitalization presents challenges due to the protective measures implemented by energy corporations for data preservation. Risks are associated with the loss of data, inadequate configuration, and unauthorized entry. Numerous companies exhibit reluctance to explore novel approaches owing to the substantial expense of errors within the energy sector (Rinku, GS 2023).

5.6 Systematic Literature Review

In order to provide a general insight into the literature on AI and energy management in organizations, we have conducted a systematic literature review. We have adopted the SPAR-4-SLR procedure suggested by Paul et al. (2021). The Scientific procedures and rationales for systematic literature reviews (SPAR-4-SLR) procedure involves three stages: (1) assembling, (2) arranging, and (3) assessing. The procedure, which was conducted in this study, is given in Fig. 5.1 (Table 5.1).

5.6.1 *The Results of Descriptive Analysis and VOSviewer*

The general summary of the studies on AI and energy management in organizations in business and management literature can be seen in Fig. 5.1.

As seen in Fig. 5.1, the very first study in that area was published in 2009, and the annual growth rate of the literature on this subject is 28.64%., and, in total, 372 authors gave contributions to this area. Table 5.2 displays the production per year.

Fig. 5.1 The general summary of the studies

As seen in Table 5.2, there is a growing trend on that subject in business and management literature.

The keyword network can be seen in Fig. 5.2. The most frequently used keywords were AI, decision support systems, decision making, energy management, energy efficiency, energy utilization, and sustainable development (Table 5.3).

As seen in Table 5.4, the production of countries over time is increasing; India and China are the two leading countries studying that subject.

Table 5.1 SPAR-4-SLR systematic literature review procedure

Assembling	
Domain	AI in energy management in organizations
Source type	Academic sources
Source quality	Scimago (Scopus)
Search period	Undefined–December 2023
Keywords	"AI," "energy," "management," and "organizations"
Total documents	5300
Arranging	
Inclusion criteria	Subject areas, language, source type
Subject area	Business, management, and accounting in Scopus
Language	English
Source type	Journals and book chapters
Document type	Articles and reviews
Total documents	116
Assessing	
Total documents	116
Analysis	Descriptive analysis, VOSviewer

Table 5.2 Production trends by years

Year	Articles	Year	Articles
2009	1	2017	5
2010	2	2018	10
2011	0	2019	7
2012	1	2020	8
2013	2	2021	8
2014	4	2022	29
2015	5	2023	34
2016	0		

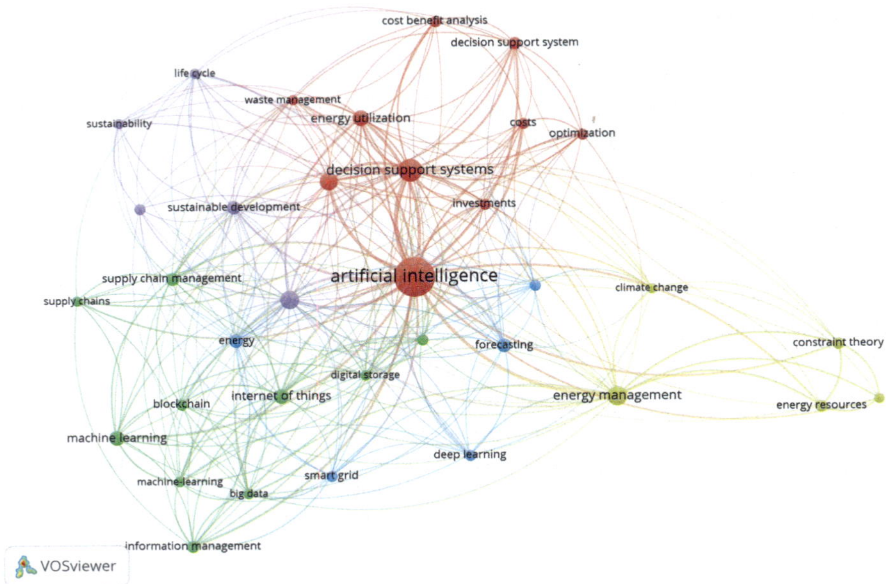

Fig. 5.2 Keyword network

Table 5.3 Most frequent keywords

Terms	Frequency	Terms	Frequency
AI	54	smart grid	5
decision support systems	27	supply chains	5
decision making	17	waste management	5
energy management	16	climate change	4
energy efficiency	15	energy management systems	4
energy utilization	13	real time systems	4
sustainable development	10	renewable energies	4
energy	9	renewable energy resources	4
forecasting	8	risk management	4
information management	8	smart power grids	4
internet of things	8	technology adoption	4
supply chain management	8	algorithm	3
constraint theory	7	ant colony optimization	3
energy resources	7	battery management systems	3
machine learning	7	benchmarking	3
cost benefit analysis	6	block-chain	3
deep learning	6	blockchain	3
investments	6	business development	3
big data	5	charging (batteries)	3
costs	5	commerce	3

(continued)

Table 5.3 (continued)

Terms	Frequency	Terms	Frequency
data analytics	5	competition	3
digital storage	5	decision making process	3
energy conservation	5	decision support system (dss)	3
energy development	5		
life cycle	5		
machine-learning	5		
optimization	5		

Table 5.4 Countries' production over time

Country	Year	Articles	Country	Year	Articles
United Kingdom	2013	6	USA	2014	2
United Kingdom	2014	6	USA	2015	4
United Kingdom	2015	7	USA	2017	7
United Kingdom	2017	7	USA	2018	7
United Kingdom	2018	7	USA	2019	9
United Kingdom	2019	7	USA	2020	9
United Kingdom	2020	10	USA	2021	16
United Kingdom	2021	10	USA	2022	20
United Kingdom	2022	19	USA	2023	25
United Kingdom	2023	36	India	2015	2
China	2014	3	India	2017	2
China	2015	6	India	2018	5
China	2017	6	India	2019	12
China	2018	24	India	2020	17
China	2019	34	India	2021	25
China	2020	34	India	2022	46
China	2021	42	India	2023	65
China	2022	64			
China	2023	74			
Italy	2014	2			
Italy	2015	2			
Italy	2017	2			
Italy	2018	2			
Italy	2019	7			
Italy	2020	7			
Italy	2021	12			
Italy	2022	16			
Italy	2023	27			

The top three authors in that subject area are Diamantoulakis P.D. (2015), Quariguasi Frota Neto J. (2010), and Vikhorev K. (2013). The other top cited authors are given in Table 5.5.

Table 5.5 Most globally cited documents

Paper	Total citations	Total Citations (TC) per year
Diamantoulakis Pd, 2015, Big Data Res	253	28.11
Quariguasi Frota Neto J, 2010, Int J Prod Res	163	11.64
Vikhorev K, 2013, J Clean Prod	159	14.45
Cubric M, 2020, Technol Soc	95	23.75
Mattiussi A, 2014, Decis Support Syst	83	8.30
Zhang L-Y, 2019, J Clean Prod	82	16.40
Abdulbaki D, 2017, J Clean Prod	76	10.86
Gonzalez Edrs, 2015, J Clean Prod	56	6.22
Herath Hmkkmb, 2022, Int J Inf Manag Data Insights	51	25.50
Flisberg P, 2012, Joperressoc	47	3.92
Yalcinkaya S, 2020, J Clean Prod	46	11.50
Javaid M, 2021, J Ind Integr Manag	41	13.67
Hu Z-H, 2014, Decis Support Syst	39	3.90
Kim Ss, 2017, J Knowl Manag	36	5.14
Majumdar D, 2018, Technol Anal Strateg Manage	34	5.67
Wang H, 2019, J Clean Prod	33	6.60
Pulselli Rm, 2019, J Clean Prod	33	6.60
Yan J, 2018, Technol Forecast Soc Change	32	5.33
Chang K-H, 2014, Decis Support Syst	32	3.20
Hunt Jd, 2013, Decis Support Syst	32	2.91
Banaszak Za, 2009, Int J Prod Econ	28	1.87
Cao M-T, 2015, J Constr Eng Manage	28	3.11
Raval M, 2021, Internet Thing	26	8.67
Wu T, 2021, Electr J	25	8.33
Torregrossa D, 2017, J Clean Prod	24	3.43

AI in energy management has emerged as a powerful tool for optimizing energy consumption, reducing costs, and improving sustainability. By leveraging AI's capabilities in data analysis, machine learning, and real-time control, organizations can achieve significant benefits and contribute to a more sustainable future.

5.7 Implications for Practice and Theory

Integrating AI into energy management offers significant implications for both practical applications and theoretical advancements. In practice, organizations can leverage AI-powered tools to optimize energy consumption in real time, automate tasks, improve decision-making, and enhance operational agility. This leads to increased efficiency, cost savings, and the ability to adapt to changing market conditions. Additionally, AI paves the way for the development of innovative business models and services, empowering energy consumers and facilitating regulatory compliance.

Theoretically, AI's integration necessitates a reevaluation of the existing energy management models, prompting researchers to refine them by incorporating real-world dynamics and diverse data sources. This opens up new research areas focused on human–AI interaction, the ethical implications of AI in decision-making, and its impact on energy markets and policy. Furthermore, AI fosters a deeper understanding of energy systems, allowing researchers to analyze the complex relationships between production, consumption, and environmental factors. This leads to the development of more effective energy policies and strategies. Ultimately, AI challenges traditional energy management paradigms, emphasizing data-driven decision-making, automation, and continuous optimization. This necessitates interdisciplinary collaboration between researchers and practitioners from diverse fields, facilitating cross-pollination of ideas and accelerating innovation in the energy sector.

5.8 Conclusion

The integration of AI into energy management applications represents a transformative breakthrough toward a more sustainable, efficient, and adaptable future. The energy sector is undergoing rapid change as it faces disruptions due to reasons such as decentralization, decarbonization, and digitalization. Developments in the field of AI can help the energy sector optimize the electricity grid and successfully maintain durability and reliability (Ahmad et al., 2021). The benefits outlined in this article show that AI is not just a technological luxury but also a strategic necessity for organizations looking to optimize energy use, reduce costs, and contribute to global environmental goals. As AI technologies continue to evolve, the synergy between AI and energy management will play an increasingly important role in shaping the future of corporate sustainability. AI is ushering in a new era of smart energy management for organizations. By leveraging the power of AI, businesses can optimize energy consumption, reduce costs, and contribute to a more sustainable future. As technology continues to advance, the integration of AI into energy management applications is poised to become a standard and indispensable part of corporate strategy in the coming years. Adopting these innovations will not only increase operational efficiency but also position organizations as leaders in the global movement toward a greener and more sustainable economy.

While the integration of AI into energy management holds great promise, it is imperative that the challenges and drawbacks are acknowledged and addressed for responsible and sustainable implementation. The energy sector can leverage the transformative potential of AI while mitigating risks by prioritizing data security, carefully guiding infrastructure integration, addressing energy consumption concerns, advocating for standardization, and ensuring ethical considerations. Balancing innovation with rigor is critical to developing a future in which AI positively contributes to energy management while protecting security, privacy, and societal well-being.

References

Aalami, H., Yousefi, G. R., & Moghadam, M. P. (2008, April). Demand response model considering EDRP and TOU programs. In *2008 IEEE/PES transmission and distribution conference and exposition*. IEEE. pp. 1–6

Abdalla, A. N., Nazir, M. S., Tao, H., Cao, S., Ji, R., Jiang, M., & Yao, L. (2021). Integration of energy storage system and renewable energy sources based on AI: An overview. *Journal of Energy Storage, 40*, 102811.

Acquisti, A., & Fong, C. M. (2018). An Experiment in Hiring Discrimination via Online Social Networks. *Proceedings of the National Academy of Sciences, 115*(16), 3996–4001.

Ahmad, T., Zhang, D., Huang, C., Zhang, H., Dai, N., Song, Y., & Chen, H. (2021). AI in sustainable energy industry: Status Quo, challenges and opportunities. *Journal of Cleaner Production, 289*, 125834.

Alassery, F., Alzahrani, A., Khan, A. I., Irshad, K., & Islam, S. (2022). An AI-based solar radiation prophesy model for green energy utilization in energy management system. *Sustainable Energy Technologies and Assessments, 52*, 102060.

Antonopoulos, I., Robu, V., Couraud, B., Kirli, D., Norbu, S., Kiprakis, A., et al. (2020). AI and machine learning approaches to energy demand-side response: A systematic review. *Renewable and Sustainable Energy Reviews, 130*, 109899.

Ballamudi, V. K. R. (2019). AI: Implication on Management. *Global Disclosure of Economics and Business, 8*(2), 105–118.

Berente, N., Gu, B., Recker, J., & Santhanam, R. (2021). Managing AI. *MIS Quarterly, 45*(3).

Çakmak, R., & Altas, I. H. (2020). A novel billing approach for fair and effective demand side management: Appliance level billing (AppLeBill). *International Journal of Electrical Power & Energy Systems, 121*, 106062.

Caruana, R., Lou, Y., Gehrke, J., Koch, P., Sturm, M., & Elhadad, N. (2015, August). Intelligible models for healthcare: Predicting pneumonia risk and hospital 30-day readmission. In *Proceedings of the 21th ACM SIGKDD international conference on knowledge discovery and data mining*. pp. 1721–1730.

Chai, D. S., Wen, J., & Nathwani, J. S. (2014). *Concept development for smart energy networks*.

Delipetrev, B., Tsinaraki, C., & Kostic, U. (2020). *Historical evolution of AI*.

Faruqui, A., Sergici, S., & Sharif, A. (2010). The impact of informational feedback on energy consumption-A survey of the experimental evidence. *Energy, 35*, 1598–1608.

Farzaneh, H. (2019). *Energy systems modeling*. Springer.

Farzaneh, H., Malehmirchegini, L., Bejan, A., Afolabi, T., Mulumba, A., & Daka, P. P. (2021). AI evolution in smart buildings for energy efficiency. *Applied Sciences, 11*(2), 763.

Göde, A., Doğan, A., & Özköse, H. (2023). *AI within digital transformation of the energy sector*.

Groves, P., Kayyali, B., Knott, D., & van Kuiken, S. (2013). *The "big data" revolution in healthcare [WWW Document]*. McKinsey Co.

Haefner, N., Wincent, J., Parida, V., & Gassmann, O. (2021). AI and innovation management: A review, framework, and research agenda☆. *Technological Forecasting and Social Change, 162*, 120392.

Hu, Y., Li, J., Hong, M., Ren, J., & Man, Y. (2022). Industrial AI-based energy management system: Integrated framework for electricity load forecasting and fault prediction. *Energy, 244*, 123195.

Irshad, K., Khan, A. I., Irfan, S. A., Alam, M. M., Almalawi, A., & Zahir, M. H. (2020). Utilizing artificial neural network for prediction of occupant's thermal comfort: A case study of a test room fitted with a thermoelectric air-conditioning system. *IEEE Access, 8*, 99709–99728.

Jose, R., Panigrahi, S. K., Patil, R. A., Fernando, Y., & Ramakrishna, S. (2020). AI-driven circular economy as a key enabler for sustainable energy management. *Materials Circular Economy, 2*, 1–7.

Kermany, D. S., Goldbaum, M., Cai, W., Valentim, C. C., Liang, H., Baxter, S. L., et al. (2018). Identifying medical diagnoses and treatable diseases by image-based deep learning. *Cell, 172*(5), 1122–1131.

Kolbjørnsrud, V., Amico, R., & Thomas, R. J. (2016). *The promise of AI*. Dublin, Ireland.
Lee, S., Yoo, S., Kim, J., & Hong, T. (2017). AI for building energy modeling and optimization: A review. *Automation in Construction, 83*, 23–33.
Li, J., Herdem, M. S., Nathwani, J., & Wen, J. Z. (2023). Methods and applications for AI, Big Data, Internet of Things, and Blockchain in smart energy management. *Energy and AI, 11*, 100208.
Maad, M. M. (2015). *History of AI*. Computer Science, College of Science, University of Baghdad.
Mell, P., & Grance, T. (2011). *The NIST definition of cloud computing* (NIST Special Publication 800-145). National Institute of Standards and Technology.
Mohsenian-Rad, A., Wong, V. W., Jatskevich, J., Schober, R., & Leon-Garcia, A. (2010). Autonomous demand-side management based on game-theoretic energy consumption scheduling for the future smart grid. *IEEE Transactions on Smart Grid, 1*(3), 320–331.
Musa, A. (2023). The impact of AI on energy management: A revolutionary shift in the power industry. *IJCRT2305381 International Journal of Creative Research Thoughts (IJCRT) www.ijcrt.org c882, 11*(5).
Nutakki, M., & Mandava, S. (2023). Review on optimization techniques and role of AI in home energy management systems. *Engineering Applications of AI, 119*, 105721.
Oyucu, S., & Herdem, M. S. (2022). Veri ve Yapay Zeka Perspektifinde Yeni Nesil Enerji Sistemleri. In *5th international conference on data science and applications (ICONDATA'22), September 7–11, 2022, Fethiye, Turkey*.
Paul, J., Lim, W. M., O'Cass, A., Hao, A. W., & Bresciani, S. (2021). Scientific procedures and rationales for systematic literature reviews (SPAR-4-SLR). International Journal of Consumer Studies, 45(4), O1-O16.
Rong, A., Zhang, C., Sun, Y., Zhang, K., & Wang, Y. (2019). Smart building energy management: A review. *Renewable and Sustainable Energy Reviews, 104*, 442–454.
Russell, S. J., & Norvig, P. (2010). *AI a modern approach* (3rd ed.). Prentice Hall.
Saleem, H., & Karmalkar, S. (2009). An analytical method to extract the physical parameters of a solar cell from four points on the illuminated J-V curve. *IEEE Electron Device Letters, 30*, 349–352.
Sharma, K., & Saini, L. M. (2015). Performance analysis of smart metering for smart grid: An overview. *Renewable and Sustainable Energy Reviews, 49*, 720–735.
Rinku, G. S. (2023). AI In sustainable energy industry: Status quo, challenges, and opportunities. *EPRA International Journal of Multidisciplinary Research (IJMR), 9*(5), 234–237.
Sinoplu, M., & Ceyhan, E. B. (2023). Akıllı Binalarda Yapay Zekâ Destekli Enerji Yönetimi: Bir İnceleme. *Uluslararası Batı Karadeniz Mühendislik ve Fen Bilimleri Dergisi, 5*(1), 39–63.
Stone, P., Brooks, R., Brynjolfsson, E., Calo, R., Etzioni, O., Hager, G., et al. (2022). AI and life in 2030: The one hundred year study on AI. *arXiv preprint arXiv:2211.06318*.
Strubell, E., Ganesh, A., & McCallum, A. (2019). Energy and policy considerations for deep learning in NLP. *arXiv preprint arXiv:1906.02243*.
Thompson, R. F., Valdes, G., Fuller, C. D., Carpenter, C. M., Morin, O., Aneja, S., et al. (2018). AI in radiation oncology: A specialty-wide disruptive transformation? *Radiotherapy and Oncology, 129*(3), 421–426.
Turing, A. M. (2012). Computing machinery and intelligence (1950). In *The essential turing: The ideas that gave birth to the computer age* (pp. 433–464).
Wang, Y., Zhang, Y., & Wang, Y. (2018). Integrating legacy systems with cloud-native applications: Challenges, principles, and practices. *IEEE Cloud Computing, 5*(5), 46–54.
Yılmaz, A., & Soylu, İ. (Eds.). (2022). *Yapay zekâ (10. bs.)*. KODLAB.

Chapter 6
Exploring Energy Management Practices in BIST 50 Enterprises: Insights and Implications

Aykut Arslan

Nomenclature

Ts	Tons
GWh	Gigawatt hour
BIST	Borsa Istanbul
CO_2	Carbon dioxide

6.1 Introduction

In today's world, energy is an important component in households, businesses, and our daily activities. Individuals with this perspective have a high sense of responsibility for energy in their economic and social areas. When it comes to energy, concepts such as energy management, energy security, energy independence, and energy resources come to mind. Energy management is defined as the control and improvement of energy consumption (Capehart et al., 2003). Energy security explains the existence and reliability of energy resources. Energy independence refers to the management and supply of energy without the need for any external influence. Energy management plays an important role in the process of achieving energy security and independence. Energy management has a wide range of functions including monitoring, controlling, and improving energy use and consumption. The broad scope of energy management includes the effective, efficient, and

A. Arslan (✉)
Piri Reis University, Tuzla, İstanbul, Turkiye
e-mail: aarslan@pirireis.edu.tr

M. Z. Sogut, M. Koray (eds.), *Energy Rationality and Management for Decarbonization*, Green Energy and Technology,
https://doi.org/10.1007/978-3-031-68497-5_6

sustainable use of energy. Energy management also includes the management of waste in energy production, the control of energy costs, and the monitoring and improvement of all kinds of emissions (ASRAE, 2015). As explained in this study, energy management has financial and environmental effects. These impacts affect the sustainability of the business and the environment. In the energy management process, energy security and energy independence are tried to be achieved through strategies such as energy conservation, energy efficiency, renewable energy, and energy storage. Energy conservation defines the reduction of energy use and consumption through efficient technology and practices. Energy can be conserved by methods such as insulation and especially individual measures such as closing unused areas at the household level and being sensitive in the use of water and electricity. At the business level, a professional energy management perspective is required. Energy efficiency refers to the improvement of energy consumption and the reduction of energy waste. Strategies such as the use of technologies such as LEDs to improve energy consumption and the use of efficient tools and equipment in production processes can provide energy efficiency. Within the framework of the renewable energy strategy, alternatives with no or low emissions are considered. Thanks to renewable energy sources, other forms of energy can also be produced efficiently and effectively. Renewable energy sources also contribute to energy security and independence. Energy storage describes the use of technologies for the long-term storage of energy produced through renewable energy sources. With energy storage technologies, energy is stored safely and continuously. Energy storage technology also reduces dependence on fossil and other types of fuels.

In this investigation, an endeavor was made to gauge the awareness levels pertaining to the aforementioned energy management strategies within the milieu of BIST 50 enterprises, denoting entities listed on the Istanbul Stock Exchange Market. BIST 50 enterprises represent entities with the highest market capitalization in the Istanbul Stock Exchange Market, a classification subject to periodic changes contingent upon their operational volumes. The BIST 50 index, encompassing diverse businesses from various sectors, serves as a distinctive parameter for the selection of enterprises examined in this study. The adoption of international standards for energy management, such as ISO 50001, has gained prominence as a strategic initiative for organizations aiming to enhance energy efficiency. However, the literature devoted to the ISO 50001 standard is still limited (Marimon & Casadesús, 2017). It has been argued (Marimon & Casadesús, 2017) that the main drivers behind ISO 50001 implementation are three-dimensional: social requirements (stakeholder pressures), ecology drivers (greenhouse gas emission reduction), and competitive advantage. The tangible benefits of ISO 50001 implementation in terms of competitive advantage are shaped by cost management and overall economic viability. Noteworthy among these instances is Cummins, a prominent player in power generation productions, which realized an annual saving exceeding USD 3 million through ISO 50001 adoption. Similarly, LG Chem, a Korean chemical company, achieved a commendable reduction of 10% in energy consumption, translating into cost savings amounting to USD 9 million. These illustrative cases serve as but a glimpse into a broader spectrum of comparable success stories characterized by substantial outcomes (Lazarte, 2016). Another example is a program spearheaded by the US Department of Energy that substantiates the standards' efficacy. This program revealed that facilities adhering to ISO 50001 not only witnessed heightened energy savings but also surpassed noncompliant counterparts by a significant margin, with performance differentials reaching up to 65%. The applicability of ISO 50001 extends beyond developed nations, as evidenced by Mexico's National Programme for the Sustainable Use of Energy (PRONASE), forecasted to yield a consequential 25% reduction in electricity consumption and a noteworthy 37% decrease in natural gas usage over a span of 2 years (Lazarte, 2016). Lazarte (2016) confirms that in the initial 5 years of ISO 50001 implementation, a consistent pattern of realized savings emerged, ranging between 5% and 30% of prevailing energy costs. This trend was observed across various sectors, including the pioneering adoption by the Brazilian electric and power company WEG. João Alfredo Silveira, Manager of the Department of Training and Continual Improvement at WEG, underscores the integral role of ISO 50001 in fostering a culture of energy efficiency. Silveira notes that beyond the tangible benefits, the standard serves to cultivate awareness among employees, aligning them with the overarching imperative of energy conservation.

In academic literature, Chiu et al. (as cited in Marimon & Casadesús, 2017) analyzed how ISO 50001 implementation improved energy performance indicators, serving as a motivation for competitors. Karcher and Jochem (as cited in Marimon & Casadesús, 2017) found positive results of ISO 50001, in terms of monetary savings, motivating other organizations to implement it, with positive staff attitude being a key enabler. Various studies, such as Cholette and Venkat's (as cited in Marimon & Casadesús, 2017), focus on the logistics chain in the USA, and Jabbour et al.'s analysis of ISO 50001's impact on CO_2 emissions contributes to the technical understanding of Energy Management Systems (EnMS) energy savings.

6.1.1 Sustainability

The effect of energy management on the sustainable competitive advantage of enterprises is clearly seen in the literature. Although the concept of sustainability has been widely used in recent years, the concept was first used in the 1700s within the framework of the supply and demand of forest industry products (Wiersum, 1995). However, in traditional agricultural production, methods such as leaving the soil fallow, producing as much as needed, and hunting as much as needed can be evaluated within the framework of sustainability. Sustainability is not a problem in periods when individuals adapt to nature in traditional lives. The negative results of discoveries, colonialism, and industry revolution have revealed sustainability problems. In the twentieth century, consumption awareness increased the sustainability issue, and it began to be emphasized more strongly (Meadows et al., 1972). In particular, the United Nations Environment and Development Report, known as the Brundtland Report, has increased awareness of sustainability. The concept of sustainability is discussed comprehensively in the Brundtland Report, which emphasizes that increasingly irresponsible consumption and production behaviors directly and indirectly threaten the world's resources. Sustainability is defined as meeting today's needs with today's resources without harming the future. The concept of sustainability, with three subheadings: social, economic, and environmental sustainability, is questioned in all areas of life. However, in subsequent studies, it has been observed that different subdimensions of sustainability are also the subject of studies. According to Belchior-Rocha (2018), sustainability has seven subdimensions. These dimensions, which are defined as economic, ecological, cultural, spatial/territorial, political, social, and environmental sustainability, show the impact of sustainability in all areas of life.

The concept of sustainable competitive advantage is derived from the concept of sustainability for businesses and refers to the achievement of the goals of enterprises by using their existing resources effectively and efficiently. The energy management of enterprises directly affects their ecological and environmental sustainability, especially economic sustainability, and indirectly affects their social, cultural, spatial/territorial, and political sustainability.

6.1.2 The Current State of Energy Management

Energy management plays a key role in many business goals, especially sustainable competitive advantage goals (Henze, 2001). Inefficient energy management increases many costs, especially financial costs. Energy management is closely related to all organizations at micro- and macro-levels. Almost all production businesses need energy. Energy is used in the entire production process of products. Apart from this, trade and service enterprises also need energy in their business activities. From the Industrial Revolution to the present day, the energy needs of businesses have generally been met from fossil fuels. The need for energy, which emerged with the Industrial Revolution, has played a role in many crises in the world, especially in the world wars. While energy plays a decisive role in macro-level crises in the world, it has often played a role as a threat instrument. More importantly, the Gross National Product (GNP) and national income of countries with energy resources have been higher than other countries. This effect of energy at the macro-level also shows its effect at the microlevel. All businesses, starting from small businesses to international businesses, carry out their activities dependent on energy. To give a quantitative example, according to the data of the Ministry of Energy and Natural Resources, natural gas consumption in Turkey in 2021 was 36,006 (10^6 Sm^3; 21.604 × 106 Sm^3 individual consumption), and petroleum product consumption was 43.505 thousand tons (800 thousand tons individual consumption). Approximately 90% of oil and 85% of natural gas are imported. According to the data of the Ministry of Energy and Natural Resources, Turkey's detailed energy balance tables for 2021 are presented at the end of this study. The fact that Turkey's energy imports are so high has also been the reason for the country's foreign trade deficit that has lasted for years. This effect at the macro-level also applies to businesses at the microlevel. The energy costs of the enterprises have a significant proportion in the production costs. Energy consumption data by sector are presented in the Table 6.1. It is seen that most of the oil consumption occurred in the transportation sector, and a large part of the natural gas consumption occurred in energy production and residences.

Table 6.1 Energy supply distribution of the sectors

Energy supply distribution	Hard coal	Lignite	Asphaltite	Coke	Coal tar	Raw petrol	Petroleum products	Natural gas	Bioenergy and waste	Electricity
	(1000 tons)	(1000 tons)	(1000 tons)	(1000 tons)	(1000 tons)	(1000 tons)	(1000 tons)	(106 Sm^3)	(1000 tons)	(GWh)
Sum of sectors	**10.019**	**11.573**	**333**	**5.143**	**15**	**0**	**42.665**	**36.006**	**8.792**	**284.271**
Industrial consumption	**7.070**	**5.987**	**74**	**5.143**	**15**	**0**	**3.282**	**13.120**	**3.090**	**134.969**
Mining activities	0	1					133	199		2.580
Food, beverage, tobacco production	474	1.010		51			18	1.536	231	9.913
Textile, leather products	421	1.317		0			12	1.656	21	20.286
Wood and product manufacturing	14	0					9	74	1.081	2.765
Manufacture of paper and printing	46	324					5	279	22	4.257
Manufacture of chemicals, petrochemicals	496	339		0			16	3.266	10	16.839
Nonmetallic mineral products	4.129	2.946	22				2.953	2.492	1.701	14.603
Basic metal industry	1.487	41		5.091	15		26	2.496	5	36.506
Machinery, electrical, electronics	0	4					3	167	8	3.109
Transportation vehicle product	0	3					4	329	1	3.124
Furniture manufacturing	0						9	48	10	779
Construction	1						13	394		3.985

(continued)

Table 6.1 (continued)

Energy supply distribution	Hard coal	Lignite	Asphaltite	Coke	Coal tar	Raw petrol	Petroleum products	Natural gas	Bioenergy and waste	Electricity
Other industries			51				78	184		16.220
Transportation	**0**	**0**	**0**	**0**	**0**	**0**	**28.520**	**313**	**266**	**1.571**
Railways							102			1.298
Seaways							357			
Airlines							895			
Pipelines								233		273
Highways							27.166	80	266	
Other sectors	**2.949**	**5.586**	**259**	**0**	**0**	**0**	**3.893**	**21.604**	**5.437**	**147.731**
Household	2.845	3.855	259				387	16.679	5.437	61.499
Trade and services	104	1.732					421	4.773		72.460
Agriculture and livestock							3.085	151		13.772

When Table 6.1 is examined, it is seen that energy management is important at the household level, especially in these sectors. The amount paid for total energy imports was 50.7 billion dollars in 2021. In 2021, Turkey's total exports were $225 billion, and imports were $271 billion. As can be seen, energy imports have an important role in the foreign trade deficit. This macro-level outlook is also felt at the microlevel. Although the situation is not very different when we look at countries other than Turkey, it is obvious that the foreign trade deficits of countries with energy resources or countries that produce value-added products are not negative in this way.

6.1.3 The Impact of Energy Consumption on Sustainability

While environmental awareness is increasing, energy-related factors have been seen as the main causes of environmental negativities. Since the early 1990s, the relationship between energy management and environmental outcomes has aroused both scientific and social interest (Dincer & Ve Rosen, 1999). There is a direct relationship between energy efficiency and environmental impacts, which is emphasized in almost every study on this subject. Within the framework of this fact, many alternatives such as effective energy management, efficient energy use, energy conservation, and orientation to renewable energy sources are expressed as solution suggestions. In addition, as described above, direct and indirect effects of energy management are observed on economic, ecological, cultural, spatial/territorial, political, social, and environmental sustainability.

Economic sustainability describes the protection of financial, natural, social, and human resources while meeting economic needs (Jeronen, 2020). More generally, economic sustainability can be defined as responsible and balanced economic management in the long term with the optimal use of available resources. In terms of its relationship with other dimensions of sustainability, economic sustainability refers to the execution of economic activities with the effective and efficient use of existing resources without jeopardizing social and environmental sustainability. On the other hand, in a broader framework, it also refers to meeting the economic expectations of the stakeholders of organizations without jeopardizing future resources. It is important for economic sustainability that costs and earnings are balanced for businesses and that they do not create problem areas in the future. Energy management can also positively affect the economic costs of businesses by ensuring that the energy consumption of businesses is kept under control and optimized. It is thought that enterprises with high awareness of energy management strategies will have optimal results in all performance indicators in the long term by using energy resources effectively and efficiently while not adversely affecting other sustainability dimensions. As the first topic of the awareness of businesses on economic sustainability, the optimal use of energy sources has led to the creation of a standard called ISO 50001. The ISO 50001 energy management standard, which was developed within the framework of ISO management system standards, was created by the cooperation of nearly 60 countries in accordance with the ISO 9001 quality management system and ISO 14001 environmental management standard. ISO 9001 is used as the standard to meet the expectations of the stakeholders, while ISO 14001 is used as the standard to meet environmental expectations. ISO 50001 develops goals and objectives for the improvement of the energy performance of businesses or institutions. In the ISO 50001 energy management process, energy consumption behaviors are tried to be shaped by establishing concrete relationships between the energy use activities of the enterprises and the energy performance indicators. Corrective measures are also taken when necessary in the energy management process. The improvement of processes such as the procurement, storage, use, and preservation of different energy sources, which are presented in detail in the table above, is done in the energy management process. The main purpose of the ISO 50001 energy management system is to establish systematic approaches to energy use, efficiency, and consumption in order to improve the energy performance of enterprises. The energy management process is carried out within the framework of a management system standard established by the International Organization for Standardization with the plan-do-check-act approach. The five basic components of the energy management system are energy policy, energy planning, implementation and execution, energy control, and internal evaluation. In the internal evaluation process, the evaluation process is carried out within the framework of internal evaluation standards specific to each enterprise.

6.2 Methods

The present investigation constitutes an exploratory research study designed to elucidate the extent of awareness among enterprises regarding energy management, as inferred from the themes and codes outlined in the 2022 annual reports of the BIST 50 enterprises within the study sample. In this study, it was tried to reach the objectives of the research by using the codes determined in the previous literature review in the determined documents. The results obtained in the reviews made with the codes determined in the sample documents were analyzed. The study was designed using the qualitative research method. The 2020 version of the MAXQDA software was used in the method used. The tendencies of the enterprises in the BIST 50 index such as energy efficiency, renewable energy use, energy storage, and energy saving have been tried to be revealed through the determined codes. The above-mentioned codes were scanned in the declarations of the BIST 50 businesses set out in the

Table 6.2 The designated themes, associated codes, and explanatory insights for reference

Theme	Code
Conservation of energy	Energy conservation, security, energy supply, energy investment, energy supply
Energy efficiency	Energy efficiency, savings, energy costs, energy policy, energy consumption, energy expenditure
Use of renewable energy	Renewable energy, solar, wind, energy transition, energy portfolio, power projects, photovoltaic, geothermal
Storage of energy	Energy conservation, storage

documents they published. In this respect, the study fully complies with the qualitative research method.

In qualitative research methods, instead of quantitative indicators, the depth and scope of the targeted information are tried to be revealed. Like every scientific research method, the research process in qualitative research methods is based on theoretical foundations. In this study, based on the theory, the energy orientations of BIST 50 enterprises were reviewed with the codes and themes determined in the BIST 50 annual documents, and the results were classified. Another advantage of qualitative research over quantitative research is that it enables different dimensions of the research problems to be revealed. In this study, the energy orientations of BIST 50 organizations have been tried to be revealed in a way that will reveal all their dimensions from an analytical point of view.

Several softwares are used in qualitative research methods. Although many of these software have similar functions, in terms of usefulness, MAXQDA software also provides systematic classification of quantitative data. In addition, MAXQDA helps researchers in organizing the content of documents and establishes meaningful relationships between related themes. On the other hand, MAXQDA also provides frequency verification of the codes determined for reliability and accuracy in qualitative research. In this study, through the thematic analysis method of qualitative research, the energy orientations of BIST 50 enterprises were revealed analytically with MAXQDA software.

In the study, the themes determined according to the literature review and the codes associated with the themes are presented in the Table 6.2.

6.3 Results

6.3.1 Descriptive Statistics

The descriptive statistics of the themes identified in the research are presented in Table 6.3.

Table 6.3 Descriptive statistics

	Use of renewable energy	Conservation of energy	Energy storage	Energy efficiency
N	50	50	50	50
Mean	32,76	5,98	3,60	14,42
Std. dev (sample)	54,281	7763	17,544	16,566
Std. dev (universe)	54,832	7842	17,722	16,735
Variance (sample)	2.946,422	60,260	307,800	274,444
Variance (universe)	3.006,553	61,489	314,082	280,044
Minimum (min.)	0,00	0,00	0,00	0,00
1st Quarter	7,00	0,00	0,00	2,00
Median	18,00	3,00	0,00	8,00
3rd Quarter	36,75	8,50	1,00	21,75
Maximum (max.)	364,00	34,00	125,00	80,00
Gap	364,00	34,00	125,00	80,00
Total	1.638,00	299,00	180,00	721,00
Standard error	7,75	1,11	2,51	2,37
Mean (%95) in down	17,18	3,75	−1,44	9,66
Mean (%95) in up	48,34	8,21	8,64	19,18
Missing	0	0	0	0
Missing (%)	0,00	0,00	0,00	0,00

6.3.2 *Total Code Statistics*

In the study, the total code statistics obtained by scanning the codes related to the themes presented above through content analysis are presented in Table 6.4.

Looking at Table 6.3, it is observed how many documents the themes are used in. The theme of "use of renewable energy" was used as the theme with the highest frequency in the annual reports of 48 of the 50 businesses. This shows that the awareness of renewable energy use in BIST 50 enterprises is very high. However, the theme of "Storage of Energy" was used in 15 of the 50 reports. The level of awareness in the world is also low in the theme of "Storage of energy," which is seen as the theme with the lowest frequency among the themes of the research. The theme of "Energy Efficiency" was the theme used in most documents after the

Table 6.4 Total theme statistics

	Documents	Frequency	%	% (valid)
Use of renewable energy	48	1638	96.00	100.00
Energy efficiency	44	721	88.00	91.67
Energy storage	15	180	30.00	31.25
Conservation of energy	34	299	68.00	70.83
Analyzed documents	50		100.00	–

theme of "Use of Renewable Energy." It can be said that the level of awareness about the theme of energy efficiency is very high. The theme of energy conservation was also used 299 times in 34 documents.

6.3.3 *Distribution of Codes by Firms*

The usage frequencies of the themes of the research on the basis of enterprises are presented in Table 6.4. The first of the columns in Table 6.4 shows the sectors in which the enterprises operate. The second column shows whether it has ISO 50001 energy management certificate. The third column shows the frequency of the energy code in general, the fourth column shows the frequency of the "Use of Renewable Energy" theme, the fifth column shows the "Energy Conservation" theme, the sixth column shows the frequency of the "Energy Storage," and the seventh column shows the frequency of the "Energy Efficiency" theme.

As mentioned above, the study is based on annual reports of firms in a sample group. We just used the information from these annual reports. It is observed that only 29 of the top 50 enterprises in the research sample have ISO 50001 energy management certificate. What is even more surprising is that some of the energy businesses on the list do not have an energy management certificate. It is observed that the themes of "Conservation of Energy" and "Storage of Energy" have less frequency than other themes. In particular, the storage of energy is seen as the most problematic area all over the world. The development of energy storage technologies is not at a sufficient level and constitutes a key point in research on energy.

An examination of Table 6.5 reveals that companies operating within the energy sector demonstrate the most substantial engagement across energy-related thematic categories.

Table 6.5 Detailed usage frequencies of themes

Name of firm	Sectors	ISO 50001 Certificate	Energy	Renewable Energy Usage	Conservation of Energy	Storage of Energy	Energy Efficiency
ENERJISA	Enery	+	2160	91	34	4	35
AKSA	Energy	+	580	15	18	1	8
SMART	IT	-	570	364	27	3	2
KONTROL	Energy	+	525	61	28	125	6
TUPRAS	Energy	+	423	39	15	2	80
Arcelik	White Goods	+	305	33	6	2	51
TURKCELL	Communication	+	257	63	16	0	32
SAHOL	Holding	+	241	38	5	0	5
YAPIKREDI	Banking	+	240	93	10	0	31
VESTEL	White Goods	+	232	25	4	12	32
CIMSA	Cement	+	226	63	9	0	35
ERDEMIR	Iron-Steel Production	+	218	18	6	2	38
PETKIM	Energy - Petro Chemicals	+	208	21	11	1	12
OYAK	Cement	+	207	16	1	0	10
ISBANK	Banking	+	202	73	14	0	22
KOC	Holding	+	191	28	7	0	18
CWENE	Energy	-	183	108	3	8	2
HALKBANK	Banking	+	181	55	11	0	36
Akbank	Banking	+	161	69	9	0	20
ODAS	Energy	-	158	6	7	0	0
GARANTI	Banking	-	155	58	5	0	20
Alarko	Holding	+	153	19	0	4	2
MIGROS	Energy	-	153	29	3	0	50
SISECAM	Glass Production	-	153	23	6	0	8
TTKOM	Communication	+	123	22	3	0	22
ENKAINS	Construction	+	114	8	6	0	10
DOHOL	Holding	+	105	20	3	0	6
FORD	Automotive	+	93	21	3	0	21
KARDEMIR	Iron Steel Ind	-	82	8	0	0	10
TOFAS	Automotive	+	79	4	2	0	26
ASELSAN	IT-Defense Ind	+	78	8	15	1	7
DOGUS	Holding	-	76	25	4	0	5
TEKFEN	Holding	-	68	18	5	1	1
CANTE	Termik Energy	-	66	5	0	0	4
THY	Airlines	+	52	7	0	0	14
EUROPOWER	Energy	-	51	13	1	12	1
Anadolu Efes	Alcohol Beverage	+	41	8	0	0	8
Astor	Energy	-	39	7	0	0	1
QNBFinansbank	Banking	-	36	7	0	0	6
BİM	Retail	-	35	13	0	0	8
HEKTAS	Fertilizer	-	35	4	1	0	7
TAV	Air Ground Services	+	32	11	0	0	4
PEGASUS	Airlines	-	16	4	0	0	3
SASA	Polyester	+	16	3	1	0	0
ISBIR	Synthetic	-	15	5	0	0	1
EMLAKGYO	Real Estate-Construction	+	10	3	0	0	0
GUBRETAS	Fertilizer	-	10	5	0	2	1
KOZAME	Mining	-	2	0	0	0	0
QNBLEASING	Banking Leasing	-	1	1	0	0	0
KOZAAL	Gold Mining	-	0	0	0	0	0

6.4 Discussion

The energy management certification system in businesses provides businesses with high awareness in seven main topics, also called 7Cs. The seven main topics are defined as consumption, carbon, cost, competence, competitiveness, continuity, and compliance. For businesses, energy management certification and, therefore, awareness lead to more positive results than expected. In the literature, it is reported that Wyeth Nutrition spent 400,000 dollars in the energy management certification process between 2012 and 2018 but earned as much as seven times higher spending (WYETH, 2017). In addition to financial gains, the reduction in emissions and energy consumption is an important gain in terms of sustainability. This study is an important proof in terms of predicting the results of the awareness of the enterprises. Having an energy management certificate requires a strategic perspective. A strategic perspective requires the involvement of senior management. The strategic point of view is emphasized in business texts, reports, and activities. This is also true for the awareness of energy management. The energy management awareness of businesses is also supported by business activities. This study is based on this idea. The awareness of BIST 50 enterprises on energy management can be seen in their business reports. It is thought-provoking that only 29 of the 50 enterprises with the highest business volume in a country have energy management certificates. In this respect, it is inevitable for businesses to have a responsible perspective. This study is exploratory and contains many clues for future studies. For example, it can be a starting point for the awareness of the enterprises with the highest volume on energy management. In addition, the fact that there are energy businesses that do not have energy management certification draws a worse scenario. Its main field of activity is energy, and sustainability becomes difficult or even impossible for businesses that are not aware of the most valid certification in the sector. In this respect, this study has an important potential for energy enterprises to see their own situation. The qualitative analysis methods used in the study are very functional tools, especially for grounded theory studies. The results of the qualitative method and tool used in this study show the problem areas for businesses in energy management. Businesses can develop and implement energy management strategies in the problem areas that arise as a result of this exploratory study.

6.5 Limitations

In terms of the limitations of the study, the study covers only the 50 enterprises with the highest volume in the Turkish stock market. The application of this study in the annual reports of companies on stock exchanges in different countries will increase the validity of the study. In addition, such studies also provide the opportunity for cross-border comparison.

6.6 Conclusions for Practitioners

The results of the study have implications for the managers of the BIST 50 companies within the scope of the research, which will affect their energy strategies. Analytical information is provided to increase awareness of energy management. Through the results of the study, comparisons can also be made with BIST 50 enterprises in the sector or in different countries. In addition, it can be ensured that the practices of businesses with good practices can be seen. The study can provide businesses with an idea about sustainability, energy efficiency, energy storage and saving, and general energy policies. In addition, references are also offered to serve as a guide for business managers in their energy strategies. In businesses that have sustainability awareness, optimization can be achieved in improvements to be made in dimensions such as energy consumption, efficiency, and storage. Through optimization, costs can be reduced in the short and medium term and sustainability can be achieved in the long term.

6.7 Theoretical Results

The theoretical framework that reveals the impact of energy management of enterprises on sustainability is limited. The increase in theoretical frames can enrich the studies. The use of this study, which was conducted using the qualitative research method, can reveal new theoretical approaches. Theoretical approaches that take into account the increasing importance of sustainability can be developed. In-depth information is provided by the study, as the research analytically takes into account all the energy-related discourses of enterprises. The comprehensive and in-depth presentation of qualitative research methods used in grounded theory studies also stands out in this study. The deep and comprehensive knowledge presented in the study can enable new studies to be carried out and develop and diversify the theoretical approaches. The results of the research could also shape future studies.

6.8 Conclusion

In this study, the corporate energy strategies of BIST 50 enterprises were tried to be examined in depth with an analytical approach. The study conducted using qualitative research methods is an exemplary study that is rarely encountered in the literature. In addition, the analysis software used in the study has enabled the energy management of the enterprises to be revealed in a multifaceted way. Energy management plays an important role in enabling businesses to gain sustainable competitive advantage. It is important in terms of awareness that the study is carried out in the enterprises with the highest turnover in Turkey because these enterprises are seen as locomotives and references, both in their industries and in other fields. These results can also enable other businesses to take proactive measures. Apart from the practice, it has been emphasized in the literature review that theoretically, energy and sustainability issues are effective in the competitive advantage of companies. When the results of the study are considered in general, it can be said that the energy strategies of BIST 50 enterprises are insufficient. From the point of view of both cost and other sustainability dimensions, a comprehensive consideration of energy strategies can be presented as a recommendation. While almost all of the competitive advantages of businesses develop depending on their energy strategies, the low level of awareness is thought-provoking. In this respect, both readers and business managers should take these conclusions into account.

Appendix A: 2021 Türkiye National Energy Balance Table.

2021 TÜRKİYE NATIONAL ENERGY BALANCE TABLES (Ministry of Energy and Natural Resources)

ENERGY SUPPLY DISTRIBUTION	Hard Coal	Lignite	Asphaltite	Coke	Coal Tar	Raw Petrol	Petroleum Products	Natural Gas3	Bioenergy and Waste7	Hydraulic	Wind	Electricity	Other Heat	Jeotermal 9	Sun
	(1000 Tons)	(1000 Tons)	(1000 Tons)	(1000 Tons)	(1000 Tons)	(1000 Tons)	(1000 Tons)	(106 Sm3)	(1000 Tons)	(GWh)	(GWh)	(GWh)	(Accord to Tep)	(Accord to Tep)	(Accord to Tep)
Domestic Production (+)	1.236	83.559	1.671			3.442		415	14287	55.927	31.437			11.234	2.059
Import (+)	37.189			1.170	6	31.418	18.725	58.704				2.334			
Export (-)	304	33		17	128		7.841	383				4.186			
Stock Change (+/-)	-40	1.304	26	89	14	-344	128	938							
SUPPLY OF ENERGY PR	**38.081**	**84.830**	**1.698**	**1.241**	**-107**	**34.516**	**7.817**	**59.674**	**14287**	**55.927**	**31.437**	**-1.852**	**0**	**11.234**	**2.059**
Statistical Difference (+/-)	**-8**	**-110**	**0**	**233**	**7**	**0**	**840**	**0**	**0**	**0**	**0**	**0**	**0**	**0**	**0**
CYCLE & ENERGY SEC	**-28.069**	**-73.367**	**-1.36**	**4.134**	**128**	**-34.516**	**35.688**	**-23.668**	**-5.495**	**-55.927**	**-31.437**	**286.123**	**3.977**	**-9.280**	**-1.199**
Electricity and Heat Gen.	-21.470	-71.448	-1.36				-283	-21.084	-5.488	-55.927	-31.437	334.723	2.065	-9.280	-1.199
Main Activity Producers5	-19.042	-70.839	-1.36				-203	-17.464	-5.404	-55.889	-31.326	304.617	1.495	-9.280	-913
Auto producers	-2.428	-610					-81	-3.619	-8[illegible]	-37	-110	30.106	570		-286
Heat Generation6	-606	-1.824					-37	-1.849	-6				2.265		
Coke Ovens	-5.620			4.134	128										
Blast Furnaces															
Oil Refineries						-34.215	38.776	-676				-2.420	-353		
Domestic Consumption and Loss	-373	-94				-300	-2.768	-60				-46.180			
TOTAL FINAL ENERGY CONSUMPTION	**10.011**	**11.463**	**333**	**5.376**	**21**	**0**	**43.505**	**36.006**	**8.792**	**0**	**0**	**284.271**	**3.977**	**1.954**	**860**
SUM OF SECTORS	**10.019**	**11.573**	**333**	**5.143**	**15**	**0**	**42.665**	**36.006**	**8.792**	**0**	**0**	**284.271**	**3.977**	**1.954**	**860**
INDUSTRIAL CONSUMPTION	**7.070**	**5.987**	**74**	**5.143**	**15**	**0**	**3.282**	**13.120**	**3.090**	**0**	**0**	**134.969**	**3.914**	**0**	**301**
Mining Activities(07,08,09)	**0**	**1**					**133**	**199**				**2.580**	**41**		
Food, Beverage, Tobacco Products Manufacturing (10,11,12)	**474**	**1.010**		**51**			**18**	**1.536**	**234**			**9.913**	**550**		
Food Products Manufacturing(10)	474	969		0			13	1.438	234			8.387	550		

2021 TÜRKİYE NATIONAL ENERGY BALANCE TABLES (Ministry of Energy and Natural Resources)

Beverage Products Manufacturing(11)							1	51				584			
Tobacco Products Manufacturing(12)							0	9				214			
Sugar Production(10.81)		40		51			4	37	0			728			
Textile, Leather Products Manufacturing(13,14,15)	**421**	**1.317**		**0**			**12**	**1.656**	**21**			**20.286**	**118**		
Textile Products Manufacturing(13)	220	1.234					11	1.367	20			16.904	118		
Manufacture of Apparel(14)	200	83		0			1	277	1			2.920			
Manufacture of Leather and Related Products(15)	1						0	12				462			
Wood and Products Manufacturing(16)	**14**	**0**					**9**	**74**	**1.081**			**2.765**	**95**		
Manufacture of Paper and Products(17,18)	**46**	**324**					**5**	**279**	**22**			**4.257**	**228**		
Manufacture of Chemical, Petrochemical Products(20,21,22)	**496**	**339**		**0**			**16**	**3.266**	**10**			**16.839**	**357**		
Manufacture of Chemical Products(20)	491	229		0			12	1.962	7			7.868	338		
Fertilizer(20)	0						1	1.021				662			
Pharmaceutical Products Manufacturing(21)	4	1					2	73				640			
Manufacture of Rubber, Plastic Products(22)	0	110					2	211	3			7.671	19		
Non-Metallic Mineral Products Manufactures (23)	**4.129**	**2.946**	**22**				**2.953**	**2.492**	**1.701**			**14.603**	**64**		
Glass Products Manufacturing(23)							1	962				2.039			
Ceramic Products Manufacturing(23)	53	383					60	1.221				2.741			
Cement Products Manufacturing(23)	4.076	2.563	22				2.892	309	1.701			9.823	64		
Basic Metal Industry (24,25)	**1.487**	**41**		**5.091**	**15**		**26**	**2.496**	**5**			**36.506**	**341**		
Iron & Steel Products (24)	1.438			5.084	15		14	1.944	1			28.808	285		

2021 TÜRKİYE NATIONAL ENERGY BALANCE TABLES (Ministry of Energy and Natural Resources)

Manufacture of Non-Ferrous Metal Products(24)	48			7			7	362				5.305	55		
Fabricated Metal Products Manufacturing(25)	0	41					5	191	3			2.392			
Machinery, Electrical, Electronics Manufacturers (26,27,28)	**0**	**4**					**3**	**167**	**8**			**3.109**	**6**		
Transportation Vehicles Manufacturing(29,30)	**0**	**3**					**4**	**329**	**1**			**3.124**	**9**		
Motor Vehicle Manufactures (29)	0						2	299	1			2.506	9		
Other Transportation Vehicles Manufacturing(30)		3					2	30	0			618			
Furniture Manufacturing(31)	**0**						**9**	**48**	**10**			**779**			
Construction(41,42,43)	**1**						**13**	**394**				**3.985**	**0**		
Other Industry			**51**				**78**	**184**				**16.220**	**2.106**		**301**
TRANSPORTATION	**0**	**0**	**0**	**0**	**0**	**0**	**28.520**	**313**	**266**	**0**	**0**	**1.571**	**0**	**0**	**0**
Railways							102					1.298			
Seaways							357								
Airlines							895								
Pipelines								233				273			
Highways							27.166	80	266						
OTHER SECTORS	**2.949**	**5.586**	**259**	**0**	**0**	**0**	**3.893**	**21.604**	**5.437**	**0**	**0**	**147.731**	**63**	**1.954**	**559**
Household	2.845	3.855	259				387	16.679	5.437			61.499		853	559
Trade & Services	104	1.732					421	4.773				72.460	63	475	
Agriculture and Livestock							3.085	151				13.772		627	
NON-ENERGY CONSUMPTION	**0**	**0**	**0**	**0**	**0**	**0**	**6.971**	**969**	**0**	**0**	**0**	**0**	**0**	**0**	**0**
Petro-Chemical Feedstock							**1.825**								
ELECTRICITY AND HEAT GEN. INFORMA															
Electrical Power Gen (GWh)	**58.026**	**42.983**	**2.373**				**281**	**111.181**	**7.779**	**55.927**	**31.437**	**334.723**		**10.793**	**13.943**
Installed Power Capacity (MW)	**9.835**	**10.120**	**405**				**258**	**25.576**	**2.035**	**31.493**	**10.607**	**99.820**		**1.676**	**7.816**
Heat Generation (TJ)	13.882	17.842					3.576	134.452	6.108				177.711		
Heat Sold (TJ)	1.915	1.968					2.055	74.545	5.979				84.407		
Unsold Heat (TJ)	11.967	15.874					1.521	59.907	128				93.304		

References

ASRAE. (2015). *Energy use and management, (inch-pound edition)*. American Society of Heating, Refrigerating and Air-Conditioning Engineers.

Belchior-Rocha, H. (2018). Social work practices and the ecological sustainability of socially vulnerable communities. *Sustainability, 10*(5), 1312. https://doi.org/10.3390/su10051312

Capehart, B. L., Turner, W. C., & Kennedy, W. J. (2003). *Guide to energy management*. Fairmont Press, Inc.

Dincer, I., & Ve Rosen, M. (1999). Energy, environment and sustainable development. *Applied Energy, 64*(1–4), 427–440.

Henze, G. P. (2001). Building energy management as continuous quality control process. *Journal of Architectural Engineering, 7*(4), 97–106.

Jeronen, E. (2020). Economic sustainability. In S. Idowu, R. Schmidpeter, N. Capaldi, L. Zu, M. Del Baldo, & R. Abreu (Eds.), *Encyclopedia of sustainable management*. Springer.

Lazarte, M. (2016). *Does ISO 50001 still live up to its promise?* Retrieved from https://www.iso.org/2016/11/Ref2135.html.

Marimon, F., & Casadesús, M. (2017). Reasons to adopt ISO 50001 energy management system. *Sustainability, 9*(10), 1740. https://doi.org/10.3390/su9101740

Meadows, D. H., Meadows, D. L., Randers, J., & Behrens, W. W., III. (1972). *The limits to growth; Potomac associates*. New American Library.

Wiersum, K. F. (1995). 200 years of sustainability in forestry: Lessons from history. *Environmental Management, 19*, 321–329.

Chapter 7
Design and Analysis of a New Double-Flash Geothermal Power Plant Integrated with Multistage Flash Desalination and Reverse Osmosis Desalination Units for Power, Fresh Water, and Hot Water Generation

Fatih Yilmaz and Resat Selbas

Nomenclature

Ex	Specific exergy, kJ/kg
Ex	Exergy rate, kW
h	Specific enthalpy, kJ/kg
s	Specific entropy, kJ/kG-K
$\dot{Q}$	Heat transfer rate
W	Work rate

Greek Letter

η	Energy efficiency
ψ	Exergy efficiency

Abbreviations

EV	Expansion valve
MSFD	Multistage flash desalination

F. Yilmaz (✉) · R. Selbas
Mechanical Engineering, Isparta University of Applied Sciences, Isparta, Turkiye
e-mail: fatihyilmaz@isparta.edu.tr; resatselbas@isparta.edu.tr

M. Z. Sogut, M. Koray (eds.), *Energy Rationality and Management for Decarbonization*, Green Energy and Technology,
https://doi.org/10.1007/978-3-031-68497-5_7

ORC	Organic Rankine cycle
P	Pump
PR	Performance ratio
RO	Reverse osmosis
SG	Single-generation
T	Turbine

7.1 Introduction

Energy has a very important place in the development of factors such as human development, industrialization, and globalization (Yilmaz et al., 2023). Along with these developing parameters, energy use increases with industrial and global development. Therefore, the need for different and alternative energy sources to meet the increasing energy request is increasing day by day. In addition, in today's world where environmental problems are very serious, it is inevitable to use renewable energy sources (RESs) to prevent or decrease these problems. RESs are important alternatives to fossil-based, environmentally unfriendly fuels for a sustainable future (Bozgeyik et al., 2022). RESs (solar, wind, geothermal, hydroelectric, and ocean) are inexhaustible compared to conventional energy sources and offer many environmental benefits (Hepbasli, 2008). Geothermal is a kind of energy that exists as heat in the Earth's interior and is almost unaffected by environmental conditions compared to other RESs (i.e., solar and wind).

Geothermal energy is among the easiest sources to use because it uses hot water or hot steam for energy production. Moreover, it can be described as a sustainable resource type in the long term as long as there is no change in the temperature and mass flow rate of hot water or steam coming out of the ground. In this energy source, electrical energy can be produced using thermal cycles and various methods. Moreover, it is possible to classify this energy source as low, medium, and high temperatures (Guler et al., 2022). Moreover, geothermal energy has many advantages such as building heating, cooling, and electricity generation or continuous thermal energy production without emitting any dangerous pollutants (Kumar et al., 2022). Conversely, the relatively low efficiency of geothermal energy–supported power production configurations may be considered a disadvantage. However, in recent years, there have been many studies on overcoming this disadvantage by designing geothermal-supported combined systems, namely cogeneration, trigeneration, or multi-generation. In recent years, there have been many studies in the literature on this subject, i.e., geothermal-supported combined systems, and it is still among the topics of interest.

In 2022, a development of the geothermally driven integrated cycle was proposed and evaluated by Yilmaz (2022). In this work, the author has examined the thermodynamic performance of the entire schema for the generation of power, hot and clean water, and hydrogen. The analysis results indicated that the overall system had 52.01% energetic and 29.45% exergetic efficiencies. In another study, Asadi et al. (2022) developed a review study that is based on geothermal- and solar-based beneficial products. Herein, the authors have concluded that the combined plants have more efficiency than simple single-generation plants. Moya et al. (2018) looked at a recent study on direct heat applications and geothermal energy power plants. They created hybrid geothermal arrangements in addition to using an organic Rankine cycle (ORC) powered by geothermal energy. In this work, several geothermal plants, which are single-flash, double-flash, binary, advanced, and dry steam, are studied. The results indicated that the binary ORC system can play a significant role in low temperatures.

Another author, Hekmatshoar et al. (2022), suggested optimizing the thermo-economic analysis of a geothermal-assisted combination system for the generation of power, hydrogen, and clean water. In this study, the authors performed analyses by combining Multi-effect desalination (MED), ORC, and Proton-exchange membrane (PEM) units with the geothermal system. They stated that the clean water production capacity of the system they designed is between 0.28 and 1.5 kg/s . Abdolalipouradl et al. (2020) developed a geothermal-driven integrated cycle that generates power and hydrogen. They performed a thermodynamic performance analysis and an exergo-economic analysis of the developed model. Their investigation results indicated that the tri-generation cycle has better performance than the single-generation model.

In 2023, Yilmaz et al. (2023) proposed and evaluated a flash-binary power cycle, in which the production of beneficial products that are power, hydrogen, cooling, and freshwater production. The authors have used herein the transcritical CO_2 power plant as a bottom cycle for power generation. The system had a 40.04% energy efficiency and a 36.31% exergy efficiency. The study of Huang et al. (2023) created an optimization and exergy analysis for the transcritical cycle and the geothermal power plant. Their proposed model indicated that the total exergy efficiency was 34.26%. Ma et al. (2023) examined a new integrated cycle that combines solar and geothermal units. After that, the developed configuration's energetic and exergetic performances were computed as 34.2% and 66.3%, respectively.

In light of the detailed review of the literary works, it should be noted that geothermal-based combined facilities, where many different useful products are produced, are significant for a sustainable future. Geothermal is a clean and sustainable source and is also an almost uninterrupted source. It is significant to highlight the importance of geothermal plants such as cogeneration, tri-generation, and multi-generation configurations to achieve better performance. In this context, the key objective of this book chapter is to perform the thermodynamic analysis of power, hot water, and clean water production with a combined system supported by geothermal energy.

In this book chapter, a thorough energetic and exergetic efficiency analysis of the geothermal-supported combined cycle that combined a double-flash geothermal system, an MSFD unit, an RO desalination unit, and a hot water preparation apparatus, is conducted and presented for beneficial products. The main difference in the developmental configuration is the integration of both MSFD and RO units for clean water production. After that, the key importance and main motivation of this book chapter can be drawn as below:

- To project a new geothermal energy–supported combined configuration
- To examine the developed plant's thermodynamic efficiency
- To examine the freshwater generation with MSFD and RO units
- To analyze the geothermal-driven power, hot water, and freshwater generation capacity
- To determine the system performance ratio with respect to various important indicators

7.2 System Layout and Analysis

7.2.1 System Definition

In this advised book chapter, a double-flash geothermal process, an MSFD unit, an RO unit, and a hot water preparation apparatus for the production of power, hot water, and clean water with a combined geothermally driven plant are examined by integrating them as shown in Fig. 7.1. The subprocesses that make up the system are thermally integrated with each other. While the double-flash geothermal system consists of two separators and two turbines, the MSFD system consists of four

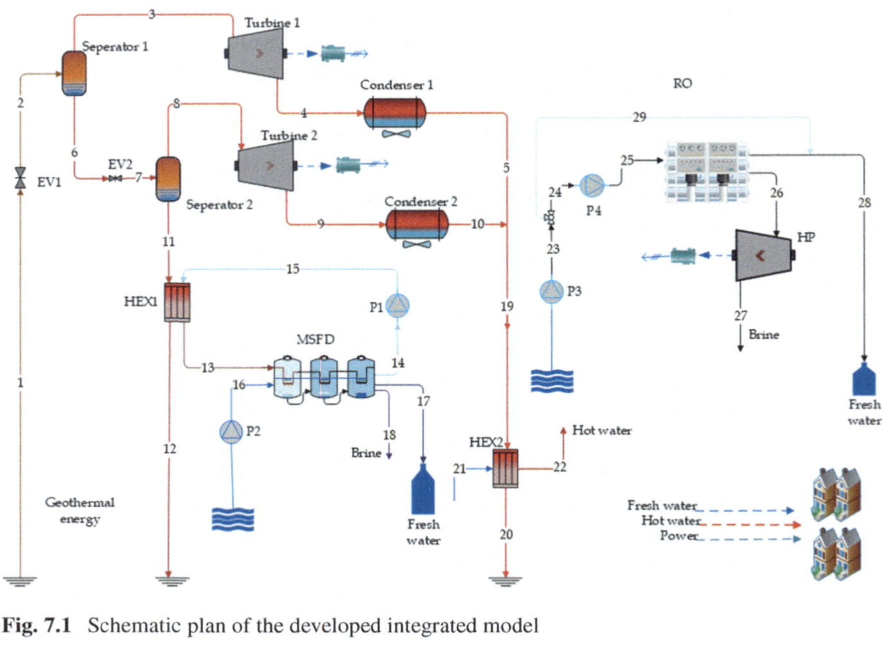

Fig. 7.1 Schematic plan of the developed integrated model

effective purification units. The electricity consumed by the pumps in the RO system is generated by the turbines.

To begin with, the geothermal source exits underground (at point 1) at 180 °C and enters expansion valve 1 (EV-1). Here, after some pressure and temperature drop, it enters separator 1 at point 2 in the liquid–vapor mixture phase. In separator 1, this mixture separates into fluid, saturated vapor, and saturated liquid phases, and while saturated steam goes to turbine 1, the saturated liquid enters the expansion valve (EV-2) again. After generating electricity with the geothermal fluid generator that expands in turbine 1, it enters the condenser at point 4, where it sheds its heat and reaches HEX2. A similar situation also applies to the geothermal fluid, which is a liquid–vapor mixture at point 7 in separator 2. The geothermal fluid, which separates into saturated liquid and saturated steam phases in separator 2, goes to turbine 2 as saturated steam at point 8, where electricity production occurs. The saturated liquid at point 11 enters HEX 1, where it supplies the thermal energy required for the MSFD subsystem.

Second, at points 16 and 23, seawater at 30 °C enters the MSFD and RO units, respectively, and after various processes, it is distilled as clean water at points 17 and 28. The electrical needs of the pumps in these systems are met by the electricity captured in the turbines.

Finally, the geothermal fluid at point 19, still at a certain temperature, enters HEX2, where hot water is produced for heating purposes. The plant continues to work uninterruptedly in this way, and eventually, both power, hot water, and clean water are produced.

7.2.2 *Modeling and Analysis*

In this book chapter, a detailed thermodynamic examination and simulation of a geothermal power plant for power and hot and clean water generation with the energy and exergy efficiencies method is done with the Engineering Equation Solver (EES) (Klein, 2021) software program. Herein, the subpart deals with the mathematical simulation of the thermodynamic study for the entire plant. For this mathematical formulation, the system analysis parameters and assumptions are tabulated in Table 7.1.

7.2.2.1 Thermodynamic Performance Analysis

In this developed book chapter, a comprehensive thermodynamic performance evaluation is conducted and presented by using energetic and exergetic efficiency along with exergy destruction rate. Before proceeding with thermodynamic calculations, the following assumptions were taken into account to make the system an ideal system:

Table 7.1 System examination data and assumptions

Indicators	Value
T_1, °C	200
P_1, kPa	1555
$\dot{m}_1$, kg^{-1}	30
$\eta_{is,T}$, %	85
$\eta_{is,P}$, %	85
T_{12}, T_{20}, °C	52.2
HEX effectiveness	0.8
MSFD and RO desalination units	
Effect number	4
T_{16}, T_{23}, °C	30
Seawater salinity rate, ppm	42
P_{16}, P_{23}, kPa	101.325
Brine salinity rate, ppm	70
Hydropower turbine isentropic efficiency, %	85
Freshwater salinity rate, ppm	0
Domestic hot water unit	
Hot water temperature, °C	44.96
Hot water pressure, kPa	101.325
Reference conditions	
Pressure, kPa	101.325
Temperature, °C	25

- Changes in kinetic and potential energy and exergy are disregarded (Nam et al., 2025).
- Chemical exergy is ignored.
- All system elements are simulated as adiabatic and continuous flow.
- Environmental temperature and pressure values were taken as 25 °C and 101.325 kPa, respectively (Yilmaz, 2025).
- Pumps and turbines operate with isentropic efficiency.
- The efficiency of HEXs is accepted as 80%.

To begin with the thermodynamic analysis, first, it should be noted that four equilibriums can be taken into account as follows (Cengel et al., 2011; Dincer, 2020):

$$\begin{cases} \sum \dot{m}_{\text{in}} = \sum \dot{m}_{\text{e}} \\ \sum \dot{Q}_{\text{in}} + \sum \dot{W}_{\text{in}} + \sum \dot{m}_{\text{in}} h_{\text{in}} = \sum \dot{Q}_{\text{e}} + \sum \dot{W}_{\text{e}} + \sum \dot{m}_{\text{e}} h_{\text{e}} \\ \sum \left(\frac{\dot{Q}}{T} \right)_{\text{in}} + \sum \dot{m}_{\text{in}} s_{\text{in}} + \dot{S}_{\text{gen}} = \sum \left(\frac{\dot{Q}}{T} \right)_{\text{e}} + \sum \dot{m}_{\text{e}} s_{\text{e}} \\ \sum \dot{\text{Ex}}^{\dot{Q}_{\text{in}}} + \sum \dot{\text{Ex}}^{\dot{W}_{\text{in}}} + \sum \dot{m}_{\text{in}} \text{ex}_{\text{in}} = \sum \dot{\text{Ex}}^{\dot{Q}_{\text{e}}} + \sum \dot{\text{Ex}}^{\dot{W}_{\text{e}}} + \sum \dot{m}_{\text{e}} \text{ex}_{\text{e}} + + \dot{\text{Ex}}_{\text{D}} \end{cases} \tag{7.1}$$

Here, the subscripts "in" and "e" indicate the inlet and outlet flows to the system components. Then, "*h*" defines specific enthalpy, "*s*" defines specific entropy, and

"$\dot{m}$" defines mass flow rate. The last terms in the equation, "$\dot{E}x^{\dot{Q}}$" and "$\dot{E}x^{\dot{W}}$," define the heat and work exergies, and these terms can be framed as:

$$\dot{E}x^{\dot{Q}} = \left(1 - \frac{T_0}{T_K}\right)\dot{Q} \tag{7.2}$$

$$\dot{E}x^{\dot{W}} = \dot{w} \tag{7.3}$$

Additionally, the term "ex" defines the flow exergy:

$$\text{ex} = (h - h_0) - T_0 (h - s_0) \tag{7.4}$$

Moreover, Table 7.2 presents comprehensive exergy balance formulations based on Eq. (7.1) for system components.

Table 7.2 Exergy destruction rate of the developed system's apparatuses

Subparts	Exergy formulation
Separator 1	$\dot{E}x_2 = \dot{E}x_3 + \dot{E}x_6 + \dot{E}x_{D,S1}$
Separator 2	$\dot{E}x_7 = \dot{E}x_8 + \dot{E}x_{11} + \dot{E}x_{D,S2}$
Turbine 1	$\dot{E}x_3 = \dot{E}x_4 + \dot{W}_{T1} + \dot{E}x_{D,T1}$
Turbine 2	$\dot{E}x_8 = \dot{E}x_9 + \dot{W}_{T2} + \dot{E}x_{D,T2}$
Condenser1	$\dot{E}x_4 = \dot{E}x_5 + \dot{E}x_{Q,CON1} + \dot{E}x_{D,CON1}$
Condenser1	$\dot{E}x_9 = \dot{E}x_{10} + \dot{E}x_{Q,CON2} + \dot{E}x_{D,CON2}$
HEX1	$\dot{E}x_{11} + \dot{E}x_{15} = \dot{E}x_{12} + \dot{E}x_{13} + \dot{E}x_{D,HEX1}$
HEX2	$\dot{E}x_{19} + \dot{E}x_{21} = \dot{E}x_{20} + \dot{E}x_{22} + \dot{E}x_{D,HEX2}$
Hydropower turbine	$\dot{E}x_{26} = \dot{E}x_{27} + \dot{W}_{HP} + \dot{E}x_{D,HP}$
Pump 1	$\dot{E}x_{14} + \dot{W}_{P1} = \dot{E}x_{15} + \dot{E}x_{D,P1}$
Pump 2	$\dot{E}x_{in,sw} + \dot{W}_{P2} = \dot{E}x_{16} + \dot{E}x_{D,P2}$
Pump 3	$\dot{E}x_{in,sw} + \dot{W}_{P3} = \dot{E}x_{23} + \dot{E}x_{D,P3}$
Pump 4	$\dot{E}x_{24} + \dot{W}_{P4} = \dot{E}x_{25} + \dot{E}x_{D,P4}$
MSFD	$\dot{E}x_{16} + \dot{E}x_{13} = \dot{E}x_{14} + \dot{E}x_{17} + \dot{E}x_{18} + \dot{E}x_{D,MSFD}$
RO	$\dot{E}x_{25} = \dot{E}x_{26} + \dot{E}x_{28} + \dot{E}x_{D,RO}$

7.2.2.2 Energy and Exergy Efficiency

In the developed book chapter, a geothermally driven combined plant is examined via the energy and exergy efficiency methods, for power, hot water, and fresh water. For both combined plants—which generate power, fresh water, and hot water—and single-generation (SG) systems, which solely generate power, energy and exergy efficiency are also examined. Basically, efficiency terms can be defined as beneficial products divided by system inputs and are formulated as below:

For SG system:

$$\eta_{SG} = \frac{\dot{W}_{net}}{\dot{m}_1 h_1 - (\dot{m}_{12} h_{12} + \dot{m}_{20} h_{20})} \tag{7.5}$$

$$\psi_{SG} = \frac{\dot{W}_{net}}{\dot{m}_1 \mathrm{ex}_1 - (\dot{m}_{12} \mathrm{ex}_{12} + \dot{m}_{20} \mathrm{ex}_{20})} \tag{7.6}$$

For combined plant:

$$\eta_{sys} = \frac{\dot{W}_{net} + \dot{Q}_{hotwater} + \dot{m}_{fw} h_{fw}}{\dot{m}_1 h_1 - (\dot{m}_{12} h_{12} + \dot{m}_{20} h_{20})} \tag{7.7}$$

$$\psi_{sys} = \frac{\dot{W}_{net} + \mathrm{Ex}_{Q,hotwater} + \dot{m}_{fw} \mathrm{ex}_{fw}}{\dot{m}_1 \mathrm{ex}_1 - (\dot{m}_{12} \mathrm{ex}_{12} + \dot{m}_{20} \mathrm{ex}_{20})} \tag{7.8}$$

After that, the generated net power can be written as below:

$$\dot{W}_{net} = \dot{W}_{T1} + \dot{W}_{T2} + \dot{W}_{HP} - \sum \dot{W}_{pumps} \tag{7.9}$$

7.3 Results and Discussion

In the developed book chapter, a detailed thermodynamic examination of the combined plant is fulfilled along with energetic and exergetic efficiency as well as exergy destruction rate. To investigate the system performance chancing, a parametric study is done with respect to different systems' indicators. Additionally, the exergetic devastation examines each system component. The case study results are shown in Table 7.3; the net power rate is 957.6 kW, the hot water capacity is 301.6 kW, and the total freshwater rate is 7.19 kg/s. Moreover, the total irreversibility is calculated as 2749 kW. For the whole system, energy and exergy efficiency are computed as 15.45% and 23.55%, respectively.

Figure 7.2 presents a comparison of the energy and exergy performance of the SG system and the entire cycle. While the energy performance of the SG system, i.e., only for electricity production, was 6.46%, the energetic efficiency of the

Table 7.3 Thermodynamic simulation outcomes of the suggested plant

Parameters	Values
$\dot{W}_{T1}$	848 5, kW
$\dot{W}_{T2}$	349 1, kW
$\dot{W}_{HP}$	13.99, kW
$\dot{W}_{net}$	957 6, kW
$\dot{Q}_{hotwater}$	301 6, kW
$\dot{m}_{fw}$	7.19, kg^{-1}
η_{SG}	6.46%
ψ_{SG}	18.63%
η_{sys}	15.45%
ψ_{sys}	23.55%
SI	1.308–
$Ex_{D,total}$	2749, kW

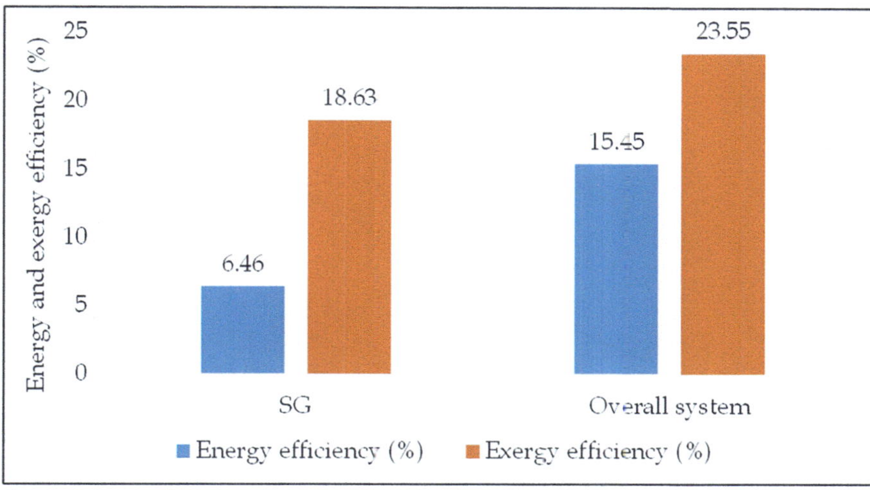

Fig. 7.2 Efficiency of the single-generation system and overall system

designed combined plant was improved and increased to 15.45%. These results show that both the thermodynamic efficiencies of combined systems are higher.

In thermodynamic analysis, irreversibility is among the most important parameters. The irreversibility happening in the plant and sub-elements is calculated and depicted in Fig. 7.3. Herein, it can be noticed that the highest irreversibility is seen in the HEX1 and MED systems, as expected. Because the heat exchange in these systems is higher than in other systems, this means high irreversibility. Total irreversibility is determined as 2749 kW for the entire system.

The change in reservoir temperature is among the greatest significant variables in the developed system. Figures 7.4 and 7.5 show the produced useful outputs and the

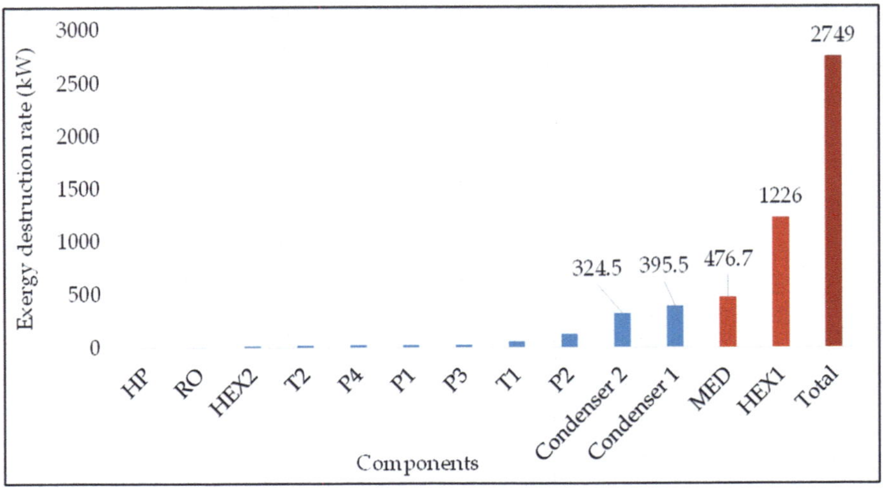

Fig. 7.3 Exergy destruction rate of the developed plant components

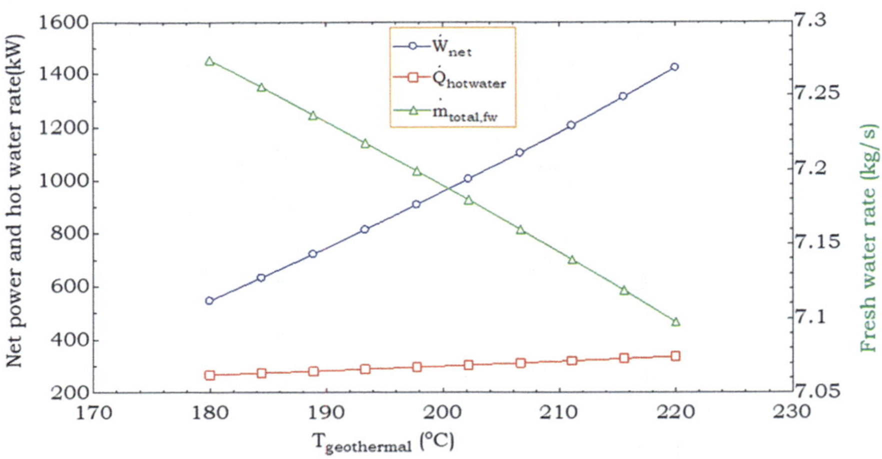

Fig. 7.4 Effect of reservoir temperature on the beneficial products

efficiency change of the cycle as the geothermal source temperature rises from 180 to 220 °C. According to Fig. 7.4, as a result of the increase in source temperature, the beneficial products obtained from the whole system increased, except for fresh water. During this interval, net power production increased from approximately 500–1400 kW. This also linearly increased the system performance as shown in Fig. 7.5. However, as can be seen, the whole irreversibility value in the plant increased as the temperatures were increased.

Figures 7.6 and 7.7 show the change in the useful outputs and system performance obtained from the developed system, depending on the rise in the amount of

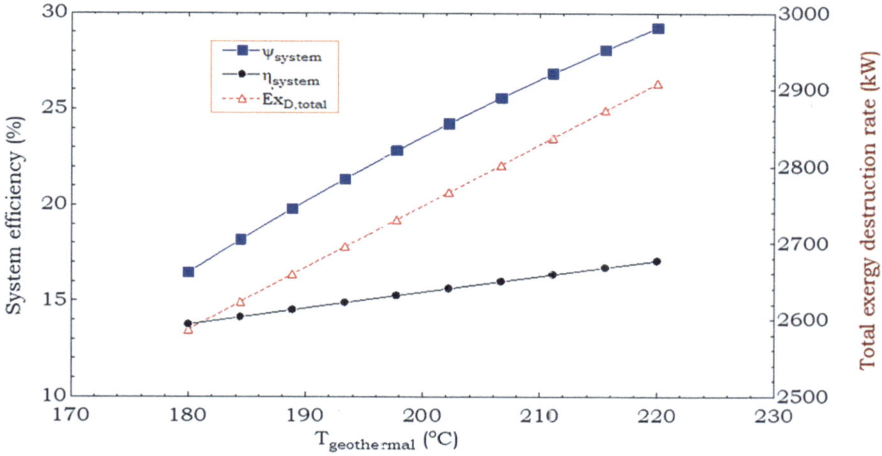

Fig. 7.5 Effect of reservoir temperature on plant efficiency and total irreversibility

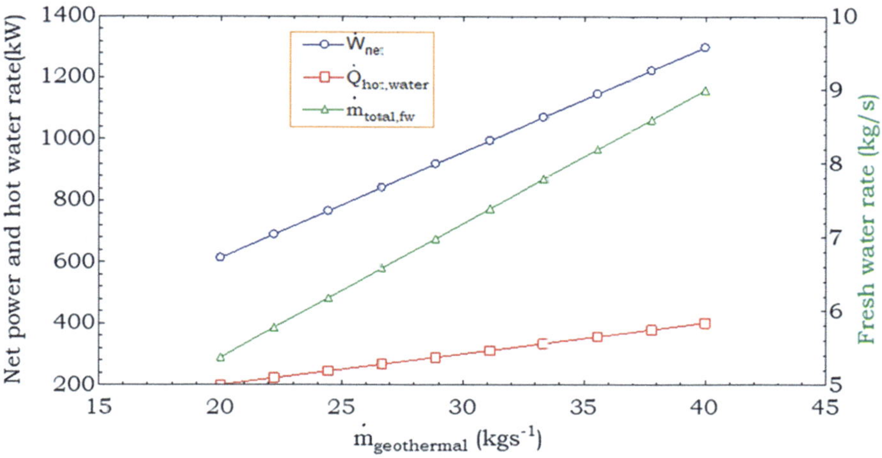

Fig. 7.6 Relationship between geothermal mass flow rate and useful products

geothermal mass rate. It can be remarked from both figures that, with the increase in geothermal mass from 20 to 40 kg^{-1}, both the advantageous outputs obtained from the plant (Fig. 7.6) and the system performance (Fig 7.7) increased linearly. The foremost reason for this situation is that the increased mass flow rate can be interpreted as an increase in the amount of energy entering the cycle and then a rise in the generated products.

Figure 7.8 examines the influence of the pressure ratio change of expansion valve 1 (EV-1) on the system energy and exergy performance as well as the power produced. As the EV-1 pressure ratio increases, while the inlet pressure is constant, the amount of pressure entering separator 1 decreases, and accordingly, the network

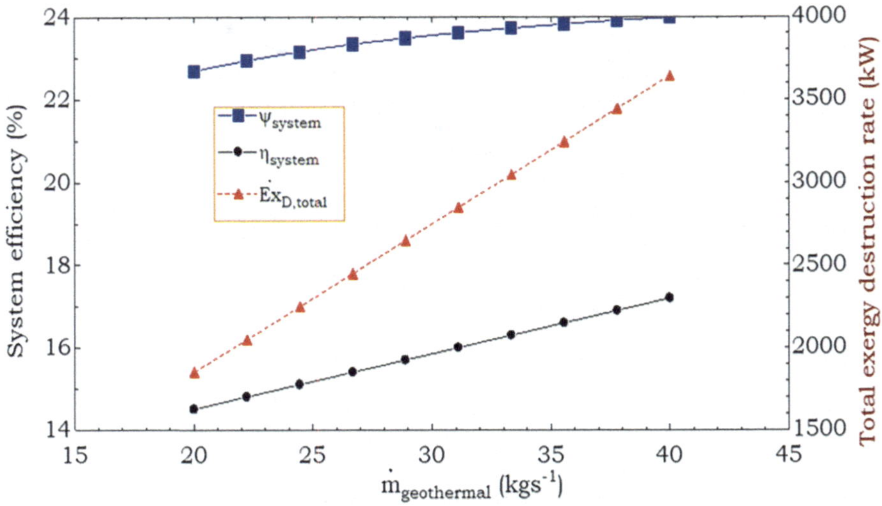

Fig. 7.7 Relationship between geothermal mass flow rate and system performance ratio and total irreversibility rate

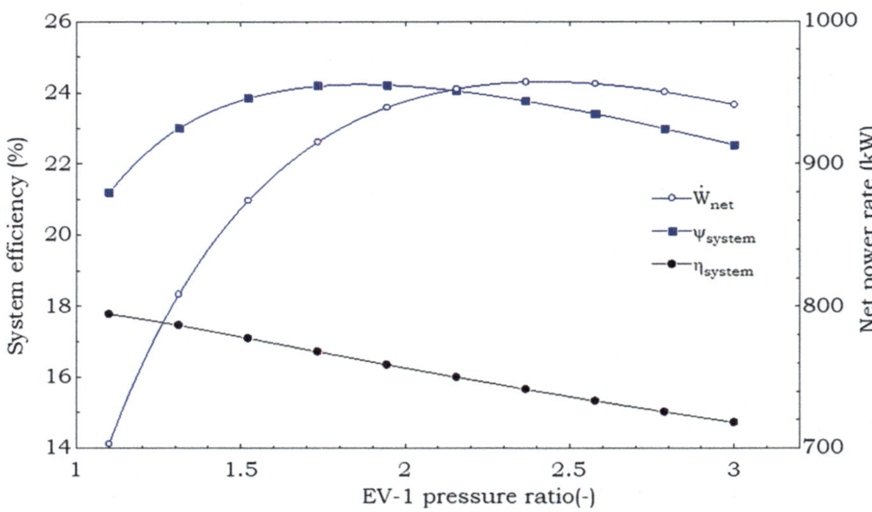

Fig. 7.8 Relationship between EV pressure ratio and system performance ratio and net power rate

production, which initially increases, then begins to decrease. The entire system's efficiency is negatively affected by this situation. Here, the ideal pressure ratio can be interpreted as up to 2 in terms of work increase. For this reason, the pressure ratio is an important parameter in flashing systems.

Figure 7.9 examines the effect of the change in steam quality at the separator 1 (state 2) inlet on the system efficiency and net power production. The increase in

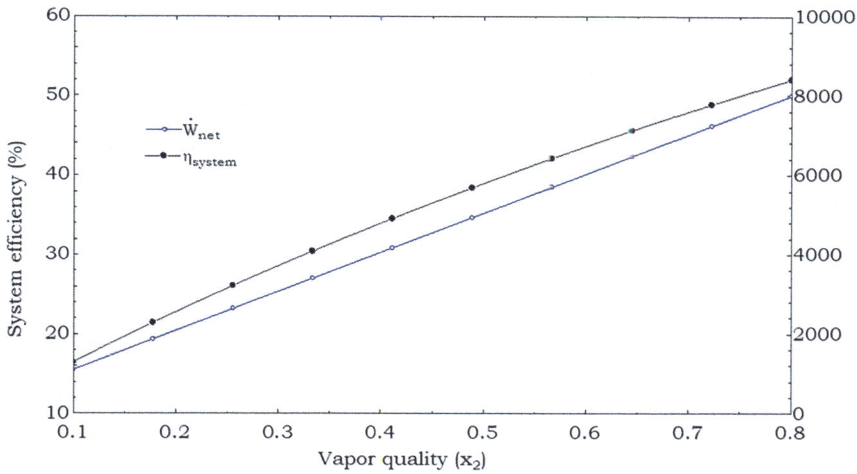

Fig. 7.9 Effect of vapor quality (X_2) on the energy efficiency and net power rate

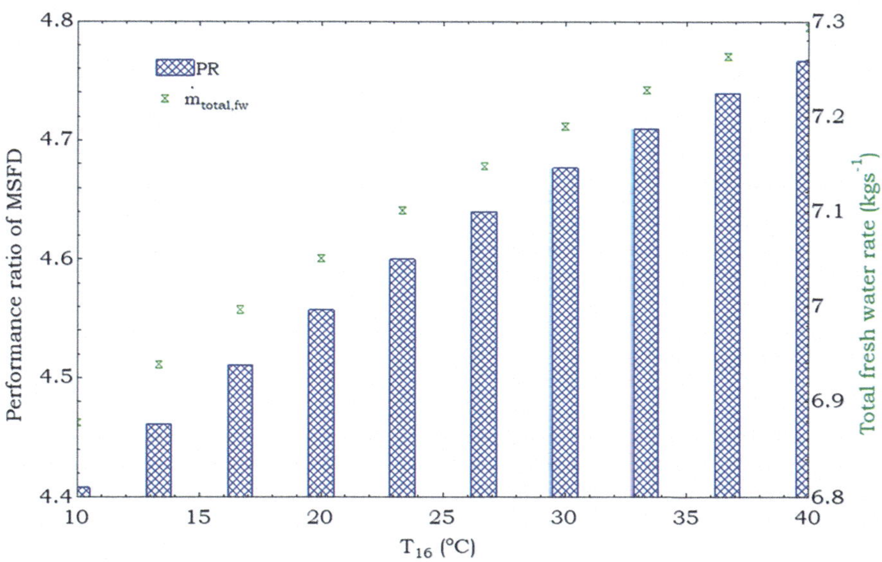

Fig. 7.10 Impact of seawater inlet temperature on the MSFD performance ratio and freshwater rate

steam quality that occurs here increases the net power rate obtained from the plant, from 0.1 to 0.8 and, accordingly, increases the energy efficiency of the developed plant.

In this developed book section, both MSFD and RO units were preferred for clean water production. According to the inlet acceptance at the same temperatures, Fig. 7.10 shows that with the rise in seawater inlet temperature, from 10 to 40 °C, both the MSFD performance ratio (PR) value and the total amount of fresh water are

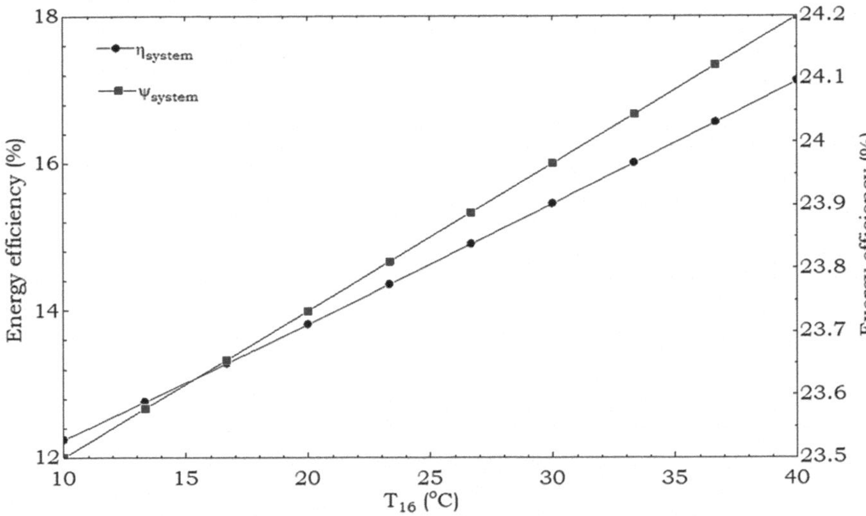

Fig. 7.11 Influence of seawater input temperature on the efficiency of the overall plant

enhanced. Here, with the enhancement in the PR value of MSFD, the amount of clean water obtained from the system also increases.

The last chart of this developed book chapter is the effect of the increase in seawater input temperature on system efficiency. Figure 7.11 depicts the change in both energy and exergy performance of the entire cycle as a result of the increase in seawater temperature from 10 to 40 °C. Here, as depicted in Fig. 7.10 above, plant efficiency increases linearly depending on the increase in the amount of clean water.

7.4 Conclusion

Geothermal energy is a clean energy source that is used in many areas around the world because it is a renewable energy source. Therefore, these types of energy will play a key role in overcoming environmental problems such as global warming. In this developed book chapter, a geothermal combined system for clean water, power, and hot water production is designed and proposed. This work deals with a detailed theoretical assessment of the energetic and exergetic performance analysis of the geothermal energy-based combined system. Furthermore, the exergetic destruction rate in the total cycle and subsystem are investigated. The analysis results indicated that the developed study offers advantages in terms of performance. Also, the prominent results can be summarized as:

- The developed plant power generation capacity is 957.6 kW, hot water capacity is 301.6 kW, and total fresh water is 7.16 kg^{-1}.

- The energy and exergy performance ratio of the suggested book chapter is computed as 15.45% and 23.55%, respectively. Also, the sustainability index of the developed chapter is 1.308.
- Based on the exergy destruction results, the total irreversibility is 2749 kW, while HEX1 has the maximum rate among the components.
- The increase in mass and temperature of the geothermal resource has increased the performance of the developed system.
- Rise in the seawater inlet temperature gives rise to system performance and generates freshwater load.

To sum up, the developed plant has offered many advantages, especially a higher performance ratio than a single-generation system. It should be noted that the combined plant has advantages based on high performance, more products, etc. Due to the increase in global warming and environmental problems, these systems continue to gain importance day by day. Moreover, it is thought that these studies will guide the experimental setup in the future.

References

Abdolalipouradl, M., Mohammadkhani, F., Khalilarya, S., & Yari, M. (2020). Thermodynamic and exergoeconomic analysis of two novel tri-generation cycles for power, hydrogen and freshwater production from geothermal energy. *Energy Conversion and Management, 226*, 113544.

Asadi, R., Assareh, E., Moltames, R., Olazar, M., Nedaei, M., & Parvaz, F. (2022). Optimisation of combined cooling, heating and power (CCHP) systems incorporating the solar and geothermal energy: A review study. *International Journal of Ambient Energy, 43*(1), 42–60.

Bozgeyik, A., Altay, L., & Hepbasli, A. (2022). A sub-system design comparison of renewable energy based multi-generation systems: A key review along with illustrative energetic and exergetic analyses of a geothermal energy based system. *Sustainable Cities and Society, 82*, 103893.

Cengel, Y. A., Boles, M. A., & Kanoğlu, M. (2011). *Thermodynamics: An engineering approach* (Vol. 5, p. 445). McGraw-hill.

Dincer, İ. (2020). *Thermodynamics: A smart approach*. Wiley.

Guler, O. F., Sen, O., Yilmaz, C., & Kanoglu, M. (2022). Performance evaluation of a geothermal and solar-based multigeneration system and comparison with alternative case studies: Energy, exergy, and exergoeconomic aspects. *Renewable Energy, 200*, 1517–1532.

Hekmatshoar, M., Deymi-Dashtebayaz, M., Gholizadeh, M., Dadpour, D., & Delpisheh, M. (2022). Thermoeconomic analysis and optimization of a geothermal-driven multi-generation system producing power, freshwater, and hydrogen. *Energy, 247*, 123434.

Hepbasli, A. (2008). A key review on exergetic analysis and assessment of renewable energy resources for a sustainable future. *Renewable and Sustainable Energy Reviews, 12*(3), 593–661.

Huang, J., Abed, A. M., Eldin, S. M., Aryanfar, Y., & García Alcaraz, J. L. (2023). Exergy analyses and optimization of a single flash geothermal power plant combined with a trans-critical CO2 cycle using genetic algorithm and Nelder–Mead simplex method. *Geothermal Energy, 11*(1), 1–20.

Klein, S. (2021). *Engineering equation solver (EES), Academic Commercial, V11.199*. 2021 F-Chart Software.

Kumar, L., Hossain, M. S., Assad, M. E. H., & Manoo, M. U. (2022). Technological advancements and challenges of geothermal energy systems: A comprehensive review. *Energies, 15*(23), 9058.

Ma, T., Lan, L., & Marefati, M. (2023). Assessment of a new multigeneration system based on geothermal plant and a linear Fresnel reflector-based solar unit: An effort to improve performance. *Process Safety and Environmental Protection, 171*, 896–913.

Moya, D., Aldás, C., & Kaparaju, P. (2018). Geothermal energy: Power plant technology and direct heat applications. *Renewable and Sustainable Energy Reviews, 94*, 889–901.

Nam, A. K., Khabbazi, H., Ghaebi, H., & Minaei, A. (2025). Designing and multi-objective optimization of a novel multigenerational system based on a geothermal energy resource from the perspective of 4E analysis. *Applied Thermal Engineering*, 258, 124673.

Yilmaz, F. (2022). Development and modeling of the geothermal energy based multigeneration plant for beneficial outputs: Thermo-economic and environmental analysis approach. *Renewable Energy, 189*, 1074–1085.

Yilmaz, F. (2025). A comprehensive energy, exergy, economic, and environmental (4E) assessment of a geothermal-driven polygeneration plant with energy storage using compressed hydrogen. *Journal of Energy Storage*, 114, 115796.

Yilmaz, F., Ozturk, M., & Selbaş, R. (2023). Proposed and assessment of a sustainable multigeneration plant combined with a transcritical CO2 cycle operated by flash-binary geothermal energy. *International Journal of Hydrogen Energy, 48*(60), 22818–22833.

Chapter 8
The Role of Innovative Ship Design in Decarbonization

Ahmet Taşdemir and Mehmet Akman

Nomenclature

CII	Carbon Intensity Indicator
CO_2	Carbon dioxide
EEDI	Energy Efficiency Design Index
EEXI	Energy Efficiency Existing Ship Index
HFO	Heavy fuel oil
IMO	International Maritime Organization
LPG	Liquefied petroleum gas
LNG	Liquefied natural gas
MeOH	Methanol
NO_X	Nitrogen oxides
SEEMP	Ship Energy Efficiency Management Plan
SO_X	Sulphur oxides

8.1 Introduction

The conventional ship design approach refers to the traditional methods and principles historically employed in ship design and construction. This approach has been the foundation of maritime engineering for many years and has evolved based on practical experience, established regulations, and available technologies

A. Taşdemir (✉)
Piri Reis University, Tuzla, İstanbul, Türkiye
e-mail: atasdemir@pirireis.edu.tr

M. Akman
İzmir Katip Çelebi University, İzmir, Türkiye
e-mail: mehmet.akman@ikc.edu.tr

M. Z. Sogut, M. Koray (eds.), *Energy Rationality and Management for Decarbonization*, Green Energy and Technology,
https://doi.org/10.1007/978-3-031-68497-5_8

(Taşdemir et al., 2023). During the design stage, key characteristics like the shape of the hull, propulsion systems, materials and construction, navigational and operational systems, regulatory compliance, and challenges and limitations of the ship influence the performance of the ships in the building and service phases. The hull's shape is carefully designed to ensure optimal resistance through water, stability in different sea conditions, and efficient propulsion. Therefore, the prediction of the ship's performance plays an important role in the early design stage of a ship (Bertram, 2000). Conventional ship design places significant emphasis on the hydrodynamic performance of the hull because this influences the resistance and propulsion, sea-keeping, and manoeuvring behaviour of ships. Therefore, the hull's shape is carefully designed to ensure optimal resistance to water, stability in different sea conditions, and efficient propulsion. Ensuring the stability and seaworthiness of the ship are paramount considerations. Stability calculations, including metacentric height and righting arm, are used to design ships that can withstand the dynamic forces of waves and maintain stability in various operating conditions.

Conventional ship design adheres to the international regulations and standards set by organizations like the International Maritime Organization (IMO). These regulations cover safety, environmental impact, and other aspects of maritime operations. While the conventional ship design approach has proven effective for many decades, it is not without its challenges and limitations, particularly in the context of the contemporary emphasis on sustainability and environmental responsibility. The process has started with the enforcement of the Energy Efficiency Design Index (EEDI) and Ship Energy Efficiency Management Plan (SEEMP), and the scope has enlarged with the adoption of the Energy Efficiency Existing Ship Index, Carbon Intensity Indicator (CII), and enhanced SEEMP (IMO, 2023b). In this context, IMO has developed a strategy to decrease the carbon intensity by at least 20% by 2030 and 70% by 2040 compared to the carbon intensity status of 2008 (IMO, 2023a). In recognizing these challenges, the maritime industry is undergoing a paradigm shift. There is a growing emphasis on innovative ship design, incorporating alternative propulsion systems, energy-efficient technologies, and sustainable materials. While the conventional ship design approach has laid the groundwork for maritime engineering, the industry's future lies in the integration of cutting-edge solutions that address the pressing issues of emissions, energy efficiency, and environmental impact. As the maritime sector navigates these uncharted waters, a blend of tradition and innovation is essential to ensure a sustainable and resilient future for global shipping. On the other hand, there has been a shift towards adopting a holistic design approach, which entails comprehensive, multidisciplinary, and multi-objective design optimization (Papanikolaou, 2010) to be used in the energy-efficient ship design process.

Historically, ships have been powered by internal combustion engines, typically running on heavy fuel oil or diesel. These engines provide the necessary thrust for propulsion and are central elements in the conventional ship design, which has been a significant contributor to global carbon emissions. Considering that more than

80% of world trade is done by seaways (Ling-Chin & Roskilly, 2016), the environmental impacts of maritime activities become increasingly more important. According to the International Maritime Organization (IMO), the maritime sector accounts for approximately 2–3% of global greenhouse gas emissions, a figure projected to rise without intervention (IMO, 2020). The challenge is not only rooted in the sheer volume of emissions but also in the longevity of maritime assets. Ships often operate for decades, so design decisions have long-lasting effects. Addressing the maritime carbon footprint requires a holistic approach that considers both existing fleets and the vessels yet to be constructed. In this study, factors affecting the energy efficiency of ships are investigated under the regulatory framework. The hull design for energy efficiency, developments on propulsion systems, alternative fuels, and energy management for sustainable shipping with measures applicable during the design and operational phases are presented. The role of ship design in energy efficiency and the environment is discussed using the numerical data of previous studies and present applications.

8.2 Energy Efficiency of Ships

In an era marked by a growing global awareness of environmental sustainability, the maritime industry stands at the forefront of the efforts to mitigate its carbon footprint. As the world navigates the challenges of climate change, the role of innovative ship design emerges as a pivotal force in the pursuit of decarbonization. Traditionally reliant on fossil fuels, the shipping sector is undergoing a transformative shift towards cleaner and more sustainable practices. This paradigmatic evolution underscores the critical importance of cutting-edge technologies and creative design solutions to revolutionize the way ships are conceived, constructed, and operated. This exploration analyzes the multifaceted aspects of innovative ship design and its crucial role in promoting energy-efficient and environmentally conscious maritime transportation. From novel propulsion systems to advanced materials, this narrative navigates the intricate waters of maritime innovation, illustrating how forward-thinking design principles are integral to charting a course towards a greener and more sustainable maritime future. Figure 8.1 shows the efficiency-influencing possible parameters for a ship.

The vast expanse of the world's oceans, once perceived as an endless frontier for exploration and trade, now functions as a focal point in the global dialogue on environmental sustainability. With climate change asserting its undeniable presence and the imperative to reduce carbon emissions becoming more urgent, industries reliant on fossil fuels are under growing scrutiny. In this context, the maritime sector finds itself at the nexus of challenge and opportunity, compelled to steer towards decarbonization while maintaining the intricate web of global trade and transportation. At the heart of this maritime transformation lies the profound impact of innovative ship design. Ships, traditionally propelled by fossil fuels, are undergoing a revolutionary transition. The design of vessels is changing from conventional paradigms to

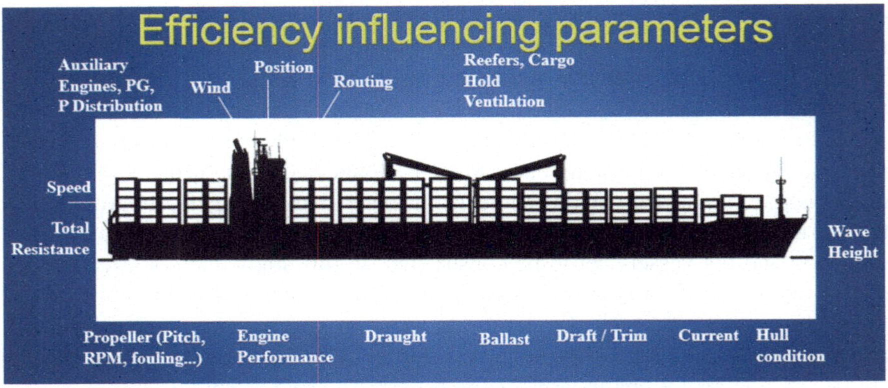

Fig. 8.1 Possible efficiency-influencing parameters for a ship

embrace cutting-edge technologies and sustainable practices, reflecting a collective effort to mitigate the environmental impact of shipping activities. This exploration delves into the multifaceted dimensions of innovative ship design, examining its pivotal role in the overarching mission of decarbonizing the maritime industry.

Several pioneering initiatives and projects exemplify the potential of innovative design in driving decarbonization. One facet of innovative ship design focuses on the optimization of hull forms to enhance energy efficiency. A prime example is the development of bulbous bows and streamlined hulls that reduce hydrodynamic resistance, allowing ships to glide more efficiently through water. Additionally, advanced coating technologies, such as low-friction paints, contribute to minimizing drag, thereby reducing fuel consumption and emissions. A pivotal element in the quest for decarbonization is the exploration of alternative propulsion systems. Traditional internal combustion engines are giving way to a new generation of power sources. Liquefied natural gas (LNG) propulsion is gaining traction as a cleaner alternative to conventional fuels, offering lower emissions and improved efficiency. Furthermore, there is a burgeoning interest in electrification, with electric and hybrid propulsion systems demonstrating their potential to revolutionize maritime powertrains.

Harnessing the power of the wind, a time-tested method in maritime history, is experiencing a renaissance with modern technological enhancements. Innovative sail designs, such as rigid sails and automated kite systems, are being integrated into ship designs to capitalize on wind energy during voyages. This hybrid approach not only reduces reliance on fossil fuels but also demonstrates the integration of traditional and contemporary maritime technologies. Solar power, abundant and sustainable, is finding its way onto the decks of ships. Solar panels integrated into the superstructure or even on sails contribute to the generation of auxiliary power, reducing the overall demand on traditional propulsion systems. This integration of renewable energy sources exemplifies the versatility and adaptability of innovative ship design in the pursuit of sustainability.

In the context of innovative ship design, advancements in materials science and digitalization play a pivotal role. The adoption of lightweight and high-strength materials, such as carbon fibre and advanced composites, contributes to the reduction of a ship's weight, enhancing fuel efficiency without compromising structural integrity, trim and ballast optimization, planned maintenance, and on-board energy management or design-oriented practices such as hull form optimization and weight reduction by improved structural design (Tasdemir & Nohut, 2012). Moreover, digital technologies, including artificial intelligence (AI) are instrumental in optimizing ship operations. Smart navigation systems, predictive maintenance, and real-time monitoring of fuel consumption enable data-driven decision-making, empowering shipowners and operators to minimize environmental impact while maximizing operational efficiency.

8.3 Hull Design for Energy Efficiency

Unlike other vehicles (e.g. aircraft, cars, and trucks) that are produced with serial production, each ship (except sister ships) must be designed from the beginning separately concerning its purpose and shipowners' requirements. Therefore, the design and manufacturing flexibility of ships and ship parts are laborious in shipbuilding (Taşdemir & Nohut, 2021). The design stage is a critical phase that plays a pivotal role in shaping the entire life cycle of a vessel in technical and economic terms. This stage involves the comprehensive planning, conceptualization, and detailed specification of a ship's architecture, systems, and features. The importance of the design stage in shipbuilding encompasses various aspects, ranging from safety and performance to technical and economic efficiency and environmental sustainability.

The ship design is a multidisciplinary process comprising a wide range of iterant analyses, which must be carried out in great detail and carefully. Concept, contract, detail, and production design are the main phases of ship design where the dimensional characteristics, stability, strength manoeuvrability, and propulsion properties are defined (Turan, 2023; Turan & Akman, 2021). The ship to be designed should fulfil especially the mission requirements of shipowners like type and amount of cargo, navigation area, and speed as well as comply with the IMO regulations. In order to achieve these goals or a nearly perfect ship design, a naval architect traces the design process through each stage.

The evaluation of hull forms in conventional shipping is a critical aspect of the ship design stage, involving a comprehensive assessment of various hydrodynamic and performance-related factors, like draught, freeboard, metacentric height, resistance, etc. The goal is to optimize the hull shape to achieve a balance between efficiency, stability, and sea-keeping characteristics. The hull optimization also serves to improve the fuel efficiency, propulsion, and manoeuvring properties of ships. Hull designs must comply with international regulations set by the IMO. These standards cover aspects like stability, safety, and environmental performance. In summary, the evaluation of hull forms in merchant shipping is a multifaceted process that involves a combination of theoretical analyses, numerical simulations, physical testing, and practical experience. The ultimate goal is to achieve an optimal balance between safety, efficiency, and environmental sustainability in the design of ships for diverse operational conditions and purposes. The design phase lasts in a so-called design process until the desired objectives are achieved. Presently, there is an increased emphasis on the development of eco-friendly vessels, aiming to curtail emissions and mitigate the effects of global warming. Prioritizing sustainability holds significant importance in preserving our planet, ensuring that forthcoming generations can responsibly utilize its resources. In addition to hull optimization during the design phase, there are many other efforts to increase the performance of the ship in terms of its technical and economic aspects. Here are some examples of efforts regarding increasing the performance:

8.3.1 Lightweight Design and Eco-friendly Materials

In shipbuilding, incorporating lightweight design principles and eco-friendly materials is a progressive approach that aligns with industry trends towards sustainability, fuel efficiency, and environmental responsibility. Advances in materials science and technology continue to drive innovation in this field, to create ships that are both efficient and environmentally friendly. Additionally, the flexible design technique is also used to ensure the optimal distribution of loads to elements during design.

8.3.2 Techniques to Reduce Friction Resistance in Shipbuilding

Reducing friction resistance in shipbuilding is crucial for improving fuel efficiency, increasing speed, and minimizing wear and tear on the vessel. Various techniques and technologies are employed to achieve this goal. Common methods to reduce friction resistance in shipbuilding are hull form optimization, low surface energy antifouling coatings, air lubrication systems as shown in Fig. 8.2, hull cleaning systems, and systems increasing propulsion efficiency. Additionally, ongoing research and technological advancements continue to contribute to innovative solutions for reducing friction resistance in shipbuilding, promoting greater energy efficiency and environmental sustainability in the maritime industry.

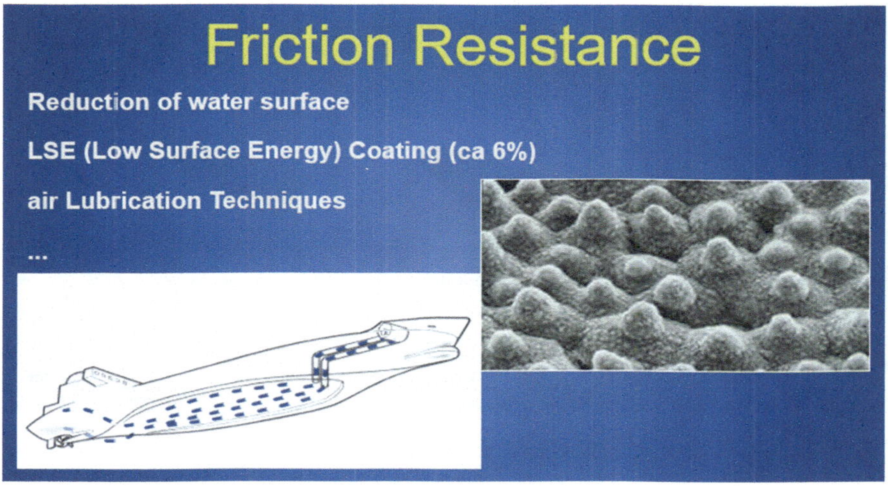

Fig. 8.2 Air lubrication and surface roughness

8.3.3 Techniques to Reduce Wind and Pressure Resistance in Shipbuilding

Reducing pressure resistance, also known as wave-making resistance or hydrodynamic resistance, is crucial in ship design to enhance overall efficiency and performance. Pressure resistance refers to the force opposing the forward motion of a ship caused by the formation of waves as it moves through the water. Several techniques are used to minimize pressure resistance in shipbuilding. It is important to note that a holistic approach involving a combination of techniques is needed to achieve optimal results in reducing pressure resistance and improving the overall hydrodynamic performance of a ship (Taşdemir & Akman, 2023). Figure 8.3 shows the pressure fields of bluff objects including square and container ship.

8.3.4 Techniques to Reduce Wave Resistance in Shipbuilding

Reducing wave resistance is a key goal in ship design to enhance fuel efficiency and overall performance. Wave resistance, also known as wave-making resistance, is the energy required to generate waves as a ship moves through the water. Minimizing this resistance is crucial for improving a ship's hydrodynamic efficiency (Fig. 8.4).

During the design stage, there are a lot of methods or temptations like reducing tip vortex losses, reducing rudder resistance, etc. to minimize the total resistance, which is crucial for fuel efficiency and operational cost savings. With the same goal, attempts have recently been made to use solar and wind energy. For this purpose,

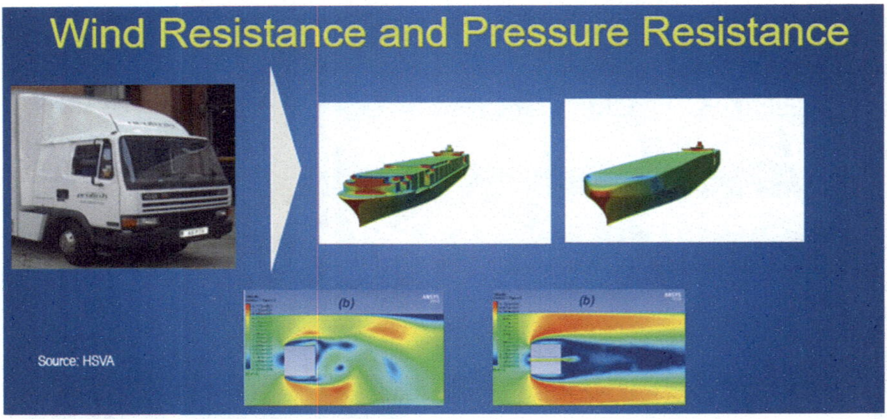

Fig. 8.3 Pressure fields of bluff bodies and a container ship (He et al., 2016; Taşdemir & Akman, 2023)

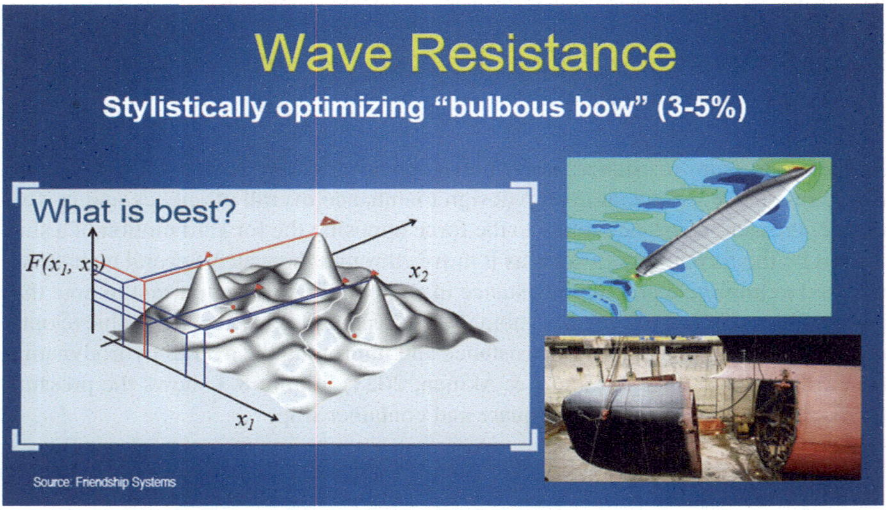

Fig. 8.4 Optimization of hull form and bulbous bow (FSYS, 2023)

computational fluid dynamics (CFD) and model testing are often used to predict the resistance a hull will experience as it moves through the water. Minimizing resistance is crucial for fuel efficiency and operational cost savings. The evaluation includes economic considerations such as the total cost of ownership, including construction costs, fuel consumption, and maintenance expenses. Shipowners seek an optimal balance between initial investment and operational savings over the vessel's life cycle (Fig. 8.5).

Fig. 8.5 Examples of reducing the tip vortex (MAN, 2016a)

8.4 Developments in Propulsion Systems

The powering stage in the ship design process includes the selection of suitable propulsion and electric power generation systems. Marine diesel engines are the traditional and widely used internal combustion engines (ICEs) for generating power on-board a vessel. The main advantages of marine diesel engines are a high power-to-weight ratio, high thermal efficiency, experienced and well-established procedures for operation, readily available spare parts, and extensive global fuel bunkering networks (Akman, 2023; Inal et al., 2022b; MAN, 2016b). Nevertheless, growing environmental concerns have started to challenge their dominance in recent years based on the efforts of the International Maritime Organization regarding reducing harmful emissions. However, the shipping industry has limited options for selecting on-board power generation technologies. While the predominant choice is internal combustion engines, there are substantial developments in fuel cells (FCs) and renewable energy sources (RESs), which can be integrated into hybrid systems for on-board powering. Hybrid systems offer multiple powering options and exhibit higher flexibility compared to traditional propulsion systems. These systems can be in serial or parallel configurations (Inal et al., 2022a) where internal combustion engines are separated to drive the propeller via a gearbox or are connected to the bus bar via alternators as shown in Fig. 8.6.

Wind energy can be directly used for propulsion as it has been used for centuries. Kites and rotors have been started to be used for commercial ships. Kites can be customized for use on nearly any type of commercial ship, and they are attached to the bow and governed by computerized controls. The kite operates at altitudes

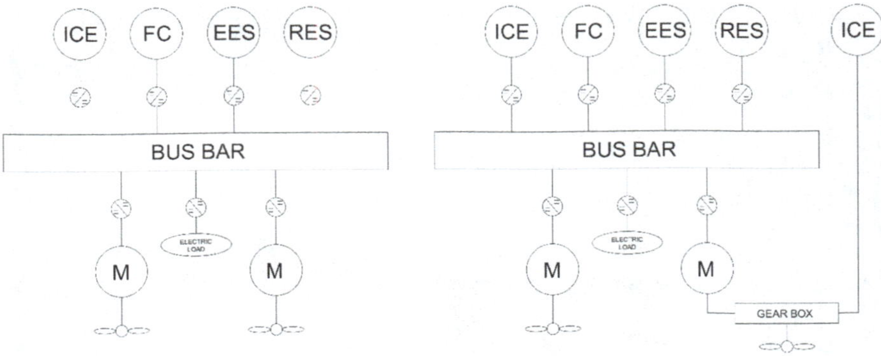

Fig. 8.6 Series hybrid (**a**) and parallel hybrid (**b**) propulsion configurations

Fig. 8.7 Kite- (**a**) and rotor- (**b**) assisted propulsion (Bass, 2021)

ranging from 300 to 1200 ft above the sea surface, taking advantage of higher wind speeds to generate significant propulsive power (Bass, 2021). It is reported that fuel savings for MS Beluga SkySails (shown in Fig. 8.7a), which is a 10,000-tonne container ship built in 2007 with a capacity of 474 TEUs, is at least 5.5% (Bass, 2021).

Rotor sails, alternatively referred to as Flettner rotors, are elevated cylinders that, when rotating, supply additional propulsive force to vessels, leading to a reduction in on-board fuel consumption. It is reported that implementing six rotor sails measuring 5 × 30 m on-board a 210,000-DWT bulk carrier (as shown in Fig. 8.7b) could result in a 29% reduction in its Energy Efficiency Design Index (Wärtsilä, 2022).

All-electric or fuel-cell-integrated propulsion systems have been commercialized and installed not only for small crafts but also for merchant ships. The Viking Lady, which is fuelled by liquefied natural gas (LNG), is stated as the pioneer merchant ship incorporating a fuel cell generating 330 kW of electric output and has

Fig. 8.8 Fuel cell and battery hybrid "Viking Lady" (Wärtsilä, 2023) and fully electric "MF Ampere" (Corvus EES, 2020)

operated successfully for over 18,500 h (Wärtsilä, 2023). Moreover, it is reported that the world's first all-electric-powered ferry MF Ampere was delivered in October 2014, and its commercial operations started in May 2015 (Corvus EES, 2020). The ferry has zero emissions and minimal noise, is powered by a 1090 kWh energy storage system, and is capable of charging in 9 min (Corvus EES, 2020) (Fig. 8.8).

8.5 Alternative Fuels for Powering

Heavy fuel oil (HFO) had been the primary fuel in global maritime shipping, accounting for 79% of total fuel consumption by energy content in 2018 based on voyage-based allocation (IMO, 2020). Alternatives to heavy fuel oil (HFO), marine diesel oil (MDO), or marine gas oil (MGO); liquefied natural gas (LNG), liquefied petroleum gas (LPG); and methanol (MeOH) are considered the short- and medium-term fuel options for complying with the strict limitations on ship-based emissions and ship energy efficiency. Table 8.1 shows some of the properties of marine fuels. According to the numerical data, the energy content of LNG and LPG is higher and the energy density of these fuels is less than that of MGO. Moreover, alternative fuels are more environmentally friendly than HFO causing more SO_X, NO_X, CO_2, and PM. The carbon content of LNG, LPG, and MeOH is less than HFO, and these fuels are naturally compliant with SO_X limits (Herdzik, 2021). Ammonia, on the other hand, has a huge potential for decarbonization based on its carbon-free structure, producibility using renewable energy, and global availability (DNV-GL, 2020). Besides, the storage and delivery pressures of fuels are remarkably different, which affects the tank and piping size during outfitting. Hence, benchmarks and considerations in the ship design process vary based on the thermochemical and physical properties of the fuels.

There are dual-fuel (DF)and tri-fuel engine options available in the market. These engines can burn LNG, LPG, methanol, or ethanol as the primary fuel to comply with the SO_X and NO_X limits; MDO is used as the pilot fuel to control knocking and misfiring (Woodyard, 2004). HFO and very-low-sulphur fuel oil

Table 8.1 Some of the key properties of marine fuels (MAN, 2022)

Fuel	Energy content (MJ/kg)	Energy density (MJ/l)	Fuel tank size relative to MGO	Supply pressure (bar)	Emission reduction compared to HFO (%)			
					SO_X	NO_X	CO_2	PM
LNG	50	21.2	1.7	300	90–99	20–30	24	90
LPG	46	26.7	1.3	50	90–100	10–15	13–18	90
Methanol	19.9	14.9	2.4	10	90–97	30–50	11	90
Ammonia	18.6	10.6 (@45 °C)	3.4 (@45 °C)	80	100	Compliant	~90	~90
MGO	42.7	35.7	1	7–8	Ref	Ref	Ref	Ref

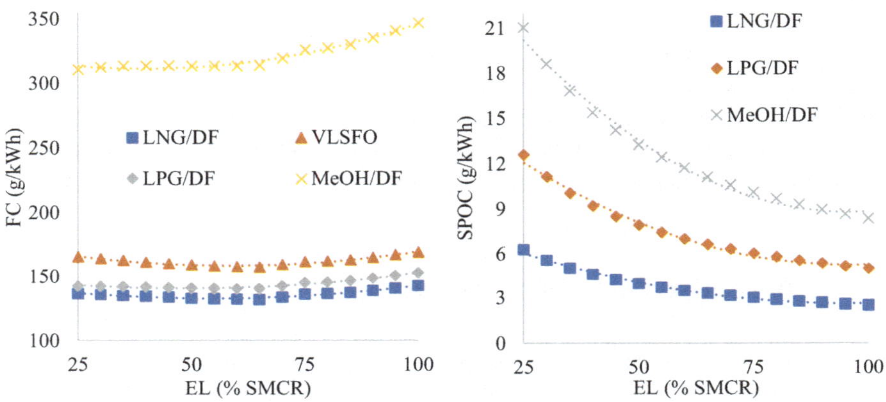

Fig. 8.9 Fuel (FC) and pilot oil consumptions (SPOC) of marine dual-fuel engines (Akman, 2023)

(VLSFO) can also be burned in these engines. However, based on the energy density of fuels as presented in Table 8.1 fuel consumption differs. Figure 8.6 shows the main fuel (a) and specific pilot oil consumptions (b) of dual-fuel (DF) engines at various engine loads with the same brake power (Akman, 2023). According to the calculations, the main fuel consumption of LNG-fuelled engines is about 16.6%, 8.6%, and 140% less compared to that of VLSFO-, LPG-, and MeOH-fuelled engines, respectively (Fig. 8.9).

Alternative fuels have different environmental impacts considering GHG emissions. Figure 8.7 shows the change in CO_2 emissions with respect to engine load at tier II (a) and tier III (b) modes of LNG-, LPG-, VLSFO-, and MeOH-fuelled engines (Akman, 2023). Based on the findings, the VLSFO-powered engine has a higher CO_2 emission rate, measuring 3.82 tons/h at 85% MCR and in tier II mode. Under the same operational conditions, engines fuelled by LNG, LPG, and MeOH emit 2.82 tons/h, 3.35 tons/h, and 3.53 tons/h, respectively. In tier III mode, CO_2 emissions increase for VLSFO, LNG, LPG, and MeOH engines at 85% MCR, with values of 3.84 tons/h, 2.83 tons/h, 3.37 tons/h, and 3.64 tons/h, respectively. However, it is worth noting that methane slip causing a greenhouse effect 28 times as strong as an equivalent amount of CO_2 is a problem in LNG-fuelled engines and

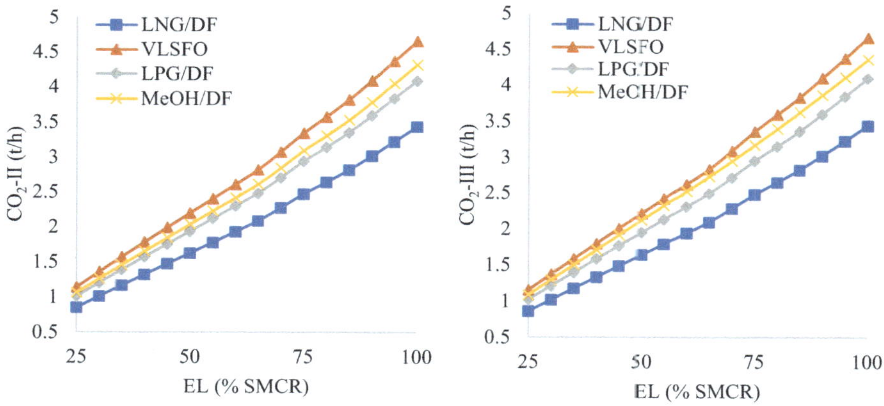

Fig. 8.10 CO_2 emission rates of dual-fuel engines at tier II (**a**) and tier III modes (**b**) (Akman, 2023)

that there are developments in combustion technologies to minimize it (MAN, 2021) (Fig. 8.10).

The Energy Efficiency Design Index is directly related to the fuel consumption and carbon content of the fuel used for powering. The engines to be installed on the designed ship must fulfil the EEDI limits defined in International Convention for the Prevention of Pollution from Ships -(MARPOL) Annex VI. This study points out that the attained EEDI of a 32,000-DWT Handymax bulk carrier can be less than the reference EEDI when LNG- and methanol-fuelled engines are selected as main and auxiliary power suppliers as shown in Fig. 8.8 (Akman et al., 2023). The attained EEDI can be decreased by using wind-assisted propulsion, solar power integration to the ship grid, and waste heat recovery; these measures have substantial potential for reducing the attained EEDI by about 15% (Barreiro et al., 2022) (Fig. 8.11).

8.6 Energy Management for Sustainability

The increasing trend of greenhouse gases sourced from international maritime transportation, fluctuations in fuel prices, and the latest technological developments increase the importance of fuel-saving and energy efficiency on-board ships. The International Maritime Organization (IMO) has enacted many regulations with MARPOL Annex VI to reduce ship-based emissions and increase ship energy efficiency. The regulatory process started with the Energy Efficiency Design Index (EEDI), Ship Energy Efficiency Management Plan (SEEMP), and Energy Efficiency Operation Indicator (EEOI); then, the scope of the regulations was expanded, and regulations such as Energy Efficiency Existing Ship Index (EEXI) and the Carbon Intensity Indicator (CII) have been introduced for more effective inspection and control. In addition, activities within the scope of the fuel consumption control, the Data Collection System (IMO-DCS) on a global basis in 2019, and the regulation

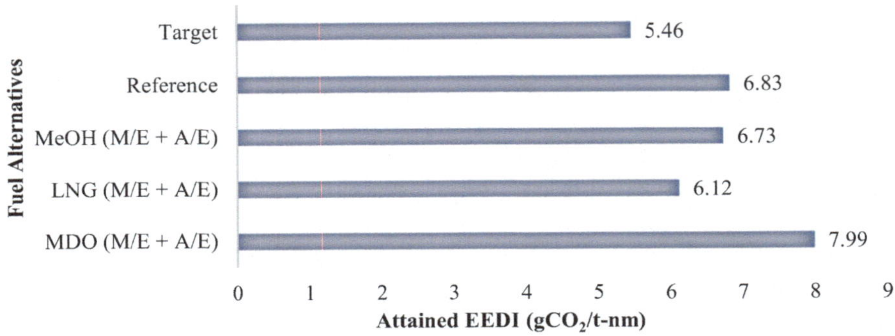

Fig. 8.11 Attained EEDI of Handymax bulk carrier with dual-fuel engines powered by alternative fuels (Akman et al., 2023)

Fig. 8.12 Concept of ship energy efficiency management in SEEMP

on the Monitoring, Reporting, and Verification of carbon dioxide (CO_2) emissions from ships (EU-MRV) on a European Union scale in 2018 started. Furthermore, the SEEMP, enhanced in 2023, requires shipowners to take effective energy efficiency and environmental measures including monitoring of data for further improvements on-board to increase the operational performance of ships (DNV, 2023). The concept of enhanced SEEMP is shown in Fig. 8.9. According to this concept, the ship's energy efficiency is based on the energy-efficient design, EEDI; plan for the energy-efficient operation, SEEMP; monitoring energy efficiency, EEOI; and collecting the data, DCS/MRV (Fig. 8.12).

In parallel with the latest developments, effective energy management systems have been critical for ships; based on that, the ships operate under dynamic conditions, and the energy production, distribution, and consumption systems on-board ships are interconnected and complex. Therefore, measurable, verifiable, and controllable energy efficiency management systems capable of monitoring key performance indicators regarding energy efficiency and the effective control of fuel consumption and emissions have been developed for ships. The latest studies and applications (Bui & Perera, 2021; Gupta et al., 2022; Kongsberg, 2013) show that the main and auxiliary engines' performance parameters and indicators such as the ship's position, speed, and surface quality of the hull and propeller are measured instantly or periodically and processed using software for condition monitoring, maintenance planning, and data reporting. Figure 8.10 shows the structure of an energy efficiency monitoring system where the primary fuel consumers are main and auxiliary engines and boilers. The parametric data are measured by sensors, and the data is processed for obtaining key performance indicators such as EEOI and

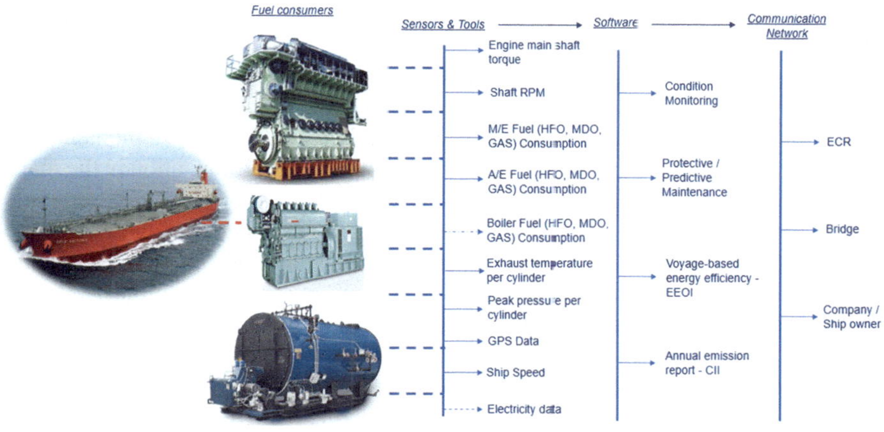

Fig. 8.13 An on-board energy efficiency monitoring system structure

CII; then, the data is reported to the engine control room, bridge, and company. The performance monitoring systems on-board are considered to be essential for continuous improvement and sustainable shipping; based on that, these systems are capable of decreasing human-based errors, facilitating decision-making process, and providing meaningful and verifiable data for taking preventive measures regarding energy efficiency and environmental impacts (Fig. 8.13).

8.7 Conclusion

The International Maritime Organization has set ambitious targets to reduce greenhouse gas emissions from international shipping. Beyond regulatory pressures, there is a growing recognition within the maritime industry that sustainability is not merely a compliance issue but a strategic imperative. Investors, consumers, companies, and stakeholders are increasingly pushing the industry to adopt a more sustainable and socially responsible approach. In this context, this study investigates the impact of regulations and factors influencing ship energy efficiency from a design perspective. Energy-efficient hull design parameters, propulsion system advancements, alternative fuels, and energy management strategies for sustainable shipping are presented. The discussion on the role of ship design in energy efficiency and environmental considerations is supported by numerical data from previous research and current applications. According to the findings, the following conclusions are drawn:

- There are various factors to design an energy-efficient ship. Rather than implementing the traditional design steps, the holistic design approach has been used recently. This methodology offers multidisciplinary and multi-objective design optimization, which is established on risk, safety, life cycle, zero emission, and efficiency.

- Hull form design and optimization is an initiative step directly affecting energy efficiency. Parametric modelling, automated optimization, and visualization of hull forms have been notably developed, recently. Therefore, the implementation of tools on these subjects facilitates the path towards decarbonization in the maritime industry.
- The transition towards fully electric or hybrid propulsion technologies on-board has accelerated. Fuel cells, wind-assisted propulsion, solar power, and highly efficient battery utilization are remarkable steps in decarbonization. However, the power-to-weight ratio is still challenging for renewable technologies, which are aimed to be used as the primary propulsion system on-board ships.
- Alternative fuels like LNG, LPG, methanol, ammonia, biofuel, and hydrogen are promising solutions and shortcuts for complying with the IMO regulations. However, based on their chemical and physical properties, ship design considerations vary. Gas fuels require more space and safety measures than conventional fuels. Not only the design aspect but also the fuel availability and life cycle assessments are crucial. Developments in green production lower the overall environmental impacts of marine fuels.
- Ships have complex machinery systems and operate at various conditions. Therefore, effective energy management on-board is substantial for low-carbon and sustainable shipping. Holistic energy management technologies should be adopted during the design phase of ships.

References

Akman, M. (2023). A techno-environmental and energy efficiency investigation of marine dual-fuel engines. *Marine Science and Technology Bulletin, 12*(2), 128–141. https://doi.org/10.33714/masteb.1247489

Akman, M., Turan, B. İ., Taşdemir, A., & Sogut, M. Z. (2023). A concept design study for the energy-efficiency of bulk carriers. In *12th international exergy, energy and environment symposium (IEEES-14), İstanbul.*

Barreiro, J., Zaragoza, S., & Diaz-Casas, V. (2022). Review of ship energy efficiency. *Ocean Engineering, 257*, 111594. https://doi.org/10.1016/j.oceaneng.2022.111594

Bass, C. (2021). *Harnessing the power of the Magnus effect*. Wärtsilä. Retrieved from https://www.wartsila.com/insights/article/how-rotor-sail-technology-can-boost-maritime-decarbonisation-efforts.

Bertram, V. (2000). *Practical ship hydrodynamics* (2nd ed.).

Bui, K. Q., & Perera, L. P. (2021). Advanced data analytics for ship performance monitoring under localized operational conditions. *Ocean Engineering, 235*, 109392. https://doi.org/10.1016/j.oceaneng.2021.109392

Corvus EES. (2020). *MF AMPERE—The world's first all-electric car ferry*. Retrieved from https://corvusenergy.com/projects/mf-ampere/.

DNV. (2023). *Ship Energy Efficiency Management Plan (SEEMP)*. Retrieved from https://www.dnv.com/maritime/hub/decarbonize-shipping/key-drivers/regulations/imo-regulations/seemp.html.

DNV-GL. (2020). *Ammonia as a marine fuel*. Retrieved from https://www.dnv.com/Publications/ammonia-as-a-marine-fuel-191385

FSYS. (2023). *CFD-driven hull form optimization of a container vessel*. Retrieved from https://www.caeses.com/industries/case-studies/container-vessel-hull-optimization/.

Gupta, P., Rasheed, A., & Steen, S. (2022). Ship performance monitoring using machine-learning. *Ocean Engineering, 254*, 111094. https://doi.org/10.1016/j.oceaneng.2022.111094

He, N. V., Mizutani, K., & Ikeda, Y. (2016). Reducing air resistance acting on a ship by using interaction effects between the hull and accommodation. In *Ocean engineering* (Vol. 111, pp. 414–423). Elsevier Ltd. https://doi.org/10.1016/j.oceaneng.2015.11.023

Herdzik, J. (2021). Decarbonization of marine fuels—The future of shipping. *Energies, 14*(14), 4311. https://doi.org/10.3390/en14144311

IMO. (2020). *Fourth IMO GHG study 2020 executive summary*. Retrieved from https://www.imo.org/en/OurWork/Environment/Pages/Fourth-IMO-Greenhouse-Gas-Study-2020.aspx.

IMO. (2023a). *2023 IMO strategy on reduction of ghg emissions from ships—MEPC.377(80)*. Retrieved from https://www.imo.org/en/MediaCentre/PressBriefings/pages/Revised-GHG-reduction-strategy-for-global-shipping-adopted-.aspx.

IMO. (2023b). *Improving the energy efficiency of ships*. Retrieved from https://www.imo.org/en/OurWork/Environment/Pages/Improving%20the%20energy%20efficiency%20of%20ships.aspx.

Inal, O. B., Charpentier, J.-F., & Deniz, C. (2022a). Hybrid power and propulsion systems for ships: Current status and future challenges. *Renewable and Sustainable Energy Reviews, 156*, 111965. https://doi.org/10.1016/j.rser.2021.111965

Inal, O. B., Dere, C., Zincir, B., & Deniz, C. (2022b). Hybrid propulsion and alternative fuels education in the course of decarbonised shipping. *Australian Journal of Maritime and Ocean Affairs, 14*(2), 97–113. https://doi.org/10.1080/18366503.2021.1940475

Kongsberg. (2013). *Ship performance system*.

Ling-Chin, J., & Roskilly, A. P. (2016). A comparative life cycle assessment of marine power systems. *Energy Conversion and Management, 127*, 477–493. https://doi.org/10.1016/j.enconman.2016.09.012

MAN. (2016a). *MAN alpha unique kappel propellers—Radical fuel savings*.

MAN. (2016b). *MAN B&W G-engines*. Retrieved from https://www.mandieselturbo.com/docs/default-source/shopwaredocuments/man-b-w-g-engines-green-ultra-long-stroke-engines15fbf9a55cda459c8d405548eea8e7e1.pdf?sfvrsn=3.

MAN. (2021). *Managing methane slip*. Retrieved from https://www.man-es.com/campaigns/download-Q1-2023/Download/managing-methane-slip/d34a34a1-cc03-4d99-a4e1-30385cf12518/Managing-Methan-Slip.

MAN. (2022). *The methanol fuelled MAN B&W LGIM engine*. Retrieved from https://www.man-es.com/docs/default-source/document-sync/the-methanol-fuelled-man-b-w-lgim-engine63e54a8470694265bdeb07314c4f0e11.pdf?sfvrsn=148c08f0_0.

Papanikolaou, A. (2010). Holistic ship design optimization. *Computer-Aided Design, 42*(11), 1028–1044. https://doi.org/10.1016/j.cad.2009.07.002

Taşdemir, A., & Akman, M. (2023). Drag reduction of bluff bodies by passive canals: A numerical and experimental study. In *4th international naval architecture and maritime symposium*.

Tasdemir, A., & Nohut, S. (2012). Fatigue analysis of ship structures with hinged deck design by finite element method. A case study: Fatigue analysis of the primary supporting members of 4900 PCTC. *Marine Structures, 25*(1), 1–12. https://doi.org/10.1016/j.marstruc.2011.09.001

Taşdemir, A., & Nohut, S. (2021). An overview of wire arc additive manufacturing (WAAM) in shipbuilding industry. *Ships and Offshore Structures, 16*(7), 797–814. https://doi.org/10.1080/17445302.2020.1786232

Taşdemir, A., Nohut, S., & Akman, M. (2023). *Gemi Türleri—Ticari Gemilerin Karakteristik Özellikleri* (1st ed.).

Turan, B. İ. (2023). Design framework for sail & rigging parameters of Bodrum Gulets: Comparison of schooner and ketch configurations. *Ships and Offshore Structures, 18*(11), 1582–1593. https://doi.org/10.1080/17445302.2022.2129918

Turan, B. I., & Akman, M. (2021). Modeling and comparison of Bodrum Gulets' Hull forms with round and transom sterns. *Journal of ETA Maritime Science, 9*(2), 120–129. https://doi.org/10.4274/jems.2021.09327

Wärtsilä. (2022). *Rotor Sail Technology*. Retrieved from https://www.wartsila.com/marine/products/propulsors-and-gears/energy-saving-technology/rotor-sail.

Wärtsilä. (2023). *Offshore references—The Viking Lady*. Retrieved from https://www.wartsila.com/marine/customer-segments/references/offshore/viking-lady.

Woodyard, D. (2004). *Pounder's marine diesel engines and gas turbines* (8th ed., pp. 48–63). https://www.sciencedirect.com/science/article/pii/B9780750658461500031.

Chapter 9
Post-COVID-19 Energy Transitions: Strategies for Global Recovery

Islam Md Rizwanul Fattah, Ibham Veza, Selçuk Sarıkoç, Shurooq T. Al-Humairi, Oyetola Ogunkunle, and Hafiz Muhammad Ali

Nomenclature

AQG	Air Quality Guidelines
CH_4	Methane
CO_2	Carbon dioxide
GAW	Global Atmosphere Watch

I. M. R. Fattah (✉)
Centre for Technology in Water and Wastewater (CTWW), School of Civil and Environmental Engineering, Faculty of Engineering and IT, University of Technology Sydney, Ultimo, NSW, Australia
e-mail: IslamMdRizwanul.Fattah@uts.edu.au

I. Veza
Department of Mechanical Engineering, Faculty of Engineering, Universitas Bung Karno, Jakarta Pusat, DKI Jakarta, Indonesia
e-mail: ibham@vezas.org

S. Sarıkoç
Department of Mechanical Engineering, Faculty of Engineering, Amasya University, Amasya, Türkiye

S. T. Al-Humairi
Chemical Engineering Department, University of Technology-Iraq, Baghdad, Iraq
e-mail: shurooq.t.ramadhan@uotechnology.edu.iq

O. Ogunkunle
Centre for Global Change, Department of Computer Science and Information Technology, School of Natural and Applied Sciences, Sol Plaatje University, Kimberley, South Africa
e-mail: oyetola.ogunkunle@spu.ac.za

H. M. Ali
Mechanical Engineering Department, King Fahd University of Petroleum and Minerals, Dhahran, Saudi Arabia
e-mail: hafiz.ali@kfupm.edu.sa

M. Z. Sogut, M. Koray (eds.), *Energy Rationality and Management for Decarbonization*, Green Energy and Technology,
https://doi.org/10.1007/978-3-031-68497-5_9

GHG	Greenhouse gas
HFC	Hydrofluorocarbon
IEA	International Energy Agency
IPCC	Intergovernmental Panel on Climate Change
IRENA	International Renewable Energy Agency
LLGHGs	Long-lived greenhouse gas emissions
N_2O	Nitrous oxide
ODS	Ozone-depleting substance
OECD	Organisation for Economic Co-operation and Development
PFC	Perfluorocarbon
RF	Radiative forcing
SF_6	Sulphur hexafluoride
STEPS	Stated Policies Scenario
STEO	Short-Term Energy Outlook
UNFCCC	United Nations Framework Convention on Climate Change
WMO	World Meteorological Organization

9.1 Introduction

From the pre-industrial era onwards, human actions, particularly the combustion of fossil fuels, have contributed to an increase in the levels of greenhouse gases (GHGs) with heat-trapping properties within the Earth's atmosphere, consequently leading to the phenomenon known as global warming. This phenomenon is commonly quantified by the escalation in the average temperature of the Earth's surface. The term 'climate change' is frequently employed as a synonymous expression, indicating a long-term alteration in the customary weather patterns that have traditionally characterised the local, regional, and global climates of our planet. These modifications have a wide variety of obvious repercussions that are equivalent to the phrase as a whole. Climate change encompasses both human- and nature-caused warming, as well as the effects it has on our planet. According to the United Nations Intergovernmental Panel on Climate Change (IPCC), rising greenhouse emissions, such as methane, nitrogen oxides, and carbon dioxide, have rapidly increased global warming. More than a hundred million lives are expected to be put in jeopardy if global temperatures rise by more than 2 °C (IPCC, 2000).

Population growth, which is expected to increase by 1.7 billion people to roughly 9.2 billion by 2040, helps support global energy output (BP, 2018). The production of energy, primarily through the combustion of fossil fuels, accounts for approximately 75% of the global emissions of greenhouse gases (GHGs) (C2ES, 2021). The total energy demand is composed of three primary elements, namely electricity, transportation, and heat supply, as indicated by Ram et al. (2022). Not only does the production of energy represent the single most important driver of climate change, but the combustion of fossil fuels and biomass also has a significant bearing on the health of the global population (Kalair et al., 2021; Wilkinson et al., 2007; Zamri et al., 2022). As a consequence of this, the global community must make the switch away from the use of fossil fuels and towards an energy mix that is predominated by low-carbon sources of energy such as nuclear power and renewable technologies (Mofijur et al., 2021; Mohamed et al., 2022).

Energy demand plays an important role in economic growth. As the global economy expands, the power sector absorbs nearly 70% of the increase in primary energy (BP, 2018). As a direct consequence, the global surface temperature is also rising. The worldwide average surface temperature has risen by more than 0.9 °C (1.6 °F) since 1906 (National Geographic, 2018). This change had a more subtle effect on the polar areas. The rate of temperature increase has nearly doubled in the past 50 years (NASA, 2010). The consequences of increasing temperatures will not happen in the distant future—the repercussions of global warming are seen right now. Rising temperatures are leading to the melting of glaciers and sea ice, inducing alterations in precipitation trends, and prompting the migration of various species. A large portion of the melting ice leads to sea level rise. Global sea levels are increasing at a pace of 3.2 mm (0.13 in) each year, and this rate has accelerated in recent years. Energy security, pollution, sustainability, and the effects of climate change are the most pressing issues confronting the global economy today.

Global economies have been impacted by the COVID-19 pandemic, which was sparked by widespread public health responses aimed at containing the virus. About 100 countries, states, and provinces—primarily in Europe, Asia, and North America—implemented complete lockdown measures to combat the pandemic from February to mid-May 2020, with another 100 implementing partial lockdowns. Lockdowns lasted 4–10 weeks on average, and they were gradually lifted beginning in the second half of April. The spread of new infections worldwide slowed and plateaued in April and May as a result of these containment efforts. Prices at the beginning of the crisis mirrored the plummeting demand for fuel in several important markets. By the end of March 2020, the price of gas had plummeted to a level not seen in 20 years, while the price of oil was also influenced by supply shocks (Statista, 2023). As the world's economies have begun to expand again, the prices of energy commodities have increased.

Different papers have examined the effect of COVID-19 on various factors, including environmental and energy issues and the global economy, to date. For instance, Forster et al. (2020) evaluated the worldwide emission reductions for ten species from February to June 2020 during the early phase of the pandemic. They concluded that the immediate impact of the response to the pandemic on the climate of the whole planet would be insignificant. Shakil et al. (2020) reviewed the studies on environmental causes and consequences of COVID-19 published between November 2019 and May 2020. Madurai Elavarasan and Pugazhendhi (2020) focussed on potential technological strategies to control the COVID-19 pandemic and emphasised deploying tech-based solutions along with management and governance to deal with it.

In another study, Khan et al. (2021) reviewed the positive impacts of COVID-19 on environmental quality throughout the world. Kumar et al. (2022) reviewed the impact of COVID-19 on GHG emissions and provided an analysis of several factors on CO_2 emission. Thus, a comprehensive review of the short-term and long-term effects of the COVID-19 global energy scenario in the next few years, along with information on GHG emissions, as well as an appraisal of the renewable energy scenario, is still lacking. The article conducts a thorough examination of the effects of using fossil fuels on greenhouse gas emissions in various global scenarios, both before, during, and after a pandemic. It also provides insight into the short- and long-term impacts that COVID-19 will have on energy and renewable energy. The article concludes with suggestions for reducing the effects of COVID-19 on energy and the environment and highlights key findings about the potential for future energy. We anticipate that this current and relevant study, which includes much critical information, will help dictate future energy policies.

9.2 Pre- and Post-pandemic Energy Scenario

9.2.1 *Fossil Fuel Scenario*

The years before the COVID-19 pandemic saw a dominant fossil fuel landscape across the globe, fuelling economies and industries despite growing concerns about climate change. Demand for fossil fuels was steadily increasing, driven by economic growth and population expansion, particularly in developing countries. This growth was projected to continue for several years, with the International Energy Agency (IEA) predicting a 30% increase in global energy demand by 2040. However, pre-existing trends towards cleaner energy sources suggested that a shift was underway, even if it was happening at a slow pace. The COVID-19 pandemic would ultimately lead to further changes and disruptions in the energy landscape, accelerating some of these pre-existing trends and creating new challenges and opportunities. According to the World Energy Outlook 2020, the global economy was projected to recover to pre-COVID-19 levels by the second quarter of 2021. However, in the long run, it was expected to be approximately 7% smaller than previously forecasted in the World Energy Outlook 2019, assuming the implementation of the Stated Policies Scenario (STEPS) (IEA, 2020b). STEPS takes into account the effects of current policy frameworks as well as today's policy intentions. According to the report, total energy demand is expected to recover to pre-crisis levels by early 2023, though trends and timing will vary by country. The pandemic's negative effects on energy demand are expected to last the longest in a number of low-income countries, where governments are less able to cushion the blow. After the crisis, Western countries' energy demand has risen slightly but not to pre-pandemic levels. The Asia Pacific region, particularly China and India, will account for the majority of the increase in energy demand. According to IEA, after dropping by approximately 1% in 2020 due to the effects of the COVID-19 pandemic, worldwide energy demand is expected to rise by close to 5% in 2021 and 4% in 2022, owing to the global economic recovery (IEA, 2021a). McKinsey & Company's Global Energy Perspective 2021 (McKinsey & Company, 2021) predicted that the demand for oil will reach its peak in the late 2020s and demand for gas will peak in the 2030s, while demand for coal will continue to fall. The reference scenario is characterised by a view that takes a prospective 'continuation of present patterns' stance. However, reflecting on the year 2021, the global economic recovery from the ramifications of the COVID-19 pandemic witnessed a resurgence in energy demand worldwide (McKinsey & Company, 2022). This, in combination with limitations in the supply side of the energy market, resulted in significant price hikes.

Figure 9.1 shows that in the forecast leading up to the year 2035, it is projected that, under all conceivable circumstances, gas demand will experience an augmentation of an additional 10–20% compared to present demand levels. However, after the year 2035, the demand for gas will likely be subject to greater fluctuations, specifically influenced by its dynamic relationship with hydrogen. The culmination of oil demand is anticipated to materialise between the years 2024 and 2027, principally propelled by the proliferation of electric vehicles (EVs)—a trend that has already commenced. Demand for coal reached its zenith in 2013 and, notwithstanding a temporary resurgence in 2021, is projected to maintain a continuous decline.

A report by IEA Oil 2021—Analysis and Forecast to 2026—contains Table 9.1, which presents the worldwide oil demand for 2019–2026 (mb/day) (IEA, 2021e). The global oil demand experienced a remarkable decline of 8.7 million barrels per day (mb/day) in 2020, driven by the widespread pandemic and severe economic downturn. However, a substantial acceleration in growth is anticipated for 2021, with a projected increase of 13.1 mb/day until 2026. Nonetheless, due to the enduring impacts of the pandemic and the growing imperative to address the transition

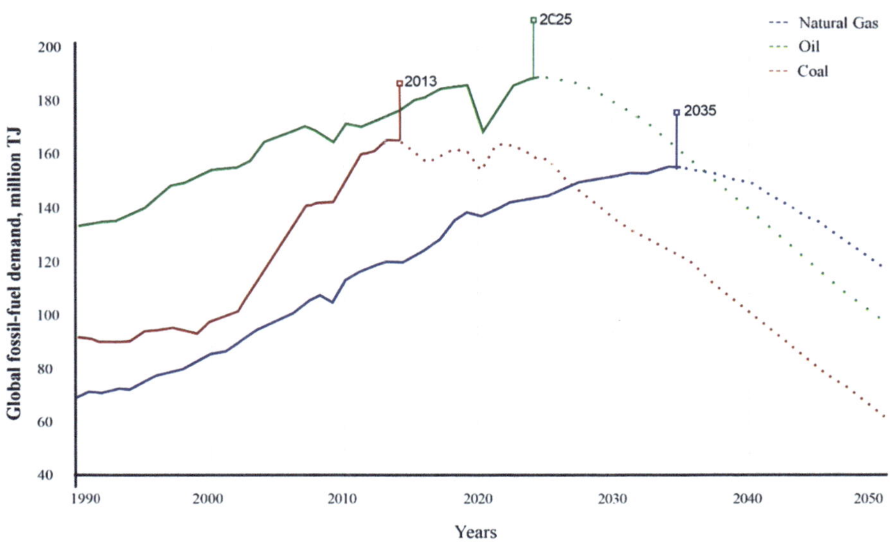

Fig. 9.1 Global fossil fuel demand per fossil fuel projected until 2050, million TJ. (Adopted from McKinsey and Company (2022))

Table 9.1 Worldwide oil demand for 2019–2026 (million barrels/day) (IEA, 2021e)

	2019	2020	2021	2022	2023	2024	2025	2026	2019–2026 growth
North America	25.3	22.2	23.8	24.5	24.7	24.7	24.6	24.6	−0.7
Central and South America	6.6	5.9	6.3	6.6	6.7	6.7	6.8	6.9	0.3
Europe	15.7	13.8	14.6	14.8	15	15	14.9	14.9	−0.8
Africa	4.2	3.8	4	4.2	4.4	4.5	4.7	4.8	0.6
Middle East	8.3	7.6	7.9	8.2	8.4	8.5	8.7	8.9	0.6
Eurasia	4.4	4.2	4.3	4.4	4.5	4.6	4.6	4.7	0.3
Asia Pacific	35.2	33.4	35.6	36.9	37.7	38.2	38.9	39.3	4.1
World	99.7	91	96.5	96.5	101.2	102.3	103.2	104.1	4.4

towards low-carbon energy systems, it is highly unlikely that oil demand will ever fully revert to its pre-pandemic trajectory. By 2026, world oil demand is projected to exceed the 2019 levels by a mere 4.4 mb/day, with the entirety of the increase expected to stem from non-Organisation for Economic Co-operation and Development (OECD) nations. Nevertheless, a significant recovery in oil demand within the OECD is anticipated in the near future. The implementation of prolonged and repeated lockdowns, coupled with stringent social distancing measures such as widespread remote work arrangements and significant restrictions on international travel, resulted in a notable decline in OECD demand by approximately 5.6 mb/day in 2020. However, as vaccination efforts progress and restrictions are gradually eased, the demand is expected to rebound.

The oil market has undergone a stark transformation since 2020. COVID-19 lockdowns triggered a significant oil surplus, plunging prices to an average of USD 44 per barrel. However, this has given way to a new reality marked by tight global supply struggling to meet resurgent demand, leading to capacity constraints among producers. The recent Russian invasion of Ukraine further exacerbated market tightness, driving oil prices to an average of USD 105 per barrel in 2022. Global oil utilisation exhibits contrasting trends: sectors like aviation are still recovering from the pandemic's impact, while others like the chemical industry remain resilient. These high prices and their economic ramifications generate additional near-term uncertainties.

Unlike other fossil fuels, natural gas demand displayed remarkable resilience during the pandemic's initial year, registering a 5% increase in 2021, doubling the average growth rate of the previous decade. However, a confluence of factors—limited new projects, weather-related demand fluctuations, LNG outages, and reduced Russian exports—strained the global gas supply balance starting mid-2021, pushing prices upwards. The Ukraine conflict further disrupted the already fragile gas market (Kammer et al., 2022). Russia's strategic withholding of gas hampered the EU efforts to replenish storage, raising the risk of winter supply shortages and causing significant market volatility and exceptionally high prices. This gas supply strain also triggered energy shortages in gas-dependent developing regions like Pakistan and Bangladesh (Varadhan et al., 2023). Notably, major gas growth markets like India and China significantly reduced their LNG imports in 2022.

Contrary to predictions of a decline, global coal demand experienced a robust rebound in 2021, approaching its historical peak. Advanced economies, previously reducing coal use, witnessed nearly a 10% increase. Emerging markets and developing economies, now accounting for over 80% of global coal demand, saw a 5% surge. However, coal production struggled to match this significant demand increase, and limited fuel-switching options further intensified pressures. Consequently, global coal prices reached unprecedented highs in the first half of 2022.

9.2.2 Renewable Energy Scenario

Renewable energy is the fastest-growing energy source, accounting for 40% of all primary energy growth. The share of renewables in global energy production increased to 29% in 2020, up from 27% in 2019. The movement restrictions, lockdowns, and economic downturn caused by COVID-19 significantly decreased global power consumption. For example, electricity consumption in the United States was down by 5% compared to April 2019, Germany was down by 12%, Spain was down by 18%, and India was down by 23% (IEA, 2020c). Despite this, renewable energy usage rose by 3% in 2020, while demand for all other fuels fell. The main driver was a nearly 7% increase in electricity generation from renewable sources (IEA, 2021b). This was primarily due to long-term contracts, priority grid access, and ongoing plant installation that supported the development of renewables, despite reduced power demand, supply chain difficulties, and building delays in many regions of the world. Bioenergy usage in the industry increased by 3%, although this was mainly offset by a drop in biofuels, as decreased oil demand reduced the use of blended biofuels. Since the beginning of the pandemic, crude oil prices have fallen, making biofuels less competitive with traditional transportation fuels. The average crude oil price for 2020 is expected to be approximately USD 40 per barrel, down from USD 64 per barrel in 2019 (IEA, 2020c). Bioenergy represents a type of sustainable energy harnessed from organic matter derived from recently living organisms, referred to as biomass. This renewable energy form possesses the capacity to generate various forms of energy including transportation fuels, heat, electricity, and other valuable commodities. Biomass serves as a renewable resource that can be converted into biofuels for transportation, including ethanol, biodiesel, and renewable hydrocarbon fuels. These biofuels can be used in aeroplanes and most vehicles, reducing the carbon intensity of transportation. Biopower technologies utilise biomass to generate heat and electricity through burning, bacterial decay, or conversion into fuels, providing a flexible and reliable energy source while lowering carbon emissions. Additionally, biomass can be used to manufacture various everyday products like plastics and chemicals, offering a renewable alternative to petroleum and natural gas. Integrated biorefineries can produce both bioproducts and biofuels, improving resource efficiency and economic viability.

Renewable energy sources utilise natural resources to generate advantageous forms of energy, primarily in the form of electricity (Hasan et al., 2021; Kalogirou, 2004). Alternative energy sources have the capacity to contribute to energy generation with minimal or no emissions of air pollutants and greenhouse gases. These non-conventional energy technologies make use of solar energy in both direct and indirect forms, as well as the characteristics of the Earth such as wind, biomass, and the downward flow of water; gravitational forces like waves and currents; and the Earth's core temperature for geothermal energy. These energy sources offer sustainable options for energy that are based on availability and local resources (Demirbas, 2005; Herzog et al., 2001). In recent years, there has been a significant global increase in the adoption of renewable energy sources, primarily driven by factors such as escalating energy demand, increasing oil prices, and environmental concerns. Renewable energy production expansion is propelled by key factors such as environmental protection and economic development (Çelik et al., 2021). Conversely, the expansion of renewable energy production is also driven by various factors, including the level of renewable energy integration into the utility grid, the cost of renewable energy technologies, feed-in tariffs, customer demand, and other related aspects. Moreover, internal drivers such as the effects of the COVID-19 crisis on investment projects, talent and technology availability, renewable energy infrastructure, and government policies play crucial roles in shaping the growth of renewable energy production.

However, the growth of the renewable energy industry may be hindered as a result of the impact of the COVID-19 pandemic on each of these driving factors (Çelik et al., 2021). Figure 9.2 illustrates various significant barriers including social and economic challenges, technological limitations, issues related to stability and dispatchability of renewable energy, geographic restrictions, and capacity factors, which impede the expansion of renewable energy production.

The substantial growth in renewable energy is primarily attributed to several factors including low variable costs, favourable regulations, long-term contracts, and the continuous expansion of wind and solar photovoltaic (PV) technologies. These two sources, wind and solar PV, are expected to account for approximately two-thirds of the overall increase in renewable energy (IEA, 2021b). According to projections, renewable energy sources are expected to contribute to approximately 33% of the total electricity generation by 2025, surpassing the capacity of coal-fired power plants. Additionally, it is estimated that until 2025, renewables will account for approximately 95% of the global increase in energy capacity (IEA, 2020c). Hydropower will remain a crucial source of renewable energy, but it will no longer make up more than half of all renewable electricity generation starting in 2024. During the predicted period, the amount of electricity generated by wind and solar PV combined will increase significantly to around 4000 TWh.

The International Energy Agency (IEA) states that global investment in clean energy must increase threefold by 2030, reaching nearly $4 trillion, in order to achieve the goal of net-zero emissions by 2050 (IEA, 2021d). The energy mix by

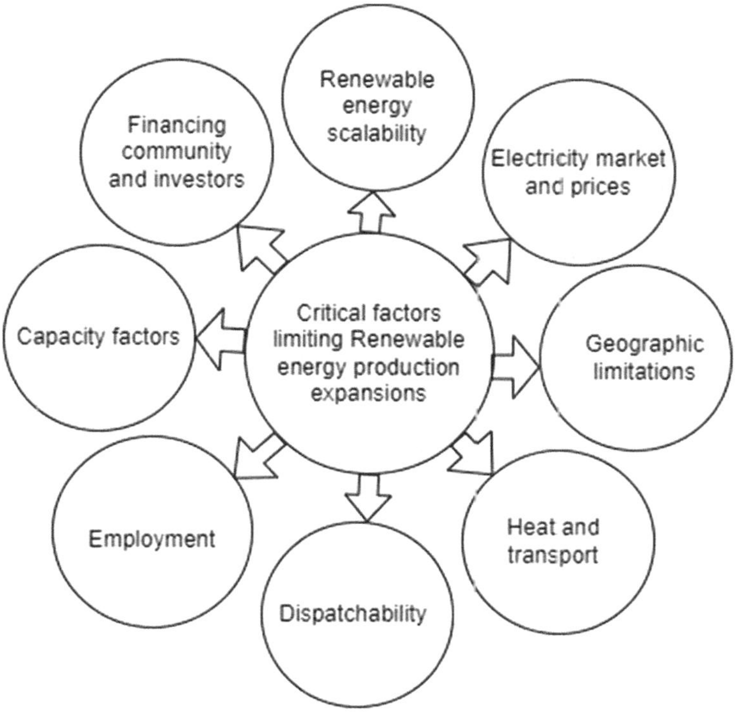

Fig. 9.2 Limiting forces for renewable energy production expansion during the pandemic

2040 will be the most diversified the world has ever seen (BP, 2018). In the years to come, there will be a dramatic change in the energy mix towards electricity, synfuels, and hydrogen, which will account for 32% of the global energy mix by the year 2035 and 50% by the year 2050 (McKinsey & Company, 2022).

9.3 Perspective and Future Energy Trends

The United Nations estimates that the planet would warm by 2.5 °C by the turn of the century if these policies are carried out in their current form. Although this is an improvement over the path that the world was on before the Paris Agreement was reached, which was a rise of 4 °C, it is still very dangerous. This shows that much more work is needed to reduce emissions even further, and more ambitious programmes and legislative measures must be developed and executed. The Paris Agreement was officially adopted on December 12, 2015, at the 21st Conference of the Parties (COP21) to the United Nations Framework Convention on Climate Change (UNFCCC) held in Paris, France. However, it is important to note that the agreement itself had been under development for several years prior to its official

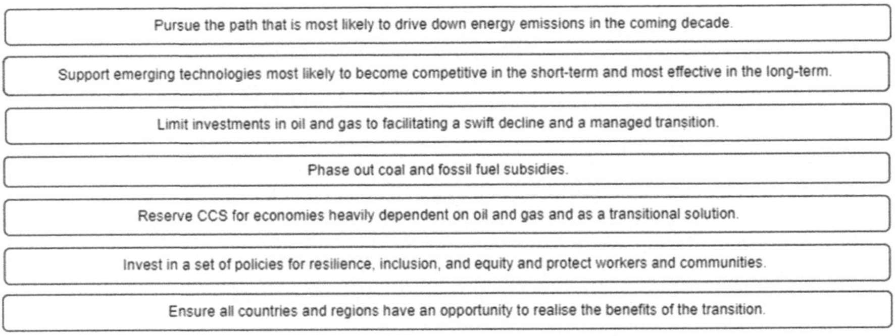

Fig. 9.3 A framework for energy transition route compatible with the climate goal of 1.5 °C

adoption. The COP28, which took place from November 30 to December 12, 2023, in Dubai, United Arab Emirates. COP28 concluded with several important outcomes, including:

- A renewed commitment to the Paris Agreement goal of limiting global warming to well below 2 °C and ideally to 1.5 °C
- A new work programme (https://enb.iisd.org/united-arab-emirates-climate-changeconference-cop28-summary) to develop a global roadmap for net-zero emissions by 2050
- Increased pledges of financial support for developing countries
- The establishment of an inclusive framework for the Loss and Damage Fund aimed at reducing vulnerability and enhancing climate adaptation

According to the International Renewable Energy Agency (IRENA), the window of opportunity to accomplish the IPCC's 2030 emission milestone is narrow, and the decisions that are taken in the following years will decide whether or not a future with a temperature rise of 1.5 °C stays within reach (IRENA, 2021). They proposed some guidelines to keep the rise in world average temperature to 1.5 °C by the end of this century in comparison to levels that existed before industrialisation. It gives top priority to technological solutions that are readily accessible and can be ramped up at the appropriate speed to meet the 1.5 °C objectives. The prerequisites are shown in Fig. 9.3.

In their endeavour to transform energy systems, policymakers, corporate leaders, investors, and consumers are displaying increasing ambition. However, if present patterns persist, global greenhouse gas (GHG) emissions are projected to decrease by approximately 25% by 2050, leading to a trajectory of 3.5 °C warming, significantly surpassing the 1.5 °C targets, despite the temporary 7% reduction in global CO_2 emissions caused by the COVID-19 pandemic (McKinsey & Company, 2021). To achieve the 2050 goal, CO_2 emissions must decline at a rate similar to that observed in 2020 over the next three decades. Moreover, although there will be a peak in oil consumption in the late 2020s and a peak in gas demand in the mid-2030s, fossil fuels will continue to hold relevance. While it is improbable for oil and gas consumption to return to pre-COVID-19 levels in the near future, they are not expected to follow their pre-pandemic growth trajectory. Despite long-term declines in demand, all three fossil fuels will retain a substantial role in the global energy system. If no further decarbonisation measures are implemented, fossil fuels are projected to contribute to over half of the world's energy consumption by 2050.

9.4 Recovery Strategies for Energy Landscape Post COVID-19

The COVID-19 pandemic significantly impacted the global energy landscape, disrupting supply chains, altering consumption patterns, and accelerating pre-existing trends. As we move towards recovery, several key strategies can guide us towards a more sustainable and resilient energy future:

9.4.1 Short-Term Recovery

In response to the damages caused by COVID-19, the governments of a number of different nations are formulating strategies for economic stimulation. They are simultaneously faced with three challenges: (1) improving public health, (2) regaining economic ground, and (3) adapting to the effects of climate change (Shang et al., 2021). The highest focus would be placed on investments in the public health system that are both long-term and sustainable. Nevertheless, a compromise between reviving the economy and addressing climate change would be something to think about. A revival of the economy based on the use of fossil fuels might bring some temporary respite; nevertheless, efforts to combat climate change would be hampered. As a result of these conditions, the significance of renewable energy technologies ought to be understood in terms of energy security and climate change rather than the economic gains that will be reaped in the short term. The use of renewable energy technologies should be an important focus of ongoing decarbonisation efforts. During the lockdown, when most individuals worked from home and received their education via an online platform, there was a significant increase in the quantity of power that was used (Li et al., 2020). A growth in the need for energy is something that may be anticipated for the future, which emphasises the importance of a production system that is environmentally friendly (Omer, 2008).

To meet the expected increase in demand for power in the foreseeable future, renewable energy technologies may be successfully combined with digital technology (Gielen et al., 2019). Investing in three new developing clean technologies—hydrogen, batteries, and carbon capture—should be part of the strategy for economic recovery (IEA, 2020a; Siddique et al., 2021). Countries all over the world are currently in the process of transitioning to a new energy system, one that looks to favour the use of electricity as the most desired form of ultimate energy carrier. This is a positive development from the point of view of energy efficiency as well as renewable sources of energy. When it is generated from renewable resources, electricity transforms into a zero-emissions kind of energy while also maintaining its reputation as an effective energy transporter (Kabeyi & Olanrewaju, 2022). Around one-fifth of the world's total final energy consumption is accounted for by electricity; however, this percentage is significantly greater in nations with high incomes and is quickly increasing in those with lower incomes (IEA, 2021c). In addition, preparations for future renewable energy technologies should be aligned with wider aims for sustainable energy, which can help encourage employment and economic recovery. The use of renewable energy sources like solar and wind power can provide electrical networks more flexibility and make them more environmentally friendly. Both bioenergy with carbon capture and storage (BECCS) (Bui et al., 2018) and direct carbon capture from the atmosphere (Kumar et al., 2015; Sanz-Pérez et al., 2016) have the potential to play an important part in the transition from conventional energy sources to renewable energy technologies. It is important to encourage better land management techniques as well as the use of automated agriculture since they will help sequester carbon in the soil rather than releasing it into the atmosphere as is the case with traditional agricultural methods.

Recycling, as a method of resource reclamation, possesses significant under-exploited potential, necessitating further exploration and support (Ferronato & Torretta, 2019). Presently, recycling contributes to only approximately 31% of waste management (Zanoletti et al., 2021). Alongside landfill mining, there is a requirement for heightened research and promotion of recycling opportunities for electronic devices prior to their disposal into the environment (SDG 12). Disused electronics and discarded batteries harbour an intriguing volume of precious and rare metals. The reclamation of these valuable resources from such waste could potentially achieve economic viability, establishing it as a pivotal mechanism for global energy recovery transformation. Given the current circumstances, there is an immediate necessity to formulate a definitive policy blueprint for the foreseeable future to assuage investor apprehension. In order to achieve these pressing goals, the proposed strategies should incorporate aggressive and economically efficient energy conservation measures, which will form the foundational response to the COVID-19 crisis.

9.4.2 Long-Term Recovery

It is important to quickly shift to clean energy sources for long-term energy stability, affordability, and national resilience. This can be achieved by increasing the use of renewable energy, making energy usage more efficient, using electric vehicles and heat pumps, utilising clean hydrogen and bioenergy with carbon capture and storage (CCS), and implementing CCS at the final stages of energy production. This will reduce the dependence on foreign energy and lessen the impact of fluctuating fossil fuel prices. However, moving away from coal can be challenging for nations that heavily rely on it, particularly in terms of ensuring a fair and equitable transition for affected workers and communities.

- The imperative of transitioning from coal-based power sources to more ecologically friendly alternatives, predominantly renewable energies, is paramount in mitigating emissions and confronting climate change. Nonetheless, this shift poses significant challenges for nations substantially dependent on coal, particularly when taking into account the necessity of equitable and just transitions for the employees and communities adversely affected. Therefore, a concerted effort from governments is indispensable to gradually phase out coal. Moreover, it is crucial to scrutinise the use of coal in certain industries, such as the iron and steel sector, cement production, and other analogous sectors, due to their sizable contribution to overall coal consumption.
- The upcoming years will be of critical importance for technological innovations, industry initiatives, and global cooperation in these areas. To facilitate this energy transition, it is essential to rectify market distortions and encourage energy-transition solutions through incentivisation. This involves the abolition of fossil fuel subsidies and the incorporation of the comprehensive costs associated with fossil fuel utilisation into their pricing structures, which extend beyond the financial sphere to include environmental, health, and societal repercussions. The adoption and refinement of fiscal measures like carbon pricing is recommended to foster transition-oriented solutions. Nevertheless, attention must be given to the social and equity considerations of these measures. The potential implications on disadvantaged populations should be carefully considered to preclude the amplification of energy poverty or the onset of other adverse consequences.
- To reach long-term climate neutrality, large annual reductions in emissions will be required in the future. The European Climate Law has set a goal of at least a 55% decrease in net greenhouse gas emissions by 2030, compared to the 1990 levels (Environment Agency (EEA), 2022). Currently, the European Parliament and member states are discussing the Fit for 55 packages, which include the proposed RePower EU plan for 2022. In order to achieve the 55% net GHG emission goal for 2030, emissions must decrease by an average of 134 million metric tons of CO_2 per year from the 2021 levels, which is more than twice the average annual reduction achieved between 1990 and 2020 (Environment Agency (EEA), 2022). Every sector must make significant efforts to reduce greenhouse gas emissions, and carbon sequestration through land use, land use change, and forestry must be increased. Additionally, energy consumption must decrease, and the proportion of renewable energy in consumption must increase by 2.5% per year in order to meet the new 45% target set in RePower EU.
- Current trends and projections from the International Energy Agency forecast that green hydrogen, presently considered a specialised market, will transform into a predominant force in the energy sector by 2030 (International Energy Agency, 2022). To actualise this transition, it necessitates substantial amplification in policy attention as well as robust, tangible actions aimed at cultivating the

global market while simultaneously diminishing costs. This effort may involve the formulation of standards and origin guarantees as well as the establishment of supportive schemes to bridge the cost disparity inherent in the implementation of green hydrogen solutions. This approach would further propagate green hydrogen use, solidifying its role in the mainstream energy market (Camacho et al., 2022).

- To meet energy needs, including the supply of raw materials, it is estimated that bioenergy's contribution will need to triple by 2030. Additionally, cleaner options should replace traditional forms of biomass, such as firewood. While there is potential for growth in the biomass industry, it is important to proceed with caution to ensure sustainable practices and avoid negative consequences. Policies supporting the use of bioenergy should also be accompanied by rigorous, evidence-based sustainability measures and regulations. To reach the goal of net-zero energy emissions by 2050, it is crucial to significantly raise the proportion of renewable energy in the overall energy supply. Specifically, this would mean raising the renewable share from the 18% recorded in 2019 to well over 30% of the total final energy consumption by 2030 (Blyth, 2022). In order to accomplish this objective, government support in all areas would need to be increased, and efficient instruments would have to be put into place to further mobilise private capital. This needs to be done in less developed countries, landlocked developing countries, and small island developing nations.
- By 2030, it is recommended that most cars sold be electric vehicles (EVs) as the market share of EVs in global car sales was only 8.3% in 2021 (Irle, 2022). While electromobility plays a crucial role in the transition to clean energy, the widespread adoption of electric vehicles (EVs) also depends on the availability of adequate charging infrastructure, financial and tax incentives, and policies such as bans or restrictions on combustion engine vehicles. To further decrease the need for personal transportation, there should be a focus on developing and promoting alternative modes of transportation, such as public transit and cycling.
- In order to enhance energy efficiency, it is crucial for new construction projects to prioritise this aspect while also significantly accelerating the rate of renovations. This can be achieved through the adoption of alterations to building codes, the establishment of energy performance standards for appliances, and the integration of renewable energy solutions like solar water heaters, heat pumps, and geothermal systems for heating and cooling purposes. While long-term endeavours to decrease carbon emissions in heating and cooling are necessary, it is essential to promptly implement these measures without delay.

9.5 Conclusion

Following a decade characterised by rapid technical and regulatory advancements within the energy sector, the year 2020 marked the commencement of a significant metamorphosis in the energy scenario. Direct repercussions of the COVID-19 pandemic have led to governments' policies playing an increasingly pivotal role in the energy transition. Given the substantial cost associated with numerous stimulus packages, their influence on the transformation of the country's energy infrastructure over the ensuing decades is of paramount importance. In the long term, profound changes within the energy framework are anticipated to persist, bringing about swift alterations to the energy landscape in future decades. Decisions made by several countries to reach net-zero emissions are projected to strongly propel the utilisation of renewable energy across all sectors, potentially exerting a major impact on international markets. While the precise outcomes of net-zero strategies remain early for a comprehensive evaluation, considerable uncertainty exists about the configuration of the global energy system within a net-zero society. The demand for an unequivocal long-term policy plan is pressing, to assuage the concerns of investment stakeholders. The strategies responding to COVID-19 should incorporate ambitious yet economically viable energy efficiency objectives as foundational components, facilitating the realisation of these urgent objectives. While the future is uncertain, we are cautiously optimistic about the potential for a positive transformation in the global energy landscape. By harnessing the momentum, embracing innovation, and prioritising both environmental and social sustainability, we can build a cleaner, more resilient, and equitable energy future for all.

References

Blyth, L. (2022). *COVID-19 slows progress toward universal energy access*. PRESS RELEASE NO: 2022/067/EEX. Retrieved 17 January, 2023, from https://www.worldbank.org/en/news/press-release/2022/06/01/report-covid-19-slows-progress-towards-universal-energy-access.

BP. (2018). *BP Energy Outlook 2018 edition*. Retrieved from https://www.bp.com/content/dam/bp/en/corporate/pdf/energy-economics/energy-outlook/bp-energy-outlook-2018.pdf.

Bui, M., Fajardy, M., & Mac Dowell, N. (2018). Bio-energy with carbon capture and storage (BECCS): Opportunities for performance improvement. *Fuel, 213*, 164–175. https://doi.org/10.1016/j.fuel.2017.10.100

C2ES. (2021). *Global emissions*. Retrieved from https://www.c2es.org/content/international-emissions/.

Camacho, M. d. l. N., Jurburg, D., & Tanco, M. (2022). Hydrogen fuel cell heavy-duty trucks: Review of main research topics. *International Journal of Hydrogen Energy, 47*(68), 29505–29525. https://doi.org/10.1016/j.ijhydene.2022.06.271

Çelik, D., Meral, M. E., & Waseem, M. (2021, 25–27 March 2021). Restrictions and driving forces for renewable energy production development and electrical energy demand in general and during COVID-19. In *2021 12th international symposium on advanced topics in electrical engineering (ATEE)*.

Demirbas, A. (2005). Potential applications of renewable energy sources, biomass combustion problems in boiler power systems and combustion related environmental issues. *Progress in Energy and Combustion Science, 31*(2), 171–192.

Environment Agency (EEA). (2022). *Trends and projections: Limited rebound in EU emissions amid post-pandemic recovery and energy crisis*. Retrieved 17 January, 2023, from https://www.eea.europa.eu/highlights/trends-and-projections-limited-rebound.

Ferronato, N., & Torretta, V. (2019). Waste mismanagement in developing countries: A review of global issues. *International Journal of Environmental Research and Public Health, 16*(6). https://doi.org/10.3390/ijerph16061060

Forster, P. M., Forster, H. I., Evans, M. J., Gidden, M. J., Jones, C. D., Keller, C. A., Lamboll, R. D., Quéré, C. L., Rogelj, J., Rosen, D., Schleussner, C.-F., Richardson, T. B., Smith, C. J., & Turnock, S. T. (2020). Current and future global climate impacts resulting from COVID-19. *Nature Climate Change, 10*(10), 913–919. https://doi.org/10.1038/s41558-020-0883-0

Gielen, D., Boshell, F., Saygin, D., Bazilian, M. D., Wagner, N., & Gorini, R. (2019). The role of renewable energy in the global energy transformation. *Energy Strategy Reviews, 24*, 38–50. https://doi.org/10.1016/j.esr.2019.01.006

Hasan, M. H., Mahlia, T. M., Mofijur, M., Rizwanul Fattah, I. M., Handayani, F., Ong, H. C., & Silitonga, A. S. (2021). A comprehensive review on the recent development of ammonia as a renewable energy carrier. *Energies, 14*(13). https://doi.org/10.3390/en14133732

Herzog, A. V., Lipman, T. E., & Kammen, D. M. (2001). *Renewable energy sources*. Encyclopedia of Life Support Systems (EOLSS) Forerunner Volume "Perspectives and Overview of Life Support Systems and Sustainable Development".

IEA. (2020a). *The coronavirus crisis reminds us that electricity is more indispensable than ever*. Retrieved 5 August, 2022, from https://www.iea.org/commentaries/the-coronavirus-crisis-reminds-us-that-electricity-is-more-indispensable-than-ever.

IEA. (2020b). *Key fuel trends in the stated policies scenario, 2019–2030*. Retrieved form: https://www.iea.org/data-and-statistics/charts/key-fuel-trends-in-the-stated-policies-scenario-2019-2030.

IEA. (2020c). *Renewables 2020. Analysis and forecast to 2025*. Retrieved from https://www.iea.org/reports/renewables-2020.

IEA. (2021a, July). *Electricity market report*. Retrieved from https://www.iea.org/reports/electricity-market-report-july-2021.

IEA. (2021b). *Global energy review 2021*. Retrieved from https://iea.blob.core.windows.net/assets/d0031107-401d-4a2f-a48b-9eed19457335/GlobalEnergyReview2021.pdf.

IEA. (2021c). *Key world energy statistics 2021*. Retrieved 5 August, 2022, from https://www.iea.org/reports/key-world-energy-statistics-2021.

IEA. (2021d). *Net zero by 2050*. Retrieved 5 August, 2022, from https://www.iea.org/reports/net-zero-by-2050.

IEA. (2021e). *Oil 2021*. Retrieved from https://www.iea.org/reports/oil-2021.

International Energy Agency. (2022). *Global hydrogen review 2022*. Retrieved 17 January, 2023, from https://www.iea.org/reports/global-hydrogen-review-2022.

IPCC. (2000). *The IPCC Special Report on Emissions Scenarios*. Retrieved from https://www.ipcc.ch/site/assets/uploads/2018/03/sres-en.pdf.

IRENA. (2021). *World energy transitions outlook: 1.5°C pathway*. Retrieved 5 August, 2022, from https://www.irena.org/publications/2021/Jun/World-Energy-Transitions-Outlook.

Irle, R. (2022). *Global EV Sales for 2022 H1*. Retrieved 17 January, 2023, from https://www.ev-volumes.com/country/total-world-plug-in-vehicle-volumes/.

Kabeyi, M. J. B., & Olanrewaju, O. A. (2022). Sustainable energy transition for renewable and low carbon grid electricity generation and supply [review]. *Frontiers in Energy Research, 9*. https://doi.org/10.3389/fenrg.2021.743114

Kalair, A., Abas, N., Saleem, M. S., Kalair, A. R., & Khan, N. (2021). Role of energy storage systems in energy transition from fossil fuels to renewables. *Energy Storage, 3*(1), e135. https://doi.org/10.1002/est2.135

Kalogirou, S. A. (2004). Solar thermal collectors and applications. *Progress in Energy and Combustion Science, 30*(3), 231–295.

Kammer, A., Azour, J., Selassie, A. A., Goldfajn, I., & Rhee, C. Y. (2022). *How war in Ukraine is reverberating across world's regions*. Retrieved 25 June, 2026, from https://www.imf.org/en/Blogs/Articles/2022/03/15/blog-how-war-in-ukraine-is-reverberating-across-worlds-regions-031522.

Khan, I., Shah, D., & Shah, S. S. (2021). COVID-19 pandemic and its positive impacts on environment: An updated review. *International journal of Environmental Science and Technology, 18*(2), 521–530. https://doi.org/10.1007/s13762-020-03021-3

Kumar, A., Madden, D. G., Lusi, M., Chen, K.-J., Daniels, E. A., Curtin, T., Perry Iv, J. J., & Zaworotko, M. J. (2015). Direct air capture of CO2 by Physisorbent materials. *Angewandte Chemie International Edition, 54*(48), 14372–14377. https://doi.org/10.1002/anie.201506952

Kumar, A., Singh, P., Raizada, P., & Hussain, C. M. (2022). Impact of COVID-19 on greenhouse gases emissions: A critical review. *Science of the Total Environment, 806*, 150349. https://doi.org/10.1016/j.scitotenv.2021.150349

Li, C., Mirosa, M., & Bremer, P. (2020). Review of online food delivery platforms and their impacts on sustainability. *Sustainability, 12*(14), 5528. https://www.mdpi.com/2071-1050/12/14/5528

Madurai Elavarasan, R., & Pugazhendhi, R. (2020). Restructured society and environment: A review on potential technological strategies to control the COVID-19 pandemic. *Science of the Total Environment, 725*, 138858. https://doi.org/10.1016/j.scitotenv.2020.138858

McKinsey & Company. (2021). *Global energy perspective 2021*. Retrieved from https://www.mckinsey.com/industries/oil-and-gas/our-insights/global-energy-perspective-2021.

McKinsey & Company. (2022). *Global energy perspective 2022*. Retrieved 20 June, 2023, from https://www.mckinsey.com/industries/oil-and-gas/our-insights/global-energy-perspective-2022.

Mofijur, M., Fattah, I. M. R., Alam, M. A., Islam, A. B. M. S., Ong, H. C., Rahman, S. M. A., Najafi, G., Ahmed, S. F., Uddin, M. A., & Mahlia, T. M. I. (2021). Impact of COVID-19 on the social, economic, environmental and energy domains: Lessons learnt from a global pandemic. *Sustainable Production and Consumption, 26*, 343–359. https://doi.org/10.1016/j.spc.2020.10.016

Mohamed, B. A., Fattah, I. M. R., Yousaf, B., & Periyasamy, S. (2022). Effects of the COVID-19 pandemic on the environment, waste management, and energy sectors: A deeper look into the long-term impacts. *Environmental Science and Pollution Research, 29*, 46438. https://doi.org/10.1007/s11356-022-20259-1

NASA. (2010). *Global warming*. Retrieved 20 March, 2021, from https://earthobservatory.nasa.gov/features/GlobalWarming/page2.php.

National Geographic. (2018). *Effects of global warming*. Retrieved from https://www.nationalgeographic.com/environment/article/global-warming-effects. https://www.nationalgeographic.com/environment/global-warming/global-warming-effects/.

Omer, A. M. (2008). Energy, environment and sustainable development. *Renewable and Sustainable Energy Reviews, 12*(9), 2265–2300. https://doi.org/10.1016/j.rser.2007.05.001

Ram, M., Osorio-Aravena, J. C., Aghahosseini, A., Bogdanov, D., & Breyer, C. (2022). Job creation during a climate compliant global energy transition across the power, heat, transport, and desalination sectors by 2050. *Energy, 238*, 121690. https://doi.org/10.1016/j.energy.2021.121690

Sanz-Pérez, E. S., Murdock, C. R., Didas, S. A., & Jones, C. W. (2016). Direct capture of CO2 from ambient air. *Chemical Reviews, 116*(19), 11840–11876. https://doi.org/10.1021/acs.chemrev.6b00173

Shakil, M. H., Munim, Z. H., Tasnia, M., & Sarowar, S. (2020). COVID-19 and the environment: A critical review and research agenda. *Science of the Total Environment, 745*, 141022. https://doi.org/10.1016/j.scitotenv.2020.141022

Shang, Y., Li, H., & Zhang, R. (2021). Effects of pandemic outbreak on economies: Evidence from business history context [original research]. *Frontiers in Public Health, 9*. https://doi.org/10.3389/fpubh.2021.632043

Siddique, A., Shahzad, A., Lawler, J., Mahmoud, K. A., Lee, D. S., Ali, N., Bilal, M., & Rasool, K. (2021). Unprecedented environmental and energy impacts and challenges of COVID-19 pandemic. *Environmental Research, 193*, 110443. https://doi.org/10.1016/j.envres.2020.110443

Statista. (2023). *Natural gas prices worldwide*. Retrieved 25 May, 2023, from https://www.statista.com/study/135240/global-natural-gas-prices/.

Varadhan, S., Bukhari, M., & Paul, R. (2023). *Gas shortage exposes fragile south Asian economies to more pain*. Retrieved 25 June, 2023, from https://www.reuters.com/markets/asia/gas-shortage-exposes-fragile-south-asian-economies-more-pain-2023-02-20/.

Wilkinson, P., Smith, K. R., Joffe, M., & Haines, A. (2007). A global perspective on energy: Health effects and injustices. *The Lancet, 370*(9591), 965–978. https://doi.org/10.1016/S0140-6736(07)61252-5

Zamri, M. F. M. A., Milano, J., Shamsuddin, A. H., Roslan, M. E. M., Salleh, S. F., Rahman, A. A., Bahru, R., Fattah, I. M. R., & Mahlia, T. M. I. (2022). An overview of palm oil biomass for power generation sector decarbonization in Malaysia: Progress, challenges, and prospects. *WIREs Energy and Environment, e437*. https://doi.org/10.1002/wene.437

Zanoletti, A., Cornelio, A., & Bontempi, E. (2021). A post-pandemic sustainable scenario: What actions can be pursued to increase the raw materials availability? *Environmental Research, 202*, 111681. https://doi.org/10.1016/j.envres.2021.111681

Chapter 10
Solar Photovoltaic Systems: Assessing Their Impact on the Advancement of Marine Decarbonization

Gülşen Akın Evingür and Büşra Osma

Nomenclature

AC	Alternative current
BMS	Battery management system
c-Si	Crystalline silicon
CdTe	Cadmium telluride
CH_4	Methane
CII	Carbon intensity indicator
CIGS	Copper indium gallium selenide
CO_2	Carbon dioxide
DCDE	Decentralizing and coevolving differential evolution
DPP	Discounted payback period
ESS	Energy storage system
EU	European Union
EVA	Ethylene-vinyl acetate
FPV	Floating photovoltaic technologies
GDEA	General Directorate of Energy Affairs
GHG	Greenhouse gas
HDPE	High-density polyethylene
IMO	International Maritime Organization
k	Extinction coefficient
MPPT	Maximum power point tracking
MOPSO	Multi-objective particle swarm optimization

G. A. Evingür (✉)
Pîrî Reis University, Tuzla, İstanbul, Türkiye
e-mail: gulsen.evingur@pirireis.edu.tr

B. Osma
Yıldız Technical University, Beşiktaş, İstanbul, Türkiye
e-mail: busra.osma@std.yildiz.edu.tr

M. Z. Sogut, M. Koray (eds.), *Energy Rationality and Management for Decarbonization*, Green Energy and Technology,
https://doi.org/10.1007/978-3-031-68497-5_10

n	Refractive index
N_2O, NO_x	Nitrous oxide
NPV	Net present value
NSGA-II	Dominated sorting genetic algorithm
OPCD	Operating point controlling device
PCTC	Pure car and truck carrier
PM	Particulate matter
PV	Photovoltaic
PVC	Polyvinyl chloride
SEA	The Solar Energy Atlas
SEEMP	Ship Energy Efficiency Management Plan
SIU	Ship's illumination unit
SPG	Ship's photovoltaic group
SPP	Simple payback period
SSS	Short sea shipping
UNCTAD	United Nations Conference on Trade and Development
VA	Vinyl acetate
ε_i	Imaginary parts of the dielectric constant
ε_r	Real parts of the dielectric constant
σ	Optical conductivity

10.1 Introduction

Renewable energy systems have become promising technologies for research to decrease the emissions of fossil fuel consumption because of global warming. According to the Review of Maritime Transport 2023 report of the United Nations Conference on Trade and Development (UNCTAD), the marine industry, which carries 80% of the world's commercial products, produces 3% of GHG emissions worldwide (Review of Maritime Transport 2023 Report, UNCTAD, 2023). Therefore, the International Maritime Organization (IMO) has developed global rules for decarbonization for use in the commercial marine industry. The aim was to decrease the amount of carbon dioxide emitted per unit of transport work in international shipping. The target is to achieve a minimum reduction of 40% by 2030, with further efforts to reach a reduction of 70% by 2050, compared to the levels in 2008 (Initial IMO strategy on reducing GHG emissions from ships, IMO, 2018). However, the overall greenhouse gas (GHG) emissions from shipping, which include carbon dioxide (CO_2), methane (CH_4), and nitrous oxide (N_2O) represented in CO_2, have risen from 977 million metric tons in 2012 to 1076 million metric tons in 2018, representing a 9.6% increase. In 2012, the total CO_2 emissions were 962 million metric tons. By 2018, this figure had increased by 9.3% to reach 1056 million metric tons of CO_2 emissions. The proportion of shipping emissions worldwide increased from 2.76% in 2012 to 2.89% in 2018 (Fourth IMO GHG Study 2020 Executive Summary, IMO, 2021). These results require the development of rules and policies in marine industries; thus, the IMO Annex 8, Resolution MEPC.346(78), 2022

Guidelines for the Development of a Ship Energy Efficiency Management Plan (SEEMP), was adopted on 10 June 2022. Part I of the SEEMP document has the "Guidance on Best Practices for Fuel Efficient Operation of Ships" title, and under this, the "Other measures" title is significant to explain and adopt renewable energy sources, such as solar (or photovoltaic) cell technology for onboard application in recent years (Resolution MEPC.346(78), Annex 8, IMO, 2022).

Floating photovoltaic (FPV) systems, which have pontoons, photovoltaic (PV) modules, mooring systems, and cables, are moored in position and float on water (Shi et al., 2023). Floating PV systems were divided into different categories by the MIRARCO research group: PV plants mounted on pontoons and rafts (using galvanized steel and plastic). Installed PV modules with mooring systems are secured by managing ropes and chains, which are fixed to bollards or mooring blocks onshore. FPV systems have many advantages, such as cooling, tracking, storage, and land-saving. To take advantage of the benefits of FPV, effective design of FPV is to be carried out with the following requirements: modularity, flexibility, robustness, safety, optimal raft size, simplicity of launching, and minimum cost (Rosa-Clot & Tina, 2017). In addition, PV systems were installed on the vessels in short sea shipping (SSS) to decrease GHG emissions (Karatuğ & Durmuşoğlu, 2020) (Martínez-López et al., 2023). According to the SEEMP proposal and PV system applications on vessels in the SSS, PV systems are a promising option in marine industries.

The choice of photovoltaic technologies from among the many types of solar cells is another important issue in the market. PV technologies consist of first generation (silicon technology), second generation (thin film technology), and third generation (multi-junction technology) (El Hammoumi et al., 2022). Because crystalline silicon (c-Si) technology is stable and has a high operating efficiency, the first generation of PV cells—which are fully commercial—are the most extensively used solar cells and make up over 90% of the PV cell market (Srivastava, 2016). Although c-Si-based photovoltaic cells are becoming increasingly popular, their cost is comparatively high owing to challenges in the industrial manufacturing of high-grade silicon (Kim et al., 2021). Because thin-film semiconductor technologies, such as amorphous Si, Cadmium telluride (CdTe), and copper indium gallium selenide (CIGS), employ extremely thin layers of semiconductor materials, second-generation PV cells can be mass-produced inexpensively but with low efficiency (Lee & Ebong, 2017). However, third-generation photovoltaic cells, such as polymer-fullerene, hybrid polymers, and perovskites, are still in the developmental stage or have not yet gained widespread market acceptance. Their aim is to enhance the electrical performance of the second generation while maintaining low production costs (Krebs-Moberg et al., 2021). In addition, we investigated and developed polymeric materials for third-generation photovoltaic cells. In this study, we examined ethylene-vinyl acetate (EVA) polymers, which have various vinyl acetate (VA) percentages, for their optical characteristics for use in solar cell applications. The optical parameters, refractive index (n), optical conductivity (σ), extinction coefficient (k), and real (ε_r) and imaginary (ε_i) parts of the dielectric constant were calculated in the range between 200 and 800 nm by analyzing the Ultra-violet visible (UV-Vis) spectroscopy. Furthermore, refractive index (n), extinction coefficient (k), real (ε_r) and imaginary (ε_i) dielectric constants were calculated and observed. Results suggest that 12% of EVA, 28% of EVA, and 40% of EVA polymers were compared among them and show promise in using solar cells as third-generation materials (Evingür & Osma, 2023).

In this chapter, FPV technologies, applications, and research, especially in marine industries, with new design concepts and using new types of materials will be discussed. The importance of FPV technologies will be reviewed and discussed with their advantage and disadvantage aspects in marine industries for being a potential renewable energy system, decreasing global warming effects and GHG.

10.2 Types of Floating Photovoltaic (FPV) Technologies

Pontoons, PV modules, mooring systems, and cables are all part of FPV systems, which are moored onshore and float on water (Xu Han et al., 2023). The MIRARCO research group classified floating photovoltaic systems into two categories: rafts (made of galvanized steel and plastic) and PV plants installed on pontoons. The controlling ropes and chains that attach to bollards or mooring blocks onshore are used to secure installed PV modules with mooring systems (Fig. 10.1).

Pontoons have been applied in small and medium FPVs. These straightforward solutions are derived from the extensive knowledge of port and dock managers. Nonetheless, a floating plant's final goal is to generate the most energy at the lowest possible cost while paying close attention to the robustness of the plant. Therefore, although pontoon is a well-established technology, its aim is not to improve the PV floating system. Because of this, numerous attempts have been made to create special solutions, mostly using modular rafts that can be put together to construct extended structures that are appropriate for the installation of PV panels (Rosa-Clot & Tina, 2017).

PV modules mounted on rafts have three panels, each tilted 35 degrees and equipped with metal flat reflectors at the back as seen in Fig. 10.2. This construction has been utilized in multiple concepts with slight modifications made utilizing

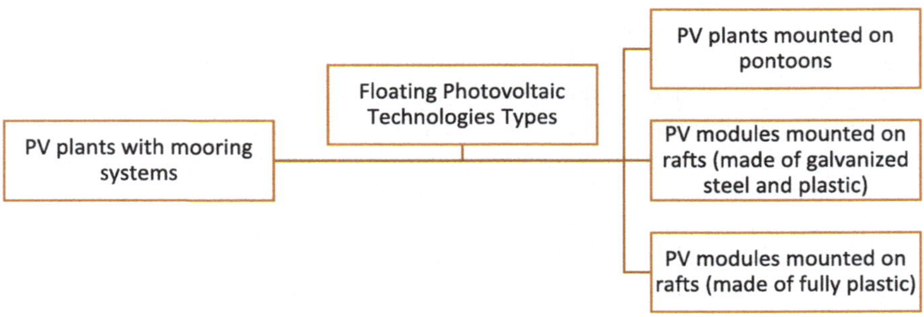

Fig. 10.1 Floating photovoltaic technology types

Fig. 10.2 Korean pilot case for PV modules on rafts in 2012 (Rosa-Clot & Tina, 2017)

Fig. 10.3 Three rafts were assembled with PVC in the Oselis system in France (Rosa-Clot & Tina, 2017)

various materials and extruded beams made of plastic materials (Yoon et al., 2014). A variety of materials, including high-density polyethylene (HDPE), polyvinyl chloride (PVC), and, as in the Korean pilot case, a combination of HDPE and galvanized steel, have been used to construct rafts holding fewer panels (Choi, 2014).

French developers have proposed a new approach. The Osesol (Nantes) proposed to construct tiny rafts entirely out of plastic (PVC) (Trapani & Redón Santafé, 2015). The business was successful in creating a PVC raft that could hold two PV panels as seen in Fig. 10.3.

In the mooring system, according to location morphology, FPV plants are fixed and tracked. Companies consider several aspects of the main characteristics: average depth of water, change of water level, structure of the basin edge, and structure of the basin bottom. According to these main characteristics and knowledge about technologies from boating and yachting industries for centuries, they decide how

FPV panels can be fixed to bollards with managing chains and ropes as seen in Fig. 10.4.

10.3 Photovoltaic (PV) Technology Applications in Marine Systems

In the past decade, FPV applications have been adapted for marine systems. In one study, the ideal sizes of hybrid PV, diesel, and energy storage system (ESS) generators for a ship's power system were determined using a revolutionary methodology. Five different operating situations are included in the model for hourly loads: regular cruising, full-speed sailing, docking, loading/unloading, and anchoring. To assign the sizes of PV and ESS here, decision-makers can use a navigation route from Dalian, China, to Aden, Yemen, considering four distinct seasons. To find the ideal sizes for the PV system and ESS and optimize the diesel generator's outputs to lower overall costs and emissions, the multi-objective particle swarm optimization (MOPSO) algorithm integrated with the non-dominated sorting genetic algorithm (NSGA) II was created. According to the simulation results, hybrid PV/diesel/ESS power generation has a lower acquired net present cost than PV/diesel power generation. The following conclusions can be represented: (a) The correction coefficient for PV generation in a ship's power system is most affected by the local time and time zone, (b) summertime irradiation has a greater effect on PV generation on board than other seasons, and (c) as PV panel costs decrease, more solar power can be added to a ship's power system. The suggested approach can be improved to investigate other mobile microgrids, like high-speed trains and container ships (H. Lan et al., 2015).

In another research, a cost–benefit analysis of photovoltaic panels placed on ships for a particular route in the Northwest Black Sea region was carried out. The results show that using photovoltaic technology is a solution for the future, even in less ideal circumstances. Higher energy output may result from modifications to the modules' ideal tilt, particularly concerning latitude and time of year, when coupled with a sizable area of installed modules. This energy can be used to lower the demand for diesel generators or provide low-energy backup for critical systems like

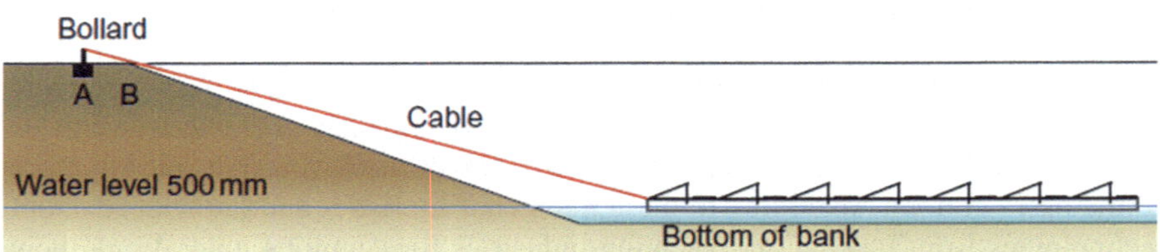

Fig. 10.4 FPV panels fixed to the bollard onshore (Rosa-Clot & Tina, 2017)

communications, GPS, emergency lights, main engine automation, etc. (Atodiresei et al., 2017).

The goal of another study is to lower fuel consumption and pollution emissions by installing a large-scale photovoltaic system onboard an oceangoing ship. First, a novel large-scale PV array structure based on the ship's photovoltaic group (SPG), the ship's illumination unit (SIU), operating point controlling device (OPCD), and batteries is presented to alter the conventional ship power grid structure to consider the structural characteristics of oceangoing ships and their navigation environment. Subsequently, the OPCD introduces the bidirectional Chuk converter to regulate the operating point of every SIU. Last, the maximum power point tracking (MPPT) of each SIU is represented as a large-scale global optimization (LSGO) problem, which may be solved offline using the decentralizing and coevolving differential evolution (DCDE) algorithm. The experimental result indicates that the suggested SPG-SIU-OPCD-based structure can resolve the "hot spot" issue in dynamic and partially shaded areas, as well as the additional power loss that may be present in some conventional constructions. Additionally, the DCDE algorithm's competitive performance is evaluated against several cutting-edge algorithms on benchmarks of 1000 dimensions. Eventually, the MPPT issue of the new structure with 3950 panels is effectively resolved by using the DCDE (Tang, 2017).

A 5000-vehicle space pure car and truck carrier (PCTC) has been chosen as the research subject for this work. Next, a unified grid-tied/stand-alone solar system with integrated battery energy storage is constructed based on the current power system. A diesel generator set, a battery storage system, grid-tied/stand-alone controlled inverters, a solar energy generation unit, a battery management system (BMS), and an energy management system are all part of the system. The use of solar energy hybrid power has the potential to cut fuel consumption by 4.02% and carbon dioxide (CO_2) emissions by 8.55% annually, based on an examination of the experimental data (Yuan et al., 2018).

A work presents a novel method for PV array arrangement on a Roll-on, Roll-off (Ro-Ro) type of vessel and does a practical calculation on the problem of power analysis of the built PV system. The vessel was navigated between Pendik (Turkiye) and Trieste (Italy), and 52 navigation times were conducted annually. In the case study, several analyses are conducted under subheadings including power, fuel, environmental, and economical using the marine vessel's 2018 trip plan. In power analysis, monthly power generation is determined by ship's trips during the pertinent months. In July, 46.88 MWh of power is produced, marking it as the most efficient month, but December generates only 9.5 MWh, indicating the lowest efficiency. During navigation, 198.91 MWh/year of power is produced, whereas 135.16 MWh/year is created at the port. At the conclusion of the year, 334.06 MWh of power is supplied to the vessel's primary power system. Consequently, 7.76% energy efficiency is attained by the installed solar system. In fuel analysis, during the voyage, shaft generators consume 710.03 tons of low sulfur fuel oil, whereas diesel generators utilize 286 tons of diesel oil at ports. The estimates indicate that the generators save 47.43 tons of low sulfur fuel oil and 26.10 tons of diesel oil annually compared to the intended solar system. It indicates that 7.38% of the annual fuel use could be potentially conserved by solar energy. In environmental analysis, significant environmental implications have been identified within the calculation. Implementing a solar system on the vessel results in a reduction of 0.23 tons of SO_x, 1.59 tons of NOx, 93.26 tons of CO_2, and 0.05 tons of particulate matter (PM) emissions during port operations, while during the voyage, the reductions are 0.09 tons of SO_x, 2.35 tons of NO_x, 139.23 tons of CO_2, and 0.06 tons of PM. In economic analysis, the investment cost of the planned solar system is predicted to be $192,060, with an O&M cost of $960.30. The NPV of the solar system is calculated to be $3,362,397.75 based on these data. The investment payback period is estimated to be around 7.2 years, as per Simple Payback Period(SPP). Nonetheless, if a discounted evaluation is performed, the investment's payback period could be around 11.2 years (Karatuğ & Durmuşoğlu, 2020).

With SSS accounting for up to 30% of all onboard power, there is a large demand for electricity supplies. With PV systems for onboard electricity production, this article discusses SSS-fleet compliance with the carbon intensity indicator (CII) regulation and market- and goal-based measures enforced by the EU. The medium-term techno-economic viability of this solution from the shipowners' perspective is examined in this research. To achieve this goal, the PV performance, technical ramifications for the vessel, and financial effects of greenhouse gas mitigation within the framework of the EU are all simulated for an SSS carrier between the Canary Islands and the Iberian Peninsula. The findings show that PV reduces overall CO_2 emissions by 3.38%, which, in turn, reduces fuel usage for power generation by 15.5%. While the improvement is not significant enough to alter the vessel's CII score, the PV investment's internal rate of return attained 55% over 10 years, with a 3-year payback period. This information, together with the reliability of the outcomes, points to PV systems as a viable mitigation strategy for SSS ships (Martínez-López et al., 2023).

All this research shows that PV adaptation on marine ships has the potential to give maritime businesses more ecologically friendly and efficient management. In addition, this research proves the correction of IMO Annex 8, Resolution MEPC.346(78), 2022 Guidelines for the Development of a Ship Energy Efficiency Management Plan (SEEMP), under the "Other Measures" title, which is significant to explain and adopt renewable energy sources, such as solar (or photovoltaic) cell technology for onboard applications (Resolution MEPC.346(78), Annex 8, IMO, 2022).

10.4 Categorization of Photovoltaic (PV)-Technology Materials

In the market, the selection of photovoltaic technology among the various types of solar cells is a tough issue. Photovoltaic technologies (PV) are divided into three categories: first generation (silicon technology), second generation (thin film technology), and third generation (multi-junction technology) as seen in Fig. 10.5 (El Hammoumi et al., 2022).

The operational efficiency values of these technologies have been analyzed in two settings: the laboratory environment and the renewable energy industry. First generation (silicon technology) has approximately 16% and 26% efficiency in commercial and laboratory settings, respectively. Second generation (thin film technology) has approximately 10% and 16% efficiency in commercial and laboratory settings, respectively. Third generation (multi-junction technology) has approximately 45% efficiency in laboratory settings (El Hammoumi et al., 2022). For instance, the porous silicon layer transfer technology was used to create a sides-contacted thin film crystalline silicon (c-Si) solar cell with a validated AM1.5 efficiency of 19.1% in the literature. The cell's aperture area is 3.98 cm^2. This is the highest efficiency for transplanted Si cells ever documented. Implementing current developments for Si wafer cells, such as surface passivation with aluminum oxide and laser ablation for contacting, resulted in an efficiency improvement of 16.9% over the prior state of the art. The short circuit current density of the cell is 37.8 mA cm^{-2}, the open circuit voltage is 650 mV, and the fill factor is 77.6% (Petermann et al., 2012). In addition, according to our research, EVA polymers, which have various vinyl acetate percentages were examined with their optical characteristics to show that they are promising materials in solar cell applications. The optical parameters, refractive index (n), optical conductivity (σ), extinction coefficient (k), and real (ε_r) and imaginary (ε_i) parts of the dielectric constant were calculated according to the analysis of the UV-vis spectroscopy. The refractive index (n), extinction coefficient (k), real dielectric constant (ε_r), and imaginary dielectric constant (ε_i) have comparable behavior concerning wavelength. Various types of EVA polymers, which have 12%, 28%, and 40% weight percent of VA, are compared. The optical conductivity increases with increased energy values, and its maximum value is found to be $4.53 \times 10^{11}\ s^{-1}$ at 6.2 eV. These results show that EVA polymers are promising potential materials for use in solar cell applications (Evingür & Osma, 2023).

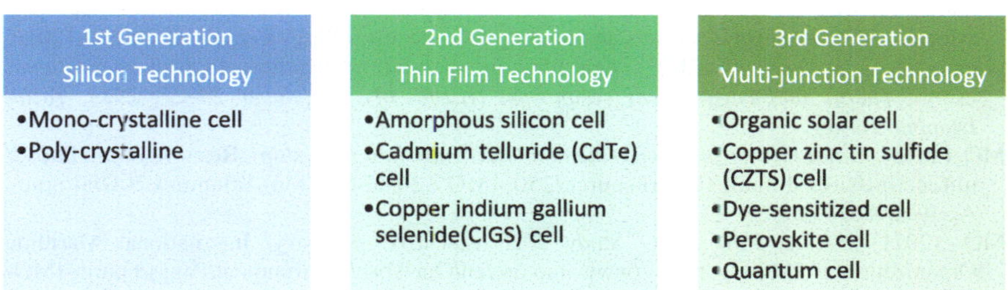

Fig. 10.5 Categorization of photovoltaic (PV) technologies (El Hammoumi et al., 2022)

10.5 Conclusion

Renewable energy development has become crucial to reduce GHG and global warming influence. On 10 June 2022, IMO Annex 8, Resolution MEPC.346(78), 2022 Guidelines for the Development of a Ship Energy Efficiency Management Plan (SEEMP), was adopted. Part I of the SEEMP document is titled "Guidance on Best Practices for Fuel Efficient Ship Operation," and the "Other Measures" section is significant for explaining and adopting renewable energy sources, such as solar (or photovoltaic) cell technology for onboard application in recent years (Resolution MEPC.346(78), Annex 8, IMO, 2022). In the past decade, FPV applications have been adapted for marine systems, especially in vessels. All this research indicates that PV adaptation on marine ships can prevent the release of tons of SOx, NOx, CO_2, and PM into the atmosphere. Furthermore, these studies demonstrate the requirements of development policy and regulations, such as IMO Annex 8, Resolution MEPC.346(78), for the next generation. For these reasons, the selection of photovoltaic technology among the various types of solar cells is an important point to contribute to these applications of marine systems. In addition, material development should be continued to increase the efficiency of PV technologies. Furthermore, the investment in PV systems is evaluated using three distinct economic indicators (net present value (NPV), discounted payback period (DPP), and simple payback period (SPP)) and determined to be economically profitable to install on marine systems (Karatuğ & Durmuşoğlu, 2020).

References

Atodiresei, D., Nicolae, F., & Cotorcea, A. (2017). Cost-benefit analysis of photovoltaic systems installed on ships on the trade routes in the Northwest Black Sea basin. *Journal of Environmental Protection and Ecology, 18*, 40–45.

Choi, Y. K. (2014). A study on power generation analysis of floating PV system considering environmental impact. *International journal of software engineering and its applications, 8*(1), 75–84.

El Hammoumi, A., Chtita, S., Motahhir, S., & El Ghzizal, A. (2022). Solar PV energy: From material to use, and the most commonly used techniques to maximize the power output of PV systems: A focus on solar trackers and floating solar panels. *Energy Reports, 8*, 11992–12010.

Evingür G. A., & Osma, B. (2023). Thin films in solar cell applications. In *12th international exergy, energy and environment symposium (IEEES-14), December 24–27, 2023, Tuzla, Istanbul, Turkey*.

IMO. (2018). *Initial IMO strategy on reduction GHG emissions from ships*. Retrieved from https://unfccc.int/sites/default/files/resource/250_IMO%20submission_Talanoa%20Dialogue_April%202018.pdf.

IMO. (2021). *Fourth IMO GHG Study 2020 executive summary*. International Maritime Organization (IMO). https://www.imo.org/en/OurWork/Environment/Pages/Fourth-IMO-Greenhouse-Gas-Study-2020.aspx.

Karatuğ, Ç., & Durmuşoğlu, Y. (2020). Design of a solar photovoltaic system for a Ro-Ro ship and estimation of performance analysis: A case study. *Solar Energy, 207*, 1259–1268.

Kim, S., Van Quy, H., & Bark, C. W. (2021). Photovoltaic technologies for flexible solar cells: Beyond silicon. *Materials Today Energy, 19*, 100583.

Krebs-Moberg, M., Pitz, M., Dorsette, T. L., & Gheewala, S. H. (2021). Third generation of photovoltaic panels: A life cycle assessment. *Renewable Energy, 164*, 556–565.

Lan, H., Wen, S., Hong, Y. Y., David, C. Y., & Zhang, L. (2015). Optimal sizing of hybrid PV/diesel/battery in ship power system. *Applied Energy, 158*, 26–34.

Lee, T. D., & Ebong, A. U. (2017). A review of thin film solar cell technologies and challenges. *Renewable and Sustainable Energy Reviews, 70*, 1286–1297.

Martínez-López, A., Ballester-Falcón, P., Mazorra-Aguiar, L., & Marrero, A. (2023). Solar photovoltaic systems for the Short Sea Shipping's compliance with decarbonization regulations in the European Union. *Sustainable Energy Technologies and Assessments, 60*, 103506.

Petermann, J. H., Zielke, D., Schmidt, J., Haase, F., Rojas, E. G., & Brendel, R. (2012). 19%-efficient and 43 µm-thick crystalline Si solar cell from layer transfer using porous silicon. *Progress in Photovoltaics: Research and Applications, 20*(1), 1–5.

Resolution MEPC.346(78), Annex 8, IMO. (2022). *Guidelines for the Development of a Ship Energy Efficiency Management Plan (SEEMP)*. Retrieved from https //wwwcdn.imo.org/localresources/en/KnowledgeCentre/IndexofIMOResolutions/MEPCDocuments/MEPC.346(78).pdf.

Rosa-Clot, M., & Tina, G. M. (2017). *Submerged and floating photovoltaic systems: Modelling, design and design,tudies*. Academic Press.

Shi, W., Yan, C., Ren, Z., Yuan, Z., Liu, Y., Zheng, S., et al. (2023). Review on the development of marine floating photovoltaic systems. *Ocean Engineering, 286*, 115560.

Srivastava, H. (2016). *Crystalline silicon PV market size*.

Tang, R. (2017). Large-scale photovoltaic system on green ship and its MPPT controlling. *Solar Energy, 157*, 614–628.

Trapani, K., & Redón Santafé, M. (2015). A review of floating photovoltaic installations: 2007–2013. *Progress in Photovoltaics. Research and Applications, 23*(4), 524–532.

UNCTAD. (2023). *Review of maritime transport 2023 report*. United Nations Conference on Trade and Development (UNCTAD). Retrieved from https://unctad.org/publication/review-maritime-transport-2023.

Yuan, Y., Wang, J., Yan, X., Li, Q., & Long, T. (2018). A design and experimental investigation of a large-scale solar energy/diesel generator powered hybrid ship. *Energy, 165*, 965–978.

Chapter 11
Carbon Management Based on Circular Chain in Decarbonization and Economic Sustainability

M. Ziya Sogut and Oral Erdogan

11.1 Introduction

Today, it is important to develop effective competition, particularly for low-carbon technologies, and prioritize energy and economic sustainability based on circular chains, due to the increasing demand for decarbonization. In fact, many nations, including developed countries such as the G-7, have revealed development frameworks toward achieving net-zero emissions in their action strategies, in order to play a role in addressing this issue. It is important to prioritize the creation of supply chains for these developed frameworks and the development of circular processes for carbon management. While this situation may be beneficial for market development and the supply chain, companies may face managerial difficulties and deficiencies caused by competition that challenge economic sustainability. In the context of national strategies, it is recommended that political strategies such as controlling product demand and improving supply security be prioritized. It is important to take criteria into account. Concerning low-carbon technologies, it is suggested that the second action process for carbon management involves identifying risks associated with cyclical processes, developing solutions, addressing market concentration, and managing technology. The third stage entails the measurement, evaluation, and dissemination of supply chains, which have become an economic necessity. This process is directly influenced by the development of zero-carbon technologies and circular chains for decarbonization, as well as the development of

M. Z. Sogut (✉)
Maritime Faculty, Piri Reis University, Tuzla, Istanbul, Türkiye
e-mail: m.zsogut@pirireis.edu.tr

O. Erdogan
Faculty of Economics and Administrative Sciences, Piri Reis University, Tuzla, Istanbul, Türkiye
e-mail: oerdogan@pirireis.edu.tr

M. Z. Sogut, M. Koray (eds.), *Energy Rationality and Management for Decarbonization*, Green Energy and Technology,
https://doi.org/10.1007/978-3-031-68497-5_11

financial instruments for their dissemination, particularly to the public. The management of this process is defined by carbon management principles and presented as a tool for cyclical chains.

The demand for social change caused by global climate change has gained international prominence. The Paris Agreement, which was implemented in 2015, supports these efforts and provides a significant achievement by committing to keep the global temperature increase below 1.5 °C. This attitude, which is particularly valuable in terms of controlling emissions that cannot be prevented, is also perceived as an important step toward global decarbonization. However, despite all efforts, the pollution caused by the consumer economy and uncontrolled production processes negatively affect the efforts to prevent global warming and lead to significant deviations from the defined targets (UNEP, 2021). A new economic cycle, in which economic roles are changing globally and environmental impacts are gaining importance, is rapidly affecting social life. These processes, defined as circular economy, have emerged as strong factors in achieving the global goal of keeping the temperature rise below 1.5 °C. The IPCC's 2022 report also sees circular economy approaches as an effective tool in this regard. In fact, the process of transforming raw materials into waste and scrap, which is the hallmark of the classical economy, has been identified as one of the main causes of global pollution. However, the development of fundamental tools such as recycling, the establishment of circular processes for them, and the identification of new economic tools related to them have enabled the formation of a new economic structure. Thus, the circular economy, which supports the development of production frameworks and related new models, has demonstrated the material or production's more environmentally friendly competence as a tool for sustainability. In particular, the reuse and recycling of products or outputs that become waste or scrap, or the creation of these cycles, have provided new opportunities. The new economic model or framework that is developing in society is known as the circular economy (Ghisellini et al., 2016). This model or approach has become an effective tool for sustainability, which has emerged as a strategic value (Hobson, 2016; Stewart & Niero, 2018). In classical approaches, financial instruments, which are evaluated in a framework different from production, have developed a new perspective in which production and service are defined in accordance with the requirements of the circular economy. Throughout this process, the commodity has been ensured to have high benefits related to the environment at all stages. Thus, this indicates that regardless of the level of the product or output, it can be considered a commodity in the economy while also providing an ambient for the continuous and sustainable protection of the environment (EMF, 2012, 2014).

It can be observed that the management tools of the industry have also changed depending on the processes revealed by the circular economy. However, a new area that uses material flow and supply arising from circular processes still presents manageable difficulties that are yet to be fully addressed. In terms of sustainable logistic flows, the compatibility of possible supply chain management with decarbonization goals is an important problem area, and solving this problem will provide a valuable perspective in meeting demands. Indeed, managing circular flows and controlling material supply will greatly contribute to resource continuity and minimize potential losses. These new areas emerging in circular production will directly affect the supply chain processes and enhance the economic and environmental sustainability of production. This new perspective, which has also become a focus of logistics today, requires a change in classical supply models (DE Angelis et al., 2018). The improvement of material continuity, the recycling of waste outputs back into production, and the emergence of new areas such as raw material mining have been particularly influential in the management plans of manufacturing companies.

After the Paris Agreement, transformation policies such as the "green call" that found a place in national strategies make carbon management necessary for sectoral players. Carbon management has become a necessity for controlling and managing fossil fuel consumption. Indeed, carbon management has now emerged as an important need for all elements. According to both the IPCC and the International Energy Agency (IEA) predictions, the effectiveness of carbon management indicates the need for carbon-free technologies. However, their potential by 2050 could increase by approximately 200 times. These technologies, which vary depending on the needs, are briefly shaped as technologies that reduce resource emissions, carbon capture and storage technologies, and carbon removal approaches. Carbon management has become a corporate commitment for corporate structures, and the management tools of such structures have revealed solutions that express environmental requirements and commercial activities (Kolk & Pinkse, 2007). Companies are now dealing with many new factors that affect costs, such as energy supply risks, by considering their prioritization in management approaches. For instance, Lee (2012) proposed a prioritization approach and identified six categories that include emission reduction commitments. Doda et al. (2016), who evaluated the effectiveness of these measures, stated that despite all these efforts, sufficient results were not achieved. They particularly noted that the outputs did not have a sufficient data structure and that there were not enough tools to describe the performance effects of the results.

Currently, many new concepts have been developed to support decarbonization for the supply chain, such as sustainable, green, environmental, and closed-loop supply chains (Gurtu et al., 2015). These concepts aim to combine sustainability and decarbonization with the manageability of supply chains (Batista et al., 2018; Govindan & Hasanagic, 2018). The circular supply chain concept plays a key role in sustainable development. It is a new concept that aims to create value from waste flows in addition to normal supply processes. Particularly in the literature, it is expressed as creating loops that will generate value from waste flows in addition to normal supply processes (Genovese et al., 2017; Batista et al., 2018). The supply chain, which is an important branch of maritime trade, creates a significant environmental problem along with heavy truck traffic. Developing and managing a circular supply chain for industrial structures can provide a valuable contribution to the decarbonization process of the transportation sector. Considering the potential of maritime transportation, which accounts for approximately 90% of global freight transport, sectoral predictions indicate that a green transition can be achieved. This study presents a sectoral perspective, taking into account Turkey's roll-on/roll-off (RORO) transportation. The potential for industrial pollution, its economic evaluations, and the environmental impact process are discussed separately. Suggestions for circular supply chains have been developed to achieve decarbonization goals.

11.2 Carbon Management and Circular Chain

Sectoral structures consider sustainability not only as a conceptual need but also as a key to economic growth for all processes. Indeed, the sustainability of the supply chain has been developed as a part of organizational sustainability. Sustainable development has three main components in standard evaluations: environmental, social, and economic impact (Elkington, 2004; Carter & Dale, 2008). On top of this basic structure, many new concepts have been developed for supply chains today, such as sustainable, green, environmental, and closed-loop supply chains (Leszczynska & Maryniak, 2017; Govindan et al., 2015; Darom & Hishamuddin, 2016). However, the literature shows that these concepts do not develop a social structure (Ahi & Searcy, 2013).

The new economic framework presented by the circular economy offers an innovative approach, particularly considering waste management. In a carbon economy, products are expected to be included in regenerative and restorative structures through circular processes, with potential gains transformed into economic structures through renewal and reuse (EMF, 2017; Weetman, 2017). Resource management in the supply chain and new frameworks have been identified as factors affecting the sustainability of resources, including carbon management. However, the sustainability of supply chains remains a controversial issue. A literature review by Farooque et al. (2019), as shown in Table 11.1, provides an overview of this issue considering restorative* (the ability of end-of-life products/materials to become technical nutrients) and regenerative** (the ability of end-of-life products/materials to become biological nutrients) materials (EMF, 2017):

Sustainability in supply chains is developed through cyclical processes, where waste is transformed into value that can be used for future cycles. Nasir et al. (2017) explored the possibility of using textile waste as insulation material in construction and found that it can be recycled and transformed into a new supply chain. However, it is important to consider not only waste-related processes but also environmental impacts related to transportation when establishing sectoral roles in the supply

Table 11.1 Conceptual review of sustainable supply chain and carbon management

	Sustainability dimension			Integration of CE	
Sustainability in SCM	Environment	Economic	Social	Restorative	Regenerative
Sustainable supply chain management[a]	✓	✓	✓	✓	
Green supply chain management[b]	✓	✓		✓	
Environmental supply chain management[c]	✓	✓		✓	
Closed-loop supply chain management[d]	✓	✓		✓	

[a] Seuring and Müller (2008)
[b] Srivastava (2007)
[c] Zsidisin and Siferd (2001)
[d] Guide and Van Wassenhove (2006)

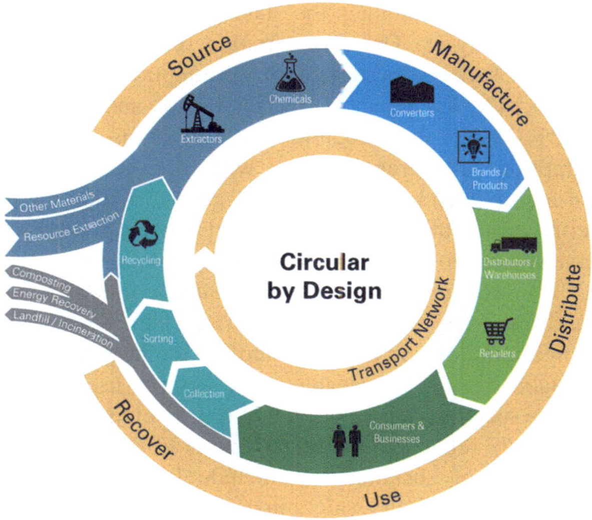

Fig. 11.1 Sustainability and chain of circular economy (Lund, 2021)

chain. This is part of the integrity that is brought about by the circular economy, as illustrated in Fig. 11.1, where all elements are interconnected and have a binding effect on each other.

In reality, transportation has intricate features in all its aspects and has tools that support each other in cycles. However, it is valuable as a framework for improving sustainability and coordinating circular tools with each other. In this regard, developing a sustainable management mechanism, especially for vehicle traffic that supports land and sea transportation, and producing cycles that will improve decarbonization in vehicles is a step that will value the sustainability of supply chains. In today's transportation, which is mostly fleet-based, managing the flow under boundary conditions and improving sustainability become important in terms of vehicle-related carbon management. As a matter of fact, it is known that consumption behavior is quite uncontrolled in road truck transportation, which burns fossil fuels and is based on low-efficiency piston engine technology. This situation points to an important problem area in terms of supply chain management that will support sectoral sustainability.

11.2.1 *Fleet Management for Circular Chain*

Today, logistics relies on three basic modes of transportation: sea, air, and land. In maritime transportation, truck fleet mobility plays a crucial role in the transfer of goods. Carbon management is a key component of energy and environmental sustainability in this process, and companies involved in this business have effective monitoring tools.

This study examines the state of transportation in Turkey as an example for impact analyses. The transportation sector, which amounts to approximately 33 GTon in Turkey, has a total fuel consumption of which road transportation accounts for 95.14%, sea transportation for 1.34%, and air transportation for approximately 3.52%. Within this structure, diesel accounts for approximately 78% of fuel consumption in road transportation, while LPG accounts for 13%, and gasoline accounts for approximately 9% (TRinsight, 2024). While highways have a high potential in terms of global emissions, incorrect driving practices due to uncontrolled or high consumption of vehicle fuel can amplify this potential.

The efficiency of the supply chain is directly related to the global impact of maritime trade and the formation of chain fleets in logistics networks. With an increase in demand and continuity in the flow of goods, monitoring has become a valuable tool. This traceable process is also useful for emission control, especially for the sectoral sustainability that has developed in recent years. Fleet management is directly related to the effectiveness of the sustainable transfer of goods. Basic indicators of sustainability include finance, maintenance, driver competence, assets, manageability, safety management, fuel management, fuel control, and efficiency (Gitahi and Ogollah, 2014). Fleet management's main expectations in terms of decarbonization are to create a circular framework and reduce costs and possible emissions while significantly reducing risks. A sustainable fleet management strategy is essential, as highlighted by Besiou et al. (2012), which establishes decarbonization targets or limits, especially in fleet management. The development of boundary conditions that support the efficiency and manageability of fleets and the development of control measures based on environmental conditions for fleet vehicles are tools that can be considered in this context. The strategic framework, together with clean fuels, creates an efficient operational and driving capacity, as well as the lowest environmental impact due to the choice of clean fuels. Fuel management, fuel control, and efficiency, in addition to finance, maintenance, driver competence, assets, manageability, and safety management, can also be considered as criteria for reducing emissions.

11.3 Methodology

This study presents a framework developed for the sustainability of circular supply chains, in particular, to support sectoral roadmaps. It is based on a dynamic approach developed for this purpose. Structurally, supply chains have defined workflows only within the raw material and sales relationship. However, this process requires a framework for cyclical continuity. As shown in Fig. 11.2, a model has been developed to support the supply chain, and sustainability criteria have been developed.

This study discusses its impact on circular supply, particularly based on RORO transport, in two aspects. In the study, this process was presented with an approach based on environmental impact assessment together with the framework of carbon management. RORO transport in Turkey has been evaluated as a potential

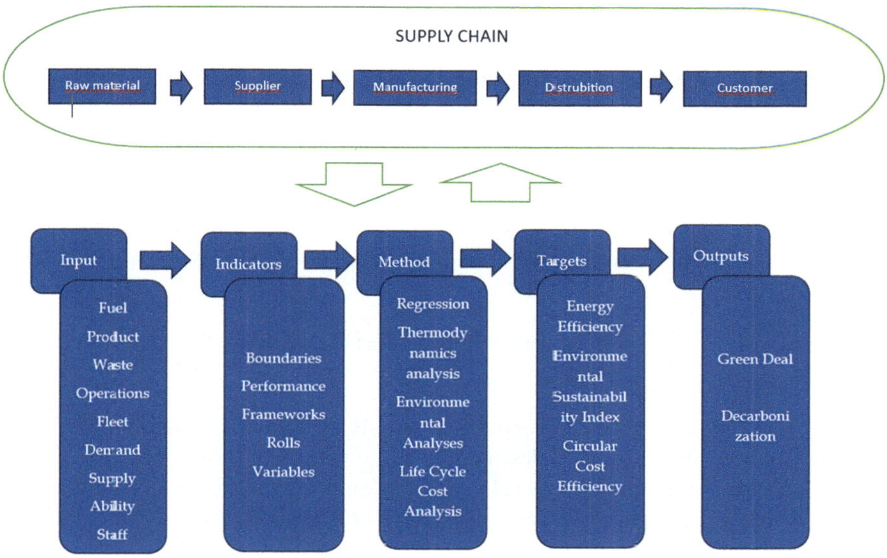

Fig. 11.2 Methodology framework of carbon management for circular chain

assessment through Western Anatolia, Black Sea, and Mediterranean logistics. In the study, linear regression was first applied and the effect of possible variables was evaluated. The standard line equation that defines the energy product relationship depends on this variable (Sermin, 2015). Controllable variables apart from production of energy systems are the variables planned by the management to minimize energy consumption energy consumption such as operating practices, system control, and maintenance standards. In general, the standard equation is a line equation showing that the energy requirement depends on specific variables. This equation is the basic equation and is defined as follows.

$$E = a + b(P) \tag{11.1}$$

Here, a is the constant, b is the coefficient, and P is the specific variable that represents the production. However, there can be many variables (x, y, and z) that affect linear variation in a process.

In the second part of the study, the irreversibilities caused by the studied transportation and the possible improvement potential for them are evaluated separately by exergy analyses. The thermodynamic analyses based on the second law in the engineering approach developed in this section are evaluated below. Within this structure, all processes are analysed on the basis of holistic analyses, as if in equilibrium. The basic mass balance for such approaches; (Moran et al., 2011):

$$\sum \dot{m}_{\text{in}} = \sum \dot{m}_{\text{out}} \tag{11.2}$$

where, $\dot{m}$ states the mass flow rates through each component of the cycle. In this context, the energy balance due to the general mass flow of the engine is handled in the same structure. General energy balance for steady-state conditions:

$$\dot{Q} - \dot{W} + \sum \dot{E}_{in} - \sum \dot{E}_{out} = 0 \tag{11.3}$$

where $\dot{Q}$ and $\dot{W}$ are the two criteria for the net transfer of heat and work. • $\dot{E}$ is the sum of energy entering and leaving system boundaries. Exergy balance due to irreversibility and dead-state temperature in thermal processes is given in Eq. (11.3).

$$\sum \left(1 - \frac{T_0}{T_k}\right) \dot{Q}_k - \dot{W} + \sum \dot{E}x_{in} - \sum \dot{E}x_{out} - \dot{E}x_{dest} = 0 \tag{11.4}$$

where $\dot{Q}_k$ and $\dot{W}$ state the net exergetic potentials of heat and work. On the other hand, Ex_{dest} is related to exergy destruction of the process considering control volume. The total exergy for the control volume related to two thermal systems is the sum of *kinetic* (Ex_{kin}), *potential* (Ex_{pot}), *physical* (Ex_{ph}),and *chemical exergy* (Ex_{ch}). The total exergy is shown in Eq. (11.5):

$$\dot{E}x = \dot{E}x_{kin} + \dot{E}x_{pot} + \dot{E}x_{ph} + \dot{E}x_{ch} \tag{11.5}$$

In this system boundary, the exergy of the flow (ψ) is derived by enthalpy and the entropy form considering the real and the dead-state conditions of the system (Cornelissen, 1997).

$$\psi = (h - h_0) - T_0 (s - s_0) \tag{11.6}$$

where h points to enthalpy, s refers to entropy, and the zero subscript is defined as a dead state based on thermophysical parameters of the surroundings. Similarly, entropy generation is a balanced condition for a steady-state, steady-flow system as given in the following equation (Moran et al., 2011; Dincer & Rosen, 2012):

$$\dot{S}_{in} - \dot{S}_{out} + \dot{S}_{gen} = 0 \tag{11.7}$$

The total entropy generation of a system is a quantitative measure of the potential for irreversibility in the environment of the control volume. Total entropy generation for all thermodynamic processes is a variable that depends on the environment and serves as a direct measure of exergy destruction. This value is determined by the extent of irreversibility for all systems, as stated in the Gouy–Stodola theorem given below (Szargut et al., 1988):

$$\dot{E}x_{dest} = T_0 \dot{S}_{gen} \tag{11.8}$$

The second law of efficiency for the processes can be expressed as a form of both exergy flows and exergy destruction. Exergy efficiency for all thermodynamic systems (Moran et al., 2011; Rosen & Dincer, 2004):

$$\eta_{Ex} = \frac{\dot{E}x_{\text{out}}}{\dot{E}x_{\text{in}}} = 1 - \frac{\dot{E}x_{\text{dest}}}{\dot{E}x_{\text{in}}} \tag{11.9}$$

Exergy destruction requires an engineering approach associated with the irreversibility of a system as a potential for improvement. Improvement potential (IP) is expressed in a form that directly creates a boundary condition in systems and is related to the efficiency of exergetic flows. Accordingly, equation of IP; (Van Gool, 1997):

$$\dot{I}P = \left(1 - \eta_{Ex}\right)\left(\sum \dot{E}x_{\text{in}} - \sum \dot{E}x_{\text{out}}\right) \tag{11.10}$$

11.4 Results and Discussion

Transportation networks are fundamental components of energy and environmental sustainability in supply chains, and they greatly impact the sectoral sustainability of supply chains. This study presents an approach based on the mobility of the RORO fleet in land transportation in Turkey, using the framework created in this context. Sectoral indicators were initially developed using regression analysis based on 9 years of data from the Western, Mediterranean, and Black Sea RORO transportation. The analysis was conducted in a holistic manner, and the subsequent environmental and economic evaluations were based on estimated fuel potential derived from a sample. Figure 11.3 provides the number of transports considered in this context and their regional distribution.

The RORO method has the highest Western mobility rate at 65.44%, with 180,000 recorded in 2020. This mobility has a significant impact on regional logistics and carries a high potential for energy consumption and environmental pollution. The Mediterranean and Black Sea regions follow with 16.9% and 17.66%, respectively. Adequate information transfer to decision-making processes in the

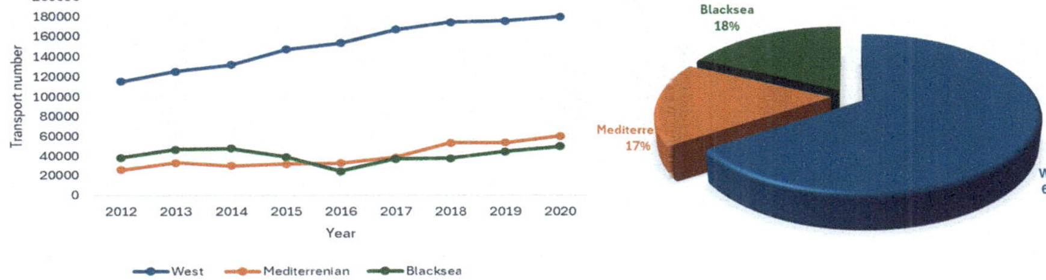

Fig. 11.3 Transport number of RORO regions and distributions

sectoral structure depends on defined values of possible variables. In this context, two basic criteria were examined for the three RORO mobility options. GDP and growth rate, which are considered independent variables in defining fleet mobility, were analyzed, and their distributions for mobility values were examined. Figure 11.4 shows the distributions and change rates over the years.

Regression analyses for both parametric values were examined separately and holistically. Multi-regression analyses based on the two independent variables obtained first are given in Table 11.2.

Although the R^2 value is 0.88 and the meaningfulness value is below 0.01, the P value for growth is significantly higher than the standard. Linear regression analyses were conducted to examine the relationship between GDP value and the distributions, as shown in Fig. 11.5.

The R^2 value for the GDP straight equation was 0.97, indicating a strong correlation. The least square difference for the actual values resulted in an equation with a low error rate of 0.16%. This parameter can be used to examine and forecast the total freight traffic of GDP effectively. However, the evaluation considered only thermodynamic analyses in terms of environmental impact and developed scenarios regarding consumption behaviors. To achieve this, a holistic model approach was created based on the total transportation numbers of unit consumption. This approach can be seen as a model that institutions and organizations can follow.

Unit modeling was conducted for a truck with a 560-kW engine power, based on the study's fleets operating for 5000 h/year. The fuel consumption limit conditions were set at a minimum of 22 kg/h (B1) and 28 kg/h (B2), with a fuel difference of 21.43% between the two borders. The annual transportation-related consumption value in Turkey was found to be 1.13 Gton/year. Based on this value, the fuel consumption rate was examined, and its distributions are presented in Fig. 11.6.

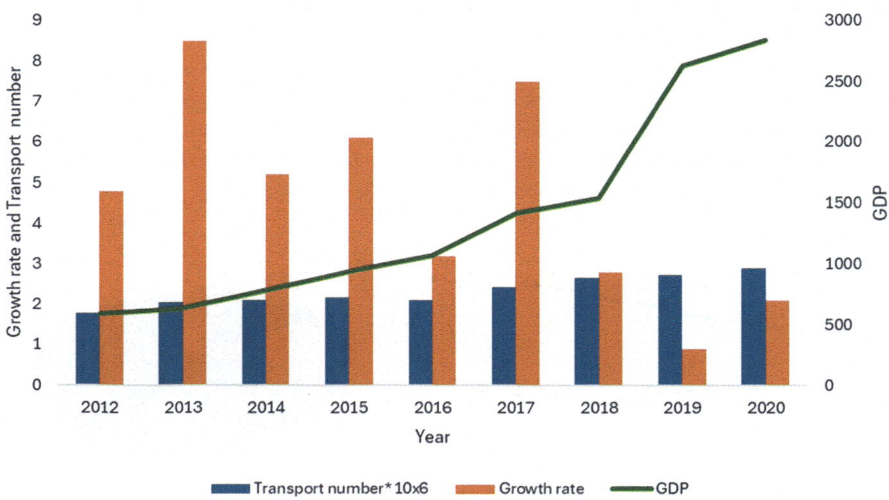

Fig. 11.4 Independent variables and their distribution over the years

Table 11.2 Multi-regression of the transport

Regression statistics	
Multi R	0.939958
R square	0.883522
Adjusted R square	0.844696
Standard error	14578.35
Observation	9

ANOVA	df	SS	MS	F	Meaningfulness F
Regression	2	9.67E+09	4.84E+09	22.75592	0.001580275
Difference	6	1.28E+09	2.13E+08		
Total	8	1.09E+10			

	Coefficients	Standard error	*t* stat	*P*-value	Low 95%	High 95%	Low 95.0%	High 95.0%
Intersection	159,193.4	24184.61	6.582424	0.00059	100015.7421	218,371	100015.7	218,371
Growth rate	2140.572	2914.119	0.734552	0.490312	−4990.020508	9271.165	−4990.02	9271.165
GDP	46.26986	8.888502	5.205586	0.002004	24.52048106	68.01924	24.52048	68.01924

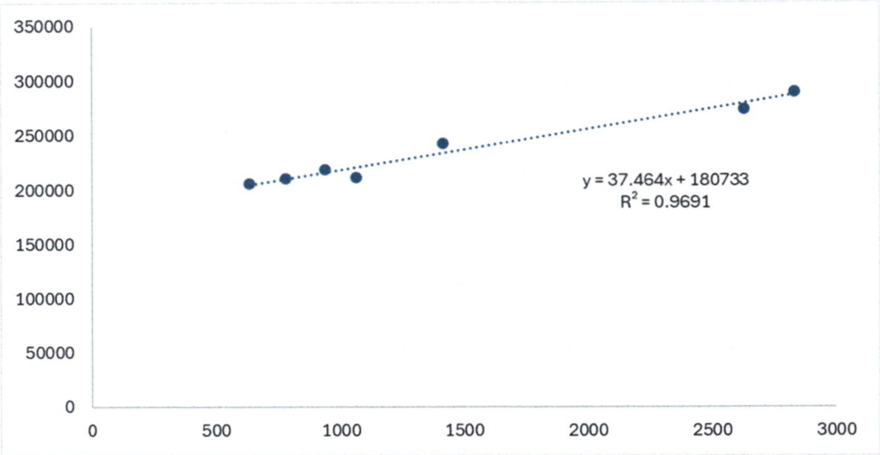

Fig. 11.5 This figure shows the regression of GDP and transport numbers

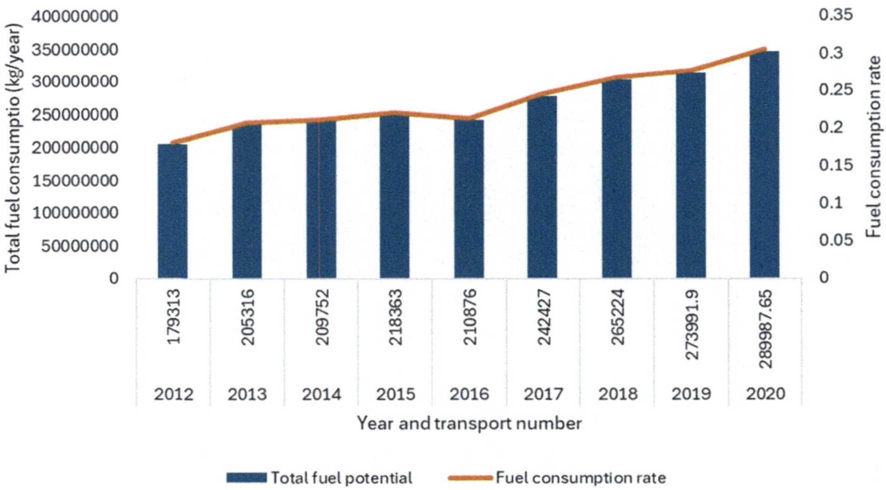

Fig. 11.6 Transport fuel consumptions and rates

When the distributions of Turkey's total transportation rate are examined in the study, the general average consumption rate shows a value of 23.69%. However, when we manage engine consumption based on the same data, we see that we can produce significant savings rates with this data. Accordingly, the distributions of efficiency potentials in terms of boundary conditions and consumption ethics were examined, and the results are given in Fig. 11.7.

For annual boundary conditions, the transportation's possible consumption efficiency has an average value of 36.95%. The efficiency distributions for consumption-dependent boundary conditions range from 31.37 to 42.90%. These two boundary features demonstrate the potential manageability that should be considered for

Fig. 11.7 Efficiency rate of fuel consumption for B1 and B2

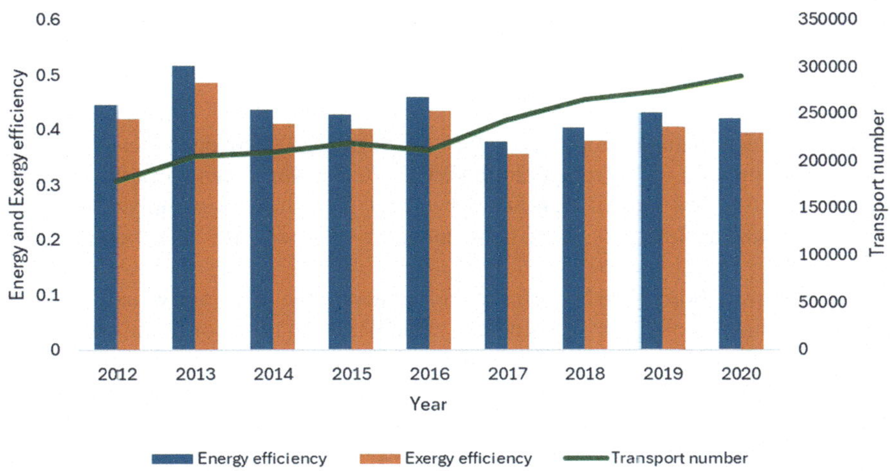

Fig. 11.8 Energy and exergy efficiencies

possible fleet management. Efficiency values were discussed with thermodynamic analysis within the scope of energy sustainability. In this context, efficiency analyses that depend on consumption conditions were conducted according to the first and second laws of thermodynamics. The distributions of these analyses are given in Fig. 11.8.

The distribution of energy efficiency boundary conditions, depending on annual consumption potentials, shows a range of 37.78–51.53%. The average efficiency potential is 43.42%. Exergy efficiency has a potential of 40.46%. It is important to

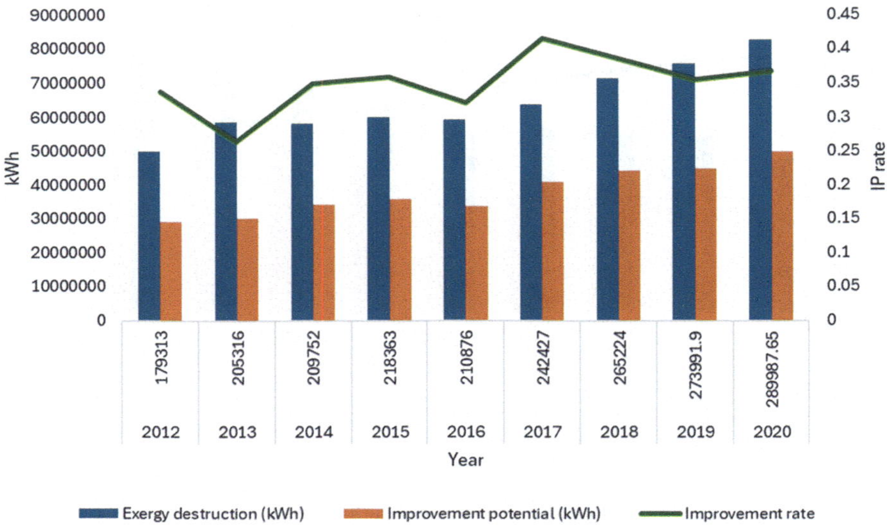

Fig. 11.9 Exergetic parameters and improvement rate

consider the framework of entropy production in terms of environmental impact. The exergy losses produce an entropy load, which averages 60%, indicating a potential for environmental pollution.

The study evaluated environmental pollution using IPCC emission factors. Exergy destruction and entropy generation were assessed using dead-state values at 1 Atm pressure and 25 °C temperature (IPCC, 2022). The potential pollution was found to be 613,908 tons of CO_2 per ton of fuel. The study also analyzed exergy destruction and resulting improvement potential (IP), with the distributions presented in Fig. 11.9.

The transportation-related consumption IP rates for the reference years suggest a potential reduction of 34.98%. Specifically, the IP rate in total consumption, which accounts for approximately 60% of exergy destruction, can serve as an indirect criterion for reducing environmental pollution. This improvement rate indicates a decrease of approximately 59.24% in total pollution potential.

These holistic approaches offer two ways of evaluating pollution and improving sectoral sustainability. The fleet management efficiency potential is, on average, 36.95%. However, the potential for pollution reduction in terms of entropy management is approximately 34.98%, which is also significant. Additionally, there is a significant potential of approximately 59.24% for total emissions in this process.

11.5 Conclusion

In this study, the development of energy and environmental sustainability in the supply chain, especially for maritime transportation, is handled with a holistic approach. In the study, a performance analysis based on prediction was made with the regression approach, an exergetic model was developed with the exergy approach, and the related energy and environmental performance analysis was discussed. The results obtained from the parametric study conducted in this context are given below.

- A forecasting model was developed for the supply chain using GDP as the independent variable. The model produced a correct equation with an R^2 value of 96.91.
- Fleet efficiency was considered in the performance evaluation, and it was found that consumption efficiency has a potential of 36.95%.
- On average, the energy efficiency of consumption potential based on thermodynamic analysis was found to be 43.42%.
- Especially for the accepted dead-state conditions, the average exergy efficiency was 40.96%, while the emission potential was 613,908 tons of CO_2 per ton of fuel.
- Considering the average IP rate of 34.98%, the system's total emission savings represent a value of 59.24%.
- Both approaches demonstrate usability as effective criteria in fleet management and sustainable development.

Supply chains, along with logistics management, have become essential tools in the modern economy. It is worth noting that fossil fuel–based fleet management has been identified as a particularly challenging area in terms of sectoral sustainability. The dimensions of irreversibility revealed indicate a potential threat to environmental sustainability that should be addressed with care and attention. When it comes to the sustainability of the supply chain in terms of energy, economics, and the environment, it appears that change management is required to turn it into a controllable and manageable framework. One potential solution could be the development of system approaches, such as electric solutions and zero-carbon fuel technologies. However, it is recommended that both approach models presented in the study be considered effective tools for enhancing boundary conditions in fleet management and promoting fuel efficiency, with the aim of improving decision-making processes. For potential performance evaluation, research areas for the future could include life process analyses, effective index studies for decarbonization, and fleet management models.

References

Ahi, P., & Searcy, C. (2013). A comparative literature analysis of definitions for green and sustainable supply chain management. *Journal of Cleaner Production, 52*, 329–341. https://doi.org/10.1016/j.jclepro.2013.02.018

Batista, L., et al. (2018). In search of a circular supply chain archetype—A content-analysis-based literature review. *Production Planning and Control, 29*(6), 438–451.

Besiou, M., Martinez, A. J. P., & Van Wassenhove, L. N. (2012). *The effect of earmarked funding on fleet management for relief & development*. INSEAD, Working Paper.

Carter, R. C., & Dale, S. R. (2008). A framework of sustainable supply chain management: Moving toward new theory. *International Journal of Physical Distribution & Logistics Management, 38*(5), 360–387.

Cornelissen, R. L. (1997). *Thermodynamics and sustainable development: The use of exergy analysis and the reduction of irreversibility*, PhD Thesis. University of Twente.

Darom, N. A. M., & Hishamuddin, H. (2016). Integration of economic and environmental aspects in sustainable supply chain management: A review. In *Paper presented at the Proceedings of the 6th International Conference on Industrial Engineering and Operations Management*. Malaysia.

De Angelis, R., Howard, M., & Miemczyk, J. (2018). Supply chain management and the circular economy: Towards the circular supply chain. *Production Planning and Control, 29*(6), 425–437.

Dincer, I., & Rosen, M. A. (2012). *Exergy: energy*. Elsevier.

Doda, B., Gennaioli, C., Gouldson, A., Grover, D., & Sullivan, R. (2016). Are corporate carbon management practices reducing corporate carbon emissions? *Corporate Social Responsibility and Environmental Management, 23*(5), 257–270.

Elkington, J. (2004). Enter the triple bottom line. In A. Henriques & J. Richardson (Eds.), *The triple bottom line, does it all add up? Assessing the sustainability of business and CSR* (pp. 1–16). Earths can Publications Ltd..

EMF (Ellen MacArthur Foundation). (2012). *Towards the circular economy*. Retrieved 20 January 2024, from https://www.ellenmacarthurfoundation.org/publications/towards-the-circular-economy-vol-1-an-economic-and-business-rationale-for-an-accelerated-transition

EMF (Ellen MacArthur Foundation). (2014). *Towards the circular economy: Accelerating the scale-up across global supply chains*.

EMF. (2017). What is the circular economy, ellen macarthur foundation, Available from: https://www.ellenmacarthurfoundation.org/circular-economy/what-is-the-circular-economy. (Accessed 29 March 2025).

Farooque, M., Zhang, A., Matthias, T., Ting, Q., & Donald, H. (2019). Circular supply chain management: A definition and structured literature review, Journal of Cleaner Production, Volume 228. *ISSN, 882-900*, 0959–6526. https://doi.org/10.1016/j.jclepro.2019.04.303

Genovese, A., Acquaye, A. A., Figueroa, A., & Koh, S. C. L. (2017). Sustainable supply chain management and the transition towards a circular economy: Evidence and some applications. *Omega, 66*, 344–357.

Ghisellini, P., Cialani, C., & Ulgiati, S. (2016). A review on circular economy: The expected transition to a balanced interplay of environmental and economic systems. *Journal of Cleaner Production, 114*, 11–32.

Gitahi, M. P., & Ogollah, K. (2014). Influence of fleet management practices on service delivery to refugees in United Nations high commissioner for refugees Kenya Programme. *European Journal of Business Management, 2*(1), 336–341.

Govindan, K., & Hasanagic, M. (2018). A systematic review on drivers, barriers, and practices towards circular economy: A supply chain perspective. *International Journal of Production Research, 56*, 1–34.

Govindan, K., Soleimani, H., & Kannan., D. (2015). Reverse logistics and closed-loop supply chain: A comprehensive review to explore the future. *European Journal of Operational Research, 240*(3), 603–626.
Guide, V. D. R., & Van Wassenhove, L. N. (2006). Closed-loop supply chains: An introduction to the feature issue (part 1). *Production and Operations Management, 15*(3), 345–350.
Gurtu, A., Searcy, C., & Jaber, M. (2015). An analysis of keywords used in the literature on green supply chain management. *Management Research Review, 38*(2), 166–194.
Hobson, K. (2016). Closing the loop or squaring the circle? Locating generative spaces for the circular economy. *Progress in Human Geography, 40*(1), 88–104.
IPCC. (2022). Climate change 2022 mitigation of climate change. In *Working group III contribution to the sixth assessment report of the intergovernmental panel on climate Change, USA*. Retrieved from https://www.ipcc.ch/report/ar6/wg3/downloads/report/IPCC_AR6_WGIII_FullReport.pdf
Kolk, A., & Pinkse, J. (2007). Towards strategic stakeholder management? Integrating perspectives on sustainability challenges such as corporate responses to climate change. *Corporate Governance, 7*(4), 370–378.
Lee, S. (2012). Corporate carbon strategies in responding to climate change. *Business Strategy and the Environment, 21*(1), 33–48.
Leszczynska, A., & Maryniak, A. (2017). Sustainable supply chain—A review of research fields and a proposition of future exploration. *International Journal of Sustainable Economy, 9*(2), 159–179.
Lund, E. (2021). *Packaging in the circular economy, Lazard asset management*. Retrieved from https://www.lazardassetmanagement.com/gl/references/fundamental-focus/packaging-in-the-circular-economy?_escaped_fragment_=
Moran, M. J., Shapiro, H. N., Boettner, D. D., & Bailey, M. B. (2011). *Fundamentals of Engineering Thermodynamics*. John Wiley & Sons Inc.
Nasir, M. H. A., Genovese, A., Acquaye, A. A., Koh, S. C. L., & Yamoah, F. (2017). Comparing linear and circular supply chains: A case study from the construction industry. *International Journal of Production Economics, 183*, 443–457.
Sermin, O. (2015). Productivity. *Energy management and savings in industry, 5th national productivity congress, ankara, 6-7*, (October).
Seuring, S., & Müller, M. (2008). From a literature review to a conceptual framework for sustainable supply chain management. *Journal of Cleaner Production, 16*(15), 1699–1710.
Srivastava, S. K. (2007). Green supply-chain management: A state-of-the-art literature review. *International Journal of Management Reviews, 9*(1), 53–80.
Stewart, R., & Niero, M. (2018). Circular economy in corporate sustainability strategies: A review of corporate sustainability reports in the fast-moving consumer goods sector. *Business Strategy and the Environment, 27*(7), 1005–1022.
Szargut, J., Morris, D. R., & Steward, F. R. (1988). *Exergy Analysis of Thermal and Metallurgical Processes*. Hemisphere Publishing Corporation.
TRinsight. (2024). Taşımacılık Türlerine Göre Yakıt Tüketimi (TR). *Taşımacılık Türlerine Göre Yakıt Tüketimi—TIRPORT Insights*. https://insights.tirport.com/kara-yolu-tasimaciliginda-yakit-turlerine-gore-yakit-tuketimi/
UNEP. (2021). *Emissions gap report 2021*. UNEP.
Van Gool, W. (1997). Energy policy: Fairy tales and factualities. *Innovation and Technology Strategies and Policies*, 93–105.
Weetman, C. (2017). *A circular economy handbook for business and supply chains: Repair, Remake, Redesign*. Kogan Page.
Zsidisin, G. A., & Siferd, S. P. (2001). Environmental purchasing: A framework for theory development. *European Journal of Purchasing & Supply Management, 7*, 61–73. https://doi.org/10.1016/S0969-7012(00)00007-1

Chapter 12
Investigation of Organic Rankine Cycle for Power Generation from Marine Diesel Generator Waste Heat

Cüneyt Ezgi, Haydar Kepekci, and Mehmet Ziya Sogut

Abbreviations

CO_2	Carbon dioxide
EES	Engineering Equation Solver
GWP	Global warming potential
IMO	International Maritime Organization
ODP	Ozone depletion potential
ORC	Organic Rankine cycle
RC	Rankine cycle

12.1 Introduction

Owing to the ever-rising expenses associated with energy, the significance of energy efficiency is amplifying daily. A method to enhance energy efficiency involves the recuperation of residual heat within systems. Given that maritime transportation stands as the most cost-effective approach in global trade, freight vessels are extensively utilized. These vessels, equipped with sizable engines, exhibit remarkably elevated energy consumption. In addition to propelling the engines of commercial ships, there is a continual demand for energy to sustain the functionality of various electronic systems. Consequently, the retrieval of energy from the ship's engine emissions, coupled with its conversion into electrical energy through the process of

C. Ezgi (✉) · M. Z. Sogut
Piri Reis University, Istanbul, Turkey
e-mail: cezgi@pirireis.edu.tr; mzsogut@pirireis.edu.tr

H. Kepekci
Istanbul Gelisim University, Istanbul, Turkey
e-mail: hikepekci@gelisim.edu.tr

M. Z. Sogut, M. Koray (eds.), *Energy Rationality and Management for Decarbonization*, Green Energy and Technology,
https://doi.org/10.1007/978-3-031-68497-5_12

recycling, will substantially augment energy efficiency (Saksa et al., 2023). A potential solution tailored for this objective is the implementation of the organic Rankine cycle (ORC), which deviates in certain aspects from the conventional Rankine cycle (Chao et al., 2024).

The Rankine cycle serves as a mechanism for the transformation of mechanical energy into electrical energy, accomplished through the utilization of a working fluid (Zhu et al., 2024). This cycle finds extensive application in diverse industrial thermodynamic contexts and energy generation. Comprising four principal elements—the boiler, turbine, condenser, and pump—the Rankine cycle operates as follows: The boiler is where water within the cycle undergoes vaporization at a high temperature and pressure. Steam, derived from the boiler, channels its flow energy into mechanical energy by propelling the turbine blades. These blades set the turbine rotor in motion. Post-turbine operation, the steam proceeds to the condenser at a reduced pressure and temperature. Here, the condenser reverts the steam to liquid water. The heat produced in this phase is exchanged with the working fluid and released into the surroundings. Responsible for transporting the condensed water back to the steam generator is the pump. Through the elevation of temperature and pressure, the condensed water undergoes vaporization once more, perpetuating the Rankine cycle (Baniam et al., 2024). Typically, the steam generator initiates the process by vaporizing water, and the resultant steam expands in the turbine, generating mechanical energy. This cycle is a prevalent method for electricity generation in thermal power plants, nuclear power plants, and certain industrial installations. The Rankine cycle assumes a pivotal role in energy production by converting heat from a primary source into mechanical and, subsequently, electrical energy. It is prominently featured in thermal power plants fueled by substances such as coal, natural gas, and nuclear energy, among others. Widely employed in industrial contexts, the Rankine cycle stands as a thermodynamic cycle primarily designated for large-scale electricity generation. Conversely, alternative refrigeration cycles are tailored for specific applications like heating, cooling, and air conditioning, selected based on distinct requirements and system applications (Liu et al., 2024).

The organic Rankine cycle (ORC) constitutes a thermodynamic process tailored for low-temperature applications, proficiently transforming heat derived from low-temperature outlets such as solar energy, geothermal energy, and recuperation of industrial waste heat into electrical energy (Tian et al., 2020). Analogous to the Rankine cycle, the ORC operates as a vapor-based thermodynamic cycle. The principal disparity between the two lies in the liquid refrigerant they employ. While the Rankine cycle commonly employs inorganic liquid refrigerants like water or water-based mixtures, the ORC opts for organic liquid refrigerants characterized by lower boiling points and distinct vaporization traits (Wang et al., 2021). The Rankine cycle traditionally taps into heat from elevated-temperature sources, employing water vaporization and finding applications in electricity generation within thermal power plants. Conversely, the ORC stands out by effectively harnessing heat from lower-temperature sources. This advantageous quality positions the ORC as an efficient converter of low-temperature heat into electrical energy, making it particularly suitable for endeavors such as industrial waste heat recovery. Given these attributes, the ORC assumes a pivotal role in fostering sustainable energy production and enhancing overall energy efficiency (Lion et al., 2020).

In the conventional Rankine cycle, the preference is for inorganic liquid refrigerants, with water being the prevalent and economically viable choice as the inorganic working fluid (Dalbakken & Ertesvag, 2024). In certain scenarios, water-based mixtures find application, particularly in the evaporator, to mitigate the risk of freezing at lower temperatures. Conversely, the organic Rankine cycle (ORC) deviates by employing organic working fluids, specifically chosen for their low-boiling-point attributes, making them suitable for low-temperature applications. Examples of these organic working fluids encompass substances such as toluene, R142b, R600a, and butane. The ORC is purposefully crafted to adeptly capture heat from low-temperature sources, aligning with the characteristics of organic working fluids (Chitgar et al., 2023). Consequently, organic working fluids are more prevalent in low-temperature applications, while inorganic working fluids like water and water-based mixtures find favor in high-temperature applications and industrial thermal power plants. The selection of optimal refrigerants for these cycles is contingent upon the requirements of the application, prevailing environmental conditions, and efficiency objectives (Xiao et al., 2024).

The working fluids employed in the Rankine cycle (RC) are generally categorized into three groups: dry, isentropic, and wet, with these classifications based on the diverse physical properties of the fluid (Zhang et al., 2020). A dry fluid denotes a refrigerant in a fully vaporized state. Simply put, if the fluid exclusively exists in the vapor phase without any presence in the liquid phase, it qualifies as a dry fluid. Typically deployed in evaporator devices, where the refrigerant remains entirely in the vapor phase, dry fluids find common usage in household cooling systems like air conditioners and refrigerators (Zhang & Li, 2023). This distinction showcases their prevalence in applications where maintaining a vaporized state is crucial for optimal functionality.

The term "isentropic" pertains to the condition in which a fluid undergoes an isentropic process, a thermodynamic process characterized by the constant maintenance of entropy (Zhang et al., 2021). An isentropic fluid exists in thermodynamic equilibrium, wherein its entropy remains steady. Such fluids are frequently employed to model ideal conditions in various applications.

On the other hand, a wet fluid signifies a state in which the fluid coexists in both the liquid and vapor phases, essentially representing a mixed state with some proportion in the liquid phase and some in the vapor phase. The concept of wet fluid often mirrors real-world scenarios encountered in engineering applications and systems (Hu et al., 2024). For instance, it may denote the condition where liquid refrigerant initiates evaporation in an evaporator or where gas begins to condense in a compressor. Systems involving wet fluids are commonly observed in expansive industrial cooling facilities, energy production plants, and large-scale air conditioning systems. The representation of both liquid and vapor phases captures the complexity of these systems in practical applications.

In recent times, there has been a surge in academic studies focused on organic Rankine cycle (ORC) systems. This shift has transformed a subject that was relatively underexplored in the past into an area of substantial research interest. Yun et al. designed a dual parallel organic Rankine cycle system to recover waste heat in marine applications. In their study using numerical methods, they examined the performance results to compare single and dual ORC systems. They found that the application of dual ORC to marine engines is much more efficient according to their annual total power output (Yun et al., 2015). In their study, Song et al. examined the waste heat recovery situation of a diesel engine with an ORC system. They designed the system by establishing two separate ORC connections fed from both jacket cooling water and engine exhaust gas. In their studies using numerical methods, they chose net power output as the evaluation criterion. As a result of their studies, they said that such a system is technically feasible and economically positive (Song et al., 2015). Tian et al. have designed a combined organic Rankine cycle system in which liquefied natural gas and dual-fuel marine engine waste heat will be used together. They used engine exhaust gas and engine jacket cooling water as the heat sources of the system. They made numerical analyses using 32 different refrigerants for the system they designed. They evaluated the results obtained from the analyses in terms of both thermal and economic efficiency. As a result, they said that the system they designed was more advantageous than the classical ORC system. They found that they increased exergy efficiency by 27.3% and energy efficiency by 15.62% in their systems. They recommended using R1150-R601a-R600a as the most suitable coolant combination for this system. The payback period of the designed system is calculated as approximately 9 years (Tian et al., 2020). Wang et al. have designed a system that will use waste heat to produce hydrogen, which they see as the largest energy source of the near future. To power the system they designed, they produced a solid oxide electrolyzer cell (SOEC) and combined it with the diesel engine of a ship used in maritime use. They used the organic Rankine cycle to generate electricity in their systems. They made extensive calculations to compare the waste heat recovery of SOEC, ORC, and diesel engines. As a result, they achieved 104.41% and 53.56% efficiency from the electrolyzer cell and ORC system, respectively. They also investigated an optimum current density for their system (Wang et al., 2021). Konur et al. have examined the ORC system, one of the waste heat recovery options in marine vessels, in detail. As a result of their investigations, they found that the waste heat recovery efficiency of classical ORC systems was 20%, the LNG waste heat recovery efficiency was 35.6%, and the effect of ORC and combined cycle arrangements on the system efficiency was 58.4%. They also stated that ORC systems can provide savings in fuel consumption (Konur et al., 2022). Lion et al. have reviewed solutions developed for low- and medium-speed diesel engines. As a result of their investigations, they stated that the use of the organic Rankine cycle would not only save fuel but also be beneficial in reducing emissions (Lion et al., 2020). Yang and Yeh performed numerical analyses using six different fluids in the ORC system to recover waste heat from the jacket water of marine engines. They derived an empirical formula to calculate the thermal efficiency and exergy rates of the ORC system. As a result, they determined that the use of R600a in the 58–68 °C range and R1234ze in the 30–35 °C range is more suitable than other fluids. They also confirmed their study with data in the literature (Yang &

Yeh, 2014). Liu et al. have proposed a system to convert the waste heat in ship exhaust gas and jacket cooling water into electrical energy. Their proposed waste heat recovery system includes three sub-cycles: Rankine cycle (RC), organic Rankine cycle (ORC), and absorption cooling cycle (ARC). They stated that the system they designed increased the thermal efficiency of ship engines by 10.5% (Liua et al., 2020). Singh and Pedersen examined different systems in their studies, investigating the suitability of waste heat recovery systems for marine vehicles. The systems they examined were the Rankine cycle, organic Rankine cycle, supercritical Rankine cycle, and Kalina cycle. As a result, they stated that the highest recovery efficiency in terms of heat quality and quantity could be obtained from the Kalina cycle (Singh & Pedersen, 2016). Mohammed et al. carried out a waste heat recovery project using the supercritical organic Rankine cycle for a cargo ship. They examined the thermal efficiency and fuel consumption of R134a and R245fa, which they used as organic fluids in the cycle, according to different evaporation pressures. As a result, they determined that the use of R134a at high pressures is more advantageous in terms of fuel consumption and thermal efficiency (Mohammed et al., 2020).

Toluene, an organic compound, finds widespread use as a component in the R-410A refrigerant mixture, which is extensively employed in air conditioning systems and various commercial and industrial cooling applications (Ahmed & Imre, 2023). The addition of toluene to this mixture serves to enhance its overall performance. R600, a natural hydrocarbon gas, is commonly utilized in household refrigeration appliances such as refrigerators and freezers. It stands out as an environmentally friendly choice among natural refrigerants, contributing to sustainability efforts (Shi et al., 2018). Isopentane, another hydrocarbon gas with refrigerant properties, is notably applied in industrial cooling systems and commercial refrigeration devices. Natural refrigerants like isopentane are esteemed for their eco-friendly attributes and energy-efficient characteristics, aligning with modern environmental considerations (Ahmed & Imre, 2023). N-hexane also recognized as R-40A, represents a hydrocarbon refrigerant variant. This refrigerant is applicable in industrial cooling systems, particularly within industrial air conditioning setups. Hydrocarbons like hexane stand out as viable alternatives among natural refrigerants, promoting eco-friendly solutions for various cooling applications (Shi et al., 2018).

In 2011, the International Maritime Organization (IMO) made amendments to MARPOL Annex VI, introducing operational and technological energy efficiency measures to reduce carbon dioxide (CO_2) emissions from international vessels. The Ship Energy Efficiency Management Plan (SEEMP) and the Energy Efficiency Design Index (EEDI) were implemented on January 1, 2013, as part of these measures. Subsequently, in 2021, member states of the IMO introduced additional energy efficiency measures, aiming to decrease the carbon intensity of international shipping by at least 40% by the year 2030 compared to the levels observed in 2008. This signifies a concerted effort to address and mitigate the environmental impact of maritime activities on a global scale.

This study introduces a Rankine cycle (RC) system engineered to capture and utilize the energy released during the full-load operation of a marine genset. The devised system is built around four dry working fluids. To derive the thermophysical properties essential for the analysis of each refrigerant within the RC, this research utilizes the Engineering Equation Solver (EES) and the NIST Reference Fluid Thermodynamic and Transport Properties Database (REFPROP) commercial software. The thermophysical properties obtained through these tools facilitate comprehensive calculations, allowing for a detailed exploration of the performance and efficiency of the RC system under consideration.

Within the context of this study, an investigation into the impact of incorporating a recuperator into the designed Rankine cycle (RC) system on efficiency has been conducted (Bellos et al., 2023). This analysis was carried out individually for each working fluid integrated into the RC design, and the results were systematically compared. Recuperators, known for their capacity to enhance efficiency in specific thermodynamic cycles, play a pivotal role in the RC system (Sun et al., 2024). In this system, a recuperator can effectively heat the feedwater by utilizing the heat from the water exiting the condenser. This strategic utilization ensures that the vapor maintains higher temperatures and, consequently, possesses more energy.

Recuperators prove particularly advantageous in systems with diminished heat emissions to the environment, facilities engaged in energy recovery processes, and continuously operating industrial applications (Bellos et al., 2023). The integration of recuperators serves to heighten system efficiency, curtail energy consumption, and mitigate environmental impacts. By recuperating waste heat from a system and transforming it into usable energy for beneficial purposes, recuperators engender significant energy savings and amplify overall system efficiency (Tian et al., 2022). Their ability to recover energy from hot gases or liquids within the system, as opposed to dissipating it as waste heat, facilitates heightened efficiency and reduced energy consumption. This not only translates into diminished operating costs but also contributes to long-term savings. The recovery of waste heat, facilitated by recuperators, holds the potential to reduce energy expenditures significantly (Sadeghi et al., 2020).

Furthermore, a distinct section of the study delves into exergy analysis and irreversibility calculations. These assessments were undertaken separately for each fluid incorporated into the organic Rankine cycle (ORC) design, and the findings were systematically compared. Exergy, a key concept in thermodynamics, serves as a measure of the quality or availability of energy (Fallah et al., 2022). It offers insights into how efficiently energy is utilized relative to the conditions of the system's surroundings. Higher exergy values signify more useful and higher-quality energy, while lower exergy values indicate energy of lower quality and lesser utility. Exergy analysis of a cooling system is instrumental in evaluating energy efficiency, identifying energy losses, and optimizing overall system performance (Fallah et al., 2022). This analytical approach proves critical for enhancing energy efficiency in cooling systems and leveraging resources more effectively.

In parallel, irreversibility calculations in thermodynamics provide insights into the effectiveness and efficiency of a thermodynamic process (Paixao et al., 2022). This parameter quantifies the inherent loss or energy waste in each process. Generally, the more irreversible a process, the less efficient it tends to be.

Irreversibility values in a cooling system offer a gauge of how effective or efficient the system's processes are, obtained through a comparative analysis with an ideal process (Paixao et al., 2022). These insights contribute to a comprehensive understanding of the system's performance and guide efforts toward greater efficiency in energy utilization.

12.1.1 The Systems for the Organic Rankine Cycle: Technical Possibilities

Less than 50% of the fuel energy is transformed into usable power production for the majority of marine diesel engines; the remaining fuel energy is primarily lost through exhaust gas, air coolers, jacket water coolers, and other mechanisms. Figure 12.1 depicts the energy balance of a typical two-stroke engine. From this, we can infer that there is great potential for heat recovery from the exhaust gas, air cooler, and jacket water. While the amount of exhaust gas produced by a two-stroke marine engine is substantial, its maximum temperature is lower—between 250 and 500 °C—than that of a four-stroke diesel engine. Thus, there is a substantial quantity of energy included in exhaust gas.

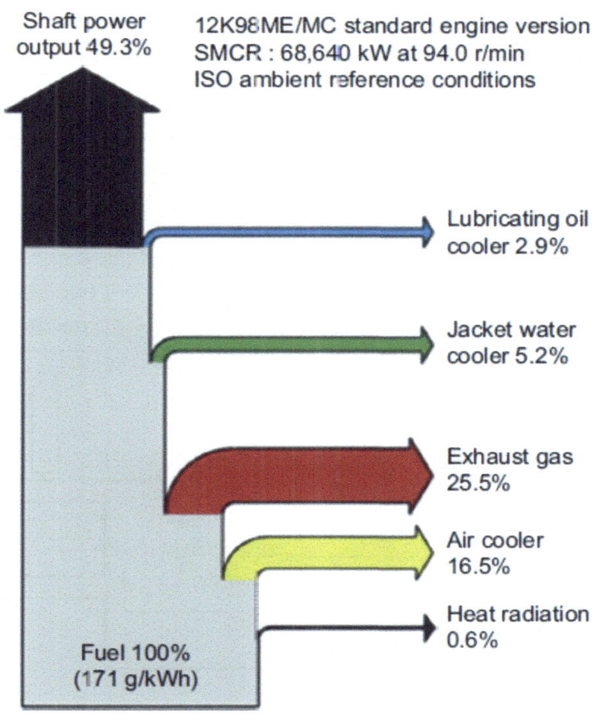

Fig. 12.1 Energy balance of typical marine diesel engine (Macchi & Astolfi 2016)

The waste heat sources for ORC systems onboard ships are as follows:

1. Exhaust gases
2. Jacket water
3. Lubricating oil
4. Scavenge air

The maximum temperature of exhaust gas produced by a diesel marine engine is in the range of 250–500 °C. At the engine output, the jacket cooling water is 70–90 °C, and the lubricating oil temperature is 80–100 °C. The scavenge air entering the charger air cooler is 130–150 °C. Heat sources onboard ships are illustrated in Fig. 12.2. Internal cooling water system for marine genset onboard ships is depicted in Fig. 12.3.

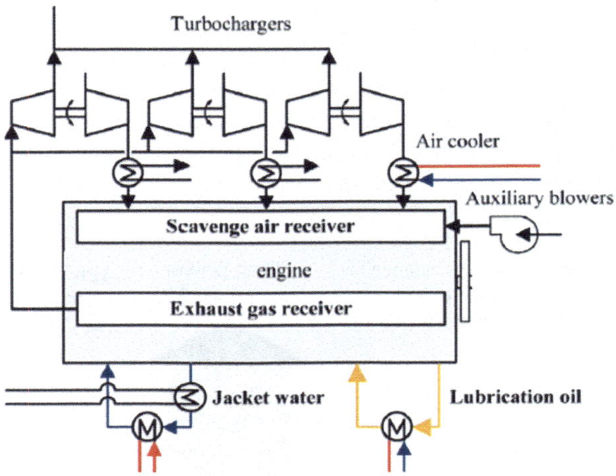

Fig. 12.2 Heat sources for marine diesel engine onboard ships

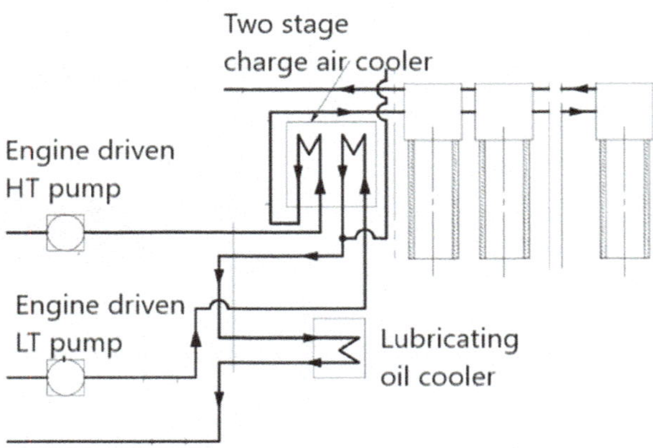

Fig. 12.3 Internal cooling water system for marine genset onboard ships

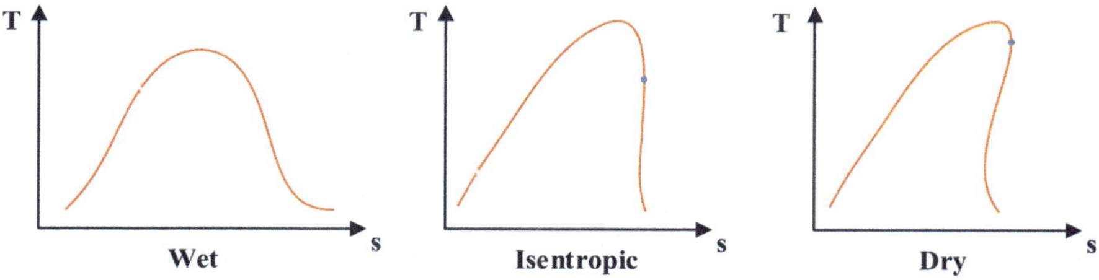

Fig. 12.4 T-S diagrams of working fluids

Table 12.1 Properties of organic dry fluids

Working fluid	Type	Critical properties			ODP	GWP (100 years)	ASHRAE 34 safety group
		Boiling point (K)	Critical temperature (K)	Critical pressure (MPa)			
Toluene	Dry	383.75	591.75	4.1263	0	3	A3
R600	Dry	266.15	418.09	4.0098	0	4	A3
Isopentane	Dry	300.98	460.35	3.378	0	11	A3
N-hexane	Dry	341.87	507.82	3.0441	0	3	A3

12.1.2 The Organic Rankine Fluids

A variety of factors can be taken into consideration while selecting an ORC's working fluid: molecular weight, complexity, and flammability. The working fluids can be classified as wet, isentropic, or dry fluids based on the slope of their saturation vapor curves on the temperature-entropy (T-s) diagram as seen in Fig. 12.4.

Within the scope of this investigation, the working fluids selected for a marine genset scenario include toluene, R600, isopentane, and n-hexane, all serving as organic fluids. The characteristics and properties of these chosen working fluids are outlined in Table 12.1 as part of this study (Lemmon et al., 2018; ASHRAE, 2019).

Temperature-entropy (T-s) and pressure-enthalpy (P-h) diagrams of selected organic fluids are given in Figs. 12.5, 12.6, 12.7, 12.8, 12.9, 12.10, 12.11, and 12.12. These figures are provided by EES software.

12.2 System Design: A Case Study

In this article, calculations have been made for the working fluid selection of the organic Rankine cycle for energy production from waste heat on ships. Technical data and heat balance of the marine genset used in the study are given in Tables 12.2 and 12.3, respectively. Values are given at 100% engine load for MAN L21/31 DF-M GenSet, Anonymous. (n.d.). A schematic of the exhaust gas flow of marine diesel engine is given in Fig. 12.13.

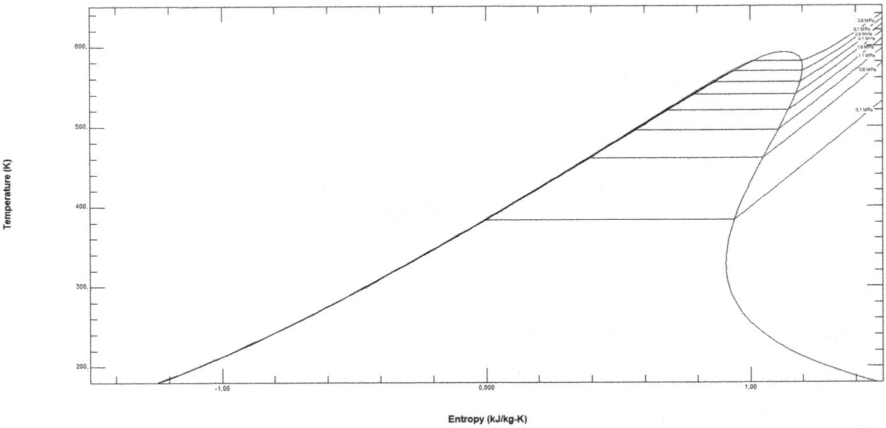

Fig. 12.5 T-S diagram of toluene

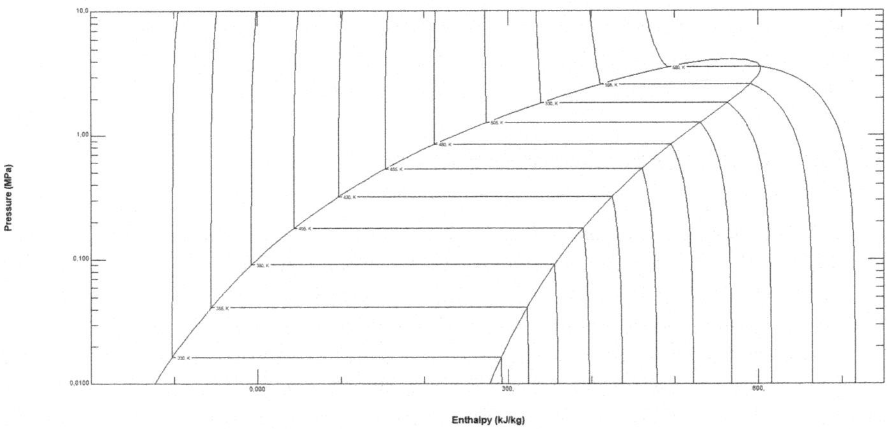

Fig. 12.6 P-h diagram of toluene

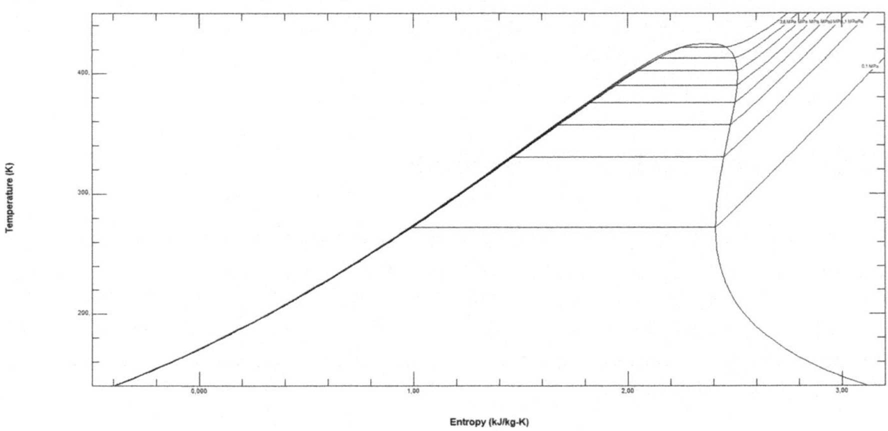

Fig. 12.7 T-s diagram of R600

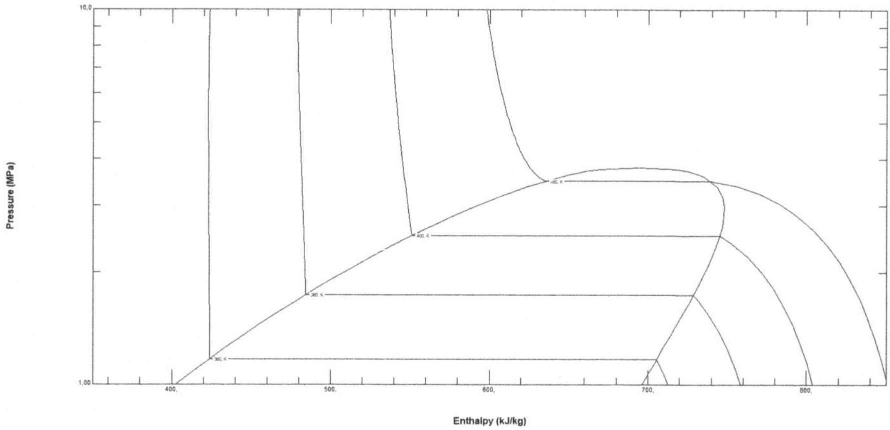

Fig. 12.8 P-h diagram of R600

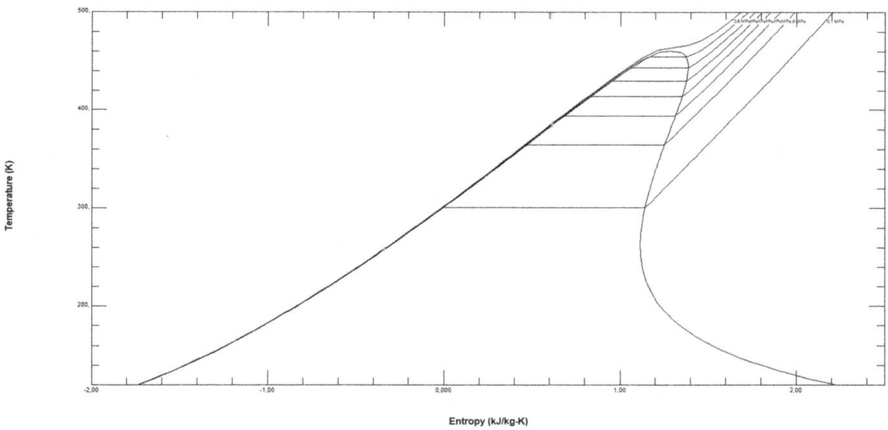

Fig. 12.9 T-s diagram of isopentane

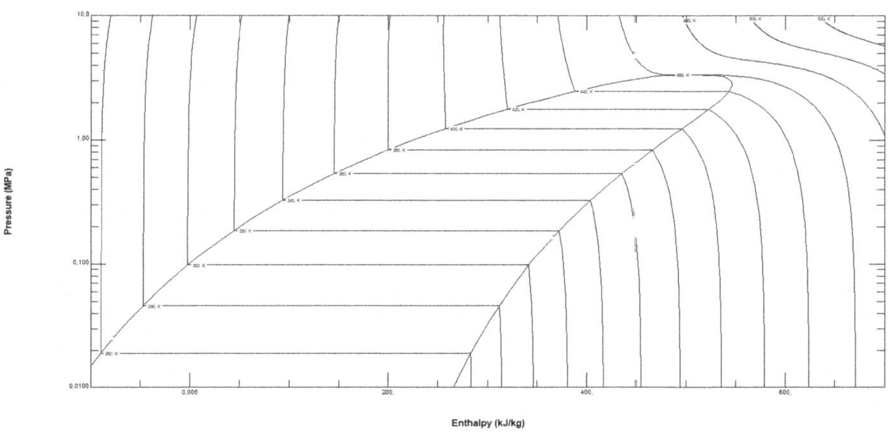

Fig. 12.10 P-h diagram of isopentane

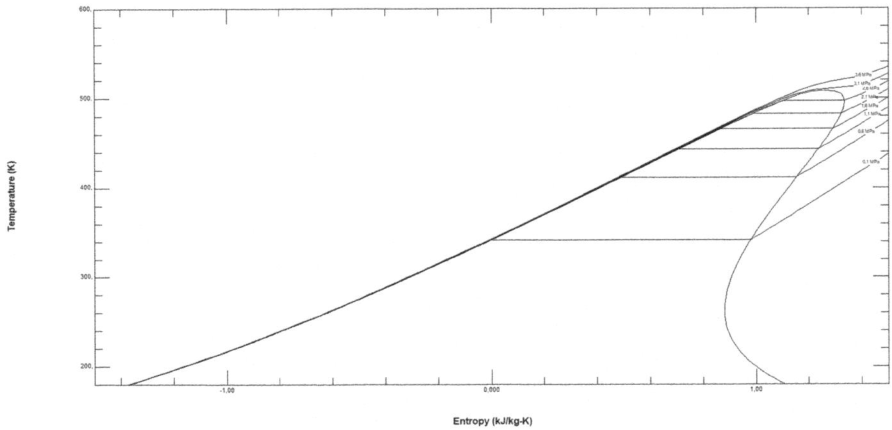

Fig. 12.11 T-s diagram of n-hexane

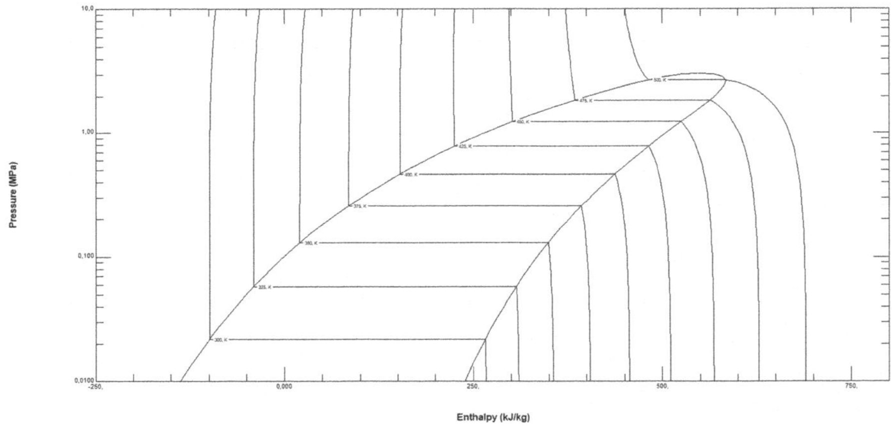

Fig. 12.12 P-h diagram of n-hexane

Table 12.2 Technical data for marine genset

Engine output	kW	1000
Speed	rpm	900
Stroke	mm	310
Cylinder bore	mm	210
Compression ratio	–	16.8

The schematic representation of the designed simple Rankine cycle (RC) system, as explored in this study, is depicted in Fig. 12.14. Simple RC systems integrated into ship engines are employed for recovering energy by analyzing the hot gases and waste heat expelled from the exhaust of the ship's engine. When the ship's engine operates, high-temperature exhaust gases are discharged. These hot gases are directed

Table 12.3 Heat balance at full load for the marine genset

Type of waste heat	Rate of heat transfer (kW)
Exhaust	366
Jacket water	208
Heat radiation engine	49
Lubricating oil	176
Charge air, HT	346
Charge air, LT	198

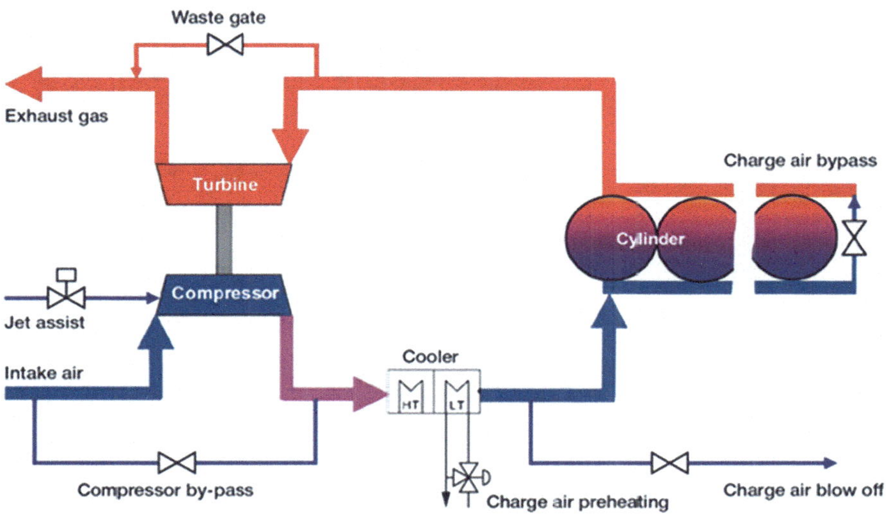

Fig. 12.13 Schematic of exhaust gas flow of marine diesel engine

into an exhaust gas boiler, where the working fluid undergoes evaporation in response to the elevated-temperature environment. The evaporation process signifies the transition of the liquid into the vapor phase, gaining expansion potential as vapor. The resulting vapor then expands within a turbine, converting the gained potential into mechanical energy. During this expansion phase, the turbine's rotation drives a generator, generating electrical energy. The steam exiting the turbine typically enters a cooling system, where the vapor is condensed back into the liquid phase. This condensation ensures that the liquid working fluid returns to the boiler. With the liquid's return to the boiler, the RC system completes a cycle, initiating the process anew.

The diagram representing the designed Rankine cycle (RC) with a recuperator system, as explored in this study, is provided in Fig. 12.15. Introducing the liquid working fluid through the recuperator is designed to elevate system efficiency and recover additional energy from waste heat. The hot fluid, emanating from the ship engine's exhaust, enters the exhaust gas boiler. In this phase, the liquid working fluid initiates vaporization, leading to a temperature increase during the process. The recuperator plays a crucial role in facilitating heat transfer between the hot vapor-containing fluid and the cold fluid earmarked for later use in the system. This exchange

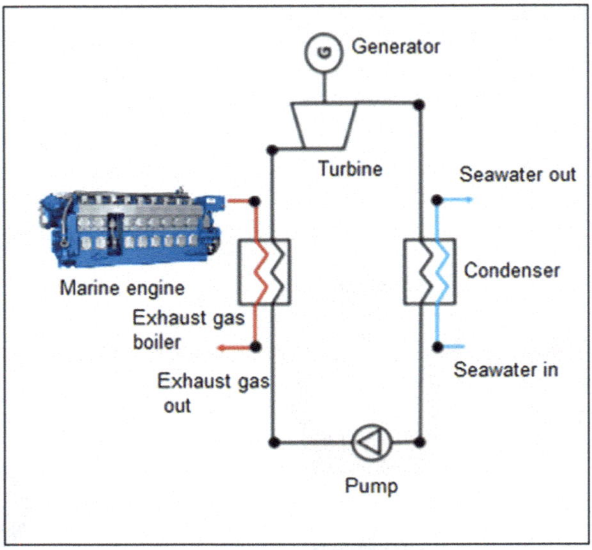

Fig. 12.14 Simple ORC system

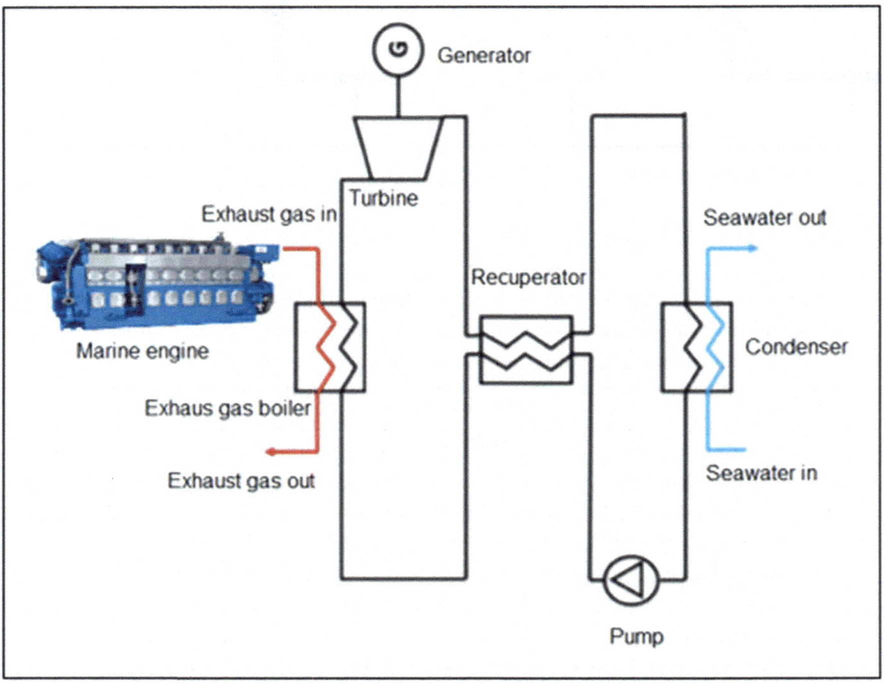

Fig. 12.15 ORC system with recuperator

enhances system efficiency by harnessing waste heat. The energy of the hot vapor transfers to the working fluid, which undergoes cooling in the recuperator, resulting in the condensation of some vapor back into a liquid state. The cooled fluid, having traversed the recuperator, is typically circulated back to the exhaust gas boiler with the aid of a pump. This fluid, now recharged with heat and carrying vaporization potential, initiates the cycle once again. This cyclic process establishes a loop within the system, and the passage of the liquid working fluid through the recuperator significantly amplifies the efficiency of the boiler. Upon return to the boiler, the liquid, having passed through the recuperator, commences the cycle at a lower temperature.

12.3 Thermodynamic Model

By using the first and second laws of thermodynamics on each component in the configurations, energy and exergy evaluations are carried out for the ORC configurations.

A six-step thermodynamic cycle is depicted in Fig. 12.16. The first stage involves pumping a working fluid from a lower pressure to a higher pressure (3–4). In order to absorb heat from the expanded vapor that is leaving the expander, the

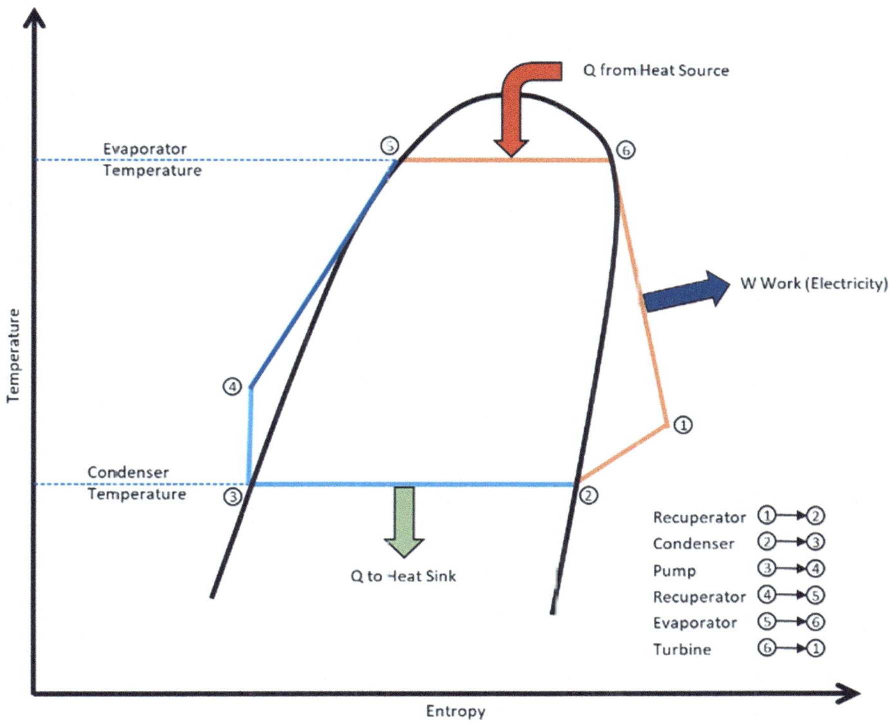

Fig. 12.16 T-s diagram of ORC

high-pressure liquid passes through the recuperator (4–5). After that, the high-pressure fluid is heated at a steady pressure within a boiler until it turns into a dry, saturated vapor (5–6). This vapor then expands and passes through a turbine, which transforms mechanical work into electrical energy (6 to 1). The liquid that emerges from the condenser (1–2) cools the hot vapor as it passes through the recuperator. The cycle then restarts when the moist vapor enters a condenser and condenses once more into a saturated liquid (2–3). To increase cycle efficiency, the cycle has a regeneration component for the majority of applications. A simplified schematic of an ORC cycle and the same ORC cycle displayed on a T-s diagram are depicted in the accompanying illustrations.

The mass balance equation for a general steady flow system is expressed as:

$$\sum_{\text{in}} \dot{m} = \sum_{\text{out}} \dot{m} \tag{12.1}$$

The energy balance equation for a general steady flow system is expressed as:

$$\dot{E}_{\text{in}} = \dot{E}_{\text{out}} \tag{12.2}$$

$$\dot{Q} - \dot{W} = \sum_{\text{out}} \dot{m}\left(h + \frac{V^2}{2} + gz\right) - \sum_{\text{out}} \dot{m}\left(h + \frac{V^2}{2} + gz\right) \tag{12.3}$$

Exergy balance for a general steady flow system is expressed as:

$$\dot{X}_{\text{in}} - \dot{X}_{\text{out}} - \dot{X}_{\text{destroyed}} = 0 \tag{12.4}$$

Net power from ORC:

$$\dot{W}_{\text{net}} = \eta_{\text{th}} \dot{Q}_{\text{exh}} \tag{12.5}$$

The rate of heat transfer from the exhaust gas is expressed as:

$$\dot{Q}_{\text{exh}} = \dot{m}_{\text{exh}} \left(h_{\text{exh1}} - h_{\text{exh2}}\right) \tag{12.6}$$

The flow exergy of a flowing fluid is expressed as:

$$\psi = \left(h - h_0\right) - T_0\left(s - s_0\right) \tag{12.7}$$

For a steady flow system, the reversible power is expressed as:

$$\dot{W}_{\text{rev}} = \Sigma \dot{m}_i \psi_i - \Sigma \dot{m}_e \psi_e \tag{12.8}$$

$$\dot{W}_{\text{rev}} = \dot{m}_{\text{exh}} \left[\left(h_{\text{exh1}} - h_{\text{exh2}}\right) - T_0\left(s_{\text{exh1}} - s_{\text{exh2}}\right)\right] \tag{12.9}$$

The exergy efficiency of the system is expressed as:

$$\eta_{\text{ex}} = \frac{\dot{W}_{\text{net}}}{\dot{W}_{\text{rev}}} \tag{12.10}$$

Irreversibility in J and W for the system, respectively, is expressed as:

$$I = W_{\mathrm{rev}} - W_{\mathrm{net}} \tag{12.11}$$

$$\dot{I} = \dot{W}_{\mathrm{rev}} - \dot{W}_{\mathrm{net}} \tag{12.12}$$

For pump:

$$\text{Mass balance}: \dot{m}_{\mathrm{in}} = \dot{m}_{\mathrm{out}} \tag{12.13}$$

$$\text{Energy balance}: \dot{m}_{\mathrm{in}} h_{\mathrm{in}} + \dot{W}_p = \dot{m}_{\mathrm{out}} h_{\mathrm{out}} \tag{12.14}$$

$$\text{Entropy balance}: \dot{m}_{\mathrm{in}} s_{\mathrm{in}} + \dot{S}_{\mathrm{gen},p} = \dot{m}_{\mathrm{out}} s_{\mathrm{out}} \tag{12.15}$$

$$\text{Exergy balance}: m_{\mathrm{in}}\, x_{\mathrm{in}} + \dot{W}_p = \dot{m}_{\mathrm{out}} x_{\mathrm{out}} + \dot{X}_{d,p} \tag{12.16}$$

For evaporator:

$$\text{Mass balance}: \dot{m}_{\mathrm{in}} = \dot{m}_{\mathrm{out}} \tag{12.17}$$

$$\text{Energy balance}: \dot{m}_{\mathrm{in}} h_{\mathrm{in}} + \dot{Q}_{\mathrm{evap}} = \dot{m}_{\mathrm{out}} h_{\mathrm{out}} \tag{12.18}$$

$$\text{Entropy balance}: \dot{m}_{\mathrm{in}} s_{\mathrm{in}} + \dot{S}_{\mathrm{gen,evap}} + \frac{\dot{Q}_{\mathrm{evap}}}{T_{\mathrm{evap}}} = \dot{m}_{\mathrm{out}} s_{\mathrm{out}} \tag{12.19}$$

$$\text{Exergy balance}: \dot{m}_{\mathrm{in}} x_{\mathrm{in}} + \dot{X}_{\mathrm{evap}} = \dot{m}_{\mathrm{out}} x_{\mathrm{out}} + \dot{X}_{d,\mathrm{evap}} \tag{12.20}$$

For steam turbine:

$$\text{Mass balance}: \dot{m}_{\mathrm{in}} = \dot{m}_{\mathrm{out}} \tag{12.21}$$

$$\text{Energy balance}: \dot{m}_{\mathrm{in}} h_{\mathrm{in}} = \dot{m}_{\mathrm{out}} h_{\mathrm{out}} + \dot{W}_t \tag{12.22}$$

$$\text{Entropy balance}: \dot{m}_{\mathrm{in}} s_{\mathrm{in}} + \dot{S}_{\mathrm{gen},t} = \dot{m}_{\mathrm{out}} s_{\mathrm{out}} \tag{12.23}$$

$$\text{Exergy balance}: \dot{m}_{\mathrm{in}} x_{\mathrm{in}} = \dot{m}_{\mathrm{out}} x_{\mathrm{out}} + \dot{X}_{d,t} - \dot{W}_t \tag{12.24}$$

For condenser:

$$\text{Mass balance}: \dot{m}_{\mathrm{in}} = \dot{m}_{\mathrm{out}} \tag{12.25}$$

$$\text{Energy balance}: \dot{m}_{\mathrm{in}} h_{\mathrm{in}} = \dot{m}_{\mathrm{out}} h_{\mathrm{out}} + \dot{Q}_{\mathrm{cond}} \tag{12.26}$$

$$\text{Entropy balance}: \dot{m}_{\mathrm{in}} s_{\mathrm{in}} + \dot{S}_{\mathrm{gen,cond}} = \dot{m}_{\mathrm{out}} s_{\mathrm{out}} + \frac{\dot{Q}_{\mathrm{cond}}}{T_{\mathrm{cond}}} \tag{12.27}$$

$$\text{Exergy balance}: \dot{m}_{\mathrm{in}} x_{\mathrm{in}} = \dot{m}_{\mathrm{out}} x_{\mathrm{out}} + \dot{X}_{\mathrm{cond}} + \dot{X}_{d,\mathrm{cond}} \tag{12.28}$$

For recuperator:

$$\text{Mass balance}: \dot{m}_1 + \dot{m}_4 = \dot{m}_2 + \dot{m}_5 \tag{12.29}$$

$$\text{Energy balance}: \dot{m}_1 h_1 + \dot{m}_4 h_4 = \dot{m}_2 h_2 + \dot{m}_5 h_5 \tag{12.30}$$

$$\text{Entropy balance}: \dot{m}_1 s_1 + \dot{m}_4 s_4 + \dot{S}_{\text{gen}} = \dot{m}_2 s_2 + \dot{m}_4 s_4 \tag{12.31}$$

$$\text{Exergy balance}: \dot{m}_1 x_1 + \dot{m}_4 x_4 = \dot{m}_2 x_2 + \dot{m}_4 x_4 + \dot{X}_d \tag{12.32}$$

12.4 Results and Discussion

The condenser temperature in this system is assumed to be 35 °C because of the maximum seawater temperature of 32 °C according to Loydu's (2022) Rules and Regulations for the Classification of Naval Ships. The efficiency of a simple cycle and a cycle with a recuperator for dry organic fluids is illustrated in Fig. 12.17.

As seen in Fig. 12.17, if the turbine inlet temperature is above 200 °C, the highest efficiency among dry fluids is achieved in the RC system with a recuperator using toluene. For the turbine inlet temperature of 300 °C, the efficiency of toluene in the recuperator system is calculated to be 41%. If the turbine inlet temperature is below 200 °C, the recuperator system with n-hexane yields the highest efficiency. For a turbine inlet temperature of 200 °C, the efficiency of n-hexane in the recuperator

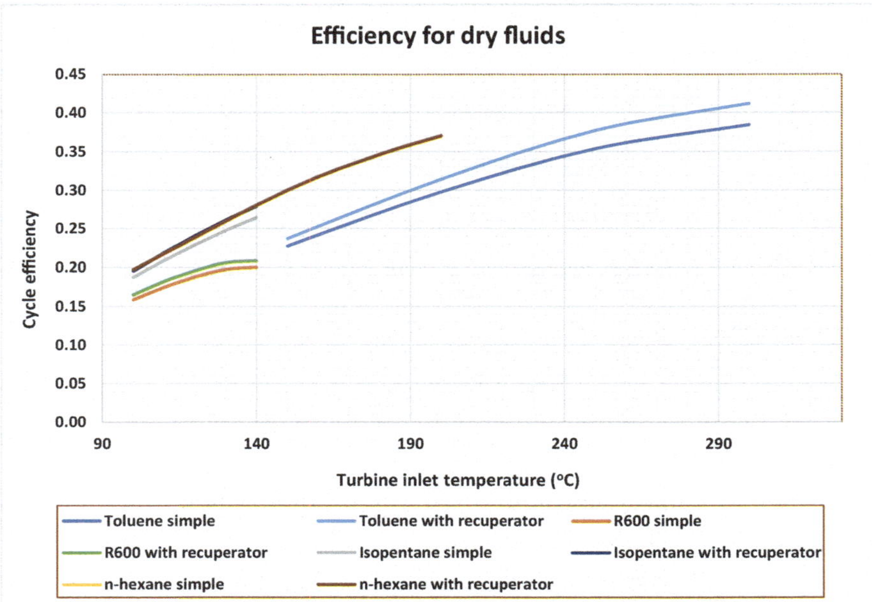

Fig. 12.17 Cycle efficiency for dry fluids

system is obtained to be 37%. It has been observed that the use of a recuperator in dry fluids increases the cycle efficiency by approximately 2%. The exergy efficiency for dry fluid is illustrated in Fig. 12.18.

As seen in Fig. 12.18, if the turbine inlet temperature is above 200 °C, the highest exergy efficiency among dry fluids is obtained from the recuperator-based RC system using toluene. For a turbine inlet temperature of 300 °C, the exergy efficiency of toluene fluid in the recuperator-based ORC system is calculated as 63%. If the turbine inlet temperature is below 200 °C, the highest exergy efficiency among dry fluids is obtained from the recuperator-based RC system using n-hexane. For a turbine inlet temperature of 200 °C, the exergy efficiency of n-hexane fluid in the recuperator-based RC system is calculated to be 60%. Additionally, it has been observed that the use of a recuperator in RC systems increases the exergy efficiency by approximately 2%. The net power and irreversibility for a simple RC with dry fluid are illustrated in Fig. 12.19.

As seen in Fig. 12.19, the highest net power among dry fluids above a turbine inlet temperature of 200 °C is obtained from toluene. For a turbine inlet temperature of 300 °C, the net power obtained from toluene is calculated to be 140.727 kW. If the turbine inlet temperature is below 200 °C, the highest net power among dry fluids is obtained from n-hexane. For a turbine inlet temperature of 200 °C, the net power obtained from n-hexane is calculated to be 135.714 kW. Additionally, as shown in Fig. 12.20, irreversibility and net power are inversely proportional. Irreversibility decreases as net power increases.

The net power and irreversibility for a dry fluid in RC with a recuperator are illustrated in Fig. 12.20. As seen in the figure, the highest net power among dry fluids with a turbine inlet temperature above 200 °C is obtained from toluene. For a

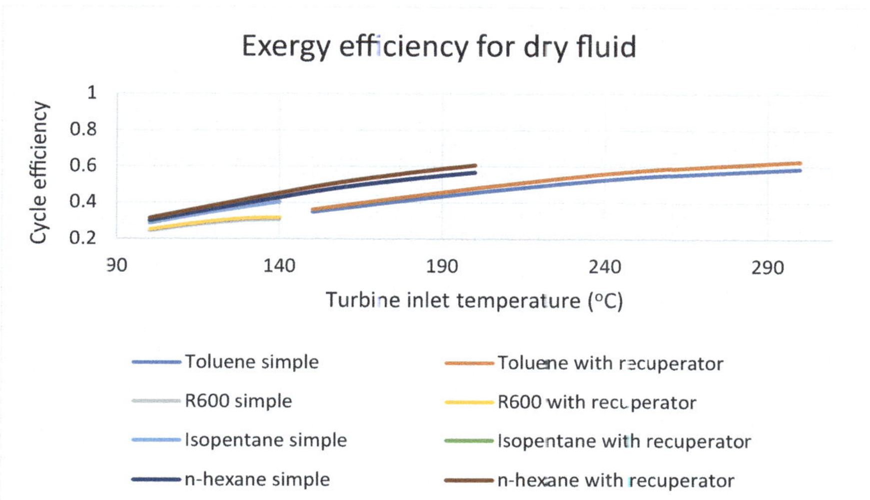

Fig. 12.18 Exergy efficiency for dry fluids

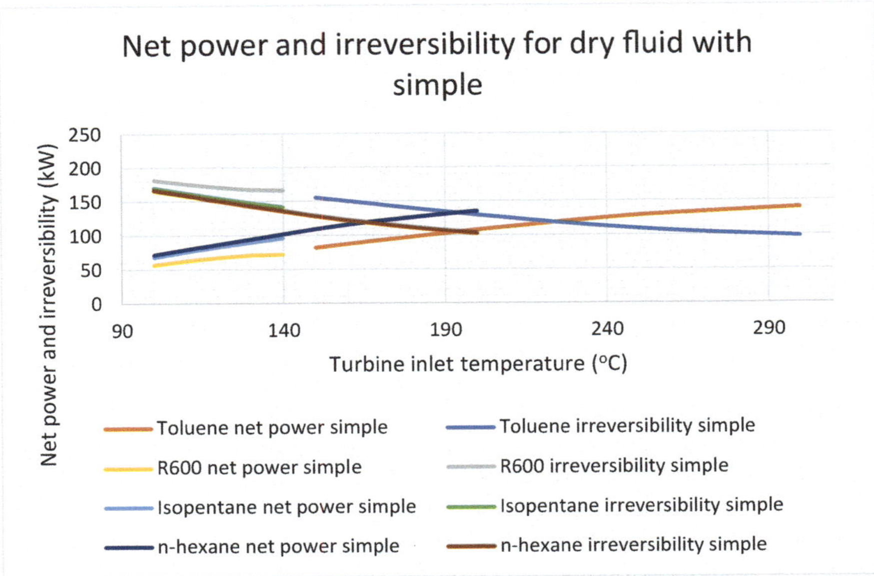

Fig. 12.19 Net power and irreversibility for dry fluids with simple ORC

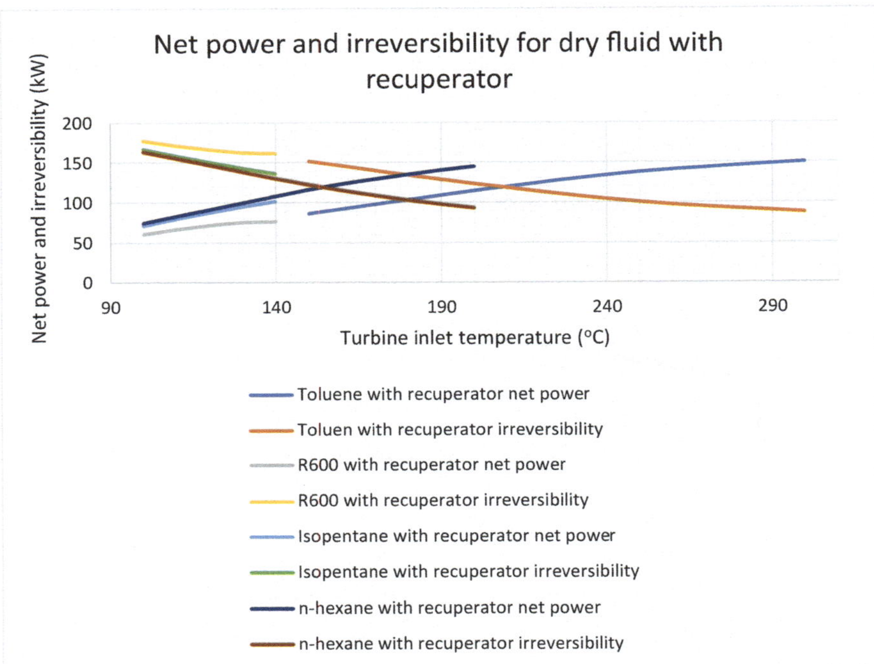

Fig. 12.20 Net power and irreversibility for dry fluids of RC with recuperator

turbine inlet temperature of 300 °C, the net power obtained from the toluene fluid is calculated to be 150.817 kW. If the turbine inlet temperature is below 200 °C, the highest net power among dry fluids is obtained from n-hexane. For a turbine inlet temperature of 200 °C, the net power obtained from n-hexane is calculated to be 145.486 kW. Additionally, as shown in Fig. 12.19, irreversibility and net power are inversely proportional. Irreversibility decreases as net power increases.

12.5 Conclusion

In this research, a thermodynamic examination has been conducted on a seawater-cooled organic Rankine cycle powered by the exhaust of a marine genset. The study delves into the cycle efficiencies of both a simple and one equipped with a recuperator configuration for the marine genset.Variations in cycle and exergy efficiencies, net power output, and irreversibility of the system have been analyzed concerning the turbine inlet temperature while maintaining a constant condensing temperature. The net power and irreversibility of the system have been precisely calculated under full-load conditions for the marine genset.

The efficiencies of the Rankine cycle (RC) and the exergy efficiencies are significantly influenced by the choice of organic fluid. In this study, the optimal fluid for low turbine inlet temperatures is found to be n-hexane, while for high turbine inlet temperatures, toluene exhibits the highest efficiency. The findings indicate that in the RC system with a recuperator, when the turbine inlet temperature surpasses 200 °C, toluene emerges as the organic fluid with the highest efficiency. Conversely, when the temperature is below 200 °C, n-hexane proves to be the most efficient organic fluid. Despite both toluene and n-hexane having an ozone depletion potential (ODP) of 0, there exists a notable contrast in their global warming potential (GWP) values. Toluene has a GWP value of 3, while n-hexane's GWP value is 11. From an environmental standpoint, considering its ODP of 0 and low GWP, toluene is recommended as the preferred organic fluid for ships operating at both low and high turbine inlet temperatures. Specifically, at a turbine inlet temperature of 300 °C, toluene demonstrates a cycle efficiency of 41%, exergy efficiency of 63%, net power output of 140.727 kW, and irreversibility of 150.817 kW under full-load conditions for the marine genset.

References

Ahmed, A., & Imre, A. (2023). Effect of critical temperature and category of zeotropic mixture working fluids on the thermal performance in subcritical ORC. *International Journal of Thermofluids, 20*, 100400.

Anonymous. (n.d.). Retrieved 30 January, 2024, from https://man-es.com/applications/projectguides/4stroke/GenSet/PG_GIMO_L2131DF-M.pdf

ASHRAE. (2019). *Standards 15&34*. Retrieved 30 January, 2024, from https://ashrae.org/technical-resources.

Baniam, M., Yari, M., & Mehr, A. (2024). Optimization waste heat recovery and power generation for industrial sustainability: A comparative study of supercritical CO2 Brayton, organic Rankine, and inverted Brayton cycles with synthesis natural gas as heat source. *Thermal Science and Engineering Progress, 47*, 102303.

Bellos, E., Lykas, P., & Tzivanidis, C. (2023). Thermodynamic investigation of a novel organic Rankine cycle including partial evaporation, dual-phase expander, flash tank, dry expander and recuperator for waste heat recovery. *Thermal Science and Engineering Progress, 46*, 102244.

Chao, Y., Wan, J., Li, Y., Deng, H., & Fu, W. (2024). Thermodynamic performance analysis of a novel integrated energy cascade system of liquid air energy storage and two-stage organic Rankine cycles. *Journal of Energy Storage, 75*, 109687.

Chitgar, N., Hemmati, A., & Sadrzadeh, M. (2023). A comparative performance analysis, working fluid selection, and machine learning optimization of ORC systems driven by geothermal energy. *Energy Conversion and Management, 286*, 117072.

Dalbakken, H., & Ertesvag, I. (2024). Safety aspects of Organic Rankine Cycles (ORC) with combustible working fluid and sub-ambient condenser pressure. *Energy Reports, 11*, 877–886.

Fallah, M., Mohammadi, Z., & Mahmoudi, S. (2022). Advanced exergy analysis of the combined SeCO2/ORC system. *Energy, 241*, 122870.

Hu, Z., Chen, Y., & Zhang, C. (2024). Role of R717 blends in ocean thermal energy conversion organic Rankine cycle. *Renewable Energy, 221*, 119756.

Konur, O., Colpan, C., & Saatcioglu, O. (2022). A comprehensive review on organic Rankine cycle systems used as waste heat recovery technologies for marine applications. *Energy Sources, Part A: Recovery, Utilization and Environmental Effects, 44*(2), 4083–4122.

Lemmon, E., Bell, I., Huber, M., & McLinden, M. (2018). *NIST standard reference database 23: Reference fluid thermodynamic and transport properties-REFPROP, version 10.0*. National Institute of Standards and Technology, Standard Reference Data Program.

Lion, S., Vlaskos, I., & Taccanic, R. (2020). A review of emissions reduction technologies for low and medium-speed marine diesel engines and their potential for waste heat recovery. *Energy Conversion and Management, 207*, 112553.

Liu, J., Zhou, T., & Yang, S. (2024). Advanced exergy and exergoeconomic analysis of a multistage Rankine cycle system combined with hydrate energy storage recovering LNG cold energy. *Energy, 288*, 129810.

Liua, X., Nguyen, M., & He, M. (2020). Performance analysis and optimization of an electricity-cooling cogeneration system for waste heat recovery of marine engine. *Energy Conversion and Management, 214*, 112887.

Loydu, T. (2022). *Rules for the classification of naval ships chapter 107*. Ship Operation Installations and Auxiliary Systems.

Macchi, E., & Astolfi, M. (Eds.). (2016). *Organic rankine cycle (ORC) power systems: Technologies and applications*. Woodhead Publishing.

Mohammed, A., Mosleh, M., El-Maghlany, W., & Ammar, N. (2020). Performance analysis of supercritical ORC utilizing marine diesel engine waste heat recovery. *Alexandria Engineering Journal, 59*, 893–904.

Paixao, V., Franco, L., & Dangelo, J. (2022). Thermoeconomic analysis of conventional and recuperative ORC for heat recovery of exothermic reactions. *Thermal Science and Engineering Progress, 33*, 101347.

Sadeghi, M., Chitsaz, A., Marivani, P., Yari, M., & Mahmoudi, S. (2020). Effects of thermophysical and thermochemical recuperation on the performance of combined gas turbine and organic rankine cycle power generation system: Thermoeconomic comparison and multiobjective optimization. *Energy, 210*, 118551.

Saksa, P., Lehtoranta, K., Kuittinen, N., Jarvinen, A., Jalkanen, J., Johnson, K., Jung, H., Ntziachristos, L., Gagn, S., Takahashi, C., Karjalainen, P., Ronkk, T., & Timonen, H. (2023).

Reduction in greenhouse gas and other emissions from ship engines: Current trends and future options. *Progress in Energy and Combustion Science, 94*, 101055.
Shi, L., Shu, G., Tian, H., & Deng, S. (2018). A review of modified Organic Rankine Cycles (ORCs) for internal combustion engine waste heat recovery (ICE-WHR). *Renewable and Sustainable Energy Reviews, 92*, 95–110.
Singh, D., & Pedersen, E. (2016). A review of waste heat recovery technologies for maritime applications. *Energy Conversion and Management, 111*, 315–328.
Song, J., Song, Y., & Gu, C. (2015). Thermodynamic analysis and performance optimization of an organic Rankine cycle waste heat recovery system for marine diesel engines. *Energy, 82*, 976–985.
Sun, S., Guo, X., Zhang, H., Li, Y., He, Z., & Li, C. (2024). Design and sensitivity analysis of a multistage solid oxide fuel cell hybrid system with an inter-cooled-recuperated gas turbine and organic Rankine cycle. *Energy, 286*, 129561.
Tian, Z., Yue, Y., Gu, B., Gao, W., & Zhang, Y. (2020). Thermo-economic analysis and optimization of a combined Organic Rankine Cycle system with LNG cold energy and waste heat recovery of dual-fuel marine engine. *International Journal Energy Research, 44*, 9974–9994.
Tian, Z., Gan, W., Qi, Z., Tian, M., & Gao, W. (2022). Experimental study of organic Rankine cycle with three-fluid recuperator for cryogenic cold energy recovery. *Energy, 242*, 122550.
Wang, F., Wang, L., Zhang, H., Xia, L., Miao, H., & Yuan, J. (2021). Design and optimization of hydrogen production by solid oxide electrolyzer with marine engine waste heat recovery and ORC cycle. *Energy Conversion and Management, 229*, 113775.
Xiao, M., Zhou, Y., Miao, Z., Yan, P., Zhang, M., & Xu, J. (2024). Multi-condition operating characteristics and optimization of a small-scale ORC system. *Energy, 290*, 130099.
Yang, M., & Yeh, R. (2014). Analyzing the optimization of an organic Rankine cycle system for recovering waste heat from a large marine engine containing a cooling water system. *Energy Conversion and Management, 88*, 999–1010.
Yun, E., Park, H., Yoon, S., & Kim, K. (2015). Dual parallel organic Rankine cycle system for high efficiency waste heat recovery in marine application. *Journal of Mechanical Science and Technology, 29*(6), 2509–2515.
Zhang, X., & Li, Y. (2023). An examination of super dry working fluids used in regenerative organic Rankine cycles. *Energy, 263*, 125931.
Zhang, X., Zhang, Y., & Wang, J. (2020). New classification of dry and isentropic working fluids and a method used to determine their optimal or worst condensation temperature used in the Organic Rankine Cycle. *Energy, 201*, 117722.
Zhang, X., Zhang, Y., & Wang, J. (2021). Evaluation and selection of dry and isentropic working fluids based on their pump performance in small-scale organic Rankine cycle. *Applied Thermal Engineering, 191*, 116919.
Zhu, H., Xie, G., & Berrouk, A. (2024). Enhancing performance of multi-pressure evaporation organic Rankine cycle/supercritical carbon dioxide Brayton cycle through genetic algorithm and machine learning optimization. *Energy Conversion and Management, 301*, 118037.

Chapter 13
Development and Performance Analysis of a Solar-Powered Thermoelectric Air Conditioning System for Space Conditioning

Manuel Fernando Ariza Taba and Koray Ülgen

Nomenclature

COP	Coefficient of performance
TEC	Thermoelectric cooler
TEC	Thermoelectric devices
TEG	Thermoelectric generators
TPV	Thermo-photovoltaic

13.1 Introduction

Solar cooling systems have taken great interest in recent decades. Since these can be used anywhere. However, integrating these devices in buildings has had major disadvantages in not having appropriate and truthful information that generate confidence using this technology for investors.

Using heating, ventilation, and air conditioning (HVAC) systems in homes and buildings consumes a considerable amount of energy. Likewise, in implementing this type of technology, several parameters must be considered, such as design, application type, and control systems. For the past decades, many researches have aimed to reduce energy consumption, and keeping optimal efficiency has been preponderant in these types of systems. Also, it generates a minor impact on the environment with the use of these devices. Different articles about the strengths and

M. F. A. Taba · K. Ülgen (✉)
Ege University, Solar Energy Institute, Bornova-İzmir, Türkiye
e-mail: koray.ulgen@ege.edu.tr

M. Z. Sogut, M. Koray (eds.), *Energy Rationality and Management for Decarbonization*, Green Energy and Technology,
https://doi.org/10.1007/978-3-031-68497-5_13

weaknesses of the systems have been published, applying different models (Afroz et al., 2018).

Likewise, different combinations have been made to improve the performance of each system and provide better comfort for users. The process of hybridization has allowed working with many technologies. However, trying to stabilize these systems requires paying attention to the control systems. For example, the use of natural ventilation combined with an HVAC system has a significant advantage while taking off the air, the energy consumption decreases. However, to maintain an average temperature within the area, the system must have a robust control system that allows its proper operation. The system compares the inside and outside of the home and determines the amount of cold or heat that the HVAC system must provide. Mathematical models have been used to evaluate system function based on data like the temperature, humidity, ppm of CO_2, solar radiation, and wind speed (Chen et al., 2018).

Different algorithms are also used to handle the power demand within the HVAC system. The algorithms allow advanced control of the system and likewise, it can work more optimally. In this document, a solution based on a system is presented in an algorithm that allows the control of the work cycles of the units of the HVAC systems. The solution for the proper functioning of this system is the use of the combination of a novel and binary search algorithm. Likewise, the HVAC system is connected to an intelligent thermostat that allows being approximated to be applied in residential buildings without making any modification in the electrical infrastructure (Adhikari et al., 2018).

HVAC systems have not only had applications in homes but also in mobility. In the city of Beijing, China, for 4 months, an HVAC system has been monitored in the metro stations. Getting results in decreasing the energy consumption of each metro station between 20 and 38% (Wang et al., 2017).

Many systems have different control systems to manage the operation of the devices. Genco et al. (2018), proposed using one model-free control which could learn and make decisions itself based on prior historical data to optimize the energy consumption of the HVAC system. Many systems have used mathematical models that analyze data and execute actions. The control system stored the data in real time and showed a saving in energy consumption of 31.17%.

Ghahramani et al. (2017), applied a mathematical model for the analysis of HVAC systems using two types of control. The first has analyzed the amount of airflow at the inlet and the second is the temperature at the air inlet of the system. The two control models analyzed the energy consumption based on humidity, heating, and cooling using fans and extractors of the system.

13.1.1 Solar-Powered HVAC Systems

Hybrid systems have been used to harness the energy that comes from the sun to generate electricity and heat at the same time. A hybrid system of solar collectors (vacuum tube or flat plate) heats the water in a storage tank and then supplies it when required. Storage tanks are charged at night and discharged during the day. The economic profitability of the system largely depends on the price of energy (Mammoli et al., 2010).

Using different refrigerants to reduce the potential for global warming has had significant consideration in recent times. Such as integrating systems such as solar-assisted heat pumps to improve efficiency in individual systems. Lee et al. (2018) have done this study on the operation of a solar-assisted heat pump and low-affected coolant (R-1233zd (E)) using a flexible solar collector. The flexible solar collector showed an average efficiency of 40% regardless of weather. In this study, the high performance of the sun-assisted heat pump using R-1233zd (E) was obtained by controlling the heat temperature source and mass flow. The heat capacity of the solar-assisted heat pump with the use of R-1233zd (E) was low compared to R-134a under the same conditions.

Deme Belafi et al. (2017) used numerical methods in the analysis of HVAC systems to predict performance and optimize control parameters. The system was installed in an educational building of an area of 7000 m^2 and analyzed throughout the year. In this study, solar assistance has represented more than 90% of the total heating. During the colder season and the hottest season, the system made considerable contributions. However, during the months when the temperature was warm, the building operated without any type of external assistance. Heating and cooling buildings were set to maintain the set temperature. Also, use excess heat to produce cold water and excess heat was stored and used when required. The numerical system has analyzed an HVAC system to predict performance and optimize the control parameters. The system took the data for 1 year in an area of 7000 m^2. In this study, over 90% of the total heating in the building was provided by the solar system. While the temperature was warm, the building did not need any help from the solar system. Meanwhile, in the coldest and hottest seasons, the system provided considerable energy. Additionally, the energy produced in excess was stored when it was needed.

Global warming and energy crises have precise objectives: reduce pollution and use energy efficiently. A solar-assisted hybrid power gas heat pump system is proposed as an alternative to reduce heating and cooling energy consumption. The system was evaluated throughout the life cycle of the system, based on the life cycle of the solar panels, since the effect of the environment directly affected them. Simulations showed that when using different photovoltaic ratios, the system achieves an improvement in energy consumption, however; it was hazardous to the environment. As a result, the return on investment for a solar-assisted hybrid power gas heat pump system was approximately 13.4 years and the photovoltaic ratio was 40% (Zhao et al., 2017).

Australia is characterized by having a climate with a lot of sun, which results in great use of air conditioners. HVAC systems produce a negative impact on the environment due to the different types of refrigerants used within the system. Different studies were carried out to analyze technologies and developments in air conditioners assisted with solar energy. The use of HVAC systems in buildings has created a problem. First the consumption of fossil fuels for the operation, and second the increase of mold, and the quality of the air inside the buildings. Also, the theories and technologies involved are discussed, such as solar energy and the different types of solar collectors Baniyounes et al., 2013).

13.1.2 Solar-Powered Thermoelectric Systems

Zeyghami et al. (2015) report on the development of thermos-mechanical technologies. Also, a detailed study of the literature on the design, configuration, and analysis of studies that had been carried out previously was made. Likewise, the improvements in thermos-mechanical solar cooling systems were shown. A system with different configurations was presented where each one of the components and the performance of the system was analyzed. The scope of the study allowed for obtaining important results for the improvement in the functioning of the system. Likewise, a comparison of the conventional solar hybrid systems with the system of this study was made. The work conducted in this research was based on a thermoelectric cogeneration system. The system uses the heat that comes from the exhaust gases to heat water and produces electricity simultaneously using the thermoelectric generators (TEG). The thermal analysis of the system was based on the position of the thermoelectric generators within the system, the TEG was in the internal and external walls of the tank and the pipes. The results showed that the power produced with the gases was 0.35 W and the hot water at a temperature of 76 °C. Six scenarios with different configurations were evaluated. In configurations 2 and 3 the water temperature had increased since the heat was trapped inside the tank by the addition of more thermal resistance to the tank wall. In case 3, the TEGs were located inside the tank wall. For cases 4 and 5, the TEGs were located inside or outside the pipe surface and the temperature was lower than in cases 1, 2, and 3. Finally, in case 6, the water temperature was similar to that in cases 4 and 5. The maximum power produced was 52 W in case 6 and the smallest was in case 2 with a value of 6.8 W. The economic analysis showed that locating the TEGs on the surface of the pipes could reduce the cost by 75% compared to if the TEGs were located on the surface of the tank. The cost of power generation for each scenario with the use of the TEGs provided by the manufacturer was ($20/TEG). To obtain greater power, the cost of the system increases its costs in great value and also the environmental impact. Likewise, the result showed that the location of the TEG does not influence the reduction of CO_2 which was found to be close to 6 tons/year. Additionally, the evaluation showed that the configuration when the TEGs are in the pipe wall the energy recovery has more profitability (Jaber et al., 2018).

The increase in the temperature of a solar cell is one of the problems affecting solar energy today. It not only affects the useful life of the photovoltaic cells but also affects the performance of the system during operation. In this article, the authors used the hybrid system that generates electricity and at the same time heat that was used for other purposes. Hence, a proposal was made for a hybrid system that involves a tube that absorbs the heat that comes from the solar cells and a thermoelectric generator (TEG) that transforms the heat into electricity. The mathematical model was based on the energy balance to evaluate the performance of the hybrid system and observe the data that came from the photovoltaic cells. The result showed that the integration of thermoelectric materials improves the performance of the photovoltaic cells to generate a greater power output. The authors concluded that the system can be used in temperate climates where electricity has a higher demand compared to thermal energy (Makki et al., 2016).

13.1.3 Application of Thermoelectric Devices

The use of thermoelectric devices (TEC) solar-assisted has increased during the past decades. This has been because of the performance of devices and their economic affordability. Some hybrid systems use thermoelectric devices for generating electricity and heating the water (Twaha et al., 2016). The solar collector can be used for different applications such as to provide energy, control systems, HVAC systems, and sensors (Fabrizio et al., 2014). The thermoelectric cooler (TEC) has been used with the solar thermoelectric generator (TEG) to analyze the performance of devices. Also, it has been used to calculate the number of devices to achieve the best performance of the system all year. The analysis of the system allows for determining the performance and the physical properties of TEC modules (Khattab et al., 2006). In hybrid thermoelectric systems such as thermoelectric cogeneration systems (TEG), both electricity generation and water heating are achieved at the same time by the system (Zheng et al., 2014). The water is used as a refrigerant on the cold side of the TEG and circulates through a heat exchanger to heat the water and generate electricity. Solar energy comes from a collector located at the top of the building. The heat coming from the collector is usually added to the boiler's exhaust pipe to supply heat. The efficiency of the system is increased, however, there are some challenges in the transfer of heat to and from the TEG. The main objective of this technology was used in heat recovery. However, it is now used in solar applications (Chen et al., 2014). Thermo-photovoltaic (TPV) cells provide the heat input to the TE to generate electricity. The TPV and TEG devices have been installed in cascade mode. The TPV acts as a power supply and TEG generates electricity from the heat. The analysis showed that these systems are suitable for some applications on a small scale (Qiu & Hayden, 2012).

During the last few years, the systems assisted with solar energy have had great importance in the improvement of the cooling systems since they use renewable energies to supply energy to the system. Even so, one of the drawbacks has been the integration of this type of system in buildings. These systems provide good opportunities for the reduction of energy consumption in buildings. The current developments have allowed a significant development, however, the integration of these systems in the buildings has generated great uncertainty. In this document, the possibilities for the integration of solar-assisted systems in the facades of buildings were discussed. The study was conducted based on the type of solar cooling technologies and cooling façade concepts. The results were obtained using a matrix that described the possibilities of several components and subsystems of the cooling process. Also, it evaluated the cooling systems based on current research, work prototypes, and developed projects (Prieto et al., 2017).

Thedeby (2014) designed a prototype for heating and cooling air using Peltier cells and a solar panel. The system was used in conjunction with an existing HVAC system that was used to preheat and pre-air the air entering the house. The objective was to reduce energy consumption and also reduce the cost of electricity. Additionally, analyze the behavior of the system and the economic feasibility to make a comparison with the Swedish heating systems. The results obtained based on theoretical and practical simulations showed that the annual energy output was 1300 kWh, 9% corresponding to the use of heating and 5% of the total energy for a Swedish house. The economic results showed that the economic viability of the system was directly related to the time of use of the heating, and the period of re-entry of the investment was between 3 and 10 years. Also, the best option was to use the system together with an existing system, which would help reduce the energy consumption of the system.

13.2 The System Design and Operation Modes

13.2.1 Design

The system is divided into two parts. The first part is constituted by the solar cells, which are responsible for supplying energy to the different devices that are part of the system (sensors, TECs, cables, fans, and all the control devices of the system). The system was designed to work with the energy that comes from the sun. The second part of the system is constituted by TEC devices that are responsible for providing heat or cold depending on the temperature required within the houses. Eight thermoelectric devices were installed in the Peltier panel and it was installed on the back of the air solar collector. The air enters through one of the collector's entrances and goes through the other entrance of the collector when having contact with the Peltier panel, the air is either cooled or heated. Finally, the air enters the home, and at the same time, the air that was inside the home is pumped outside to complete the cycle of air circulation.

The solar air collector was modified inside to be able to install the Peltier panel and in the same way, the channels through which the air circulates inside it. The materials used in the system played a fundamental role in performance. The Peltier panel is made of aluminum for the dissipation of cold or heat. The fans are responsible for the circulation of air inside the collector. The sensors are responsible for making measurements inside and outside the system. The control system is responsible for making decisions of the system based on the established parameters to maintain an average temperature inside the home. The control systems change the air velocity, closing and opening of the air inlets of the collector, and the amount of

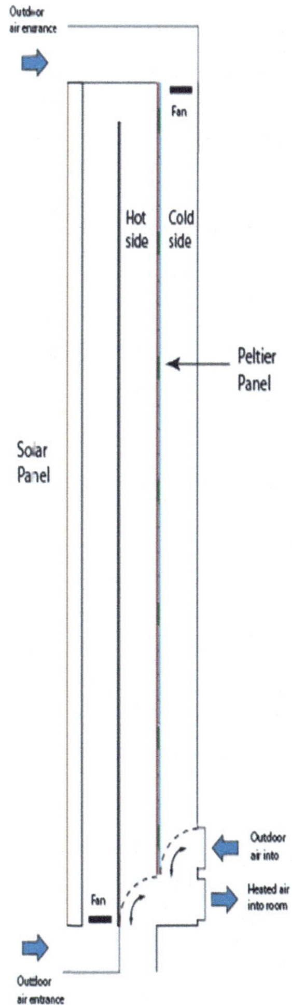

Fig. 13.1 Diagram of the system

energy that was provided to the TEC devices for the generation of cold or heat. Figure 13.1 shows a diagram of the system (Ariza Taba, 2019).

13.2.2 Operating Modes

The system consists of two operation modes, which are responsible for providing cold or heat. Figure 13.2 shows the heating mode on the left side and the cooling mode on the right side (Ariza Taba, 2019).

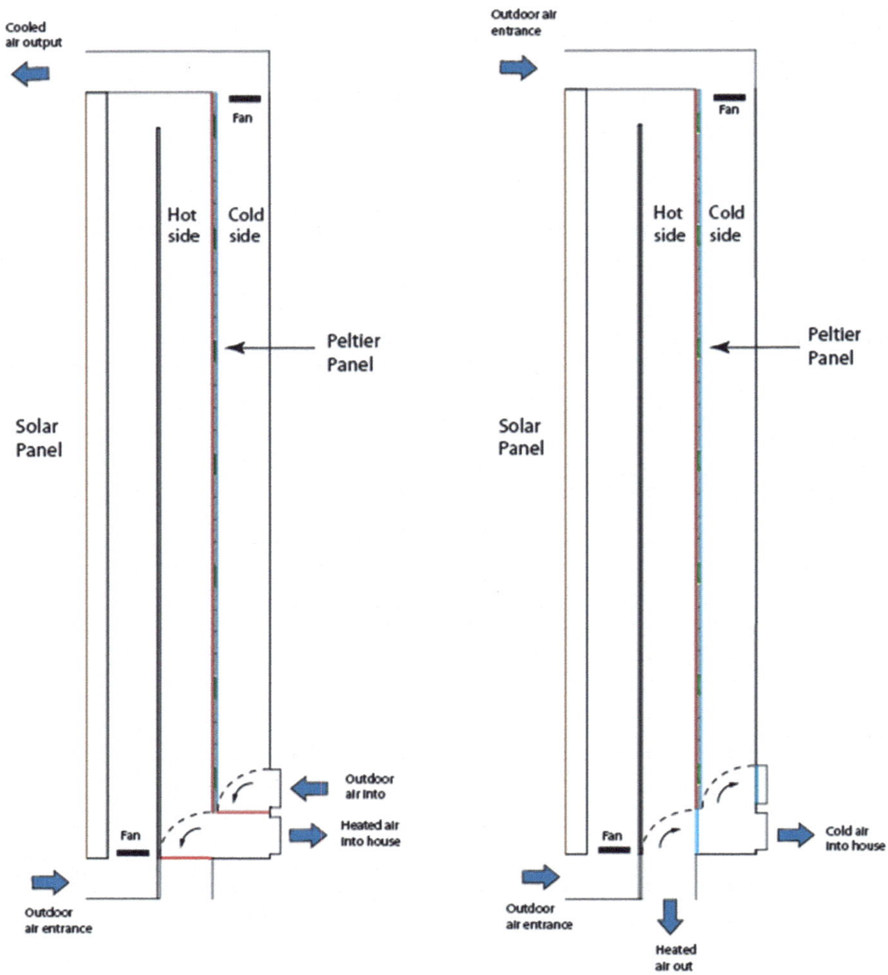

Fig. 13.2 Heating and cooling mode

13.2.2.1 The Cooling Mode

The cooling mode is designed to cool the air that enters the collector. As soon as the air enters the collector, the temperature sensors sense the air temperature. The air is cooled at the moment it has contact with the Peltier panel. Finally, the cold air enters the house while the warm air inside the room is withdrawn to the outside.

13.2.2.2 The Heating Mode

The heating mode is responsible for heating the air that enters the collector. The air enters through one of the collector inputs and is sensed by the temperature sensors. The air is heated when it has contact with the Peltier panel, then it is directed toward the house while the cold air in the room is pumped outside of the house by system fans. The objective of the two modes of operation of the system was to maintain the room temperature, humidity, and remove contaminating particles within the home. The advantages of the system are that it requires less energy for its operation since it does not have many mobile parts; its maintenance is more economical, and also the integration of this type of system is much more affordable since it does not need any type of external liquid to cool or heat the air.

13.3 Methodology

This section presents the analytical methodology considered in the project to obtain the necessary results of the system. The project was conducted at Ege University, Solar Energy Institute in Izmir-Turkey, (latitude 38°27′ and longitude 27°14′). The data was provided by the meteorological station located at the Solar Energy Institute (Ariza Taba, 2019). The data was analyzed and the amount of energy produced by the device was analyzed, and the number of light hours was also considered. Based on the data obtained and the ambient temperature, the period of use of the device was determined for both summer and winter. To find the amount of energy generated with the use of the system, the amount of solar radiation, the daily solar radiation, and the solar radiation per hour for the location was collected and analyzed. All climatic parameters were also considered since the functioning and efficiency of the system depended on it.

13.3.1 Solar Radiation Analysis

The extra-terrestrial solar radiation received on a horizontal surface located in the upper limit of the atmosphere has been calculated using the Eq. (13.1). All the equations used in this section to calculate the solar radiation analysis come from this literature (Ariza Taba, 2019; Duffie & Beckman, 2013).

$$G_o = G_{SC} * \left(1 + 0.033 * \cos\frac{360 * n}{365}\right) * \cos\phi * \cos\delta * \cos\omega + \sin\phi * \sin\delta \qquad (13.1)$$

Hence, using the equation below the daily solar radiation on horizontal radiation in the absence of the atmosphere has been evaluated using the Eq. (13.2),

$$H_o = \frac{24 * 3600 * G_{SG}}{\pi}\left(1 + 0.033 * \cos\frac{360 * n}{365}\right) * \left(\cos\phi * \cos\delta * \sin\omega_s + \frac{\pi\omega_s}{180} * \sin\phi * \sin\delta\right) \quad (13.2)$$

The amount of energy received hourly on a horizontal surface has further been evaluated using the given Eq. (13.3),

$$I_o = \frac{12 * 3600}{\pi} * G_{SG} * \left(1 + 0.033 * \cos\frac{360 * n}{365}\right) * \left[\cos\phi * \cos\delta * (\sin\omega_2 - \sin\omega_1) + \frac{\pi(\sin\omega_2 - \sin\omega_1)}{180} * \sin\phi * \sin\delta\right] \quad (13.3)$$

The monthly average clearness index is defined as the average daily amount of energy produced monthly on a horizontal surface divided by the average of extra-terrestrial radiation that has been calculated using the Eq. (13.4),

$$K_T = \frac{H}{H_o} \quad (13.4)$$

Incident angle θ, on a horizontal surface, when the radiation directly comes, the angle between normal and the surface is known as the incident angle was estimated using the Eq. (13.5),

$$\cos\theta = \sin\delta * \sin\phi * \cos\beta - \sin\delta * \cos\phi * \sin\beta * \cos\gamma + \cos\delta * \cos\phi * \cos\beta * \cos\omega + \cos\delta * \sin\phi * \sin\beta * \cos\gamma * \cos\omega + \cos\delta * \sin\beta * \sin\gamma * \sin\omega \quad (13.5)$$

Zenith angle θ_z, is the angle between the vertical at the site and the site to the sun line and can be calculated using Eq. (13.6),

$$\cos\theta_z = \cos\phi * \cos\delta * \cos\omega + \sin\phi * \sin\delta \quad (13.6)$$

Direct radiation or beam radiation, which describes the straight-line path of the solar radiation from the sun to the surface of the earth, was calculated using the Eq. (13.7),

$$R_b = \frac{\cos\theta}{\cos\theta_z} \quad (13.7)$$

Thus, total radiation on tilted surfaces was further estimated using the expression illustrated below in Eq. (13.8),

$$I_T = I_b * R_b + I_d * \left(\frac{1+\cos\beta}{2}\right) + I * \rho_g * \left(\frac{1-\cos\beta}{2}\right) \tag{13.8}$$

However, in the case of this study, the total radiation striking the surface of the panel was estimated using the Eq. (13.9) as given by:

$$I_{\beta T} = I_b * R_t + I_d * 0.5 + I_T * \rho_g * 0.5 \tag{13.9}$$

13.3.2 PV Panel Energy Balance

The characteristics and performance of the PV panel are determined by different factors. In this study, the maximum power point for the solar panel in consideration was estimated using Eq. (13.10) given below (Duffie & Beckman, 2013):

$$P_{\text{mpp}} = V_{\text{mpp}} * I_{\text{mpp}} \tag{13.10}$$

whereas the panel Fill Factor FF, for the solar PV technology was estimated using the following Eq. (13.11),

$$\text{FF} = \frac{V_{\text{mpp}} * I_{\text{mpp}}}{V_{\text{OC}} * I_{\text{SC}}} \tag{13.11}$$

The photoelectric efficiency of solar cells η, for the solar PV panel considered in this study, was estimated considering the Eq. (13.12) given below:

$$\eta = \frac{I_{\text{SC}} * V_{\text{OC}} * \text{FF}}{A * P_{\text{sun}}} \tag{13.12}$$

Thus, the total energy produced (E) by the solar PV panel considering losses due to the balance of system components and low solar radiation and dust was evaluated using the Eq. (13.13) given below.

$$E = I_{\beta T} * A * \eta * (1-\lambda)(1-\rho') \tag{13.13}$$

13.3.3 TEC Module Energy Balance

The analysis of the TEC devices allows for estimating the performance of the device in terms of the efficiency and the quantity of energy consumption, as well as the number of energy transfers in the form of heat or cold. In TEC devices both sides of the element cold side and the hot side have been considered, for the cold side, the following equations have been considered in the evaluation [23].

13.3.3.1 Hot Side Heat Transfer

The energy balance heat side is given by Eq. (13.14),

$$Q_H = 2 * N * \left(\alpha * I * T_H - \frac{I^2 * \rho}{2g} - k * g * \left(T_H - T_C \right) \right) \tag{13.14}$$

where α is Seebeck coefficient (V/K), ρ is resistivity (Ωm) and k is conductivity (W/mK).

13.3.3.2 Cold Side Heat Transfer

Peltier cooling is defined as the process of cooling components via the Peltier effect and is given by Eq. (13.15),

$$Q_{cs} = 2 * \alpha * N * I * T_C \tag{13.15}$$

Joule heating (known as resistive or ohmic heating) is the process where the electric current flows through a resistance and the energy is converted into heat theoretically as given by the Eq. (13.16),

$$Q_J = \frac{2 * N * I^2 * \rho}{g} \tag{13.16}$$

Conducted heat is giving by Eq. (13.17),

$$Q_{con} = 2 * k * N * g * \Delta T \tag{13.17}$$

where k is thermal conductivity, and ΔT is a temperature difference between the cold and hot side of the TEC device ($T_H - T_C$). The energy balance cold side is given by Eq. (13.18),

$$Q_C = 2 * N * \left(\alpha * I * T_C - \frac{I^2 * \rho}{2g} - k * g * \left(T_H - T_C \right) \right) \tag{13.18}$$

Electrical energy consumption is given by Eq. (13.19),

$$Q_E = 2 * N * \left(\frac{I^2 * \rho}{2g} + \alpha * I * \Delta T \right) \tag{13.19}$$

The coefficient of performance (COP) is defined as the amount of heat pumped divided by the amount of supplied power. COP expresses the number of units of heat pumped for each unit of power supplied is given by Eq. (13.20),

$$\mathrm{COP} = \frac{Q_C}{Q_E} = \frac{Q_H}{Q_E} \tag{13.20}$$

13.4 Result and Discussion

The results showed that Izmir has great potential to generate energy from the sun's energy for heating and cooling the building through the use of thermoelectric devices. Figure 13.3 below presents the hourly solar radiation at the project site as recorded by the Solar Energy Institute Meteorological Station in Izmir [23]. Figures 13.4 and 13.5 show the variation of the cell temperature and module efficiency throughout the year concerning project site ambient temperature. The ambient air temperature was directly related to the performance of the PV system, the maximum value reached for ambient temperature was 39 °C in summer, and the minimum value reached was near −0.1 °C in winter. The average temperature in spring was near 12 °C, and in autumn was near 20 °C.

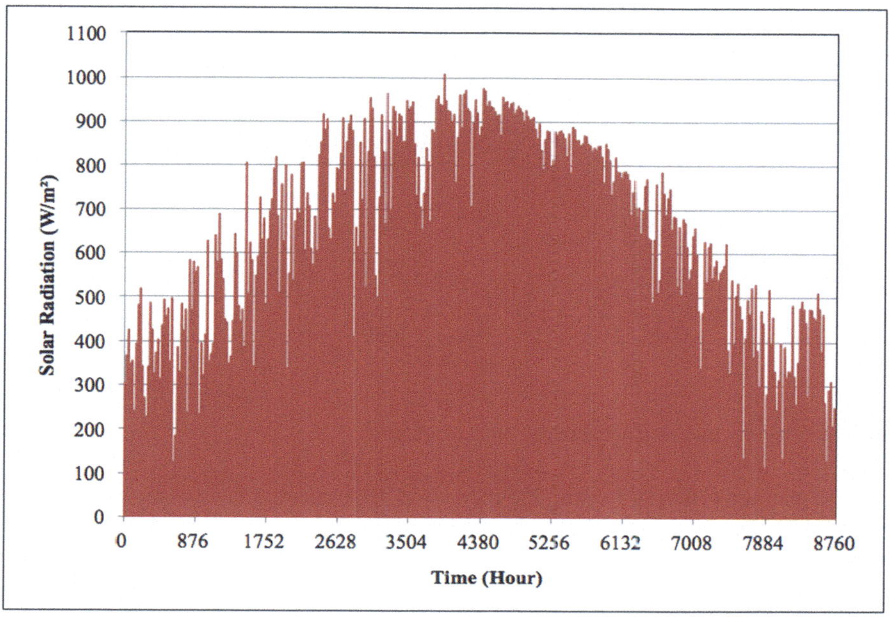

Fig. 13.3 Hourly solar radiation in Izmir, Turkey

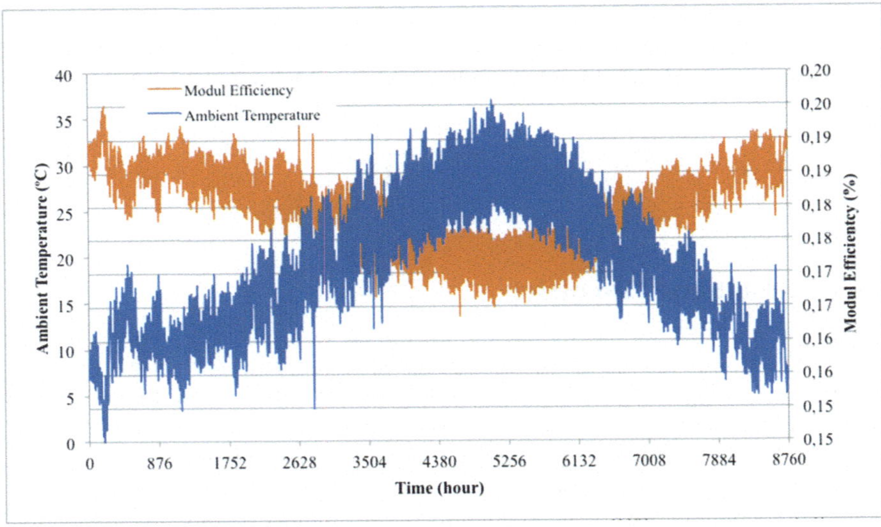

Fig. 13.4 Hourly ambient temperature and module efficiency

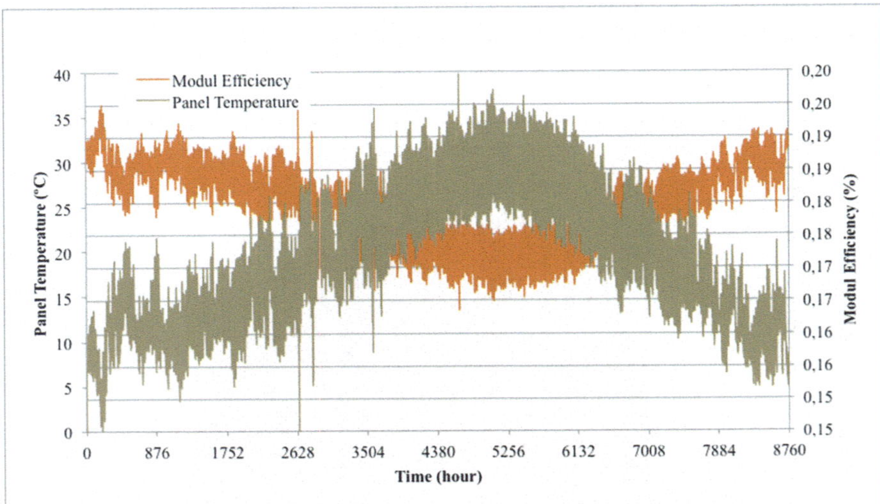

Fig. 13.5 Hourly panel temperature and module efficiency

The average efficiency of the PV panel was 18% during the year; however, there were some changes where the performance of the PV panel was high, for example in winter until the middle of spring the average module efficiency was 19%. The best performance of the PV panel was during spring and at the end of autumn.

Figure 13.6 shows the hourly energy output by the solar PV system during the year. It is worth noting that the system energy output at the beginning of the year

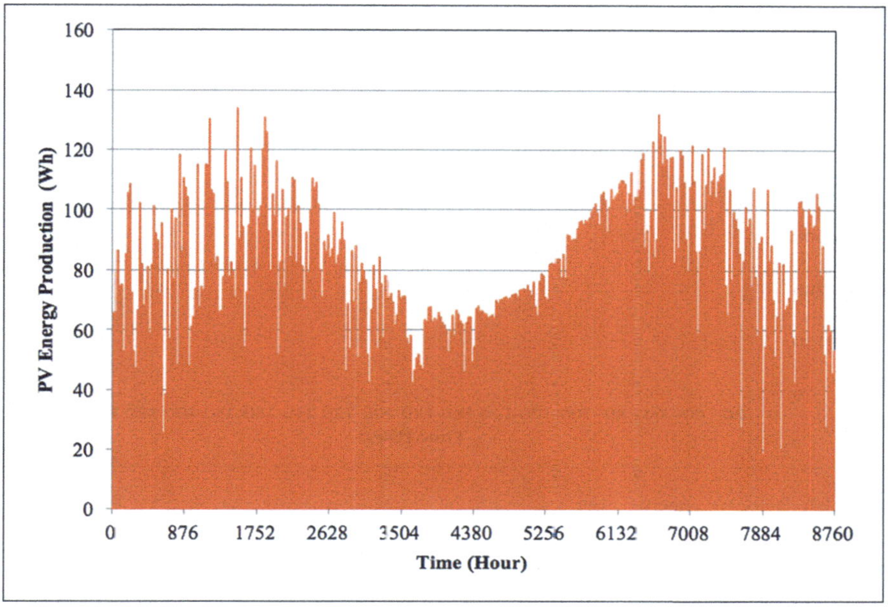

Fig. 13.6 Hourly PV energy output

was considered good due to a higher amount of solar radiation being received by the device. To make a comparison between energy output and solar radiation, the results achieved at the beginning of the year the solar radiation was low which was reflected in the solar PV panel energy output. In February, the energy output by the PV panel increased and was constant throughout the months until November. However, there was a significant change at the beginning of March when energy output increased, and there were some changes in energy output across the months until May.

Figure 13.7 illustrates the daily energy output of solar PV panels per day; thus, the results showed a good performance of the system even when the ambient air temperature was very high.

In the middle of the year, the energy output was less as compared to other months and days of the year. However, during the periods of hottest days, the performance of the system was reduced due to high cell temperatures, hence affecting the energy output and the overall system performance. The following sections highlight the system performance and energy output during the coldest day and hottest day of the year.

13.4.1 The Coldest Day

The meteorological parameters measured for the coldest day (January 9th, 2018) of the year in which the study was conducted are given in Table 13.1. Figure 13.8 shows the changes in the hourly ambient temperature, solar PV cell temperature,

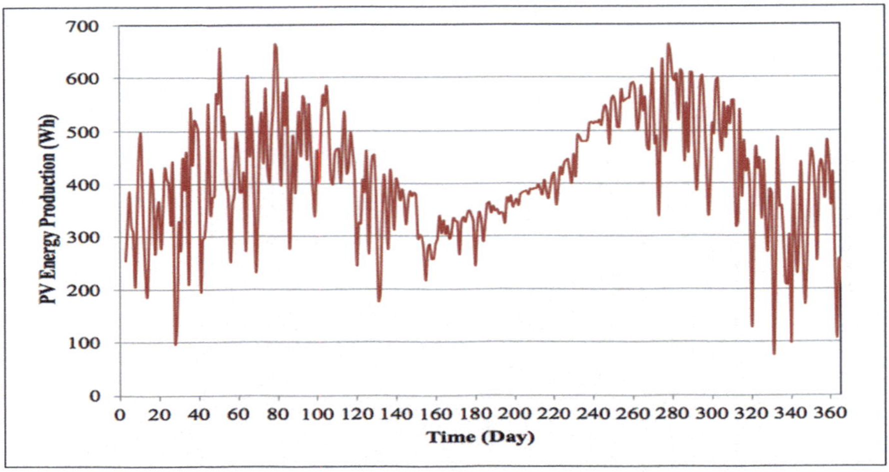

Fig. 13.7 Daily PV energy output

Table 13.1 Meteorological parameter of the coldest day

Hour	Wind speed (m/s)	Wind direction (°)	Humidity (%)	Ambient temperature (°C)	Horizontal solar radiation (W/m²)	Vertical solar radiation (W/m²)
1	3.08	81.4	62.57	0.89	0.00	0.00
2	3.14	86.3	63.27	0.69	0.00	0.00
3	3.60	84.9	63.16	0.62	0.00	0.00
4	3.83	93.6	63.06	0.57	0.00	0.00
5	4.17	79.5	62.82	0.47	0.00	0.00
6	4.06	72.5	64.34	0.23	0.00	0.00
7	4.39	65.6	65.00	0.15	0.00	0.00
8	4.45	53.9	65.06	0.12	0.00	0.00
9	3.93	87.0	66.85	−0.27	5.43	2.2
10	4.12	58.9	61.87	0.94	124.75	68.8
11	5.09	60.8	55.68	2.34	287.78	201.8
12	5.26	61.8	53.58	3.36	415.14	329.9
13	4.19	66.7	48.06	4.84	497.64	402.9
14	3.78	69.4	43.55	6.03	518.73	384.8
15	3.89	58.2	41.74	6.70	477.82	289.3
16	3.67	116.3	41.84	7.26	361.80	155.8
17	3.53	125.7	44.67	6.79	213.85	21.7
18	3.09	133.9	48.91	5.78	56.42	6.0
19	3.12	82.8	56.77	3.86	0.63	0.00
20	3.62	53.3	60.50	2.85	0.00	0.00
21	3.88	52.1	58.41	3.03	0.00	0.00
22	3.64	52.9	57.93	2.84	0.00	0.00
23	3.68	80.0	61.09	2.03	0.00	0.00
24	3.64	62.0	63.84	1.58	0.00	0.00

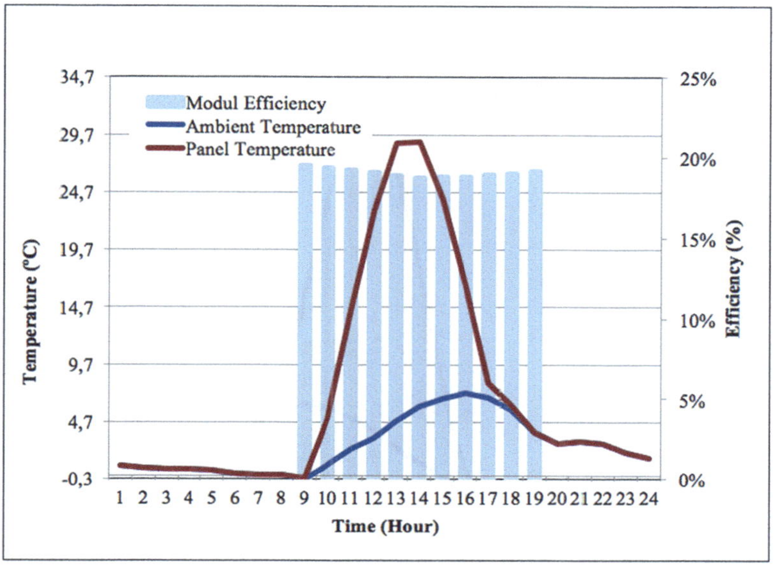

Fig. 13.8 Comparison of hourly ambient temperature, panel temperature, and module efficiency for the coldest

and the efficiency of the module during the coldest day of the year. In the night, the temperature values for the coldest day reached closer to 0 °C. However, during the daytime, the maximum solar panel temperature reached nearly 29.70 °C.

Figure 13.9 shows the relationship of values between ambient air temperature, solar PV cell temperature, and the hourly solar PV energy output for the coldest day. The production of energy was directly related to the amount of solar radiation that was present. Additionally, it was observed that a direct relationship existed between solar radiation and temperature since the increase in ambient air temperature increased energy output. The maximum energy output was observed when the sun formed a line perpendicular to the solar module.

Figure 13.10 shows the hourly system energy output concerning hourly solar radiation. The graph shows the values of solar radiation and the maximum energy output achieved by the selected PV solar panel. The variations in energy output were due to ambient air temperature changes, system losses, and panel efficiency.

13.4.2 *The Hottest Day*

The meteorological parameters measured for the hottest day (July 30, 2018) of the year in which the study was conducted are given in Table 13.2. The hottest day was considered in this study to analyze the system performance during this critical period when the system had to provide more cooling to the room. The evaluation of

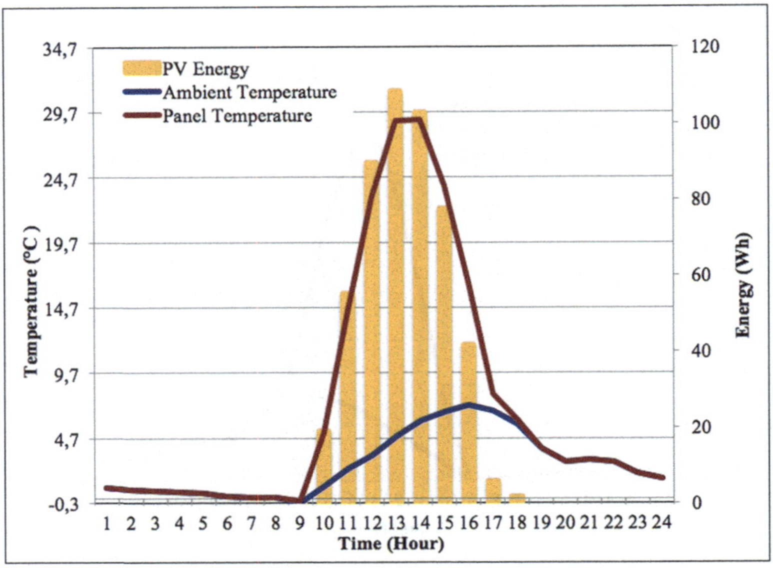

Fig. 13.9 Hourly ambient temperature, panel temperature, and energy output for the coldest day

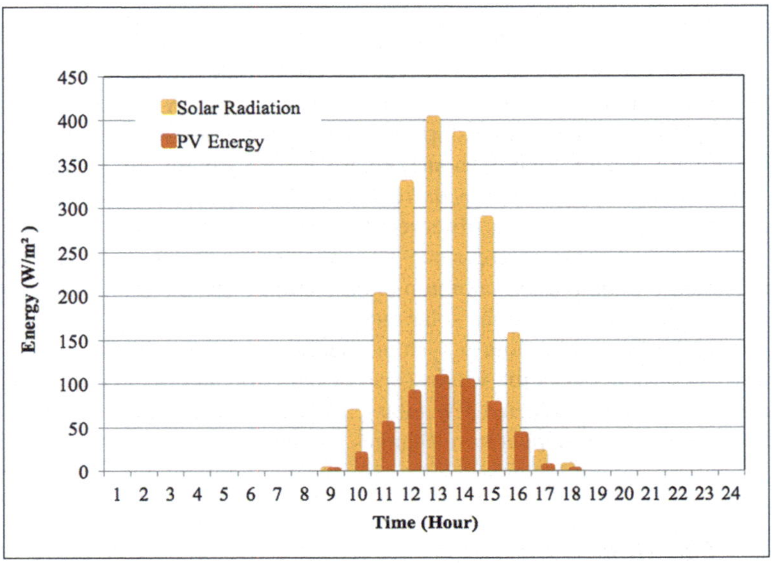

Fig. 13.10 Comparison of solar radiation and PV energy output for the coldest day

Table 13.2 Meteorological parameter of the hottest day

Hour	Wind speed (m/s)	Wind direction (°)	Humidity (%)	Ambient temperature (°C)	Horizontal solar radiation (W/m²)	Vertical solar radiation (W/m²)
1	2,75	83,43	46,67	27,16	0,00	0,00
2	2,97	102,33	49,29	26,54	0,00	0,00
3	2,55	89,34	52,82	25,84	0,00	0,00
4	1,98	134,95	56,18	25,24	0,00	0,00
5	2,12	123,87	56,84	24,89	0,00	0,00
6	2,38	67,13	57,34	24,48	0,00	0,00
7	2,54	58,61	58,55	24,26	21,28	2,28
8	1,70	124,51	52,96	26,78	148,16	14,97
9	2,71	105,31	44,78	29,43	332,72	33,42
10	2,54	130,39	43,82	30,60	526,39	96,80
11	2,76	196,46	42,74	31,95	680,30	185,70
12	4,16	192,66	40,89	32,57	804,73	270,40
13	3,52	238,00	39,80	33,54	872,76	308,69
14	4,20	250,37	38,41	34,33	883,08	283,68
15	4,94	251,70	37,05	35,09	825,90	207,00
16	5,34	244,14	36,64	35,45	743,77	123,67
17	4,84	237,32	32,65	36,19	616,48	61,80
18	3,98	232,28	28,77	36,96	448,29	44,98
19	3,72	280,46	29,85	36,35	261,21	26,46
20	3,10	230,46	35,91	34,59	92,68	9,61
21	2,27	218,85	48,11	31,17	3,85	1,02
22	1,25	190,96	55,67	29,10	0,00	0,00
23	1,13	174,28	59,21	28,15	0,00	0,00
24	1,04	99,19	58,62	27,44	0,00	0,00

the system provided significant information, which was used not only to identify the limitations of the system but also to the optimal performance of the system.

Figure 13.11 shows the variation in the efficiency of the module during the hottest day, in contrast to the ambient air temperature and cell temperature during the hours of operation of the system. The maximum efficiency of the module achieved was not greater than 18%. There was a reduction in the efficiency of the module during the 12 h between 9 am and 6 pm. The reduction of panel efficiency was due to the increase in ambient air temperature and solar PV cell temperature.

Figure 13.12 shows the various comparisons of energy generation and temperatures. Nevertheless, the maximum solar PV panel energy generation for the hottest day was observed to be 70 Wh with a temperature above 50 °C. The average ambient air temperature for the hottest day was 32 °C. The area under the curve shows the total energy generated during the working time of the module. During the hottest day, solar radiation hours were about 15 h. This provided a considerable opportunity for the application of the system due to a higher amount of solar radiation, which allows a reduction in the dependence of the grid and using solar radiation to generate self-consumption cooling in the building.

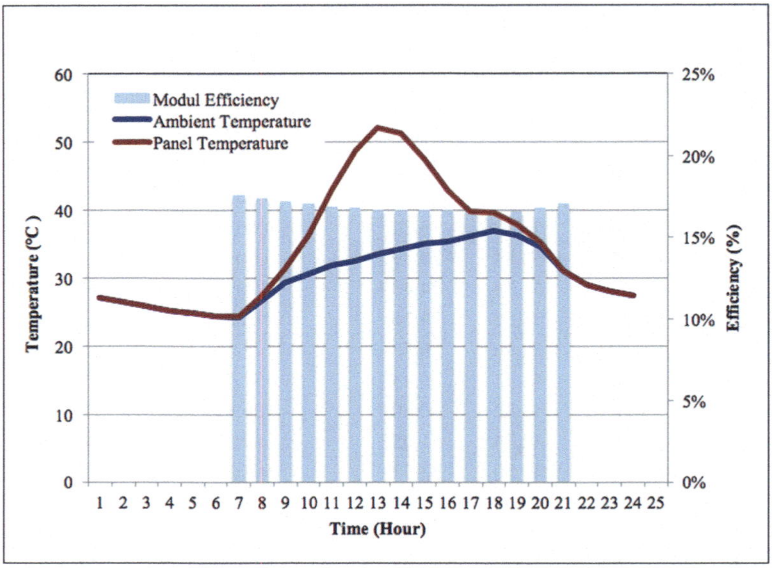

Fig. 13.11 Comparison of hourly ambient temperature, panel temperature, and module efficiency for the hottest day

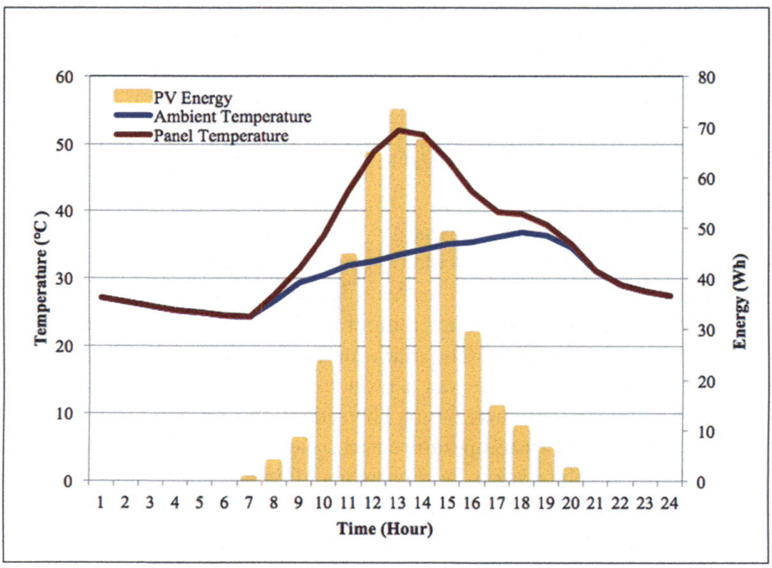

Fig. 13.12 Hourly ambient temperature, panel temperature, and energy output for the hottest day

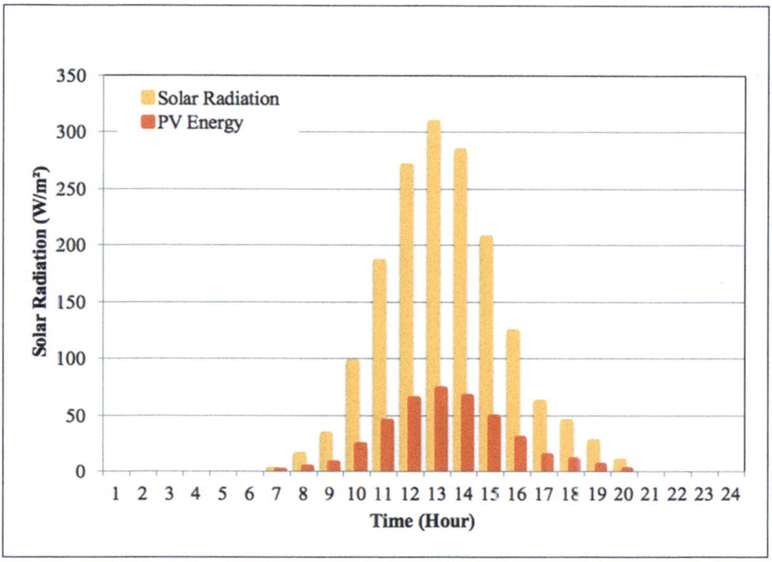

Fig. 13.13 Comparison of solar radiation and PV energy output for the hottest day

Figure 13.13 shows the relationship between hourly solar radiation and system energy generation. The highest solar radiation received at the project site between 1 pm and 2 pm, was close to 900 W/m^2 on the horizontal surface. However, the maximum solar radiation striking the surface of the solar PV panel during this period decreases because the sun is not directly on top of the panel. Hence, affecting the energy generated by the solar PV panel, which in this case was observed to decrease as compared to the coldest day despite the presence of higher solar radiation. The energy generation for this period was observed to be close to 72 Wh, slightly lesser than for the coldest day.

Figure 13.14 shows the comparisons of the hourly efficiency of the module and energy generated for the coldest and hottest days. The curves of energy generation during the hottest day and the coldest day show a reduction in energy generation for the hottest day as compared to the coldest due to increased ambient temperature and also the change in the location of the sun from the southern hemisphere during the winter to the northern hemisphere during the summer as evidenced in Fig. 13.14.

Thus, due to the vertical positioning of the solar PV panel, less solar radiation strikes the surface of the panel in summer as compared to winter. Hence, maximum generation was observed in winter for the coldest day with the value of 110 Wh as compared to 75 Wh for the hottest day. Additionally, another important factor is that during the coldest day, the amount of solar radiation hours is less, not more than 8 h despite the energy generation being greater. The efficiency of the module also shows a considerable difference between the days as highlighted in Fig. 13.14. It was observed that the difference in solar PV panel efficiency between the 2 days evaluated was close to 3%. Also, the variation of efficiency only has an important

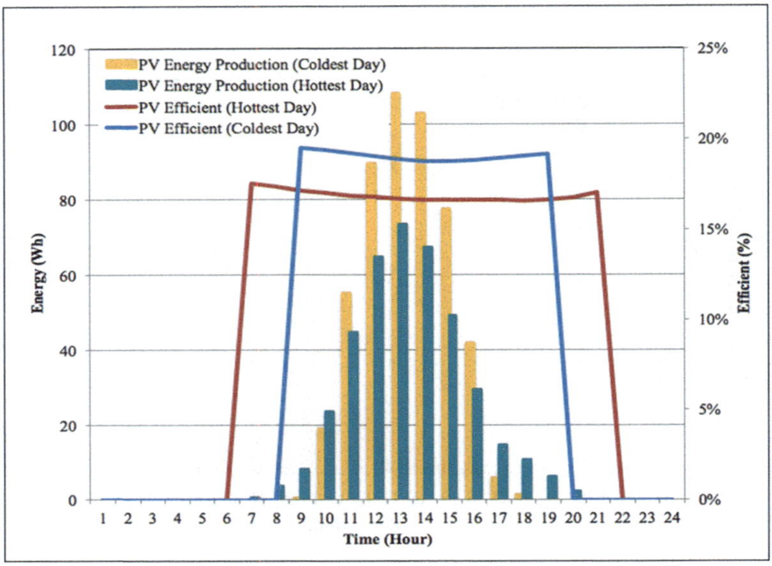

Fig. 13.14 Comparison of PV energy production and module efficiency for coldest and hottest day

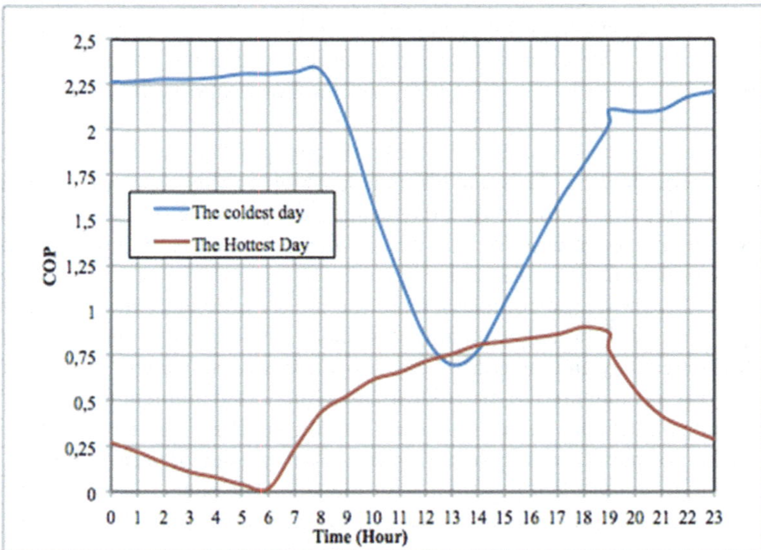

Fig. 13.15 The performance coefficient for the coldest and hottest day

variation during the period when solar radiation is present. The maximum value of solar PV panel efficiency for the coldest day was 20% and the sunniest day was 18%.

Figure 13.15 shows the performance coefficient for the coldest and hottest days of the year. The graph shows that during some hours of the day, the fluctuation of the COP does not change suddenly as it could be observed for the hottest day where the graph shows a fluctuation for some hours of the day. The maximum values of the coefficient of performance (COP) for the system during the hottest were found to be less than the value of 0.91. However, for the coldest day, the COP values were found to be more than 0.70 but less than 2.33.

13.5 Conclusion

The system operation depended on hourly solar radiation striking the solar PV panel installed vertically to the wall of the test room. The energy generated by the solar panel supplied all energy requirements for the thermoelectric modules that generated heating and cooling. The Peltier panel heated or cooled the air that was exposed to its surface, and then the cold or hot air was let to flow into the room depending on the room temperature.

The results showed that Izmir has solar energy potential suitable for using PV solar panels for powering the thermoelectric device for heating and cooling buildings. Despite higher ambient air temperature in summer across Izmir, which has negative effects on the performance of the devices, the area has a suitable climate for the development of solar energy-based thermoelectric devices for heating and cooling buildings.

However, in the design of a thermoelectric cooler HVAC system, ambient air temperature and room temperature are the most important parameters that need to be considered during the whole design and analysis process. However, installing the system vertically affected the amount of solar radiation striking the surface of the solar PV panel more, especially in summer, thus affecting the overall system performance. Nonetheless, if the system were to be installed at an optimal angle, its performance could be considerably improved.

References

Adhikari, R., Pipattanasompom, M., & Rahman, S. (2018). An algorithm for optimal management of aggregated HVAC power demand using smart thermostats. *Applied Energy, 217*, 166–177.

Afroz, Z., Shafiullah, G. M., Urmee, T., & Higgins, G. (2018). Modeling techniques used in building HVAC control systems: A review. *Renewable and Sustainable Energy Reviews, 83*, 64–84.

Ariza Taba, M. F. (2019). *Design of solar assisted thermoelectric air-conditioning system*. Master thesis, Graduate School of Natural and Applied Science, Ege University, Turkey.

Baniyounes, A. M., Ghadi, Y. Y., Rasul, M. G., & Khan, M. M. K. (2013). An overview of solar assisted air conditioning in Queensland's subtropical regions, Australia. *Renewable and Sustainable Energy Reviews, 26*, 781–804.

Chen, W. H., Wang, C. C., & Hung, C. I. (2014). Geometric effect on cooling power and performance of an integrated thermoelectric generation-cooling system. *Energy Conversion and Management, 87*, 566–575.

Chen, Y., Norford, L. K., Samuelson, H. W., & Malkawi, A. (2018). Optimal control of HVAC and window systems for natural ventilation through reinforcement learning. *Energy and Building, 169*, 195–205.

Deme Belafi, Z., Hong, T., & Reith, A. (2017). Smart Building Management Vs. *Intuitive Human Control-Lessons Learnt from an Office Building in Hungary, Building Simulation, 10*, 1–44.

Duffie, J. A., & Beckman, W. A. (2013). *Solar engineering of thermal processes* (4th ed.). Wiley.

Fabrizio, E., Seguro, F., & Filippi, M. (2014). Integrated HVAC and DHW production systems for zero energy buildings. *Renewable and Sustainable Energy Reviews, 40*, 515–541.

Genco, A., Viggiano, A., Viscido, L., Sellito, G., & Magi, V. (2018). Dynamic analysis of HVAC systems for industrial plants with different airflow control systems. *Thermal Science and Engineering Progress, 6*, 330–345.

Ghahramani, A., Karvigh, S. A., & Becerik-Gerber, B. (2017). HVAC system energy optimization using an adaptive hybrid metaheuristic. *Energy and Buildings, 152*, 149–161.

Jaber, H., Ramadan, M., Lemenand, T., & Khaled, M. (2018). Domestic thermoelectric cogeneration system optimization analysis, energy consumption and CO2 emissions reduction. *Applied Thermal Engineering, 130*, 279–295.

Khattab, N. M., Shenawy, E. I., & E.T. (2006). Optimal operation of thermoelectric cooler driven by solar thermoelectric generator. *Energy Conversion and Management, 47*(4), 407–426.

Lee, S. J., Shon, B. H., Jung, C. W., & Kang, Y. T. (2018). A novel type solar assisted heat pump using a low GWP refrigerant (R-1233zd (E)) with the flexible solar collector. *Energy, 149*, 386–396.

Makki, A., Omer, S., Su, Y., & Sabir, H. (2016). Numerical investigation of heat pipe-based photovoltaic–thermoelectric generator (HP-PV/TEG) hybrid system. *Energy Conversion and Management, 112*, 274–287.

Mammoli, A., Vorobieff, P., Barsun, H., Burnett, R., & Fisher, D. (2010). Energetic, economic and environmental performance of a solar-thermal-assisted HVAC system. *Energy and Buildings, 42*, 1524–1535.

Prieto, A., Knaack, U., Auer, T., & Klein, T. (2017). Solar cool facades: Framework for the integration of solar cooling technologies in the building envelope. *Energy, 137*, 353–368.

Qiu, K., & Hayden, A. C. S. (2012). Development of a novel cascading TPV and TE power generation system. *Applied Energy, 91*(1), 304–308.

Thedeby, A. (2014). *Heating and cooling with solar powered peltier elements*. Master Thesis, Department of Energy Planning, Division of Efficient Energy Systems, Lund University, Sweden, ISSN 0282-1990.

Twaha, S., Zhu, J., Yan, Y., & Li, B. (2016). A comprehensive review of thermoelectric technology: Materials, applications, modelling and performance improvement. *Renewable and Sustainable Energy Reviews, 65*, 698–726.

Wang, Y., Feng, H., & Xi, X. (2017). Monitoring and autonomous control of Beijing Subway HVAC system for energy sustainability. *Energy for Sustainable Development, 39*, 1–12.

Zeyghami, M., Goswami, D. Y., & Stefanakos, E. (2015). A review of solar thermo-mechanical refrigeration and cooling methods. *Renewable and Sustainable Energy Reviews, 51*, 1428–1445.

Zhao, S., Xu, G., Wang, J., Chen, T., & Zhang, X. (2017). Environmental performance of a solar assisted hybrid power gas heat pump based on life cycle assessment. *Energy Procedia, 142*, 419–425.

Zheng, X. F., Liu, C. X., Boukhanouf, R., Yan, Y. Y., & Li, W. Z. (2014). Experimental study of domestic thermoelectric cogeneration system. *Apply Thermal Engineering, 62*(1), 69–79.

Chapter 14
An In-depth Analysis of the Energy and Exergy Performance of Radiant Heating Integrated with Ground Source Heat Pumps in Building Environments

Ahmet Dogan, Nurullah Kayaci, and Hakan Demir

Nomenclature

ASHP	Air source heat pump
AT	Accumulation tank
CHP	Combined heat and power
COP	Coefficient of performance
FC	Fan coil
FCU	Fan coil unit
GHE	Ground heat exchanger
GSHP	Ground source heat pump
HP	Heat pump
PCM	Phase change material
RH	Radiant heating
RHC	Radiant heating and cooling
RP	Radiant panel
YTU	Yildiz Technical University

A. Dogan
Department of Mechanical Engineering, Erzincan Binali Yıldırım University, Erzincan, Turkey
e-mail: adogan@erzincan.edu.tr

N. Kayaci · H. Demir (✉)
Department of Mechanical Engineering, Yıldız Technical University, Istanbul, Turkey
e-mail: kayaci@yildiz.edu.tr; hdemir@yildiz.edu.tr

M. Z. Sogut, M. Koray (eds.), *Energy Rationality and Management for Decarbonization*, Green Energy and Technology,
https://doi.org/10.1007/978-3-031-68497-5_14

14.1 Introduction

The global perspective on building energy consumption reveals a significant trend, with the building sector accounting for over 30% of global energy usage and exceeding half of global electricity consumption (Energy Agency, 2020, IEA, 2022). This sector has a broader impact, leading significantly to global carbon emissions (Friedlingstein et al., 2020; Ibrahim et al., 2021). In Turkey, the building sector emerges as a dominant consumer, accounting for 33% of total energy consumption (Authority, 2015). The prominence of heating and cooling systems, which consume around 70% of total energy within buildings, is particularly noteworthy, indicating a significant opportunity to advance energy efficiency in this area (TÜRKİYE NATIONAL ENERGY PLAN, 2022). This powerful energy consumption in the building sector highlights the critical need for efficient and sustainable solutions, such as the incorporation of innovative systems such as radiant heating and cooling (RHC) (Sartori et al., 2012; Wu & Skye, 2021; Agbodjan et al., 2022). RHC represents a promising path to address the escalating energy demand within buildings, providing a potential solution to curbing energy consumption, particularly in heating and cooling applications (Rhee et al., 2017; Liao et al., 2022).

The integration of RHC systems into building structures not only promises to reduce energy consumption but also aligns with the essential of lowering carbon emissions. Its implementation is a critical strategy in global efforts to combat climate change and encourage sustainable practices within the built environment (Rhee & Kim, 2015; Wang et al., 2019). The potential of decreasing carbon footprint and dependency on traditional energy sources becomes palpable when renewable energy, such as solar or geothermal power, is integrated into RHC systems (Doğan et al., 2022; Dogan et al., 2022, 2023). The emphasis on RHC systems is an important step toward tackling the challenges posed by the significant energy demand in buildings, thus aiding in a more sustainable and environmentally conscious future (Hawks & Cho, 2024).

RHC systems, which can be used as viable alternatives to traditional air systems, significantly improve overall plant efficiency by ensuring adequate thermal comfort. These systems use radiant heat transfer across large surfaces such as walls, floors, or ceilings to save energy (Behzadi et al., 2022). The mechanism employs lower water temperatures for radiant heating (RH) and higher temperatures for radiant cooling, a configuration that inherently supports energy efficiency (Hesaraki & Huda, 2022). Furthermore, the adaptability of RHC systems extends to their integration with low-exergy sources, enhancing their eco-friendliness even further (Szabó & Kalmár, 2019).

In the current literature, there are some studies on energy and exergy analysis of radiant panels (RP) combined into a ground source heat pump (GSHP), although not many. Kilkis (2012) introduced a new engineering metric based on the Rational Exergy Management model to investigate the potential and benefits of RHC systems integrated with GSHPs, and stated that this metric provides a new perspective on the real benefits of heat pump (HP) technology in reducing CO_2 emissions. In the study, the results of case studies pertaining to various scenarios were compared with those of a base case, which includes an RP system that was integrated to a natural gas-fired condensing boiler for heating and a grid power-driven chiller for cooling. The findings indicated that if RPs were used integrated with combined heat and power (CHP) systems or GSHPs powered by renewable energy, there was a significant potential to reduce CO_2 emissions. Akbulut et al. (2016) studied exergy and exergo-economic analyses of radiant wall heating systems connected to a vertical ground heat exchanger (GHE) at Yildiz Technical University (YTU). According to the results, the energy and exergy efficiencies of the entire system were obtained to be 67.36% and 27.40%, respectively, and the energy and exergy efficiency of the wall heating system panels were determined to be 86.61% and 82.90%, respectively. In the study of Oruc et al. (2019), the energy and exergy analyses were examined in a phase change material (PCM)-embedded radiant wall heating system using GSHP and solar energy. Energy and exergy efficiencies of the radiant heating system were obtained using four different PCMs. Compared to the situation where PCM was not used in radiant systems, it was claimed that the energy efficiency of SP26E PCM, which achieved the best results, rose from 62 to 87% and the exergy efficiency increased from 14 to 56%.

Kayaci (2020) performed comprehensive evaluations of the energy, exergy, and economic aspects of the radiant wall panel unit (RW-PU), floor heating fan coil unit (FH-FCU), and radiator fan coil unit (R-FCU) systems, which were linked to GSHP systems. With the help of the code created in the MATLAB software program, the inlet-outlet water and soil temperatures of the horizontal Ground Heat Exchangers (GHE) designed for heating and cooling an office in Istanbul were determined and verified with experimental data. According to the results, the average values of coefficient of performance (COP) and second-law efficiency for RW-PU, FHFCU, and R-FCU were obtained to be 4.01, 3.37, 3.01, and 39.29%, 38.04%, and 36.63%, respectively.

In the study of Shao et al. (2021), the energy and exergy efficiencies of an air source heat pump (ASHP) system equipped with a direct-condensation radiant heating panel (DRHP) were assessed. The proposed thermodynamic model was used to examine the impacts of external, indoor, and condensation temperatures on the efficiency of the system. The findings indicated that utilizing a low compression ratio enhanced both energy and exergy efficiency, outperforming the performance of existing ASHP systems. The system's economic performance was also examined, with a reliable model suggested and hourly heating load rates offered. They obtained that the investment pay-back term of the system is 7.3 years, and its internal rate of return is 11.2%.

Based on the above-mentioned studies on RPs integrated with GSHPs and ASHPs, it appears that there is a research gap for this field in the existing literature. The originality of the present study can be expressed as follows: (i) GHE pipes were buried horizontally in the ground under the building foundation; (ii) energy and exergy analyses of wall, ceiling, and wall + ceiling RP implementations were conducted, as opposed to singular applications on either walls or ceilings; and (iii) the experimental and theoretical studies of whole systems (GSHP + RP systems) were performed in an actively used office environment.

14.2 System Description

The experimental setup was established in the YTU Laboratory building by applying the GSHP system combined RPs (wall and ceiling), depicted in a schematic and actual appearance in Fig. 14.1. The GSHP system can be outlined into three primary subgroups, namely test room (Fig. 14.1a), the heat pump unit (Fig. 14.1b), and the GHEs (Fig. 14.1c).

The test room (Fig. 14.1a) comprises RPs incorporated into the ceiling and walls of an office area, which are linked to a GSHP with a heating capacity of 10.5 kW. The system is fundamentally organized into three primary sections. Initially, there is a designated space at the entrance of the building where assessments are carried out to evaluate the level of comfort and inside temperature conditions. The walls and ceiling of this room are equipped with RPs. The southwest facade features six RPs with dimensions of 2 m × 0.6 m, while the northeast face has nine panels of identical size. Furthermore, the suspended ceiling contains a grand total of 29 RPs. Each panel has dimensions of 1 m by 0.6 m and is constructed with identical structural layers. The panels can be independently regulated using spherical valves installed on the collector group above the suspended ceiling, as well as on the panels situated

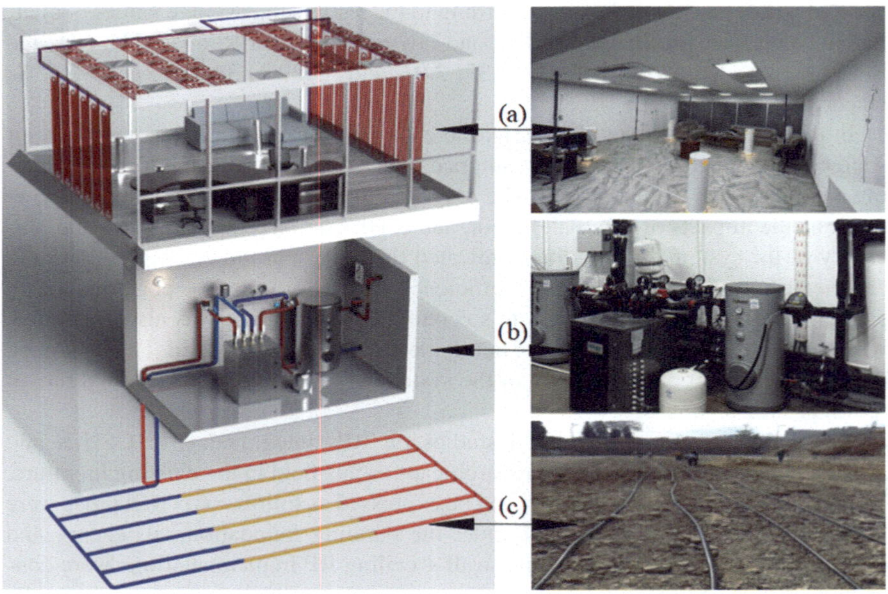

Fig. 14.1 Schematic and actual views of (**a**) test room, (**b**) heat pump system, and (**c**) GHE pipes

on the ceiling and walls. The second section (Fig. 14.1b) comprises the heat pump unit located in the mechanical room on the ground floor of the building. The key equipment in the heat pump unit includes the main block of the heat pump, the storage (accumulation) tank, balancing tank, and water circulation pumps. The third part (Fig. 14.1c) consists of parallel and horizontal GHEs buried beneath the building foundation. The heat pump control device in the mechanical room initiates the system by modifying the panel return temperature according to the heating or cooling operation. The heat pump produces hot water in the heating operation and cold water in the cooling operation, which is then kept in the accumulation tank (AT). Motorized valves at the collector inlets are controlled by thermostats on the walls, allowing the water to flow based on the requirements of the test room. In addition, the temperature sensor located on the panel return line regulates the opening of the motorized valve, so assuring the circulation of water.

14.3 Measurement System

The test room was instrumented with a comprehensive array of measurement devices, notably utilizing 51 T-type thermocouples. These thermocouples facilitated the measurement of various parameters critical for a thorough analysis, encompassing water flow rate, indoor air velocity, humidity, outdoor ambient temperature, conditioned and unconditioned surface temperatures, indoor temperature, and hallway air temperature. Surface temperatures, including those of RP surfaces, uncooled walls, inner and outer window surfaces, and the hallway glass, were meticulously recorded using 26 thermocouples. Vertical temperature profiling, an essential aspect for both thermal comfort assessment and energy efficiency considerations, was conducted through the strategic placement of 23 thermocouples distributed across five metal stands, each measuring 2.5 m in height.

Beginning with the heat pump system, as illustrated in Fig. 14.1b and c, the circulation pumps, accumulation and expansion tanks, flow switches, and the intricate piping system are showcased in both projected and real views. These components collectively contribute to the efficient operation of the GSHP system, ensuring optimal heat exchange and fluid circulation. The measurement system, constituting the second subgroup, comprises a suite of essential equipment, including flow meters, manometers, thermometers, rotameters, RTD sensors, and the Programmable Logic Controller (PLC) unit. These instruments play a pivotal role in monitoring and gauging various parameters that are critical to the performance and efficiency of the GSHP system. The third subgroup comprises the GHE, which is composed of an arrangement of 10 parallel pipes that are horizontally placed beneath the foundation of the Laboratory's building. These pipes extend to a length of 85 m, with a 0.5 m spacing between each. Water flows into the GHE pipes in a highly organized manner, eventually reaching a collector and then being transported to the heat pump. Additional insights into this experimental setup subsection can be gleaned from our previous publications (Kayaci & Demir, 2018, 2020).

14.4 Thermodynamic Analysis

This paper focuses on the energy and exergy flow of a ground source heat pump system providing energy to a facility heated by radiant heating from the walls and ceiling, with an emphasis on the analysis of the fundamental laws of thermodynamics. The conducted analyses are made considering the following assumptions:

- There is no kinetic and potential energy effect within the system, and there are no chemical or nuclear reactions. All processes are in a steady state and continuous flow.
- The environment or system surroundings are considered as a large thermal reservoir and are not affected by the local effects of sources or sinks.
- The dead state temperature has been considered as 4.2 °C, representing the average outdoor ambient temperature.
- Heat transfer through the pipes connecting the components and pressure drops in the heat exchangers within the heat pump have been neglected.
- The compressor and all pumps are assumed to operate adiabatically with an assumed isentropic efficiency of 85%.
- Air is a perfect gas with a consistent specific heat and its humidity level is regarded as negligible.

Mass Balance:

The mass balance equation (Eq. 14.1) emphasizes the conservation of mass within a system, requiring that the sum of mass entering and leaving the system equals zero.

$$\sum \dot{m}_{in} = \sum \dot{m}_{out} \tag{14.1}$$

General Energy Balance:

Equation (14.2) represents the fundamental principle that the rate of net energy entering a system equals the rate of net energy leaving the system. This ensures the conservation of energy within the system.

$$\dot{E}_{in} - \dot{E}_{out} = \Delta\dot{E}_{system} = 0(\text{steady}) \tag{14.2}$$

$$\dot{E}_{in} = \dot{E}_{out} \tag{14.3}$$

Extended Energy Balance:

Equation (14.4) provides a detailed energy balance, incorporating heat ($\dot{Q}$), work ($\dot{W}$), and enthalpy (h). with net heat input and net work output represented by $\dot{Q}_{net}$ and $\dot{W}_{net}$, respectively.

$$\dot{Q}_{net} + \sum \dot{m}_{in} h_{in} = \dot{W}_{net} + \sum \dot{m}_{out} h_{out} \tag{14.4}$$

General Exergy Balance:

Equation (14.5) is an expression of the exergy balance principle, which states that the rate of net exergy entering a system equals to the rate of exergy destruction within the system.

$$\dot{E}x_{\text{in}} - \dot{E}x_{\text{out}} = \dot{E}x_D \tag{14.5}$$

Extended Exergy Balance:

Equation (14.6) provides a comprehensive view of the exergy balance, incorporating flow exergy (ψ), heat exergy $\dot{E}x_Q$, work ($\dot{W}$), and exergy destruction ($\dot{E}x_D$).

$$\dot{m}_{\text{in}}\psi_{\text{in}} + \dot{E}x_{Q_{\text{in}}} + \dot{W}_{\text{in}} = \dot{m}_{\text{out}}\psi_{\text{out}} + \dot{E}x_{Q_{\text{out}}} + \dot{W}_{\text{out}} + \dot{E}x_D \tag{14.6}$$

The term ψ (Eq. 14.7) represents flow exergy, accounting for changes in enthalpy, entropy, velocity, and elevation from the dead state. The dead state is denoted by the subscript "0."

$$\psi = (h_i - h_0) - T_0(s_i - s_0) + \frac{V_i^2}{2} + gz \tag{14.7}$$

The Eqs. (14.8), (14.9), and (14.10) quantify exergy transfer due to heat. The term $\dot{E}x_{\text{heat}}$ represents the difference between exergy transfer in and out due to heat. $\dot{Q}_k$ represents the rate at which heat is transferred from the boundary at temperature T_k at point k.

$$\dot{E}x_{Q_{\text{in}}} = \left(1 - \frac{T_0}{T_{\text{source}}}\right)\dot{Q}_{\text{in}} \tag{14.8}$$

$$\dot{E}x_{Q_{\text{out}}} = \left(1 - \frac{T_0}{T_{\text{boundary}}}\right)\dot{Q}_{\text{out}} \tag{14.9}$$

$$\dot{E}x_{\text{heat}} = \sum\left(1 - \frac{T_0}{T_k}\right)\dot{Q}_k = \dot{E}x_{Q_{\text{in}}} - \dot{E}x_{Q_{\text{out}}} \tag{14.10}$$

The Eq. (14.11) expresses the principle of entropy balance inside the system. The principle asserts that the difference between the rate at which net entropy enters the system and the rate at which net entropy leaves the system must be equal to the rate at which entropy is generated inside the system.

$$\dot{S}_{\text{in}} - \dot{S}_{\text{out}} + \dot{S}_{\text{gen}} = 0 \tag{14.11}$$

Equation (14.12) measures the rate of exergy destruction ($\dot{E}x_D$) by multiplying the dead state ambient temperature (T_0) with the rate of entropy generation ($\dot{S}_{\text{gen}}$).

This reflects the irreversible losses within the system, highlighting the degradation of exergy associated with entropy generation.

$$\dot{E}x_D = T_0 \dot{S}_{\text{gen}} \tag{14.12}$$

The ratio of the desired exergy output to the exergy input, proposed by Kotas (2012), is called exergy efficiency, which is presented below:

$$\eta_{ex} = \frac{\dot{X}_H}{\dot{X}_{\text{used}}} \tag{14.13}$$

where $\dot{X}_H$ reflects the total exergy transfer rates envisaged by the system. $\dot{X}_{\text{used}}$ represents the exergy input rate necessary for the performance of the process. When we consider the entire heat pump system, the exergetic efficiency is represented by Eq. (14.14).

$$\eta_{ex,\text{HP}} = \frac{\dot{X}_H}{W_{a,\text{in}}} \tag{14.14}$$

where $\dot{X}_H$ represents the exergy transfer rate by heat, and $W_{a,\text{ in}}$ represents the actual work input rate to the system per unit time. The calculated power supplied to the compressor can be determined using the following equation.

$$\dot{W}_{a,\text{in}} = \frac{\dot{W}_{\text{Comp}}}{\eta_C} \tag{14.15}$$

where $\dot{W}_{\text{Comp}}$ represents the work input rate to the compressor, η_C denotes the efficiency of the compressor.

In the context of an ideal heat pump system functioning between low and high-temperature reservoirs at T_L and T_H, the maximal heating coefficient of performance, denoted as $\text{COP}_{\text{cc, HP}}$ (Eq. 14.16), is derived from the Carnot cycle.

$$\text{COP}_{\text{cc,HP}} = \frac{T_H}{T_H - T_L} \tag{14.16}$$

The Eq. (14.17) is utilized to calculate the heating COP for the heat pump unit.

$$\text{COP}_{a,\text{HP}} = \frac{\dot{Q}_H}{\dot{W}_{a,\text{in}}} \tag{14.17}$$

where, $\dot{Q}_H$ represents the rate of space heating load. Exergy efficiency, expressed as η_{ex}, can also be computed as follows:

$$\eta_{ex} = \frac{\text{COP}_a}{\text{COP}_{cc}} \tag{14.18}$$

The experiments were carried out in the GSHP system built at the YTU Laboratory building, as previously stated. The positions of the measuring instruments in the experimental setup are shown in Fig. 14.2. Measurements of temperature, pressure, and flow rate were recorded at several locations throughout the system. In addition, temperature measurements were conducted at several sites both inside and outside of the building. The mathematical expressions utilized for system analysis were extended to all components and subcomponents of the GHEs circuit, as illustrated in Table 14.1. Comprehensive mass, energy, entropy, and exergy balance equations for the ground circuit are detailed in Table 14.1. The heat pump system is comprised

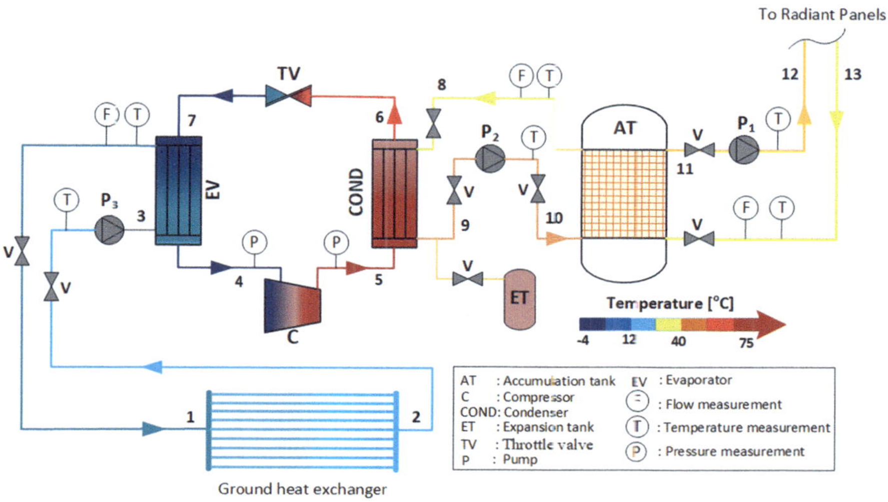

Fig. 14.2 Schematic and actual views of (**a**) test room, (**b**) heat pump system, and (**c**) GHE pipes

Table 14.1 The balance equations for the ground circuit units

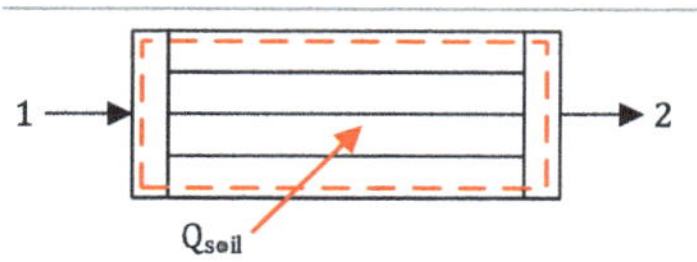 Ground heat exchanger	$\dot{m}_1 = \dot{m}_2 = \dot{m}_{\mathrm{GHE}}$ $\dot{m}_{\mathrm{GHE}} h_1 + \dot{Q}_{\mathrm{soil}} = \dot{m}_{\mathrm{GHE}} h_2 + \dot{Q}_{\mathrm{soil,GHE}}$ $\dot{m}_{\mathrm{GHE}} s_1 + \dot{S}_g + \frac{\dot{Q}_{\mathrm{soil}}}{T_{\mathrm{soil}}} = \dot{m}_{\mathrm{GHE}} s_2$ $\dot{m}_{\mathrm{GHE}} ex_1 + \dot{E}x_{\mathrm{soil}} = \dot{m}_{\mathrm{GHE}} ex_2 + \dot{E}x_{D,\mathrm{GHE}}$
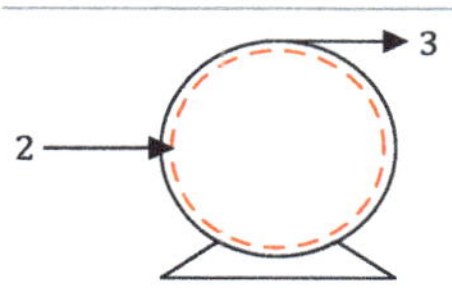 Pump1	$\dot{m}_2 = \dot{m}_3 = \dot{m}_{\mathrm{GHE}}$ $\dot{m}_{\mathrm{GHE}} h_2 + \dot{W}_p = \dot{m}_{\mathrm{GHE}} h_3$ $\dot{m}_{\mathrm{GHE}} s_2 + \dot{S}_g = \dot{m}_{\mathrm{GHE}} s_3$ $\dot{m}_{\mathrm{GHE}} ex_2 + \dot{W}_p = \dot{m}_{\mathrm{GHE}} ex_3 + \dot{E}x_D$

of essential components such as the evaporator, compressor, condenser, throttling valve, and units. The theoretical formulations employed for system analysis were extended to encompass both the heat pump circuit as a whole and its individual subunits, depicted in Table 14.2. Detailed mass, energy entropy, and exergy balance equations for the heat pump circuit are presented in Table 14.2.

The RP heating system integrated into both the wall and ceiling comprises an accumulation tank, two pumps, and room RPs. Mathematical analyses have been applied to each component within this system. In Table 14.3, detailed mass, energy,

Table 14.2 The balance equations for the heat pump circuit units

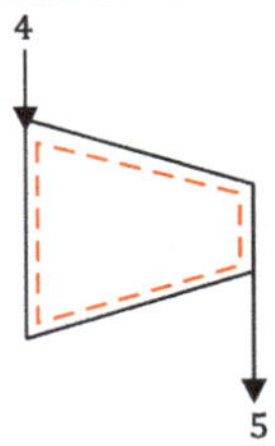 Compressor	$\dot{m}_4 = \dot{m}_5 = \dot{m}_{HP}$ $\dot{m}_{HP} h_4 + \dot{W}_{comp} = \dot{m}_{HP} h_5$ $\dot{m}_{HP} s_4 + \dot{S}_g = \dot{m}_{HP} s_5$ $\dot{m}_{HP} ex_4 + \dot{W}_{comp} = \dot{m}_{HP} ex_5 + \dot{E}x_D$
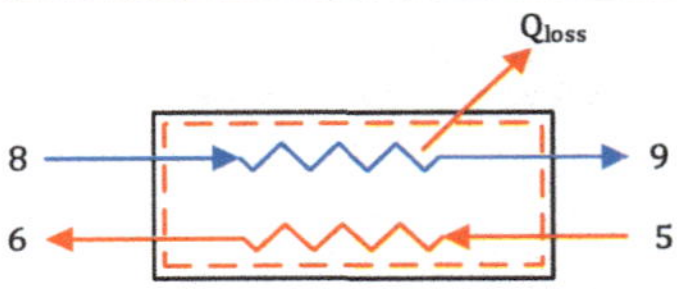 Condenser	$\dot{m}_5 = \dot{m}_6 = \dot{m}_{HP}$ $\dot{m}_8 = \dot{m}_9 = \dot{m}_{at}$ $\dot{m}_{HP} h_5 + \dot{m}_{at} h_8 = \dot{m}_{HP} h_6 + \dot{m}_{at} h_9 + \dot{Q}_{loss}$ $\dot{m}_{HP} s_5 + \dot{m}_{at} s_8 + \dot{S}_g = \dot{m}_{HP} s_6 + \dot{m}_{at} s_9 + \frac{\dot{Q}_{loss}}{T_b}$ $\dot{m}_{HP} ex_5 + \dot{m}_{at} ex_8$ $= \dot{m}_{HP} ex_6 + \dot{m}_{at} ex_9 + \dot{E}x_D + \dot{E}x_{loss}$
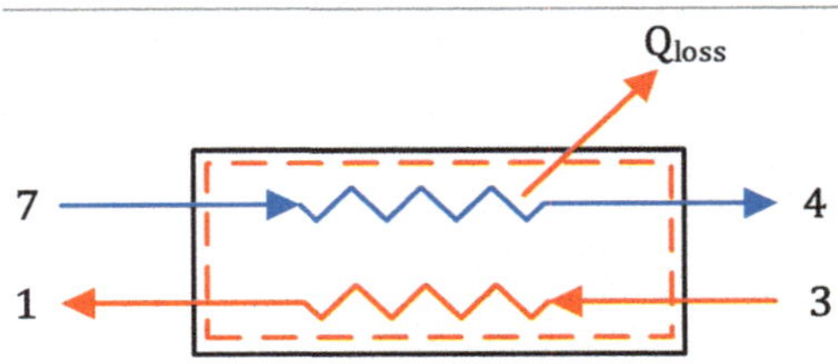 Evaporator	$\dot{m}_1 = \dot{m}_3 = \dot{m}_{GHE}$ $\dot{m}_7 = \dot{m}_4 = \dot{m}_{HP}$ $\dot{m}_{HP} h_7 + \dot{m}_{GHE} h_3 = \dot{m}_{GHE} h_1 + \dot{m}_{HP} h_4 + \dot{Q}_{loss}$ $\dot{m}_{HP} s_7 + \dot{m}_{GHE} s_3 + \dot{S}_g$ $= \dot{m}_{HP} s_1 + \dot{m}_{GHE} s_4 + \frac{\dot{Q}_{loss}}{T_b}$ $\dot{m}_{HP} ex_1 + \dot{m}_{GHE} ex_3$ $= \dot{m}_{HP} ex_2 + \dot{m}_{GHE} ex_4 + \dot{E}x_D + \dot{E}x_{loss}$
 Throttle valve	$\dot{m}_6 = \dot{m}_7 = \dot{m}_{HP} \rightarrow h_1 = h_2$ $\dot{m}_{HP} h_6 = \dot{m}_{HP} h_7$ $\dot{m}_{HP} s_6 + \dot{S}_g = \dot{m}_{HP} s_7$ $\dot{m}_{HP} ex_6 = \dot{m}_{HP} ex_7 + \dot{E}x_D + \dot{E}x_{loss}$

Table 14.3 The balance equations for the heat pump circuit units

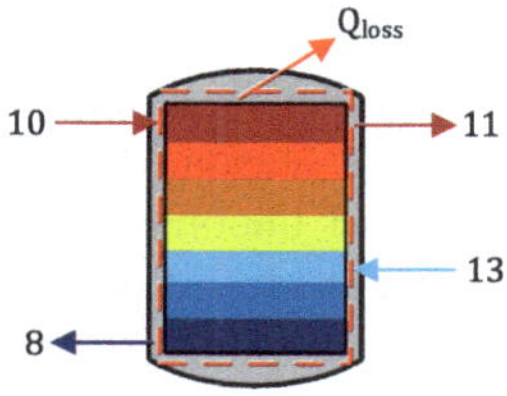 Accumulation tank	$\dot m_8 = \dot m_{10} = \dot m_{at}$ $\dot m_{11} = \dot m_{13} = \dot m_{rp}$ $\dot m_{at} h_{10} + \dot m_{rp} h_{13} = \dot m_{at} h_8 + \dot m_{rp} h_{11} + \dot Q_{loss}$ $\dot m_{at} s_{10} + \dot m_{rp} s_{13} = \dot m_{at} s_8 + \dot m_{rp} s_{11} + \frac{\dot Q_{loss}}{T_{air,room}}$ $\dot m_{at} ex_{10} + \dot m_{rp} ex_{13} = \dot m_{at} ex_8 + \dot m_{rp} ex_{11} + \dot Ex_D + \dot Ex_{loss}$
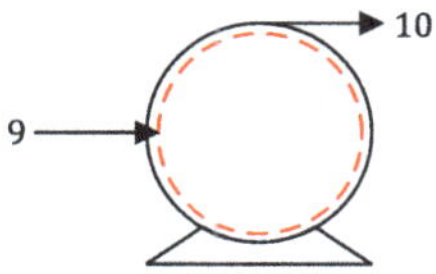 Pump 2	$\dot m_9 = \dot m_{10} = \dot m_{at}$ $\dot m_{at} h_9 + \dot W_p = \dot m_{at} h_{10}$ $\dot m_{at} s_9 + \dot S_g = \dot m_{at} s_{10}$ $\dot m_{at} ex_9 + \dot W_p = \dot m_{at} ex_{10} + \dot Ex_D$
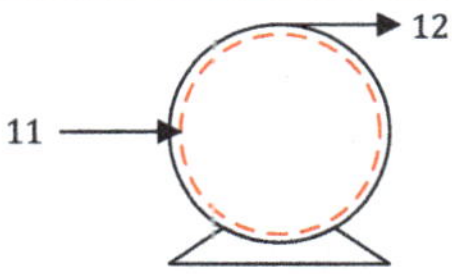 Pump 3	$\dot m_{11} = \dot m_{12} = \dot m_{rp}$ $\dot m_{rp} h_{11} + \dot W_p = \dot m_{rp} h_{12}$ $\dot m_{rp} s_{11} + \dot S_g = \dot m_{rp} s_{12}$ $\dot m_{rp} ex_{11} + \dot W_p = \dot m_{rp} ex_{12} + \dot Ex_D$
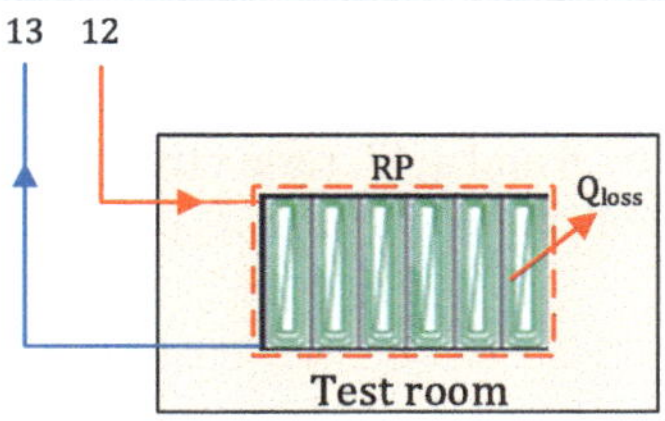 RP HX	$\dot m_{12} = \dot m_{13} = \dot m_{rp}$ $\dot m_{rp} h_{12} = \dot m_{rp} h_{13} + \dot Q_{loss}$ $\dot m_{rp} s_{12} + \dot S_g = \dot m_{rp} s_{13} + \frac{\dot Q_{loss}}{T_{air,room}}$ $\dot m_{rp} ex_1 = \dot m_{rp} ex_2 + \dot Ex_D + \dot Ex_{loss}$

entropy, and exergy balance equations are explicitly presented for each individual component, providing an in-depth insight into the system dynamics. An office room system that was combined with a GSHP and heated RPs was evaluated for its energy and exergy performance in this study.

14.5 Discussion and Conclusions

The study was conducted at the YTU Science and Technology Application and Research Center. Experiments were carried out during the heating season to measure the surface temperatures of RPs, the surface temperatures of unconditioned surfaces, the temperatures of the ground, the inlet and outlet temperatures of ground and test room water, the pressure and temperature values at various points on the heat pump unit, as well as the temperatures of the air both inside and outside the building. A presentation of the measurements and the results of the calculations for 13 locations in Fig. 14.2 can also be seen in Table 14.4. With the use of fundamental thermodynamic laws and balancing equations, some of the property data that are presented in Table 14.4 were derived directly from the results of measurements, while others were produced through calculations.

The energy ratios for each component of the GSHP system and the exergy destruction values in the exergy analysis are provided in Table 14.5. In the process of analyzing Table 14.4, it is observed that the highest exergy destruction, amounting to 0.39 kW, occurs in the evaporator, while the exergy destruction from the RP heat exchangers is 0.26 kW. The exergy destruction in the heat pump unit (compressor, condenser, evaporator, and throttle valve) constitutes approximately 67% of the total exergy destruction. The exergy destruction due to a pressure drop of the refrigerant passing through the throttle valve is 0.21 kW. The lowest exergy destruction within the GSHP system occurs in the GHE, amounting to 0.05 kW. The COP_{cc} and COP_a value of the heat pump system is calculated as 67.6 and 4.47, respectively. An exergy efficiency (η_{ex}) of the whole GSHP system is calculated as 6.6%.

Table 14.4 Measured and calculated properties data for the GSHP

Point	Mass flow (kg/s)	Working fluid	Pressure [kPa]	Temp. [°C]	Enthalpy [kj/kg]	Entropy [kj/kg.K]	Energy rate [kW]	Exergy [kj/kg]	Exergy rate [kW]
1	0.32	Water	180	12.1	51	0.18	16.32	0.28	0.12
2	0.32	Water	180	14.2	60.15	0.21	19.25	0.1	0.03
3	0.32	Water	180	14.2	60.15	0.21	19.25	0.1	0.03
4	0.021	R410A	705	−3.8	420.2	1.82	8.82	61.91	1.3
5s	0.021	R410A	2910	69.9	459.4	1.82	9.64	100.98	2.12
5a	0.021	R410A	2910	75	466.3	1.84	9.74	102.12	2.14
6	0.021	R410A	2910	47.7	281.1	1.26	5.9	82.32	1.79
7	0.021	R410A	705	−3.8	281.1	1.3	5.9	71.96	1.56
8	0.147	Water	200	39.4	165	0.56	24.2	4.1	0.6
9	0.147	Water	200	45.6	191.3	0.65	28.1	6.4	0.94
10	0.147	Water	200	45.6	191.3	0.65	28.1	6.4	0.94
11	0.175	Water	200	43.9	204.61	0.69	23	15.72	1.76
12	0.175	Water	200	43.9	204.61	0.69	23	15.72	1.76
13	0.175	Water	200	39	184.17	0.62	20.7	12.78	1.44

Table 14.5 Energy capacity and exergy destruction of system components

System components	Capacity value [kW]	Exergy destruction [kW]
Total electric power in	1.05	–
Compressor	0.85	0.12
GHE	2.99	0.05
RP HX	3.87	0.26
Condenser	3.87	0.2
Evaporator	3.01	0.39
Throttle valve	–	0.21
Accumulation tank	3.87	0.14

It is evident that the exergy destruction in the GSHP system predominantly takes place in the heat pump equipment. The implementation of thermal isolation enhancements in the evaporator and condenser, along with mechanical upgrades in the compressor, particularly in the heat pump system, will lead to a substantial decrease in energy consumption.

References

Agbodjan, Y. S., Wang, J., Cui, Y., Liu, Z., & Luo, Z. (2022). Bibliometric analysis of zero energy building research, challenges and solutions. *Solar Energy, 244*, 414–433.

Akbulut, U., Utlu, Z., & Kincay, O. (2016). Exergy, exergoenvironmental and exergoeconomic evaluation of a heat pump-integrated wall heating system. *Energy, 107*, 502–522. https://doi.org/10.1016/j.energy.2016.04.050

Authority, E. M. R. (2015). *Energy market regulatory authority*. World Energy.

Behzadi, A., Holmberg, S., Duwig, C., Haghighat, F., Ooka, R., & Sadrizadeh, S. (2022). Smart design and control of thermal energy storage in low-temperature heating and high-temperature cooling systems: A comprehensive review. *Renewable and Sustainable Energy Reviews, 166*, 112625.

Dogan, A., Kayaci, N., Demir, H., & Kemal Sevindir, M. (2022). An experimental comparison of radiant wall and ceiling cooling system integrated with ground source heat pump and direct expansion fan coil system in a highly glazed office room. *Energy and Buildings, 273*, 112412. https://doi.org/10.1016/j.enbuild.2022.112412

Doğan, A., Kayaci, N., Demir, H., & Sevindir, M. K. (2022). Experimental investigation of thermal comfort performance of a radiant wall and ceiling panel system. *Journal of Thermal Engineering, 8*(4), 551–561. https://doi.org/10.18186/thermal.1149161

Dogan, A., Kayaci, N., Kanbur, B. B., & Demir, H. (2023). Experimental investigation of mean radiant temperature trends for a ground source heat pump-integrated Radiant wall and ceiling heating system. *Buildings, 13*(10). https://doi.org/10.3390/buildings13102420

Energy Agency, I. (2020). Statistics report Key World Energy Statistics 2020. https://www.iea.org/reports/key-worldenergy-statistics-2020

Friedlingstein, P., O'sullivan, M., Jones, M. W., Andrew, R. M., Hauck, J., Olsen, A., Peters, G. P., Peters, W., Pongratz, J., & Sitch, S. (2020). Global carbon budget 2020. *Earth System Science Data Discussions, 2020*, 1–3.

Hawks, M. A., & Cho, S. (2024). Review and analysis of current solutions and trends for zero energy building (ZEB) thermal systems. *Renewable and Sustainable Energy Reviews, 189*, 114028.

Hesaraki, A., & Huda, N. (2022). A comparative review on the application of radiant low-temperature heating and high-temperature cooling for energy, thermal comfort, indoor air quality, design and control. *Sustainable Energy Technologies and Assessments, 49*, 101661.

Ibrahim, I. A., Ötvös, T., Gilmanova, A., Rocca, E., Ghanem, C., & Wanat, M. (2021). *International energy agency*. Kluwer Law International BV.

IEA. (2022). Technology and innovation pathways for zero-carbon-ready buildings by 2030. https://www.iea.org/reports/technology-and-innovation-pathways-for-zero-carbon-ready-buildings-by-2030

Kayaci, N. (2020). Energy and exergy analysis and thermo-economic optimization of the ground source heat pump integrated with radiant wall panel and fan-coil unit with floor heating or radiator. *Renewable Energy, 160*, 333–349. https://doi.org/10.1016/j.renene.2020.06.150

Kayaci, N., & Demir, H. (2018). Long time performance analysis of ground source heat pump for space heating and cooling applications based on thermo-economic optimization criteria. *Energy and Buildings, 163*, 121–139. https://doi.org/10.1016/j.enbuild.2017.12.034

Kayaci, N., & Demir, H. (2020). Comparative performance analysis of building foundation ground heat exchanger. *Geothermics, 83*, 101710. https://doi.org/10.1016/j.geothermics.2019.101710

Kilkis, B. (2012). Exergy metrication of radiant panel heating and cooling with heat pumps. *Energy Conversion and Management, 63*, 218–224.

Kotas, T. J. (2012). *The exergy method of thermal plant analysis*. Paragon Publishing.

Liao, W., Luo, Y., Peng, J., Wang, D., Yuan, C., Yin, R., & Li, N. (2022). Experimental study on energy consumption and thermal environment of radiant ceiling heating system for different types of rooms. *Energy, 244*, 122555. https://doi.org/10.1016/j.energy.2021.122555

Oruc, O., Dincer, I., & Javani, N. (2019). Application of a ground source heat pump system with PCM-embedded radiant wall heating for buildings. *International Journal of Energy Research, 43*(12), 6542–6550.

Rhee, K. N., & Kim, K. W. (2015). A 50 year review of basic and applied research in radiant heating and cooling systems for the built environment. *Building and Environment, 91*, 166–190. https://doi.org/10.1016/j.buildenv.2015.03.040

Rhee, K. N., Olesen, B. W., & Kim, K. W. (2017). Ten questions about radiant heating and cooling systems. *Building and Environment, 112*, 367–381. https://doi.org/10.1016/j.buildenv.2016.11.030

Sartori, I., Napolitano, A., & Voss, K. (2012). Net zero energy buildings: A consistent definition framework. *Energy and Buildings, 48*, 220–232.

Shao, S., Zhang, H., Fan, X., You, S., Wang, Y., & Wei, S. (2021). Thermodynamic and economic analysis of the air source heat pump system with direct-condensation radiant heating panel. *Energy, 225*, 120195.

Szabó, G. L., & Kalmár, F. (2019). Investigation of energy and exergy performances of radiant cooling systems in buildings—A design approach. *Energy, 185*, 449–462.

TÜRKİYE NATIONAL ENERGY PLAN. (2022). *TÜRKİYE NATIONAL ENERGY PLAN 2022*. Turkey.

Wang, Y., Yin, Y., Zhang, X., & Jin, X. (2019). Study of an integrated radiant heating/cooling system with fresh air supply for household utilization. *Building and Environment, 165*, 106404.

Wu, W., & Skye, H. M. (2021). Residential net-zero energy buildings: Review and perspective. *Renewable and Sustainable Energy Reviews, 142*, 110859.

Chapter 15
Energy, Exergy, Exhaust, and Economic Attributes of a CI Engine Powered with Binary Blends of Watermelon Seed and Pine Oil, with Diethyl Ether as a Fuel Additive

Manikandaraja Gurusamy, Balaji Subramaniyan, Malarmannan Subramaniyan, Olusegun D. Samuel, Christopher C. Enweremadu, and Selçuk Sarıkoç

Nomenclature

BSCO	Brake-specific carbon monoxide
$BSCO_2$	Brake-specific carbon dioxide
BSEC	Brake-specific energy consumption
BSHC	Brake-specific hydrocarbon dioxide
BSNO	Brake-specific nitrogen monoxide emissions
BSSootFT	Brake-specific soot Fischer-Tropsch
CI	Compression ignition methane

M. Gurusamy · B. Subramaniyan · M. Subramaniyan
Department of Mechanical Engineering, SRM Institute of Science and Technology, Kattankulathur, Chengalpattu district, Tamil Nadu, India

O. D. Samuel (✉)
Department of Mechanical Engineering, Federal University of Petroleum Resources, Effurun, Delta State, Nigeria

Department of Mechanical Engineering, Bioresources and Biomedical Engineering, University of South Africa, Science Campus, South Africa
e-mail: samuel.david@fupre.edu.ng

C. C. Enweremadu
Department of Mechanical Engineering, Bioresources and Biomedical Engineering, University of South Africa, Science Campus, South Africa

S. Sarıkoç
Department of Mechanical Engineering, Faculty of Engineering, Amasya University, Amasya, Türkiye

M. Z. Sogut, M. Koray (eds.), *Energy Rationality and Management for Decarbonization*, Green Energy and Technology,
https://doi.org/10.1007/978-3-031-68497-5_15

EE	Energy conversion efficiency
EG	Entropy generations
ExE	Exergy efficiency
IC	Internal combustion Fischer-Tropsch
SI	Sustainability index
TEPEloss	Energy loss
TEPExD	Exergy destruction
TEPExL	Exergy loss

15.1 Introduction

One of the better options for diesel fuel replacement in compression ignition methane (CI) engines is biofuels (Suresh & Vijayaragavan, 2022; Arulkumar & Vijayaragavan, 2023). In the last 10 years, a wide range of biofuels have been investigated and made available (Siva Chidambaram et al., 2021; Deshwar et al., 2022). Oil from the seeds of *Citrullus lanatus*, a cool, sweet watermelon, is a novel and fascinating biofuel that is becoming more well-known since it can be used in internal combustion (IC) engines (Xavier, 2020; Ogunwole, 2015; Asokan et al., 2021; Teja et al., 2022). For scientists and engineers searching for sustainable and alternative energy sources, watermelon seed oil (WSO) is an appealing option (Panneerselvam et al., 2015; Oragwu, 2020; Pittala, 2020). Gathering and cleaning the seeds is the first step in the time-consuming process of producing watermelon seed oil. To extract the highest grade oil, the seeds are first crushed or squeezed using a solvent. After completion, the product has a beautiful golden tinge and a faint, pleasant scent. Watermelon seed oil contains significant levels of antioxidants, essential fatty acids, and other health-promoting components. Besides being rich in nutrients, watermelon seed oil may also be utilized as a sustainable source of energy (Krishnaiah et al., 2016; Panneerselvam et al., 2016; Rekha et al., 2018). Due to its unique, composition, watermelon seed oil is a great fit for internal combustion engines. The oil burns with unique properties due to its high concentration of fatty acids (palmitic, oleic, and linoleic acids) (Baboli, 2010; Dumitru et al., 2017; Prasad et al., 2023). During combustion operations, antioxidants tend to improve the stability and oxidation resistance of the oil (Abdu, 2019). Watermelon seed oil is an excellent fuel for internal combustion engines because of its properties, which include increased performance and decreased pollution. The oil's high cetane number and low viscosity make it a promising blending component for traditional diesel fuels (Nanditta et al., 2020; Tanay et al., 2020). Because they have higher combustion efficiency and lower PM emissions than conventional fossil fuels, these two fuel types might be less harmful to the environment (Talibi et al., 2018; Saravanan et al., 2020). Furthermore, watermelon seed oil may be considered a renewable and sustainable resource because watermelons are grown all over the world (Garba et al., 2020; Karaye et al., 2020; Ouassor et al., 2020; Iweka et al., 2023). Growing watermelon for human consumption and biofuel production is a sustainable way to increase agricultural productivity and reduce our reliance on finite fossil fuel resources (Dumitru et al., 2017; Prasad et al., 2023). The possibility of using watermelon seed oil as a biofuel is currently being investigated. This includes the process of testing new fuel blends, improving production processes, and determining how well the oil works with various engine setups (Krishnaiah et al., 2016; Panneerselvam et al., 2016). Future energy and transportation systems may be more environmentally friendly and sustainable if watermelon seed oil is combined with biofuels (Jhalani et al., 2019; Kim & Kim, 2022; Karunamurthy et al., 2023).

Research rarely utilizes watermelon seed oil or any of its various blends. Panneerselvam et al. (2015) produced biodiesel from watermelon seed oil through transesterification. The study investigates three biodiesel and diesel blends (B20, B80, and B100) and compares their efficiency, emissions, and combustion parameters with regular diesel fuel. The data indicates that decreasing the heating value of biodiesel increases fuel consumption, mostly for the brakes, while concurrently lowering the brakes' thermal efficiency. Despite these drawbacks, using biodiesel and its blends dramatically lowers emissions of smoke and nitrogen oxide while simultaneously increasing emissions of hydrocarbons (HCs) and carbon monoxide (CO). Studying combustion has shown that when the percentage of biodiesel in a mixture is higher than conventional diesel, cylinder pressure and heat release rates drop. Asokan et al. (2018) investigated the transesterification process of utilizing methanol and KOH as catalysts to produce hybrid biodiesel from papaya and watermelon seed oil. WPB is a novel kind of biodiesel made from a blend of papaya and watermelon seed oils. In the investigation, a variety of diesel blends and WPB, including B0, WBB20, WPB30, WPB40, and B100, were used. The results of the investigation showed that WPB20 performed better than the other blends. Experiments carried out on a 4-stroke, single-cylinder diesel engine revealed that WPB20 fuel performs and burns similarly to diesel. On the other hand, WPB20 has improved emission characteristics, leading to a reduction in smoke emissions by 27.27%, CO emissions by 23.8%, and HC emissions by 8.3%. The study suggests that using WPB20, a blend of 20% biodiesel, might significantly reduce the amount of diesel used. The effects of mixing diesel fuel with watermelon seed oil (WSO) biodiesel on the performance, emissions, and combustion of a single-cylinder direct injection (DI) diesel engine were investigated in a separate study by Asokan et al. (2021). Watermelon seed oil can be converted into biodiesel through transesterification. Subsequently, three different ratios are used to create a blend of diesel fuel and biodiesel: 20%, 30%, and 40%. The B20 blend performs well overall, showing a slight increase in NOx emissions but a decrease in CO and HC emissions based on the data. Under maximum load conditions, the B20 fuel exhibits significantly low brake-specific fuel consumption (BSFC) of 0.275 kg/kWh and high brake thermal efficiency (BTE) of 32.5%. These results indicate that watermelon seed oil biodiesel has great potential as an alternative to traditional fuels.

The Nigerian watermelon (*Citrullus vulgaris*) seed oil was converted into biodiesel by Nwufo et al. (2023) utilizing a direct base-catalyzed transesterification process. A 93.68% yield was achieved at 55 °C and 1.0 wt% sodium hydroxide. The researchers further stressed that the methyl ester produced from watermelon seed oil has a lower cetane number and heating value than biodiesel produced from other feedstocks. However, science attributes an ether to the chemical formula $(C_2H_5)_2O$. The liquid known as diethyl ether (DEE) is colorless, highly flammable, and has a pleasant fruity scent. In the past, diethyl ether has been extensively used in a variety of industries, including the automotive sector, industrial solvents, and anesthesia. Diethyl ether has recently become an increasingly popular fuel substitute for internal combustion (IC) engines.

The fuel is selected for internal combustion engines due to its higher cetane number and oxygen content, which enhance ignition characteristics and reduce impurities for a cleaner, more efficient combustion. Their investigation revealed that the engine's performance has significantly increased under a range of load conditions. DEE may have a beneficial impact on the combustion parameters because it enhances the cylinder's maximum pressure and the rate at which heat is released. Because of this, DEE may be a good alternative to conventional diesel fuel, although there are a few minor issues with engine stability and combustion duration.

Bridjesh et al. (2018) used waste plastic oil (WPO) diesel blends as fuel for diesel engines by adding 2-methoxy ethyl acetate (MEA) and DEE as additives. They concluded that MEA might be used as a diesel and WPO additive. Nanthagopal et al. (2019) investigated the impacts of IC engines on different DEE fractions in diesel and calophyllum inophyllum biodiesel mixtures. Their results showed that the combustion characteristics were somewhat poorer when the diesel engine operated with all of the ternary blends of DEE.

The diesel engine performance of internal combustion engines running on biodiesel-diesel-DEE-water emulsion fuels (BDDWE) was examined by Dinesha et al. (2019). The tendency of BDDWE to improve performance and reduce emission attributes was observed. Tamilvanan et al. (2021) investigated the effects of oxygenated additives, such as diethyl ether (DE) and ethanol (ET), on a CI engine powered by biodiesel fueled by *Calophyllum inophyllum*. The DE and ET were credited for the improved performance behaviors and superior combustion characteristics. Sakthivadivel et al. (2022) evaluated the effects on engine performance characteristics of neem biodiesel-DEE-Al_2O_3 nanoparticles. The reduction of emission tails and improvement of performance parameters were the consequences of the doping of Al_2O_3 nanoparticles. Diethyl ether as a fuel additive (5%, 10%, and 15%) was tested by Rami Reddy et al. (2022) in relation to the various features of diesel engines powered by waste mango seed biodiesel (MSME) blend (MSME10, MSME20, MSME30, and MSME40). Diesel engines could be operated efficiently by using MSME20 + 5% DEE. The effects of DEE, methoxy ethyl acetate (MEA), and waste plastic oil on the combustion and emission properties of diesel engines were examined by Gnanamoorthi and Murugan (2022). According to their findings, the brake thermal efficiency was improved, and the CO, HC, and smoke emissions were comparable to those of neat diesel with test fuel D50W40DEE10, but lower than those of neat diesel with 50D40W10MEA. The efficacy of fuel additives for HCCI engines running on DME, DEE, and n-heptane was investigated by Banke et al. (2023). Fuel additives facilitate enhanced heat release and allow fuel-rich HCCI operation to be controlled. Hydrogen induction improved CO, HC, and smoke emissions and improved in-cylinder pressure and heat release in CI engines, according to Gurusamy and Ponnusamy's (2023) evaluation of the performance and emissions of diesel engines running on blends of hydrogen and DEE additives. Singh and Kumar (2020) reported that CI engines can efficiently utilize 10% of pine and eucalyptus oil combined with 80% diesel fuel. They emphasized once more that there had been actualized decreases in CO emissions of 13% and smoke of 20.14%.

Gurusamy and Subramanian (2023) reported that blends of pine oil (PO) with DEE and benzyl could improve the CI engine characteristics and have a negative impact on HC and CO emissions. The study by Mebin Samuel et al. (2020) reported an increase in brake thermal efficiency (BTE) and a decrease in carbon monoxide (CO), Hydrocarbon (HC), and smoke emissions in diesel engines powered by pine oil blended with diesel fuel containing 1,4-dioxane additive.

A review of the literature reveals that watermelon seed oil and diethyl ether have been extensively investigated as potential internal combustion engine fuel alternatives. Watermelon seed oil is recognized as a novel and promising biofuel due to its distinct composition, which is abundant in fatty acids and antioxidants. Research indicates that this particular feature can decrease emissions, increase combustion efficiency, and function as a sustainable energy source. Very little research has been done on the use of watermelon seed oil, particularly when it comes to the benefits and drawbacks of its combustion and emissions characteristics. Diethyl ether's high volatility and oxygen concentration make it a promising alternative fuel for internal combustion engines. The chemical's low boiling point and high volatility increase evaporation rate and maximize combustion efficiency. Whether used as an additive or in conjunction with conventional fuels, diethyl ether can enhance the ignition characteristics, reduce pollutants, and increase the overall combustion efficiency of diesel engines. As suggested by the literature, more investigation and analysis are needed to close the information gaps and make diethyl ether and watermelon seed oil viable options for use as biofuels in the energy and transportation sectors.

15.2 Materials and Methods

The watermelon seed oil, pine oil, and diethyl ether were purchased in Chennai in Tamil Nadu, India. The fuel properties of watermelon seed oil, pine oil, and DEE were determined in the thermal laboratory located at the college. A hydrometer was used to determine the specific gravity of the tested samples. Similarly, a bomb calorimeter was used to determine the calorific value of the fuel, and a Redwood viscometer was used to measure the viscosity of the oils. Tables 15.1 and 15.2 present the fuel properties and fuel test matrix in a summarized form.

Table 15.1 Fuel properties

Fuel properties	Diethyl ether	Watermelon seed oil	Pine oil	Diesel
Calorific value in (MJ/kg)	33.7	37.82	41.3	42.5
Viscosity @ 40 °C (cST)	0.33	30	3	2.5
Cetane number	<125	40	11	45–50
Stoichiometric air-fuel ratio	11.1	12.32	13.6	15
Density (kg/m³)	715	915	865	830
% of oxygen	21.62	11.21	10.39	0
% of hydrogen	13.51	11.78	11.68	14.28
% of carbon	64.86%	77.01	77.92%	85.71
Carbon to hydrogen ratio	0.40	0.54	0.56	0.50
Molecular weight	74	857	154	168
Chemical formula	$C_4H_{10}O$	$C_{55}H_{101}O_6$	$C_{10}H_{18}O$	$C_{12}H_{24}$

Table 15.2 Fuel test matrix

Fuel blending details	Fuels short forms
70% of watermelon seed oil with 30% of pine oil	W70
50% of watermelon seed oil with 50% of pine oil	W50
30% of watermelon seed oil with 70% of pine oil	W30
90% of W70 with 10% of diethyl ether	W70D10
90% of W50 with 10% of diethyl ether	W50D10
90% of W30 with 10% of diethyl ether	W30D10

15.3 Details of an Experimental Test Rig

An experiment was conducted on a computerized four-stroke water-cooled single-cylinder naturally aspirated compression ignition engine. An eddy current dynamometer was used to apply brake power to the engine. The load, speed, and crank angle position were measured using a load sensor, speed sensor, and crank angle encoder. A temperature sensor was used to measure the exhaust gas temperature. The fuel flow rate was measured using a fuel flow transmitter, and the airflow rate was measured using an airflow transmitter. All measured data instruments were collected using data acquisition systems. Tailpipe emissions were measured using an AVL gas analyzer and AVL smoke meter. The experiment was repeated three times and the average was taken for the calculations. The schematic diagram and engine specification are provided in Table 15.3. The same experimental test rig was used by Prakash et al. (2018) and Rajasekar et al. (2020) to test their biofuel samples (Fig. 15.1).

Table 15.3 Specification of engine test rig

Description	Specification
Make, model, and number of cylinders	TV1, Kirolskar, single-cylinder engine
Number of strokes, type of cooling	4-stroke, water-cooled engine
Type of dynamometer	Eddy current dynamometer, water-cooled, rated power of 7.5 kW @ 1500 rm
Stroke length, bore diameter, connecting rod length, and compression ratio	87.5 mm, 110 mm, 125 mm, 17.5
Injection timing and injection pressure	Before 27° BTDC and 200 bar

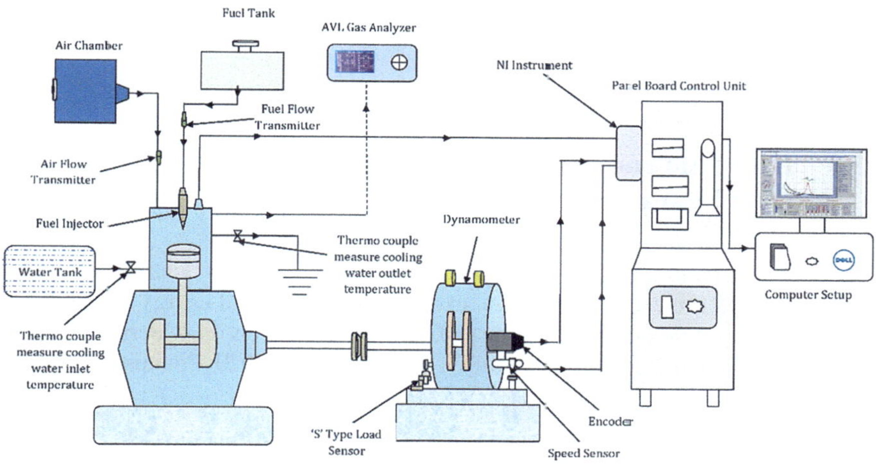

Fig. 15.1 Schematic diagram of engine layout

15.4 Uncertainty of the Experiment

The uncertainty of the experiment is calculated using the observed error and instrumental error. Firstly, the observed error is calculated from the mean and standard deviation of the three consecutive readings as discussed elsewhere (Samuel et al., 2021; Thippeshnaik et al., 2023). The 95% confidence level is set for the observed error. The accuracy of the instrument is taken as the instrumental error. Finally, the geometric mean of the square of the observed error and instrumental error is taken as the uncertainty of each parameter, and its details are given in Table 15.4. The overall uncertainty of the experiment is calculated using Eq. (15.1).

$$\text{Overall uncertanity of experiment} = \sqrt{\begin{pmatrix} \text{BTE}^2 + \text{BSEC}^2 - \text{CO}^2 + \text{CO}_2{}^2 \\ + \text{HC}^2 + \text{NO}^2 + \text{smoke}^2 \end{pmatrix}} \quad (15.1)$$

Table 15.4 Uncertainty of the experiment

Parameter	Uncertainty (±%)
BTE	0.80
BSEC	0.85
CO	1.00
HC	1.70
CO_2	0.93
NO	0.50
Smoke	0.66

$$\text{Overall uncertanity of experiment} = \sqrt{\begin{pmatrix} 0.8^2 + 0.85^2 + 1^2 + 0.93^2 \\ +1.7^2 + 0.50^2 + 0.6^2 \end{pmatrix}} = \pm 2.60\%$$

15.5 Exergetic and Thermoeconomic Parameters Based Parameter

Equations (15.2), (15.3), (15.4), (15.5), and (15.6) are used to determine the fuel exergy, exergy efficiency, energy loss, exergy destruction, and exergy loss, respectively, of internal combustion engines (IC engines) that run on watermelon seed, pine oil, and diethyl ether as fuel additives (Karagoz et al., 2020; Ponnusamy, 2023).

$$\text{Fuel exergy} = m_f * \text{CV} * \left\{ 1.0401 + 0.1728 * \frac{\text{H}}{\text{C}} + 0.0432 * \frac{\text{O}}{\text{C}} \right\} \tag{15.2}$$

$$\text{ExE} = \left(1 - \frac{A_d}{A_{in}} \right) * 100 \tag{15.3}$$

$$\text{TEPEloss} = \frac{\text{Enloss}}{\text{CI}} \tag{15.4}$$

$$\text{TEPExD} = \frac{\text{ExD}}{\text{CI}} \tag{15.5}$$

$$\text{TEPExL} = \frac{\text{Exloss}}{\text{CI}} \tag{15.6}$$

15.6 Results and Discussions

15.6.1 Performance Attributes of CI Engine

Figures 15.2 and 15.3 show the energy conversion efficiency (ECE) and brake-specific energy consumption (BSEC) in relation to engine load, respectively. Engine performance was tested at maximum speed with engine load ranging from 25% to 100% at intervals of 25%.

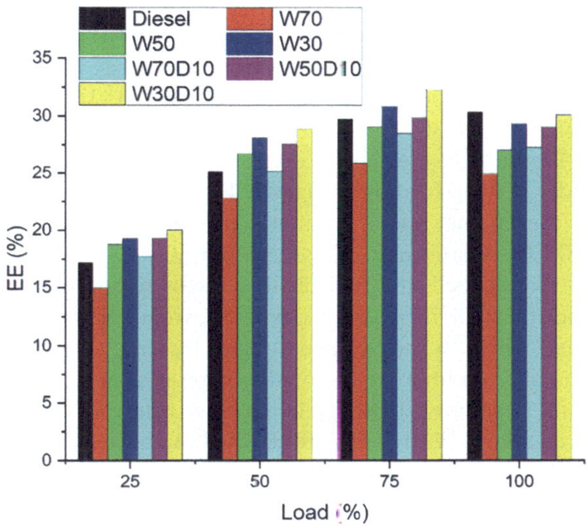

Fig. 15.2 Variation of energy conversion efficiency with load

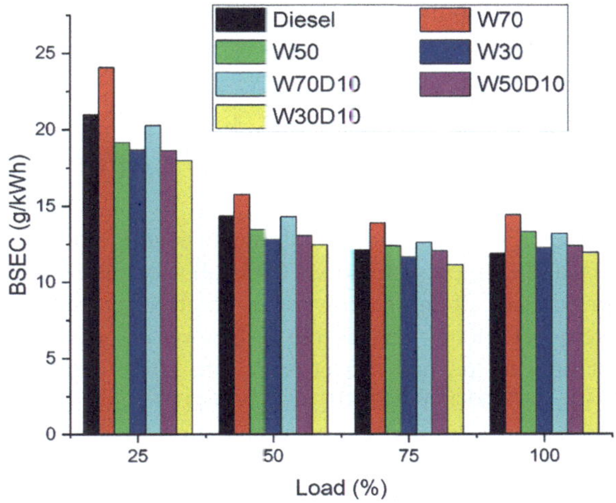

Fig. 15.3 Variation of BSEC with load

The quality of fuel energy transformed into productive labor is determined by a metric called ECE. The engine's heat and brake power are directly correlated with the energy conversion efficiency (Suresh & Vijayaragavan, 2022; Selvanathan & Vijayaragavan, 2023). Due to pine oil's lower stoichiometric air-fuel ratio requirements, blending it with watermelon seed oil results in improved atomization and a better combustible air-fuel mixture with a higher degree of homogeneity (Tanay et al., 2020; Kalra et al., 2021). As a result, W30 is anticipated to generate a greater thermal efficiency than W50 and W70. In actuality, at full load conditions, W30's energy conversion efficiency is 7.8% higher than that of W70, at 14.9%. W70D10 should be more thermally efficient than W70 due to its low viscosity and greater cetane number of DEE; similarly, W50D10 and W30D10 should be more energy-efficient than W50 and W30. In fact, W70D10 exhibits an efficiency that is 8.56% greater than W70, W50D10 exhibits an energy conversion efficiency that is 6.94% higher than W50, and at rated power, W30D10 exhibits an efficiency that is 2.58% higher than W30.

Figure 15.3 presents the diverse BSEC characteristics of the distinct fuel samples under varying loading scenarios. The heating value of the fuel, the power operating on the engine, and the amount of energy used to generate the power all affect BSFECs (Gurusamy, 2023; Ponnusamy, 2023). When pine oil and watermelon seed oil are blended, the fuel blend's viscosity and cetane number decrease while its calorific value increases. Conversely, when DEE is added, the fuel blends' calorific value, viscosity, and cetane number increase (Nanthagopal et al., 2019). DEE and pine oil both improve the oxidation process and reduce viscosity, which ultimately leads to improved energy conversion using less fuel. Regarding the justification given above, W30 and W30D10 ought to have lower BSEC values. Also, the lowest BSEC is found in W30 and W30D30 when compared to W70 by 13.30%, W50 by 12.27%, W70D10 by 6.56%, and W50D10 by 3.6%. In contrast, at full load conditions, W30D10 is closer to W70D10 and W50D10, at 21.57%, 12.10%, 3.35%, 11.16%, and 4.31% greater than W70, W50, W30, W70D10.

When combining camphor oil with Mahua oil, Tanay et al. (2020) discovered a comparable decrease in BSEC.

15.6.2 Exergy Attributes of CI Engine

Figures 15.4, 15.5, and 15.6 show the exergy efficiency (ExE), sustainability index (SI), and entropy generations (EGs) in relation to engine load, respectively. Exergy attributes of CI engine powered with all tested fuels at various loading conditions. Exergy efficiency defines the utilization of energy from the point of the highest potential of the fuel (Sheet & Yamin, 2020; Gurusamy et al., 2023). W70, W50, W30, W70D10, W50D10, and W30D10 exhibit 18.6%, 11.7%, 4.24%, 10.96%, 5.12%, and 1.7% lower at rated operating conditions in contrast to diesel. The highest ExE of all the biofuel blends is achieved by W30D10 at full load situation; this is 17.1. The above observation is consistent with the increase in ExE resulting from

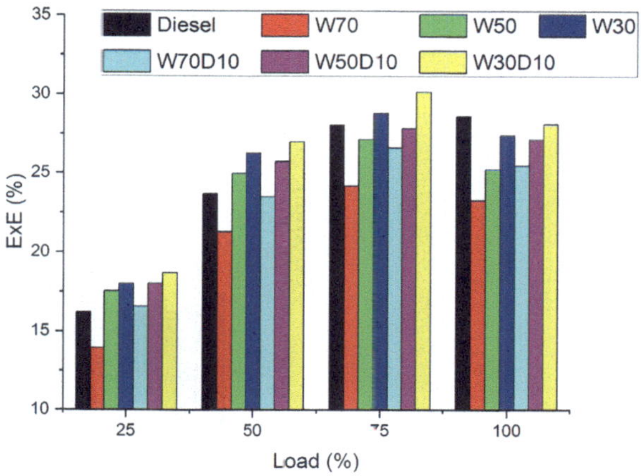

Fig. 15.4 Variation of exergy efficiency with load

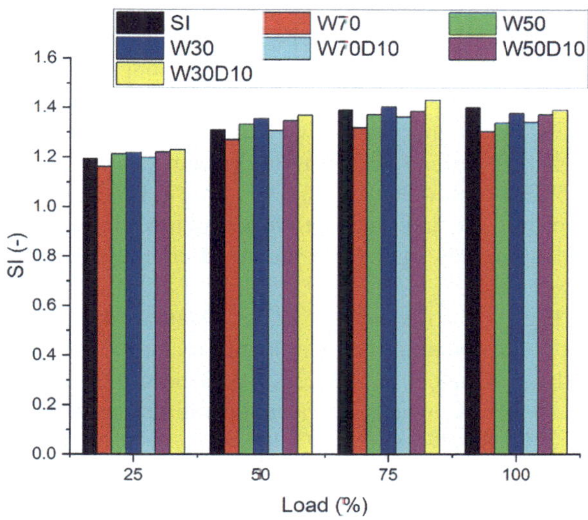

Fig. 15.5 Variation of sustainability index with load

higher fuel blend fractions of pine and DEE additives. The values were 8%, 10.18%, 2.58%, 9.42%, and 3.42% greater than W70, W50, W30, W70D10, and W50, respectively. The addition of pine oil lowers the gasoline blend's cetane number and viscosity. As a result, it decreases droplet size, raises the level of fuel and air homogeneity, and ultimately permits the premixed combustion to burn a greater quantity of fuel. Lastly, adding pine oil improves the ExE. Additional DEE addition enhances the formation of the fuel mixture, increases homogeneity, and stabilizes the combustion process.

The sustainability index (SI) variations are demonstrated in Fig. 15.5. The significant factor is SI, which is directly correlated with ExE (Yesilyurt, 2020; Tiwari et al., 2023; Doğan et al., 2023). The fuel blend with the highest ExE in this

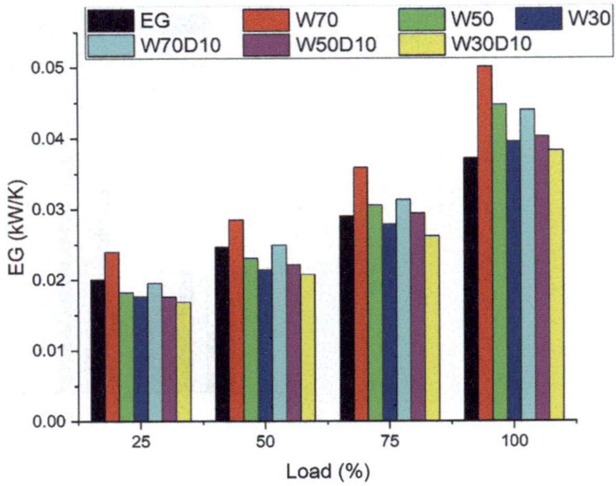

Fig. 15.6 Variation of entropy with load

situation ought to have the highest SI. Similarly, the fuel blend that has the lowest SI should also have the lowest ExE. Among the biofuel blends, W30D10 produces the highest SI during rated operating circumstances, with 6.3%, 3.83%, 1%, 3.55%, and 1.34% higher than W70, W50, W30, W70D10, and W50D10. Diesel fuel has a higher SI during rated operating circumstances for the W70, W50, W30, W70D10, W50D10, and W30D10—6.63%, 4.48%, 1.67%, 4.2%, 2.01%, and 0.67%. Through improved air-fuel mixes produced by lowering fuel blend volatility, fuel viscosity, and ultimately improving premixed combustions, the addition of pine and DEE raises the SI.

Figure 15.6 displays the EG of all tested fuels. The fuel energy degradation from the actual potential energy is defined by the EG parameter (Chintala & Subramanian, 2014; Çakmak et al., 2022; Rai & Sahoo, 2022). The EG for diesel is found to be 11.15 kW/K, which, at rated operating scenarios, is 34.56%, 20.22%, 6.27%, 18.21%, 8.28%, and 2.86% less than W70, W50, W30, W70D10, W50D10, and W30D10. The biofuel sample with the least amount of entropy generation is W30D10; at rated working circumstances, it is 30.81%, 16.88%, 3.37%, 14.92%, and 5.26% less than W70, W50, W30, W70D10, and W50D10. The improved combusted mixture generation and higher energy transfer to the piston from burning more fuel in premixed combustion result in improved performance when pine oil and DEE are added.

15.6.3 Economic Attributes of CI Engine

Figures 15.7 and 15.8 show the energy and exergy features based on thermoeconomic parameters, namely energy loss and exergy destruction (TEPEloss and TEP ExD, respectively), in relation to load.

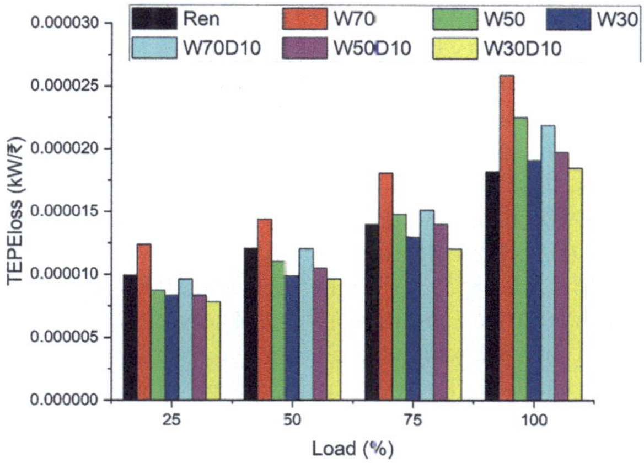

Fig. 15.7 Variation of thermoeconomic parameter based on energy loss with load

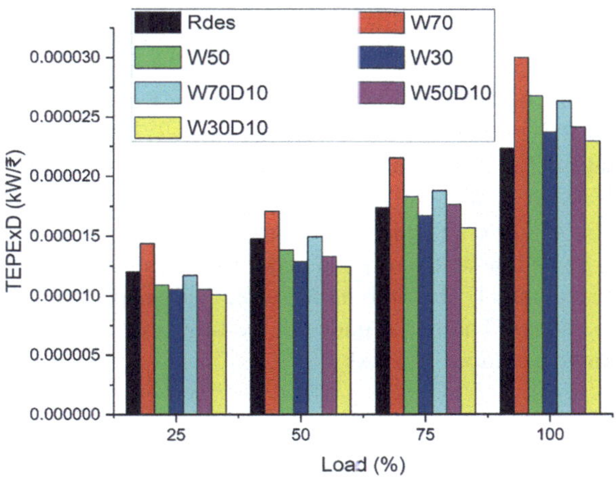

Fig. 15.8 Variation of thermoeconomic parameter based on exergy destruction with load

The ratio of energy loss to engine capital cost is known as TEPEloss. The engine setup costs a total of ₹500,000 (in Indian rupees), \$6003.87, or ₦5,388,908.61 (in Nigerian naira), and this information is utilized to calculate the thermoeconomic parameter. The ETC for diesel is obtained as 1.82×10^{-5} kW/₹ which is 42.22%, 23.93%, 5.07%, 20.47%, 8.46% and 1.71% lower than W70, W50, W30, W70D10, W50D10 and W30D10 at rated operating conditions. Regarding the biofuel samples, the least, ETC is obtained for W30D10 at rated operating conditions, in fact, that is 39.81%, 21.88%, 3.30%, 18.44%, and 6.63% lower than W70, W50, W30, W70D10, and W50D10. The TEPEloss is dependent on the energy losses that occur in the engine, so it must be as low as much as possible. With regard to biofuels, the addition of pine oil and DEE reduces energy loss and improves mechanical energy transfer, especially at premixed combustions. Therefore,

TEPEloss is lowest for W30 and W30D10 among the biofuel blends either in binary or ternary blends.

The thermoeconomic parameter variations based on TEP ExD are presented in Fig. 15.8. The parameter TEPExD is derived from Eq. (15.5). Underrated operating conditions, the diesel's TEPExD is 2.231×10^{-5} kW/₹, which is 34.5%, 20.22%, 6.25%, 18.21%, 8.3%, and 2.86% less than that of W70, W50, W30, W70D10, W50D10, and W30D10. Furthermore, in comparison to the other biofuel blends, W30D10 exhibits the lowest TEPExD. In actuality, at rated operating conditions, it is 5.26%, 16.88%, 3.31%, 14.92%, and 30.8% less than W70, W50, W30, W70D10, and W50D10. According to the observation, adding pine oil outperforms watermelon seed oil, which has a higher viscosity and raises the temperature and pressure inside the cylinder while also causing more cycle fluctuations. Conversely, DEE lowers the cyclic fluctuation by managing the quick combustion phase with a somewhat reduced ignition delay. As a result, DEE and pine oil lessen the losses and energy losses related to TEPExD.

Figure 15.9 depicts the variations in the thermoeconomic parameter depending on exergy loss (TEPExL). Equation (15.6) is used to find the TEPExL. Diesel fuel has the lowest TEPExL during full load conditions compared to W70, W50, W30, W70D10, W50D10, and W30D10. In fact, it has TEPExL values of 42.22%, 23.91%, 5.074%, 20.47%, 8.46%, and 1.71% lower. Similarly, W30D10 has the lowest TEPExL of all the biofuel blends at full load conditions, which is less than W70, W50, W30, W70D10, and W50D10 by 39.83%, 21.83%, 3.30%, 18.44%, and 6.63%. The cost related to the energy loss is called TEPExL. Consequently, it ought to be as low as feasible. in order to minimize the economies connected with loss. Pine oil and DEE combination effectively expedites the combustion phase and minimizes loss. Therefore, under all loading circumstances, the combination of DEE and pine oil reduces TEPExL.

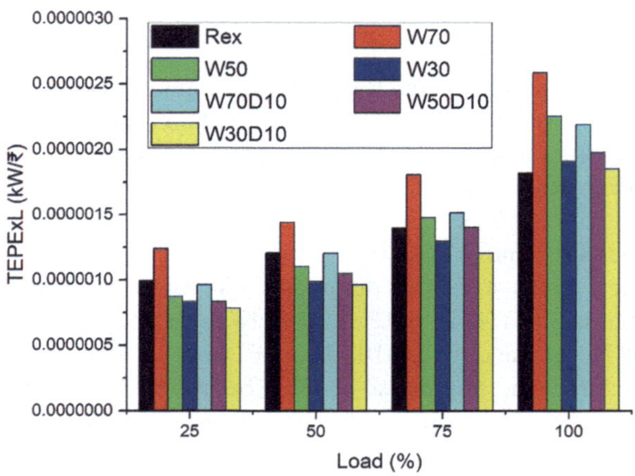

Fig. 15.9 Variation of thermoeconomic parameter based on exergy loss with load

15.6.4 Exhaust Attributes of CI Engine

Figures 15.10, 15.11, 15.12, 15.13, and 15.14, respectively, display the exhaust attributes of the CI engine, which are brake-specific hydrocarbon (BSHC), brake-specific carbon monoxide (BSCO), brake-specific carbon dioxide ($BSCO_2$), brake-specific nitrogen monoxide (BSNO), and brake-specific soot (BSSoot), in relation to load.

The BSHC emissions for each tested fuel under different loading situations are explained in Fig. 15.10. In comparison to other evaluated fuels, diesel fuel exhibits lower BSHC emissions. At full load conditions, it is higher than W70, W50, W30, W70D10, W50D10, and W30D10, respectively, by 81.5%, 44.02%, 28.78%,

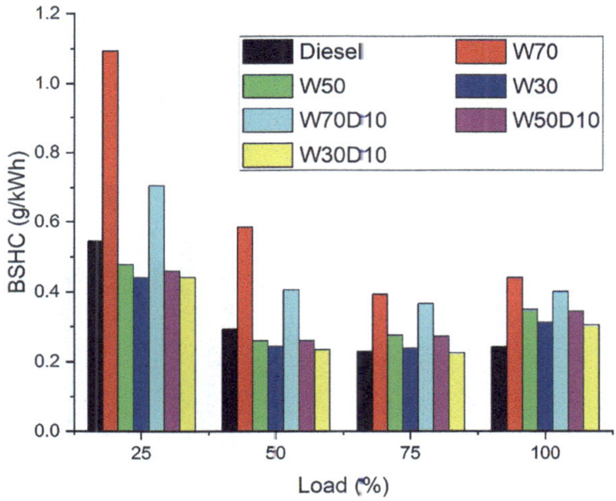

Fig. 15.10 Variation of BSHC attributes of CI engine with load

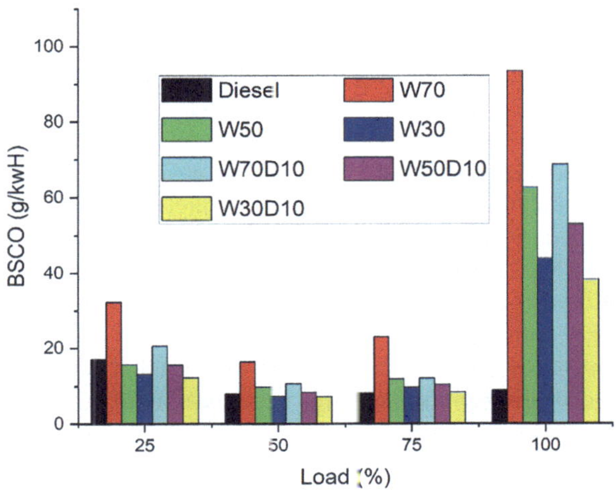

Fig. 15.11 Variation of BSCO attributes of CI engine with load

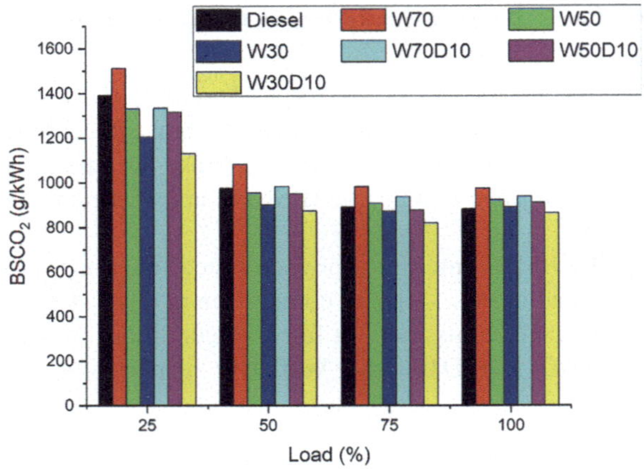

Fig. 15.12 Variation of $BSCO_2$ attributes of CI engine with load

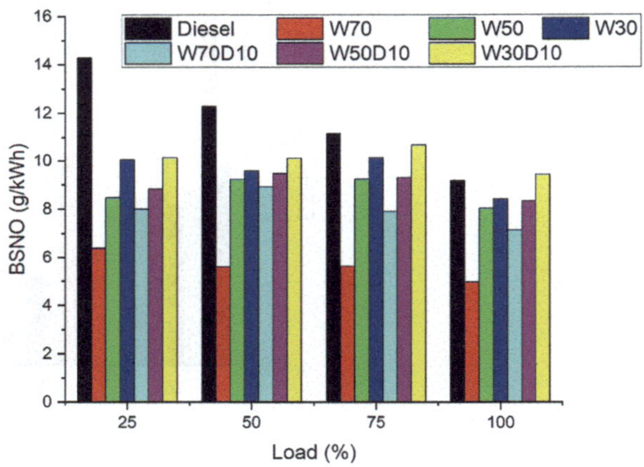

Fig. 15.13 Variation of BSNO attributes of CI engine with load

65.25%, 42.18%, and 25.60%. Additionally, W70D10 has the lowest BSHC emissions of all the biofuels that have been evaluated. W30D10, for instance, has a BSHC that, under full load conditions, is 44.71%, 14.75%, 2.53%, 31.58%, and 13.20% lower than W70, W50, W30, W70D10, and W50D10, respectively. Due to its reduced fuel droplet size, higher envelope-to-core jet length ratio, reduced sac volume fuel, improved air-fuel mixing, and decreased exhaust gas dilution with fresh entry air due to the engine exhaust's lower temperature, W30 exhibits the lowest BSHC emissions among the binary fuel blends. Similarly, the inclusion of DEE aids in combustion and lowers the emissions of BSHC. Consistent with our findings, Sathiyamoorthi and Sankaranarayanan (2017) discovered that adding diesel fuel to low-viscosity ethanol in a mixture that contains 25% lemongrass oil reduces hydrocarbon emissions. This is because fuel-born oxygen effectively participates in the combustion process.

The BSCO discrepancies for all evaluated fuels under varied loading scenarios are explained in Fig. 15.11. When compared to other studied fuels, diesel is

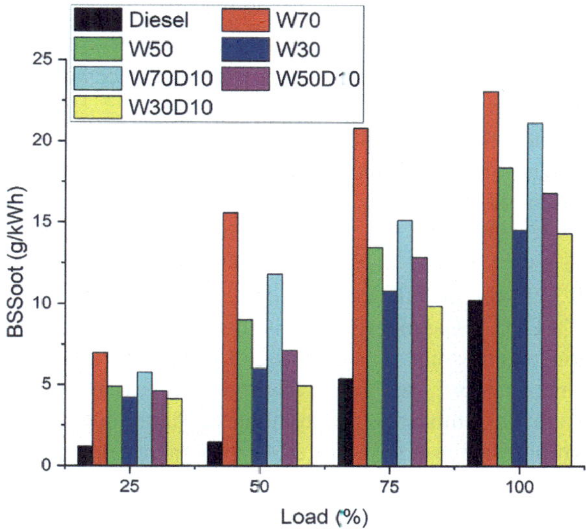

Fig. 15.14 Variation of brake-specific soot emissions with load

determined to have lower BSCO emissions at full load conditions. Most remarkably, diesel produced BSCO emissions of 8.9 g/kWh at full load, which is 34.76%, 30.02%, 24.16%, 17.65%, 12.05%, and 6% lower than the CI engine. Furthermore, with regard to biofuel blends, W30D10's BSCO is determined to be significantly lower—at full load operating conditions, it is 25.79%, 22.78%, 17.23%, 11.10%, and 5.8% lower than W70, W50, W30, W70D10, and W50D10, respectively. The reduced CO emissions for W30 and W30D10 among the blends of biofuels are confirmed by the reading above. Partially burning carbon atoms produce the CO emission. Because of improved heat transmission brought on by the high-pressure, high-temperature combustion process, the addition of pine oil and DEE accelerates the oxidation rates of carbon atoms. Similar results were reported by Jalasutram et al. (2019) when low-viscosity lemongrass oil was added to diesel fuel.

The variations in $BSCO_2$ for every fuel evaluated under different loading scenarios are explained in Fig. 15.12. The successful completion of carbon atom oxidation is measured by $BSCO_2$ emissions (Suresh & Vijayaragavan, 2022). The diesel fuel produced a BSCO2 of 882.17 g/kWh at full load, which is 10.47%, 4.58%, and 0.89% less than W70, W50, and W30, respectively. Similarly, at full load settings, diesel $BSCO_2$ emissions were determined to be 3.25% and 6.26% lower than W70D10 and W50D10, respectively. It is currently 2.06% less than W30D10. W30D10 showed reduced $BSCO_2$ emissions under full load conditions among the evaluated biofuel blends; these are, respectively, 12.8%, 6.78%, 3.02%, 8.58%, and 5.38% lower than W70, W50, W30, W70D10, and W50D10 at rated operating conditions. W30 produced the most CO_2 emissions when it came to biofuel binary blends. Proper atomization, inventive air-fuel mixing because of the smaller fuel droplet size, and a larger spray envelope than the fuel jet core are the key causes of W30's greater CO_2 emissions. A greater quantity of carbon atoms burn during the fast combustion phase due to the increased heat released by the lower cetane rating of pine oil, which also promotes effective heat transfer and accelerates the rate of carbon atom oxidation. Furthermore, W30D10 has greater $BSCO_2$ emissions based on ternary fuel mixes because of the higher cetane rating and lower DEE viscosity that control the combustion process.

The BSNO variations for all evaluated fuels under varied loading scenarios are shown in Fig. 15.13. When diesel fuel is used under full load conditions, its BSNO emissions are 45.87%, 12.26%, 8.12%, 22.07%, and 9% greater than those of W70, W50, W30, W70D10, and W50D10. The W70 exhibits lower BSNO emissions than the tested biofuel samples, with values of 62.08%, 69.72%, 43.96%, 68.09%, and 89% lower at full load settings than the W50, W30, W70D10, W50D10, and W30D10, respectively. W30 is the binary fuel blend with the highest BSNO emissions because it burns at a higher temperature when there is ambient oxygen and nitrogen present within the allotted time. Second, compared to diffused combustion, premixed combustion predominates and produces higher BSNO emissions. W30D10 exhibits increased BSNO emissions in relation to DEE addition because of fine atomization, which enhances premixed combustion by somewhat lowering ignition latency. Pine oil added to diesel at 10% and 15% increases NO_x emissions, according to Vasudevan and Vinoth Kanna (2019). This is because of the higher oxygen content and faster heat release rate of pine oil when Jatropha oil methyl ester was added to pine oil, Rajasekar et al. (2020) likewise observed comparable outcomes.

The BSSoot variations for all evaluated fuels under varied loading scenarios are interpreted in Fig. 15.14. Diesel exhibits lower CO emissions compared to all evaluated biofuel blends underrated operating circumstances. Specifically, diesel has 92%, 79%, 42%, 85%, 64.3%, and 39.9% fewer soot emissions than W70, W50, W30, W70D10, W50D10, and W30D10, respectively. Similarly, at full load conditions, the biofuel blends W30D10, W70, W50, W30, W70D10, and W50D10 show decreased BSSoot emissions, with respective values of 61.3%, 28%, 1.48%, 47.72%, and 17.41%. This observation confirms that pine oil's low cetane effect, leading to a longer ignition delay and faster heat release rate, dominates the combustion process. Consequently, less fuel is consumed in the later stages of combustion, resulting in scattered combustions. The addition of DEE to binary mixes of olive oil and watermelon seed oil improves fuel atomization and produces the best combustible air-fuel charges. Furthermore, the higher cetane rating of DEE reduces cyclic variations and decreases soot emissions, enhancing combustion efficiency.

15.7 Conclusion

The following results were obtained for the current research work and are explained below.

The EE of the W30 is found to be higher than that of W50 and W70. Similarly, W30D10 shows a higher BTE than W50D10 and W70D10. Consequently, W30D10 exhibits lower BSEC compared to other tested biofuel blends at all loading conditions.

The ExE is an important thermodynamic parameter that was found to be higher for W30D10 compared to other tested biofuel blends. Consequently, it also has lower entropy generation and a higher sustainability index at all loading conditions.

The thermoeconomic parameter is used to define the cost associated with thermodynamic irregularities. TEPEloss is found to be lower for W30D10 at all load conditions due to the reduction of cyclic irregularities caused by the DEE addition to W30. Furthermore, TEPExL is another parameter found to be lower for W30D10. Additionally, TEPExD is also found to be lower for W30D10 due to the lower thermodynamic irregularities caused by the DEE addition to W30.

In summary, a watermelon mixture with pine oil and DEE additive, specifically W30D10 fuel, can be used in a CI engine with a 90% W30 and 10% DEE combination.

References

Abdu, B. (2019). Comparative study of antimicrobial potentials of seed oils of Jatropha curcas and Tamarindus indica. *International Journal of Scientific Research in Science, Engineering and Technology, 6*, 37–49. https://doi.org/10.32628/ijsrset19653

Arulkumar, S., & Vijayaragavan, M. (2023). The influence of hydrogen as a supplementary fuel on the characteristics of CI engines running on Calophyllum inophyllum oil diesel blend. *Fuel, 351*, 128959. https://doi.org/10.1016/j.fuel.2023.128959

Asokan, M. A., Senthur Prabu, S., Kamesh, S., & Khan, W. (2018). Performance, combustion and emission characteristics of diesel engine fuelled with papaya and watermelon seed oil bio-diesel/diesel blends. *Energy, 145*, 238–245. https://doi.org/10.1016/j.energy.2017.12.140

Asokan, M. A., Prabu, S. S., Akhil, V. S., et al. (2021). Performance and emission behaviour of diesel and blends of watermelon seed oil biodiesel in direct injection diesel engine. *Materials Today Proceedings, 45*, 3274–3278. https://doi.org/10.1016/j.matpr.2020.12.469

Baboli, Z. M. (2010). Characteristics and composition of watermelon seed oil and solvent extraction parameters effects. *Journal of the American Oil Chemists' Society, 87*, 667–671. https://doi.org/10.1007/s11746-010-1546-5

Banke, K., Freund, D., Atakan, B., & Kaiser, S. A. (2023). Evaluation of fuel additives for HCCI engines operated on fuel-rich methane/air mixtures: DME, DEE, and n-heptane. *Applications in Energy and Combustion Science, 13*, 100112. https://doi.org/10.1016/j.jaecs.2023.100112

Bridjesh, P., Periyasamy, P., Krishna Chaitanya, A. V., & Geetha, N. K. (2018). MEA and DEE as additives on diesel engine using waste plastic oil diesel blends. *Sustainable Environment Research, 28*, 142–147. https://doi.org/10.1016/j.serj.2018.01.001

Çakmak, A., Yeşilyurt, M. K., Erol, D., & Doğan, B. (2022). The experimental investigation on the impact of n-octanol in the compression-ignition engine operating with biodiesel/diesel fuel blends: Exergy, exergoeconomic, environmental analyses. *Journal of Thermal Analysis and Calorimetry, 147*, 11231–11259. https://doi.org/10.1007/s10973-022-11357-w

Chintala, V., & Subramanian, K. A. (2014). Assessment of maximum available work of a hydrogen fueled compression ignition engine using exergy analysis. *Energy, 67*, 162–175. https://doi.org/10.1016/j.energy.2014.01.094

Deshwar, P., Kumar, G., Subramanian, B., et al. (2022). Simultaneous reduction of NOx and smoke emissions in gasoline-ethanol blended RCCI Engine with biogas as a ternary fuel. *International Journal of Energy for a Clean Environment, 24*, 83–101. https://doi.org/10.1615/interjenercleanenv.2022040767

Dinesha, P., Kumar, S., & Rosen, M. A. (2019). Combined effects of water emulsion and diethyl ether additive on combustion performance and emissions of a compression ignition engine using biodiesel blends. *Energy, 179*, 928–937. https://doi.org/10.1016/j.energy.2019.05.071

Doğan, B., Çelik, M., Bayındırlı, C., & Erol, D. (2023). Exergy, exergoeconomic, and sustainability analyses of a diesel engine using biodiesel fuel blends containing nanoparticles. *Energy, 274*, 127278. https://doi.org/10.1016/J.ENERGY.2023.127278

Dumitru, M. G., Badea, D. N., & Tutunea, D. (2017). Determination of physicochemical properties and emissions for different blends of biodiesel from watermelon (Citrullus lanatus L.) seeds and diesel fuel. *Revista de Chimie, 68*, 2771–2775. https://doi.org/10.37358/rc.17.12.5976

Garba, N. A., Shehu, S., & Mathew, D. C. (2020). Characterization of watermelon (Citrullus vulgaris) and calabash (Lagenaria vulgaris) seeds oil as potential feedstock for biodiesel production. *International Journal of Science for Global Sustainability, 6*, 119–124.

Gnanamoorthi, V., & Murugan, M. (2022). Effect of DEE and MEA as additives on a CRDI diesel engine fueled with waste plastic oil blend. *Energy Sources, Part A: Recovery, Utilization and Environmental Effects, 44*, 5016–5031. https://doi.org/10.1080/15567036.2019.1657206

Gurusamy, M. (2023). Experimental investigation of engine characteristics of CI engine fueled with fatty acid Millettia Pinnata oil with non-fatty acid Pinus oil blended fuel. *Environmental Progress and Sustainable Energy, 42*, e14155. https://doi.org/10.1002/ep.14155

Gurusamy, M., & Ponnusamy, C. (2023). The influence of hydrogen induction on the characteristics of a CI engine fueled with blend of camphor oil and diesel with diethyl ether additive. *International Journal of Hydrogen Energy, 48*, 24054–24073. https://doi.org/10.1016/j.ijhydene.2023.03.188

Gurusamy, M., & Subramanian, B. (2023). Study of PCCI engine operating on pine oil diesel blend (P50) with benzyl alcohol and diethyl ether. *Fuel, 335*, 127121. https://doi.org/10.1016/j.fuel.2022.127121

Gurusamy, M., Ponnusamy, C., & Varuvel, E. G. (2023). Energy, exergy, and emission analysis of biofuels in compression ignition engine with camphor oil as premixing fuel. *Energy Technology, 12*, 2300676. https://doi.org/10.1002/ente.202300676

Iweka, S. C., Falowo, O. A., Amosun, A. A., & Betiku, E. (2023). Optimization of microwave-assisted biodiesel production from watermelon seeds oil using thermally modified kwale anthill mud as base catalyst. *Heliyon, 9*, e17762. https://doi.org/10.1016/j.heliyon.2023.e17762

Jalasutram, A. K., Kiran Babu, B., Krishna, C. V., & Mekala, R. (2019). Combustion phenomena, emissions and performance characteristics of lemongrass oil in direct injection (DI) compression ignition diesel. *International Journal of Recent Technology and Engineering, 8*, 2380–2386. https://doi.org/10.35940/ijrte.A3382.078219

Jhalani, A., Sharma, D., Soni, S. L., et al. (2019). A comprehensive review on water-emulsified diesel fuel: Chemistry, engine performance and exhaust emissions. *Environmental Science and Pollution Research, 26*, 4570–4587. https://doi.org/10.1007/s11356-018-3958-y

Kalra, S., Agarwal, A., Kumar, S. A., & Mathanraj, V. (2021). Experimental investigation of IC engine performance using calophylum inophylum biodiesel. *Journal of Physics Conference Series, 2054*, 012028. https://doi.org/10.1088/1742-6596/2054/1/012028

Karagoz, M., Uysal, C., Agbulut, U., & Saridemir, S. (2020). Energy, exergy, economic and sustainability assessments of a compression ignition diesel engine fueled with tire pyrolytic oil−diesel blends. *Journal of Cleaner Production, 264*, 121724. https://doi.org/10.1016/j.jclepro.2020.121724

Karaye, I. U., Ladan, M. U., Shehu, A., et al. (2020). Biochemical characterization of seed oil in sesame and watermelon for production of biodiesel. *International Journal of Science for Global Sustainability, 6*, 32–38.

Karunamurthy, K., Janvekar, A. A., Palaniappan, P. L., et al. (2023). Prediction of IC engine performance and emission parameters using machine learning: A review. *Journal of Thermal Analysis and Calorimetry, 148*, 3155–3177. https://doi.org/10.1007/s10973-022-11896-2

Kim, S., & Kim, J. (2022). Feasibility assessment of hydrogen-rich syngas spark-ignition engine for heavy-duty long-distance vehicle application. *Energy Conversion and Management, 252*, 115048. https://doi.org/10.1016/j.enconman.2021.115048

Krishnaiah, G., Pasnoori, S., Santhoshi, P. S., et al. (2016). Ultrasonic and microwave effects on crystalline Mn(II) carbonate catalyzed biodiesel production using watermelon (Citrullus vulgaris) seed oil and alcohol (fibrous flesh) as exclusive green feedstock. *Biofuels, 7*, 735–741. https://doi.org/10.1080/17597269.2016.1192835

Mebin Samuel, P., Devaradjane, G., & Gnanamoorthi, V. (2020). Performance enhancement and emission reduction by using pine oil blends in a diesel engine influenced by 1,4-dioxane. *International journal of Environmental Science and Technology, 17*, 1783–1794. https://doi.org/10.1007/s13762-019-02455-8

Nanditta, R. V., Deenadayalan, S. R., Manikandaraja, G., & Udayakumar, P. (2020). Performance and emission characteristics of CI engine fuelled with turpentine oil-diesel blend with diethyl ether as additives. *IOP Conference Series: Materials Science and Engineering, 912*, 042075. https://doi.org/10.1088/1757-899X/912/4/042075

Nanthagopal, K., Ashok, B., Garnepudi, R. S., et al. (2019). Investigation on diethyl ether as an additive with Calophyllum Inophyllum biodiesel for CI engine application. *Energy Conversion and Management, 179*, 104–113. https://doi.org/10.1016/j.enconman.2018.10.064

Nwufo, O. C., Nzebuka, G. C., Okonkwo, B. U., et al. (2023). Watermelon (Citrullus vulgaris) seed oil as a potential feedstock for biodiesel production in Nigeria. *Biofuels, 14*, 713–720. https://doi.org/10.1080/17597269.2023.2167272

Ogunwole, O. A. (2015). Production of biodiesel from watermelon (*Citrullus lanatus*) seed oil. *Leonardo Journal of Science*, 63–74.

Oragwu, I. P. (2020). Solvent-extracted watermelon seed oil (Citrulus vulgaris) and application in skin-care products. *COOU Journal of Physical ciences, 3*. 2020.

Ouassor, I., Aqil, Y., Belmaghraoui, W., & El Hajjaji, S. (2020). Characterization of two Moroccan watermelon seeds oil varieties by three different extraction methods. *Oilseeds and Fats, Crops and Lipids, 27*, 1–8. https://doi.org/10.1051/ocl/2020010

Panneerselvam, N., Murugesan, A., Vijayakumar, C., et al. (2015). Performance, emissions and combustion characteristics of CI engine fuel with watermelon (Citrullus vulgaris) methyl esters. *International Journal of Ambient Energy, 38*, 308–313. https://doi.org/10.1080/01430750.2015.1087431

Panneerselvam, N., Murugesan, A., Vijayakumar, C., & Subramaniam, D. (2016). Optimization of biodiesel produced from watermelon (Citrullus vulgaris) using batch-type production unit. *Energy Sources, Part A: Recovery, Utilization, and Environmental Effects, 38*, 2343–2348. https://doi.org/10.1080/15567036.2015.1048389

Pittala, A. K. (2020). Performance and emission characteristics of watermelon seed oil biodiesel blend in a single-cylinder diesel engine. *AIP Conference Proceedings, 2311*, 020017. https://doi.org/10.1063/5.0033966

Ponnusamy, M. G. C. (2023). Assessment of engine characteristics of compression ignition engine fuelled with Polanga oil and camphor oil blend. *International Journal of Environmental Science and Technology, 21*, 603. https://doi.org/10.1007/s13762-023-04949-y

Prakash, T., Edwin Geo, V., Martin, L. J., & Nagalingam, B. (2018). Improved cold flow properties and combustion analysis of high viscous castor oil and its biodiesel in a CI engine. *Nature Environment and Pollution Technology, 17*, 1183–1192.

Prasad, A., Sivanraju, R., Teklemariam, A., et al. (2023). Influence of nano additives on performance and emissions characteristics of a diesel engine fueled with watermelon methyl ester. *Journal of Thermal Engineering, 9*, 395–400. https://doi.org/10.18186/thermal.1285915

Rai, R. K., & Sahoo, R. R. (2022). Effect of CNT and Al_2O_3-CNT hybrid nano-additive in water-emulsified fuels on DICI engine energetic and exergetic performances. *Journal of Thermal Analysis and Calorimetry, 147*, 3577–3589. https://doi.org/10.1007/s10973-021-10746-x

Rajasekar, V., Geo, V. E., Martin, L. J., & Nagalingam, B. (2020). The combined effect of low viscous biofuel and EGR on NO-smoke tradeoff in a biodiesel engine—An experimental study. *Environmental Science and Pollution Research, 27*, 17468–17480. https://doi.org/10.1007/s11356-019-05449-8

Rami Reddy, S., Murali, G., & Dhana Raju, V. (2022). Assessment of diethyl ether as a fuel additive on the diverse characteristics of diesel engine powered with waste mango seed biodiesel blend. *International Journal of Ambient Energy, 43*, 3365–3376. https://doi.org/10.1080/01430750.2020.1824944

Rekha, G., Rose, A. L., & Rose, D. A. L. (2018). Phytochemical, minerals and physicochemical properties of watermelon seed oil. *International Journal of Innovative Science and Research Technology, 3*, 966.

Sakthivadivel, D., Ganesh Kumar, P., Prabakaran, R., et al. (2022). A neem oil-based biodiesel with DEE enriched ethanol and Al_2O_3 nano additive: An experimental investigation on the diesel engine performance. *Case Studies in Thermal Engineering, 34*, 102021. https://doi.org/10.1016/j.csite.2022.102021

Samuel, O. D., Waheed, M. A., Taheri-Garavand, A., Verma, T. N., Dairo, O. U., Bolaji, B. O., & Afzal, A. (2021). Prandtl number of optimum biodiesel from food industrial waste oil and diesel fuel blend for diesel engine. *Fuel, 285*, 119049.

Saravanan, S., Kaliyanasunder, R., Rajesh Kumar, B., & Lakshmi Narayana Rao, G. (2020). Effect of design parameters on performance and emissions of a CI engine operated with diesel-biodiesel-higher alcohol blends. *Renewable Energy, 148*, 425–436. https://doi.org/10.1016/j.renene.2019.10.049

Sathiyamoorthi, R., & Sankaranarayanan, G. (2017). The effects of using ethanol as additive on the combustion and emissions of a direct injection diesel engine fuelled with neat lemongrass oil-diesel fuel blend. *Renewable Energy, 101*, 747–756. https://doi.org/10.1016/j.renene.2016.09.044

Selvanathan, A., & Vijayaragavan, M. (2023). Effect on minor addition of aromatic (benzyl alcohol) and diethyl ether in Calophyllum inophyllum blended diesel fuel in a CI engine operates by hydrogen energy as a secondary fuel. *Energy, 285*, 129529. https://doi.org/10.1016/j.energy.2023.129529

Sheet, E. A. E., & Yamin, J. A. A. (2020). Effect of biodiesel blending with diesel fuel on CI engine performance and energy availability. *International Journal of Exergy, 32*, 213–226. https://doi.org/10.1504/IJEX.2020.108174

Singh, K., & Kumar, N. (2020). Synergistic effect of pine oil and eucalyptus oil on the performance of a compression ignition engine. *Energy Sources, Part A: Recovery, Utilization and Environmental Effects, 47*, 2057–2068. https://doi.org/10.1080/15567036.2020.1840669

Siva Chidambaram, M., Sai Sriharsha, A. V., Srinath, A., et al. (2021). Experimental investigation of CI engine fueled with diesel blended Mahua oil with DEE and kerosene as additives. *IOP Conference Series: Materials Science and Engineering, 1130*, 012058. https://doi.org/10.1088/1757-899x/1130/1/012058

Suresh, P., & Vijayaragavan, M. (2022). Experimental investigation on use of hydrotreated Simarouba oil (green diesel) and hydrogen gas in a dual-fuel CI engine. *Environmental Science and Pollution Research, 30*, 36033. https://doi.org/10.1007/s11356-022-24829-1

Talibi, M., Hellier, P., Morgan, R., et al. (2018). Hydrogen-diesel fuel co-combustion strategies in light duty and heavy duty CI engines. *International Journal of Hydrogen Energy, 43*, 9046–9058. https://doi.org/10.1016/j.ijhydene.2018.03.176

Tamilvanan, A., Balamurugan, K., Ashok, B., et al. (2021). Effect of diethyl ether and ethanol as an oxygenated additive on Calophyllum inophyllum biodiesel in CI engine. *Environmental Science and Pollution Research, 28*, 33880–33898. https://doi.org/10.1007/s11356-020-10624-3/Published

Tanay, K. V. R., Vikram Kumar, M., Sreenath Reddy, J., et al. (2020). Experimental investigation of performance and emissions of CI engine fuelled with mahua oil blended camphor fuel. *IOP Conference Series: Materials Science and Engineering, 912*, 042071. https://doi.org/10.1088/1757-899X/912/4/042071

Teja, N. B., Ganeshan, P., Mohanavel, V., et al. (2022). Performance and emission analysis of watermelon seed oil methyl ester and n-butanol blends fueled diesel engine. *Mathematical Problems in Engineering*. https://doi.org/10.1155/2022/2456338

Thippeshnaik, G., Prakash, S. B., Suresh, A. B., Chandrashekarappa, M. P. G., Samuel, O. D., Der, O., & Ercetin, A. (2023). Experimental investigation of compression ignition engine combustion, performance, and emission characteristics of ternary blends with higher alcohols (1-heptanol and n-octanol). *Energies, 15*(18), 6582.

Tiwari, C., Verma, T. N., Dwivedi, G., & Verma, P. (2023). Energy-exergy analysis of diesel engine fueled with microalgae biodiesel-diesel blend. *Applied Sciences (Switzerland), 13*, 1–24. https://doi.org/10.3390/app13031857

Vasudevan, A., & Vinoth Kanna, I. (2019). A detailed analysis and performance analysis of diesel blended with eucalyptus/pine seed oil on a CI engine. *International Journal of Ambient Energy*. https://doi.org/10.1080/01430750.2019.1586767

Xavier, J. (2020). Research Article: Performance and emission characteristics of diesel engine using watermelon seed oil methyl ester. *Journal of Biological Sciences, 20*, 88–93. https://doi.org/10.3923/jbs.2020.88.93

Yesilyurt, M. K. (2020). The examination of a compression-ignition engine powered by peanut oil biodiesel and diesel fuel in terms of energetic and exergetic performance parameters. *Fuel, 278*, 118319. https://doi.org/10.1016/j.fuel.2020.118319

Chapter 16
Management of Energy Efficiency and the Environment for Process Cooling Through the Use of Free Cooling

M. Ziya Sogut

16.1 Introduction

The effective use of energy and energy efficiency in industrial production processes is becoming increasingly important due to rising energy costs and stricter environmental regulations. The thermal management of hydraulic fluids used in production processes offers significant potential for energy cost savings Traditionally, mechanical cooling systems have been employed to remove heat from hydraulic fluids, but these systems can consume a substantial amount of energy, particularly in companies with long-term work programs. However, by leveraging the day-to-day changes in outside air temperature caused by seasonal effects, companies can take advantage of natural cooling applications for process liquids. These natural cooling systems, known as free-cooling applications, operate on the principle of cooling the hydraulic fluids by allowing them to come into contact with cooler outside air. This is achieved through a free fall movement, which takes into account the seasonal temperature variations. By harnessing this method, companies can achieve significant energy consumption cost advantages, especially when the maximum temperature difference (ΔT) between the process liquid and the outside air is within 10 °C. Implementing free-cooling applications not only helps companies reduce their energy costs but also contributes to their overall energy efficiency goals and environmental sustainability efforts. By optimizing the use of natural cooling systems, companies can make substantial progress toward achieving their energy efficiency targets while complying with environmental restrictions.

In addition to the advantages mentioned, another key benefit of free-cooling applications is the reduction in carbon dioxide (CO_2) emissions. By utilizing the

M. Z. Sogut (✉)
Marine Engineering, Piri Reis University, Istanbul, Turkey

M. Z. Sogut, M. Koray (eds.), *Energy Rationality and Management for Decarbonization*, Green Energy and Technology,
https://doi.org/10.1007/978-3-031-68497-5_16

outdoor air temperature to cool processes instead of relying solely on mechanical systems, free cooling significantly reduces energy consumption. This, in turn, leads to a decrease in CO_2 emissions, contributing to a more sustainable and environmentally friendly approach to cooling. Furthermore, the use of Annual recurrence frequency of outdoor temperature values (BIN) values in capacity calculations has revolutionized the way cooling systems are designed. BIN values provide a more accurate representation of the actual climatic conditions experienced at a specific location. By considering the number of hours at each temperature value, engineers can better estimate the cooling requirements and optimize the system accordingly (Bulut et al., 2001; Cansevdi, 2014). The manufacturing process-cooling application discussed in the study serves as a practical example of the benefits of free cooling. By comparing the performance of a closed-circuit mechanical system-supported cooling tower system with a load-sharing closed-circuit cooling tower free-cooling application, the study highlights the energy consumption, cost, and CO_2 emission advantages of the latter. This analysis provides valuable insights for the plastics and manufacturing sector, which extensively relies on such cooling systems. Overall, the study emphasizes the importance of considering climatic data, such as seasonal averages, peaks, and BIN values, in designing and implementing free-cooling applications. By doing so, companies can achieve significant energy efficiency gains, reduce costs, and contribute to a greener future.

16.2 Free-Cooling Applications and Sustainability

In industrial applications, cooling is often a necessary function for production processes. These processes, which work especially with thermodynamic principles, have different structural features according to mechanical systems and applications EU, (2001) (Table 16.1) .

Generally, cooling processes can be classified as follows:

- Depending on the source type, dry air-cooling or evaporative water-cooling applications: Both applications operate in a thermodynamic cycle structure, dependent on latent and sensible heat transfer. While water cooling works based on the removal of latent heat within the evaporative cooling structure, the transfer of sensible heat comes to the fore in the air-cooling system.
- Open or closed cooling systems according to structural operating characteristics: In the cooling function, heat transfer is a heat transfer action that occurs depending on the ambient air. In closed systems, heat transfer occurs by the fan effect, and condensation in the cycle is directed in this structure.
- Direct or indirect cooling systems: In direct systems, heat transfer depends directly on the fluid or process within the cooling cycle. In processes that are mostly carried out through a heat exchanger, heat transfer is solved in a single circuit. In indirect processes, the cooling process is carried out in two or more cycles or system cycles. This type of cycle process is used especially in processes

Table 16.1 Technical and thermodynamic features of some cooling systems (Eurovent, 1998)

Cooling system	Cooling medium	Main cooling principle	Minimum approaches (K)[a]	Minimum achievable end temperature of the process medium[b] (°C)	Capacity of industrial process (MW_{th})
Open once-through system—direct	Water	Conduction/convection	3–5	18–20	<0.01–>2000
Open once-through system—indirect	Water	Conduction/convection	6–10	21–25	<0.01–>1000
Open recirculating cooling system—direct	Water[c] Air[d]	Evaporation[e]	6–10	27–31	<0.1–>2000
Open recirculating cooling system—indirect	Water[c] Air[d]	Evaporation[e]	9–15	30–36	<0.1–>200
Closed circuit wet cooling system	Water[c] Air[d]	Evaporation + convection	7–14[f]	28–35	0.2–10
Closed circuit dry air cooling system	Air	Convection	10–15	40–45	<0.1–100
Open hybrid cooling	Water[c] Air[d]	Evaporation + convection	7–14	28–35	0.15–2.5[g]
Closed hybrid cooling	Water[c] Air[d]	Evaporation + convection	7–14	28–35	0.15–2.5[g]

[a] Approaches of heat exchanger and cooling tower must be added
[b] End temperatures depend on the site's climate (data are valid for average middle European climate conditions 30°/21 °C dry/wet bulb temperature and 15 °C max water temperature)
[c] Water it the secondary cooling medium and is mostly recirculated. Evaporating water transfers the heat to the air
[d] Air is the cooling medium in which the heat is transferred to the environment
[e] Evaporation is the main cooling principle. Heat is also transferred by conduction/convection but in a smaller ratio
[f] Where an indirect system applies or convection is also involved the approach in this example increases with 3–5 K leading to an increased process temperature
[g] Capacity of small units—with a combination of several units or specially built cooling, systems higher capacities can be achieved

with very high-temperature ranges. While the desired cooling occurs with additional heat exchange, heat transfer produces heat that can be used in different processes. In these structures, which have a dynamic design feature, system designs are planned for multiple purposes.

Today, the cooling systems preferred in industrial applications have different models and applications depending on the cooling demand. These are:

- Single pass systems (applications with or without cooling tower)
- Open recirculation systems (Water type cooling towers)
- Closed-circuit systems (air-cooled or closed-circuit water systems)
- Combined hybrid (wet/dry) systems (open hybrid or closed-circuit hybrid towers)
- Free-cooling systems (EC, 2001)

Free-cooling applications are subject to evaluation based on various parametric and regional criteria. These evaluations take into account physical constraints arising from environmental conditions, as well as fundamental technical requirements such as cooling capacity and temperature criteria (ICS, 2024). The ambient air temperature serves as a fundamental criterion in these evaluations. For cooling towers, the airside is considered the cooling source, while the water side is the source to be cooled (Zhang et al., 2014). In a study conducted by Liu et al. (2022), critical temperatures for cooling were highlighted, specifically in data centers. The researchers emphasized the significance of climate data in determining the energy efficiency of free cooling for this cooling process. This suggests that the effectiveness of free cooling can be influenced by the prevailing climate conditions, which further underscores the importance of considering regional criteria in evaluating free-cooling applications.

Cooling towers are widely used in various industrial applications due to their special structural characteristics. Numerous studies have been conducted in the literature to investigate the energy demand and water consumption of cooling towers, employing geometry-based models (Schlei-Peters et al., 2015, 2017; Wang et al., 2013). However, it is worth noting that since regional ambient temperatures and specific process requirements influence cooling towers, there is currently no established reference model or assessment for these processes (Táboas & Vázquez, 2021). Therefore, determining the energy and water consumption of cooling towers requires a comprehensive engineering assessment, which goes beyond simply adhering to a standard level of energy efficiency. This assessment takes into account the specific conditions and demands of each cooling tower system, ensuring optimal performance and resource utilization. While there is limited documentation available on the subject of cooling towers, some empirical approaches have been developed, taking into consideration factors such as geometry and regional criteria, with a particular focus on electricity consumption. In 2001, the European Commission published a reference document that provides valuable insights into cooling systems, including current best practices and information on water and energy consumption (EU, 2019).

This document serves as a valuable resource for understanding the efficiency and performance of cooling systems. Furthermore, Table 16.2 highlights the presence of several studies in the literature that delve into the topic of cooling towers.

These studies show critical information such as thermal capacity and cooler outlet temperature, is useful the engineering evaluation, and contributes additional insights into various aspects of cooling tower design and operation. Examples from

Table 16.2 Previous research about cooling towers (Wenzel et al., 2023)

Researchers	Dry	Wet	Hybrid	Operating points	Reference object(s)	Parameters
DIN 15240 (DIN, 2019)	x	x		ns	Example values	Pf,el
EU (2001)	x	x	x	1	Example values	Qi; Pf,el; mw,i; a; Jw,o,min
Eurovent (2016)		x		3	Efficiency target	Pf,el
Hincke and Hainbach (2018)	x	x	x	Variation	Example values	mw,i; a
Qi et al. (2016)		x		Variation	Neural network	Jw,o
Schlei-Peters et al. (2017)		x		Variation	Modeling	Qi; Pf,el; mw,i
Schulze et al. (2019)		x		Variation	Simulation	Pf,el; mw,i; Jw,o
Wang et al. (2013)		x		Variation	Experiment + simulation	Qi; Pf,el; mw,i; a; Jw,o

Qi nominal thermal capacity, *Pf,el* fan power, *mw,i* freshwater consumption, *Jw,o* coolant outlet temperature. *a* approach, *Jw,i* coolant inlet temperature, *Jw,o* coolant outlet temperature, *Jw,o,min* minimum coolant outlet temperature

the literature include analyses and evaluations based on some technical evaluations under market conditions. Studies related to technological assessments, such as serial assessment criteria, especially in terms of production quality, stand out. For example, Gerke et al. (2017) presented an international benchmarking model for the traceability of the efficiency of refrigeration appliances. In their study, Monfet and Zmeureanu (2013) discussed a plant model for evaluating refrigeration processes based on manufacturer data. Tan and Fok (2008) highlighted a study on the design and selection of unique cooling systems, such as the preference for thermoelectric cooling.

Free-cooling systems are energy-efficient applications that are most often used to reduce the cost of cooling water produced by a central cooling group or to reduce energy consumption. In short, these systems are systems that provide the desired water source cooling in processes without relying on the operation of a compressor or chiller group by taking advantage of low outdoor temperatures. Water source free-cooling applications can be divided into two main categories.

1. Evaporative cooling applications
   (a) Open-circuit (direct free-cooling coil) cooling tower applications
   (b) Closed-circuit (indirect free cooling) cooling tower applications
2. Heat exchanger cooling applications
   (a) Integrated (integrated free-cooling coil) cooling coil applications
   (b) Dry and wet/dry cooler applications (Sogut et al., 2014; Acül, 2007)

When selecting free-cooling applications, it is crucial to consider the intended use of the system and the required cooling loads. These factors play a significant role in determining the effectiveness of the system. One way to evaluate the advantage of the system is by examining specific temperatures during different seasons. In free-cooling applications, the natural cooling of process water can be easily achieved by considering the temperature range expressed as BIN values. BIN values refer to the number of hours the outside air temperature remains within certain intervals over a specific period at a given location (ASHRAE, 2000). By understanding these values, it becomes easier to determine the feasibility of using free cooling for process water. When conducting a cost analysis of these systems, several factors need to be taken into account. These include the cooling capacity, operating time, and operating procedures of the chillers. Additionally, the cost of free cooling and the payback period should be considered. It is also important to assess the impact of other equipment within the system, regional values, and energy costs. By considering all these factors, a comprehensive cost analysis can be conducted to determine the viability of implementing free-cooling applications.

16.2.1 Evaporative Cooling Applications

In evaporative cooling systems, the condenser is typically cooled using water that is circulated from cooling towers. However, in situations where the tower water is insufficient due to high ambient temperatures, a hybrid chiller is employed to cool the condenser water. This hybrid chiller acts as a backup cooling mechanism. In certain applications, such as direct free cooling, the cooling group is not required when the temperature of the water used in the cooling system falls below the ambient temperature. In this case, the chiller is bypassed and the chilled water needed for the condenser is cooled by sending it directly to the tower. This approach brings the desired chilled water temperature closer to the wet bulb temperature of the environment, optimizing the cooling process. One of the main challenges faced by these systems is the potential contamination of the clean chilled water system by the dirty condenser water. To mitigate this issue, closed-circuit (indirect) systems or dry cooler applications are commonly employed in cooling tower setups. These alternative methods help maintain the cleanliness and integrity of the chilled water system, ensuring efficient and reliable cooling. For a visual representation of a direct free-cooling application, refer to Fig. 16.1, which illustrates the flow diagram of such a system.

Another advantage of closed-circuit (indirect free cooling) cooling tower applications is their ability to eliminate the risk of contamination. By separating the condenser water circuit from the chilled water circuit, these systems ensure that there is no cross-contamination between the two circuits. This is particularly important in industries where maintaining water purity is crucial, such as pharmaceutical or food processing. However, it is worth noting that closed-circuit cooling towers typically operate at higher water temperatures compared to open-circuit cooling towers. This is due to the closed flow of the system, which reduces the cooling efficiency. Despite this drawback, closed-circuit cooling towers are widely used in

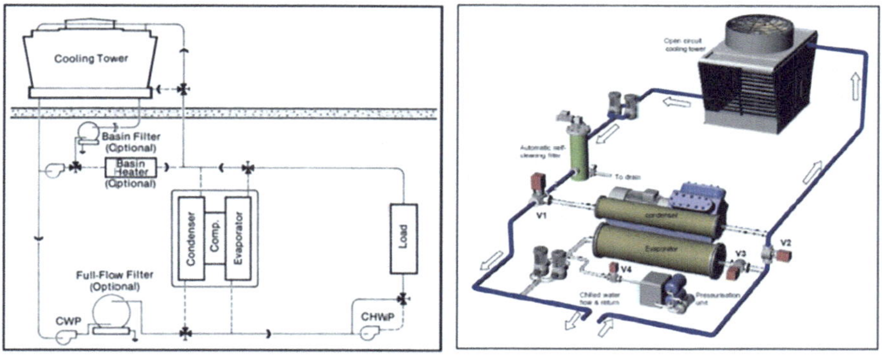

Fig. 16.1 Principle diagram of open-circuit cooling water tower system (Acül, 2009)

free-cooling applications. Three main types of systems utilize cooling towers: systems that solely rely on cooling towers, systems that incorporate additional heat exchangers, and systems that utilize refrigerant gas. Among these, free-cooling applications using closed-circuit cooling towers are the most prevalent. In free-cooling applications with closed-circuit cooling towers, both water circuits are regulated based on the outside temperature. This allows for optimal cooling efficiency by utilizing the ambient air temperature to cool the condenser water circuit, reducing the need for mechanical refrigeration. This not only saves energy but also reduces operating costs, making it an attractive option for many industries. During the summer season, when the outdoor temperature exceeds the temperature of the process water, the cooling group remains operational. In this scenario, the condenser water from the cooling circuit is circulated through the condenser in a closed circuit. This process allows the heat from the process water to be transferred to the condenser water, which is then cooled by the cooling tower before being recirculated. In contrast, during the winter season when the outdoor temperature drops below the process water temperature, a free-cooling system is employed. The water from the cooling tower is directly circulated into the chilled water circuit in a closed loop. This setup enables the cold water returning from the process to be pre-cooled before it is sent back to the process via the evaporator. By utilizing this closed-loop system, the cooling load on the chillers is reduced, resulting in significant energy savings (Sogut et al., 2014; Acül, 2007). Figure 16.2 illustrates a closed-circuit cooling tower free-cooling application with load sharing. This configuration allows multiple cooling towers to work together, sharing the cooling load and ensuring efficient operation. Load sharing ensures that the cooling towers operate optimally, preventing any individual tower from being overloaded and maintaining a consistent cooling capacity. Overall, the combination of closed-circuit cooling, free cooling, and load sharing provides an energy-efficient solution for maintaining the desired process water temperature while minimizing energy consumption.

Water-cooling towers are widely used in various industries to remove excess heat from industrial processes. These towers operate by evaporating a small portion of

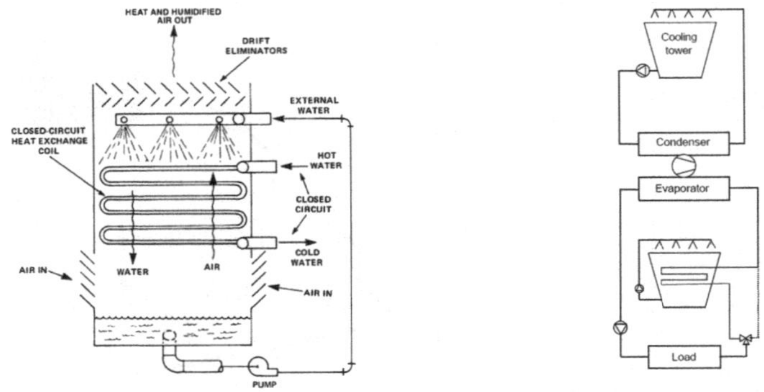

a: Indirect (Cansevdi, 2014)

b: Free Cooling Application with Load Sharing (Bulut et al, 2001)

Fig. 16.2 Closed-circuit cooling tower: (**a**) Indirect (Cansevdi, 2014). (**b**) Free-cooling application with load sharing (Bulut et al., 2001)

the water, which helps in cooling the remaining water. However, the water temperature in these towers can sometimes be higher than the wet bulb temperature, depending on their design. When it comes to closed-circuit cooling towers, the water temperature can be 2–3 °C higher than in open-circuit towers. This difference is due to the closed-circuit design, which allows for better control over the cooling process. The cooling criterion, which determines the desired temperature, can vary depending on the specific requirements of the industrial process. The wet bulb temperature, on the other hand, is influenced by the size of the tower. Generally, larger towers tend to have lower wet bulb temperatures. However, the size of the tower directly affects the investment costs, making cost analysis an essential step in the decision-making process. While water-cooling towers offer advantages in terms of capital costs, they can present operational challenges. One of the main issues is the contamination of the cooling heat exchanger by the water coming directly from the tower. This contamination can lead to the formation of scale on the surfaces, reducing the efficiency of the heat exchange process. Moreover, water loss through evaporation is another concern associated with water-cooling towers. As water evaporates, it needs to be replenished, which can lead to increased water consumption. Depending on the process characteristics, this increased water usage can cause water-related problems, such as scarcity or increased costs. To mitigate these issues, regular maintenance and water treatment are crucial. Proper water treatment can help prevent scale formation and reduce the risk of contamination, ensuring the efficient and reliable operation of water-cooling towers. Additionally, implementing measures to minimize water loss through evaporation, such as using drift eliminators, can help conserve water resources and reduce operational costs.

Another advantage of using additional heat exchangers in closed systems is that they allow for greater flexibility in controlling the temperature of the water circuit (Fig. 16.3).

By circulating the water through the heat exchanger, the system can adjust the heat load in the chilled water circuit without relying solely on a chiller. This is particularly beneficial when the ambient temperature is low, as the heat exchanger can provide the necessary heat load to maintain the desired temperature. Furthermore, the use of a plate heat exchanger in these systems offers several advantages. Plate heat exchangers are known for their high heat transfer efficiency due to their large surface area. This means that they can effectively transfer heat between the water circuit and the surrounding environment, ensuring efficient cooling or heating as required. In Fig. 16.3, the flow diagram of the system with an additional heat exchanger illustrates how the water circuit is directed through the heat exchanger. As the water passes through the heat exchanger, it exchanges heat with the surrounding environment, either gaining or losing heat depending on the desired temperature. This allows for precise control over the temperature of the water circuit without the need for a separate chiller unit. Overall, the use of additional heat exchangers in closed systems provides a cost-effective and energy-efficient solution for maintaining the desired temperature in the water circuit, especially in situations where the ambient temperature is low.

In addition to shaping the ΔT change of the process through the pressure difference of the refrigerant, other factors can influence the thermal load distribution and cooling load in heating, ventilating, and air-conditioning (HVAC) systems. One such factor is the use of variable speed drives (VSDs) in the system. VSDs can adjust the speed of the compressor and fans, allowing for more precise control of the cooling capacity and energy consumption. Furthermore, the selection of the appropriate heat exchanger type can also impact the thermal load distribution. Different types of heat exchangers, such as shell and tube, plate, or finned tube heat exchangers, have varying heat transfer characteristics and are suitable for different

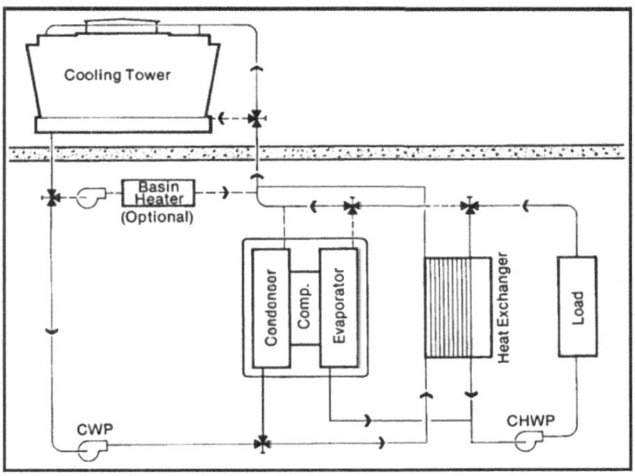

Fig. 16.3 Systems using additional heat exchangers (Cansevdi, 2014)

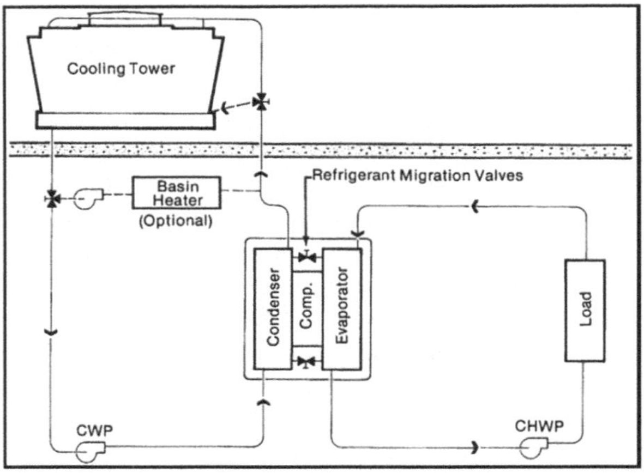

Fig. 16.4 Refrigerant circulating systems (Cansevdi, 2014)

applications. The choice of heat exchanger depends on factors such as the required heat transfer rate, fluid properties, and space limitations. In certain cases, where there is a need for precise temperature control or when dealing with sensitive processes, additional temperature sensors and control systems can be implemented. As can be seen in Fig. 16.4, these systems continuously monitor the fluid temperature and adjust the operation of the valves and other components to maintain the desired cooling load.

It is worth noting that the use of recirculating systems, although less common in the manufacturing sector, can offer advantages in terms of energy efficiency and process stability. These systems involve the recirculation of the cooling fluid, allowing for better control of the cooling process and reducing the overall energy consumption. Overall, the thermal load distribution and cooling load in HVAC systems are influenced by various factors, including the type of valves, heat exchangers, temperature control systems, and the use of recirculating systems. By considering these factors, engineers can design and optimize HVAC systems to meet the specific cooling requirements of different processes. In these systems, the condenser water can be supplied at a temperature lower than the specified chilled water temperature. This allows for efficient cooling of the condenser in the secondary circuit, which functions as a water-cooled heat pump and is cooled by a tower. The primary circuit, on the other hand, utilizes counter-flow evaporators to provide direct process cooling. The movement of heat in both fluid circuits leads to phase changes in the refrigerant, resulting in differential pressure in the condenser and evaporator. This unique characteristic allows the system to operate without the need for a compressor in the refrigerant circuit, forming a cycle based on the heat exchange process. However, it is important to note that this system may not be suitable for all process-cooling applications. Its free-cooling capacity is limited to a maximum of 30% of the chiller's design capacity. The extent of the free-cooling capacity depends on the temperatures of the primary and secondary water used in the system.

16.2.2 Cooling Applications with Heat Exchangers

In addition to air conditioning and process cooling, finned tube heat exchanger (cooling coil) systems have found another common application in the field of refrigeration. These systems are often used in commercial refrigeration units to remove heat from the refrigerated space and maintain the desired temperature. When it comes to free-cooling applications using a cooling coil, the system operates in different modes depending on the ambient temperature. In full mechanical cooling mode, the system relies solely on mechanical refrigeration to cool the space, without utilizing any free cooling. This mode is typically used when the ambient temperature is too high for free cooling to be effective. In partial free-cooling mode, also known as load sharing or pre-cooling, the system combines mechanical cooling with free-cooling. The cooling coil is activated to pre-cool the space, reducing the load on the mechanical refrigeration system and improving energy efficiency. Finally, in full free-cooling mode, the chiller does not operate at all, and the cooling coil alone is sufficient to maintain the desired temperature. This mode is employed when the ambient temperature drops significantly below the cooling water temperature, allowing for efficient cooling without the need for mechanical refrigeration. It is worth noting that, unlike evaporative cooling applications, finned coil systems do not pose a risk of contamination or reduction of the cooling water in the closed circuit. This makes them a reliable and efficient choice for various cooling applications. Furthermore, finned coil systems have two different applications (Fig. 16.5): chiller and integrated free-cooling coil applications, as well as dry and wet/dry cooling system applications. In chiller and integrated free-cooling coil applications, the cooling coil is integrated into the chiller unit, allowing for seamless operation and enhanced energy efficiency. On the other hand, dry and wet/dry cooling system applications involve the use of finned coil systems in conjunction with air-cooled condensers or cooling towers, providing effective cooling in different environments and conditions.

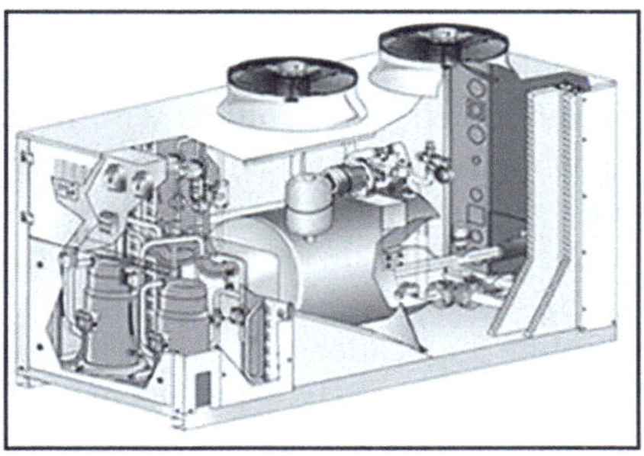

Figure 5.a

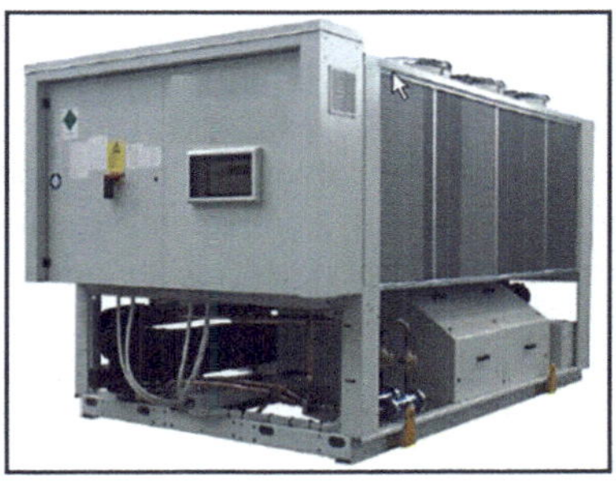

Figure 5.b

Fig. 16.5 (**a**) Schematic of an air-cooled water chiller with an integrated free-cooling coil (ASHRAE, 2000). (**b**) Air-cooled water chiller with integrated free-cooling coil (De Saulles, 2004)

Applications with chillers and integrated free-cooling coils (Fig. 16.5b) offer energy-efficient solutions by utilizing the existing chiller and incorporating free-cooling technology. The key to designing such systems lies in careful consideration of various factors. Firstly, the purpose of the system needs to be determined, whether it is for air conditioning or process cooling. This will help in establishing the specific requirements and parameters for the system. Another crucial aspect is the outdoor temperature and its comparison with the cooling water temperature. By analyzing this, one can identify the periods when free cooling can be employed effectively. Free-cooling periods occur when the outdoor temperature is lower than the cooling water temperature, allowing the system to use the ambient air to cool the water without the need for mechanical refrigeration. However, it is important to address the issue of water freezing in extremely cold weather conditions. To prevent this, a glycol-water mixture can be used in the system. The concentration ratio of the glycol-water mixture needs to be carefully determined to ensure that the system remains operational even in freezing temperatures. By considering these factors and designing the system accordingly, applications with chillers and integrated free-cooling coils can provide efficient and sustainable cooling solutions for various purposes.

Air-cooled water chillers with free-cooling coils offer efficient cooling solutions in various seasons. These systems are designed to adapt to different ambient temperatures and optimize their cooling operations accordingly. During the summer season, when the ambient temperature is high, the air-cooled water chillers utilize the free-cooling coils to provide cooling. In this scenario, the cooling group and the air cooler work independently. For instance, in an environment where the ambient temperature reaches 35 °C, the cooling water is maintained at 10 °C, and the return water temperature is 15 °C. The cooling coil directly cools the water, ensuring effective cooling. In winter conditions, when the ambient temperature drops to 5 °C, the air-cooled water chillers switch to a different cooling method. They rely on 100% outside air for cooling purposes. This approach allows the system to efficiently cool the water without the need for additional energy consumption. During the spring season, when the ambient temperature is around 13 °C, air-cooled systems with free-cooling coils utilize the ambient air for pre-cooling. The water that undergoes pre-cooling is then directed to the cooling group, ensuring the desired cooling effect. These operational variations in air-cooled water chillers with free-cooling coils enable them to provide efficient cooling solutions throughout the year, adapting to different environmental conditions. The flexibility of these systems ensures optimal performance and energy efficiency, making them a reliable choice for various cooling applications (Climaveneta, 2014).

In addition to their use in heat exchanger cooling applications, dry coolers have gained popularity in both dry and wet/dry cooler applications. These coolers operate by transferring the return water load to the air through the use of a fan. Acting as a heat exchanger, the system allows the return water to pass through the pipe while the outside air passes between the lamellae, thanks to the fan's effect. One notable advantage of dry coolers is that they maintain constant water levels during closed operation. However, it is crucial to take precautions to prevent freezing, especially in winter conditions. To safeguard against freezing, it is recommended to use antifreeze, such as ethylene-glycol, at a concentration of 25–35%, depending on the specific system design. It is worth noting that the use of antifreeze will impact the water temperature in the cooling system, and this factor should be considered during analysis and system design. Figure 16.6 provides a visual representation of different dry cooler systems, including the horizontal type, V type, and dry cooler systems with a chiller. These variations offer flexibility in terms of installation and can be tailored to meet specific cooling requirements.

Cooling in this type of system relies solely on the surrounding temperature. Generally, a cooling effect of approximately 5 °C above the ambient temperature can be attained. Wet-dry coolers are the preferred choice for achieving lower temperatures, especially when there is a cooling group involved in the system. Opting for a dry cooler (Fig. 16.7), on the other hand, can lead to reduced cooling efficiency. Dry cooling systems can be employed alongside water-cooling groups in free-cooling applications across a range of industries, including chemistry and plastics. Figure 16.7 provides a visual representation of the various methods through which free cooling can be accomplished using these types of coolers.

Wet-dry coolers are designed to function similarly to dry coolers, but they incorporate a water spray system to enhance their cooling capabilities when necessary. This water spray system plays a crucial role in creating an adiabatic cooling effect within the incoming airflow. When the fluid within the system needs to be cooled to

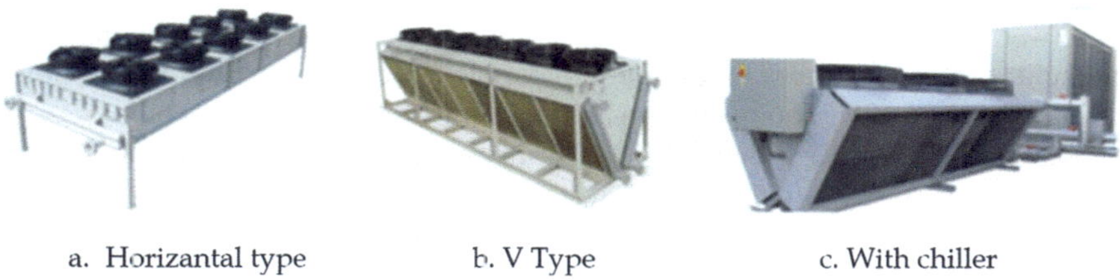

Fig. 16.6 Dry coolers (Sogut et al., 2014; Acül, 2009). (**a**) Horizontal type. (**b**) V Type. (**c**) With chiller

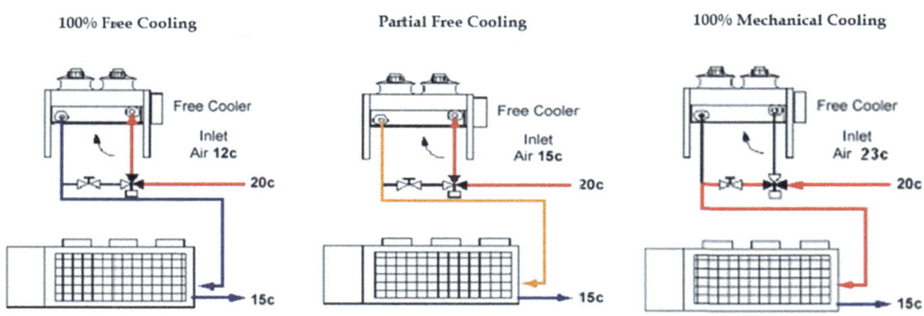

Fig. 16.7 Dry cooler application with cooling group (TPC, 2024)

temperatures lower than the outdoor temperature, the pressurized water spray system is activated. This system saturates the inlet air with moisture, resulting in a reduction of the air temperature below the ambient temperature. It is important to note that these systems are only activated during peak temperature periods throughout the year to provide additional cooling support. In addition to their primary application, wet-dry cooling systems can also be employed in condensers to lower the condensation temperature within cooling groups. This application proves to be beneficial in certain scenarios where reducing the condensation temperature is necessary. There are three distinct methods by which wet-dry cooling systems can be implemented, and the choice of implementation depends on the intended use of the system. These different approaches ensure that the wet-dry cooling system can be tailored to meet specific cooling requirements and optimize its performance accordingly.

Direct water spray systems are commonly employed in wet-dry coolers to raise the humidity of the incoming air to nearly 100%. These systems' effectiveness is influenced by the relative humidity of the surrounding air. To provide a safety margin, it is customary to spray a slightly higher amount of water than what can evaporate in the environment, allowing some of the water to remain in liquid form. However, it is crucial to consider the quality of the water used in these systems, taking into account factors such as hardness and sediment properties. To ensure optimal performance, it is necessary to eliminate water hardness and clean any sediment or impurities. The presence of accumulated lime or deposits, particularly on the heat exchanger surfaces or fins, can significantly reduce the system's efficiency. Moreover, to safeguard the system against corrosive environments containing salt and acid, epoxy paint or stainless steel materials are employed. Figure 16.8 illustrates a wet-dry cooler equipped with a direct spray system.

Fogging systems are widely used in wet-dry coolers to regulate the humidity of the inlet air and bring it to the desired wet bulb temperature. These systems operate by spraying water at a high pressure of around 70 bar through specially designed nozzles. These nozzles are engineered to produce water droplets with a size below 35 μm, ensuring efficient evaporation and cooling. The performance of the fogging system is influenced by various factors, including the quality of the water used.

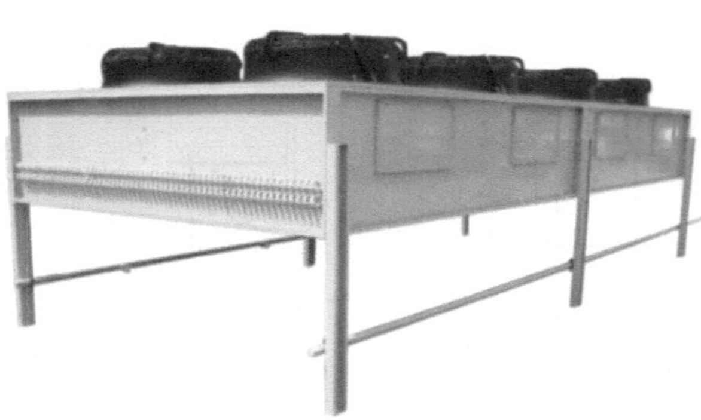

Fig. 16.8 Wet-dry cooler with direct water spray system (Acül, 2009)

Fig. 16.9 Wet-dry cooler with fogging system (Acül, 2009)

Water hardness and sediment content can have a direct impact on the system's effectiveness. High levels of water hardness can lead to mineral deposits and scaling, which can clog the nozzles over time. Similarly, sediment in the water can cause blockages and reduce the system's efficiency. To maintain optimal performance, it is crucial to prevent clogging of the nozzles. Regular maintenance and cleaning of the system, including the nozzles, are essential. Figure 16.9 provides a visual representation of how the nozzles operate in such fogging systems, highlighting the importance of keeping them clear and free from obstructions. By ensuring proper maintenance and monitoring, the fogging system can continue to provide efficient cooling and humidity control in wet-dry coolers.

The spraying method mentioned above is not only limited to dry coolers but can also be applied to wet-dry coolers. In this case, a water spraying system is installed on the network to achieve the desired cooling effect. The system works by spraying water from nozzles at specific intervals onto a large and dense structure positioned in front of the dry cooler. The frequency and amount of water sprayed can be adjusted according to the system's requirements. By spraying water onto the surface, the temperature of the inlet air in contact with the structure is reduced, thereby enhancing the cooling efficiency. This method proves to be particularly effective in dry climates, where it can lower the inlet temperature by up to 20 °C. The reduction in temperature significantly improves the overall performance of the cooling system. To ensure optimal operation, these systems are equipped with a fully automated control panel. This control panel manages various aspects, including water consumption, system performance, operating time, frequency adjustment, and performance optimization. By automating these processes, the system can operate efficiently and effectively without constant manual intervention. For a visual representation, please refer to Fig. 16.10, which displays a wet-dry cooler equipped with an on-net water spray system. This illustration provides a clear depiction of how the spraying method is implemented in practice.

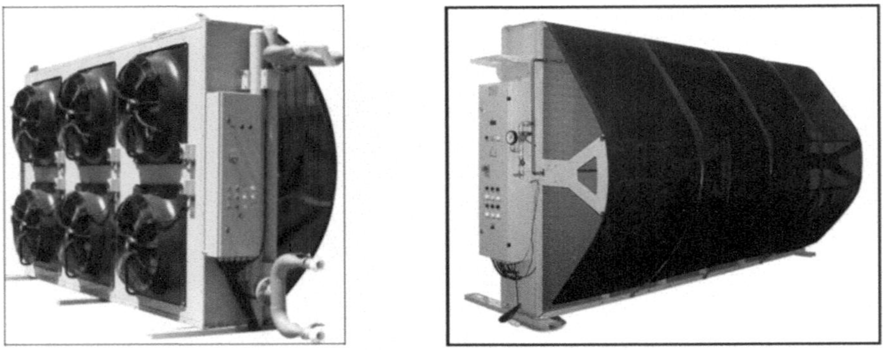

Fig. 16.10 Wet-dry cooler with over-network water spraying system (Acül, 2009)

16.3 Load and Energy Analyses in Free-Cooling Applications

Process cooling is an essential requirement across various industries, with the manufacturing sector being a prominent user. In manufacturing processes, hydraulic fluids often need to be cooled using either direct or indirect cooling methods. The cooling process is primarily designed for continuous flow operations. Evaluating the cooling within this flow can be achieved by considering the thermodynamic principles of a heat exchanger. Consequently, the energy balance in a process-cooling system becomes a crucial factor to consider. By maintaining a proper energy balance, the cooling process can be optimized to ensure efficient and effective cooling of hydraulic fluids, thereby enhancing the overall performance of the manufacturing processes (Cengel & Boles, 2006).

$$\sum E_{\text{in}} = \sum E_{\text{out}} \tag{16.1}$$

where $\sum E_{\text{in}}$ is the heat load to be absorbed in the system for process cooling, $\sum E_{\text{out}}$ is the heat load exerted by the refrigerant in process cooling. In the case where kinetic and potential effects are neglected in the unit energy, the energy distribution of the system is enthalpy-dependent. Enthalpy is a thermodynamic property that accounts for both the internal energy of a system and the work done by or on the system. Accordingly, the energy distribution of the system depending on enthalpy is:

$$\dot{m}_{\text{hot}} \cdot \left(h_{\text{hat,in}} - h_{\text{hot,out}}\right) = \dot{m}_{\text{col}} \cdot \left(h_{\text{col,out}} - h_{\text{col,in}}\right) \tag{16.2}$$

Here, $\dot{m}$ is the mass of process liquid and refrigerant for both circuits, and h is the enthalpy values defined for each point. In process-cooling systems, the flow of coolants can be continuously adjusted, providing precise control over temperature. This flexibility is a result of the energy flows resulting from the different cooling temperatures of different processes. Energy balance according to the energy load of each fluid circuit to find the load values of process fluids:

$$\dot{m}_{\text{hot}} \cdot C_{\text{p,hot}} \left(T_{\text{hat,in}} - T_{\text{hot,out}}\right) = \dot{m}_{\text{col}} \cdot C_{\text{p,col}} \left(T_{\text{col,out}} - T_{\text{col,in}}\right) \tag{16.3}$$

where $C_{\text{p,hot}}$ is the specific heat capacity of the process fluid, $C_{\text{p,col}}$ is the specific heat capacity of the cooling fluid, and T is the temperature of the process fluid and the cooling fluid for both circuits. To determine the cooling load, it is crucial to analyze the cooling machines in process cooling based on a vapor compression cycle. The power consumption of the compressor can be calculated by subtracting the load on the cold source from the load on the hot reservoir in a refrigeration cycle. Shortly:

$$W_{\text{comp}} = Q_{\text{h}} - Q_{\text{L}} \tag{16.4}$$

where W_{comp} represents compressor power consumption, Q_{h} represents hot source load, and Q_{L} represents cold source load. In free-cooling applications, savings depend on load values.

$$\sum W_{\text{free cooling}} = \sum n_{\text{h,BIN}} \cdot Q_{\text{cooling}} \tag{16.5}$$

16.3.1 Environmental Indicators

It is important to note that traditional environmental analyses often base predictions or approaches to CO_2 emissions on fossil fuels, as reflected in the emission factors published by the Intergovernmental Panel on Climate Change (IPCC). This study proposes two initiators that were confidently developed using an entropy-based approach. The approach, confidently developed by Sogut (2021), relies on analyzing the irreversibility of processes based on the second law of thermodynamics. The study uses the Environmental Performance Index (EPI) and the Sustainability Index (SI) to provide manageable results. The Environmental Performance Index (EPI) focuses on environmental impact, while the Sustainability Index (SI) measures the potential for reversibility in systems or processes. It is crucial to maintain the distinction between the two measures, as they serve different purposes. By prioritizing environmental impact, the EPI provides a comprehensive assessment of a system's sustainability, while the SI emphasizes environmental efficiency:

$$\text{EPI} = \left(\frac{\sum S_{\text{gen}}}{\sum \dot{\text{E}}\text{x}_{\text{in}}}\right) * T_{0} \tag{16.6}$$

The goal is for the impact value to approach zero or the SI, which represents the limit of reversibility. EPI is a value ranging from zero to one and is directly related to the temperature of the environment. The extent of irregularities or problems in the system can be determined through the effect of irreversibility, and its potential for improvement can be calculated using the (Eq. 16.7) equation for the irreversibility condition. When dealing with real processes, it is possible to assess the limit condition using a potential directly linked to the yield of Carnot cycles. Considering the thermal system performance, the formula of Improvement Potential (IP) developed by Van Gool can be revised with Carnot's efficiency (1997):

$$\mathrm{IP} = \left(\eta_{\mathrm{Carnot}} - \eta_{11}\right)\cdot\left(\dot{\mathrm{Ex}}_{\mathrm{in}} - \dot{\mathrm{Ex}}_{\mathrm{out}}\right) \tag{16.7}$$

The EPI value serves as a measure of the environmental impact caused by reversible processes at the dead state temperature. It is a valuable metric that allows for comparison against a predetermined boundary condition in terms of environmental impact. On the other hand, the Sustainability Index (SI) provides a means to quantify the environmental impact of a process or system under consideration, assuming reversible conditions. This index is particularly useful as it indicates the degree to which the EPI has deviated from reversibility. Consequently, the SI offers a dependable measure of the environmental impact associated with the process or system being evaluated. By considering both the EPI and SI, one can gain a comprehensive understanding of the environmental implications of a given process or system. Accordingly, SI:

$$\mathrm{SI} = \left(\frac{\sum S_{\mathrm{gen,c}}}{\sum \dot{\mathrm{Ex}}_{\mathrm{in}}}\right) * T_0 \tag{16.8}$$

The SI defines the extent of environmental pollution caused by entropy generated under reversible conditions for structures that consume direct energy (Sogut, 2021). This refers to the minimum limit or reversibility value of pollution caused by irreversibility in the systems in terms of process management and energy efficiency. In other words, it sets boundaries for potential process improvement, providing a valuable metric for decision-makers in energy management. This metric is referred to as the energy efficiency ratio and represents a corporate objective in system management. Accordingly, the energy efficiency ratio:

$$\theta = \frac{\sum\left(\mathrm{EPI} - \mathrm{SI}\right)\cdot\mathrm{IP}}{\sum \mathrm{Ex}_{\mathrm{in}}} \tag{16.9}$$

16.4 Results and Discussion

In addition to the two commonly used systems for process cooling in manufacturing technologies, there are specific applications where highly effective 100% free cooling is employed, particularly in the manufacturing industries. These industries benefit from the use of free-cooling applications, where the secondary circuit can be fully for processes with temperatures below 25 °C. This means that the cooling process can be achieved without the need for additional energy consumption, resulting in cost savings and increased efficiency. Figure 16.11 provides a visual representation of the flow diagram specifically designed for processes with temperatures below 25 °C. This diagram illustrates the various components and their interconnections, highlighting the key elements involved in the cooling process. By following this diagram, manufacturers can ensure their cooling systems' proper functioning and optimal performance for applications with temperatures below 25 °C.

The 25 °C > *T* process, which is widely employed across various industries, involves the utilization of a three-circuit flow system. This system incorporates a cooling group that adapts to the specific cooling demands, natural cooling, and crucially, the external ambient temperature. To better understand this arrangement, Fig. 16.12 illustrates the flow diagram of this structure, referred to as a free-cooling application with a load-sharing closed-circuit cooling tower. This diagram provides a visual representation of how the components interact within the system, aiding in the comprehension of its functioning.

In the study, the evaluation of the defined capacity for the sample case, namely the outside air temperature and the temperature of the process fluid, played a crucial role in assessing the energy and environmental impact of free-cooling applications. This study focused on the BIN values specific to the Bursa province to provide a comprehensive analysis. By considering the outside air temperature, the study aimed to determine performance about the optimal conditions for free-cooling application. Temperature distributions and BIN values for free cooling in Bursa province are given in Table 16.3.

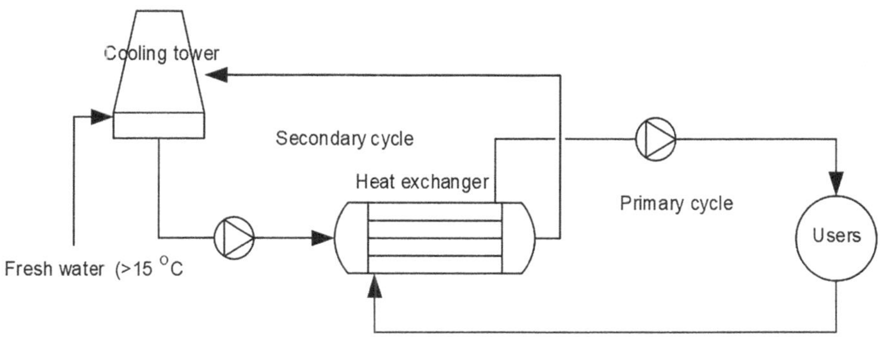

Fig. 16.11 Closed-circuit cooling tower free-cooling application

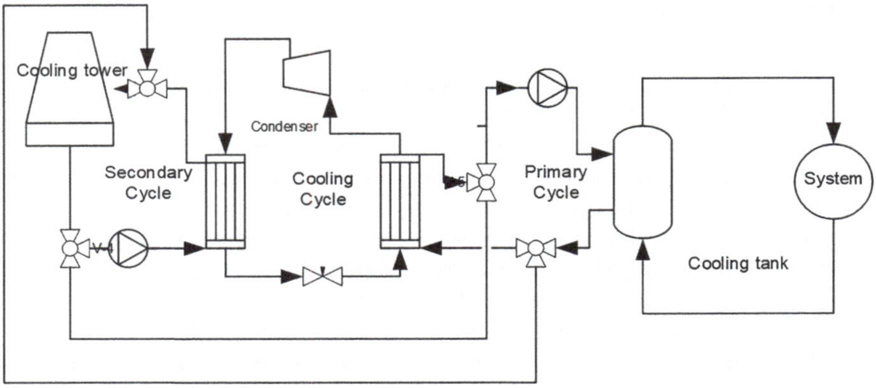

Fig. 16.12 Free-cooling application with load-sharing closed-circuit cooling tower

Table 16.3 Bursa province BIN values and working methods

Temperature range (°C)	<-15	-15/-12	-12/-9	-9/-6	-6/-3	-3/0	0/3	3/6	6/9	9/12	12/15	15/18	18<
Temperature recurrence frequency (Hour/year)	0	0	2	5	36	237	568	810	1028	961	942	985	3166
Operation preference	%100 Free cooling								Partial Cooling Zone			Mech. cooling	
									Mech. Cooling region				
									20%	45%	72%	100%	
									80%	55%	28%		
									Free cooling region				

The study discusses the use of free-cooling applications for temperatures below 17 °C, with 100% free cooling being possible when the temperature drops below 10 °C (Cansevdi, 2014; Kakı, 2014). However, for temperatures between 10 and 17 °C, hybrid applications are used depending on the load. In the province of Bursa, the total free-cooling daily rate is 63.78%, and the rate of 100% free-cooling application days throughout the year is 42%. Building upon these findings, further analyses were conducted to determine the free-cooling application rates for specific temperature ranges in Bursa. For instance, the rates for mold cooling temperature ranges of 10–16 °C and oil cooling temperature ranges of 27–33 °C were calculated. These rates indicate the extent to which free cooling can be utilized in these processes, resulting in energy savings. The distribution of these rates is illustrated in Fig. 16.13.

The study focused specifically on evaluating the use of process cooling in industrial applications, with a particular emphasis on the cooling of manufacturing mold.

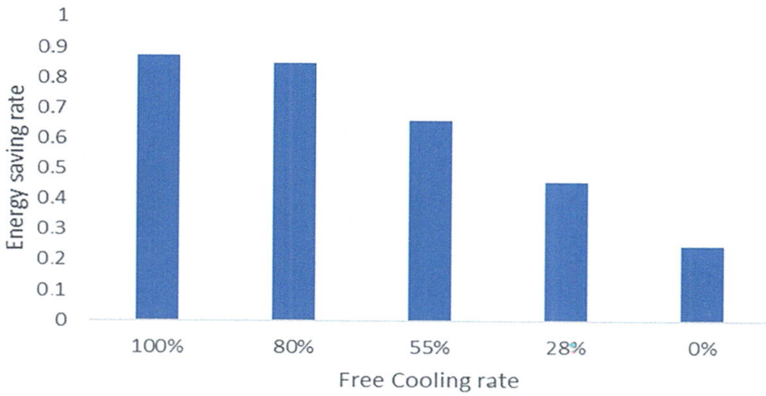

Fig. 16.13 Energy-saving rates depending on free-cooling usage rate

Table 16.4 Operating and capacity parameters of the manufacturing mold cooling process

Parameters	Input	Output
Mold cooling water temperature (°C)	10	16
Oil cooling water temperature (°C)	27	33
Working time	3 shifts (8 h) 6 days a week	
Mold number	9	
Cooling water flow rate (m^3/h)	64	
Oil flow rate (m^3/h)	89.3	
Cooling requirement (mold cooling) (kW)	160	
Cooling requirement (oil cooling) (kW)	80	
Chiller capacity (R-134a) kW (12–7 °C)	240	
Compressor consumption (kW/h)	72.67	

This application is widely used in the manufacturing sector and is known for its significant energy consumption. In industries where production processes operate at high intensity and require continuous operation, substantial cooling energies are necessary. To provide a more comprehensive understanding, the specific parameters of the manufacturing process-cooling process are outlined in Table 16.4.

The analyses are carried out by comparing a mold cooling process with an existing cooling group. For this purpose, the energy consumption and the amount of energy saved depending on the reference process, as well as the usage rate of the free-cooling application in Bursa, were evaluated and analyzed as basic criteria. Accordingly, the distributions of the achieved energy savings are shown in Fig. 16.14.

The energy saving achieved by free cooling in Bursa province is indeed impressive, amounting to 228,544.2 kWh per year. This significant reduction in energy consumption is a positive step toward sustainability and cost-efficiency. To put this into perspective, the total energy required by the system for the normal cooling process is calculated to be 544,513 kWh per year. With the energy-saving measures

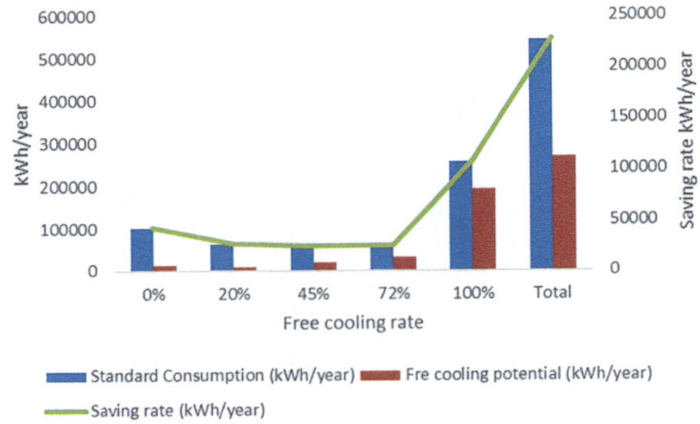

Fig. 16.14 Energy efficiency effects of using free cooling

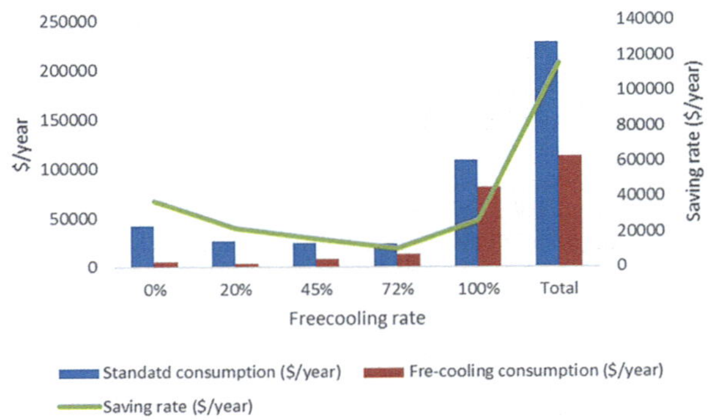

Fig. 16.15 Free-cooling application saves energy costs

implemented, the total energy saving for the company is found to be 42%. Energy cost data related to these consumption and savings amounts were also examined. The results obtained are given in Fig. 16.15.

In the mold cooling application, the annual saving cost effect for the free-cooling application, which has an annual average saving potential of 34.28% for Bursa province, shows a potential of 50.71% of the total cost. This provides a significant gain for the energy cost sustainability of the business. The environmental impact of free-cooling application is directly addressed by the second law of thermodynamics. In the analysis, which is actually considered as an entropy approach, environmental impact assessment and efficiency ratio are discussed through the Environmental Performance Index (EPI) and Sustainability Index. However, first of all, exergetic parameters related to free-cooling rates were examined and the distributions are given in Fig. 16.16.

The exergetic approach provides more realistic results about process performance under dead state conditions and also reveals information about the environmental impact of free-cooling applications due to exergy destruction. The mechanical system exergy efficiency has a potential of 12.5%, while the total process efficiency can increase to a potential of 49.3%. In exergetic analyses, the IP ratio and resulting entropy value are crucial in assessing environmental impact. Therefore, this study examined IP values, related entropy production, and exergy destruction at each stage of the process. The results are presented in Fig. 16.17.

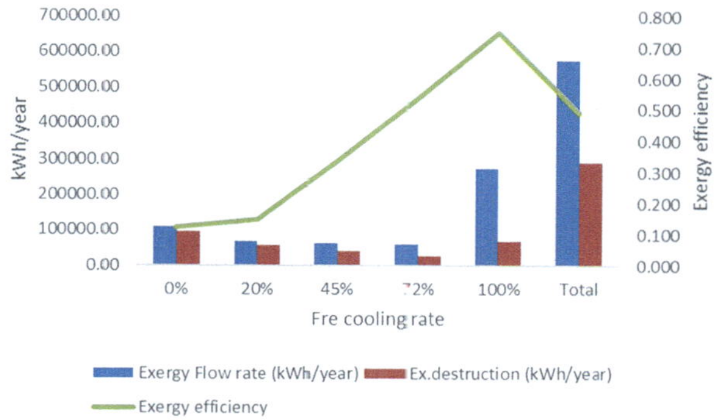

Fig. 16.16 Exergetic parameters of free-cooling application

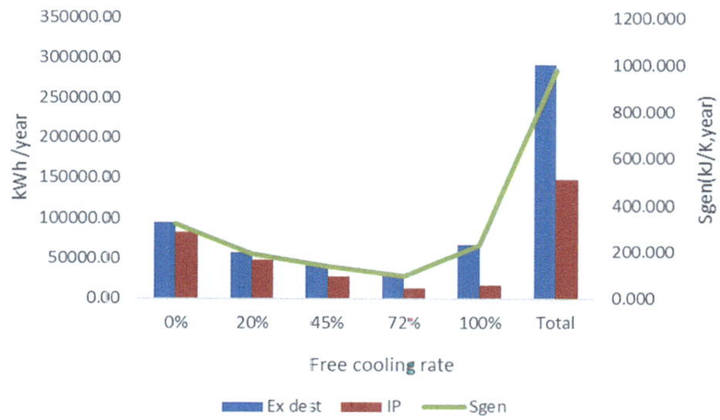

Fig. 16.17 Exergetic parameters of free-cooling application

Although the process described has a high potential for producing entropy, considering the amount of exergy destruction involved, the IP ratio can provide important information about the process's reversibility. For instance, when the examined process is operating under mechanical conditions, it has a rate of 76.51%. However, when free cooling is applied to the entire system, this value drops to an average of 35.71%. This situation is especially valuable in terms of entropy production. The study also evaluated the free-cooling effect on environmental indices in managing entropy sustainably. The results of this study are presented in Fig. 16.18.

The EPI value of a completely mechanical system was found to be 0.875, while the value was 0.25 when full free cooling was used. When the system was used in hybrid mode, the total EPI value was 0.507. Similarly, the SI value was 0.153 for the mechanical process, but it increased to 0.246 for free cooling. The average SI value of the process was 0.171. The energy efficiency potential of the process was also evaluated, and it was found to be 0.826 for the mechanical process and 0.016 for the free-cooling process. The energy efficiency ratio of the hybrid operation process was 0.662. All these values indicate that mechanical cooling produces high entropy and has a significant potential for environmental pollution. The environmental impact assessment of the process was also discussed, and the CO_2 emission savings caused by free cooling were calculated and presented in Fig. 16.19, using the IPCC approach.

Taking into account the climate conditions of Bursa province, the total CO_2 emission equivalent for the manufacturing mold cooling system, based on total

Fig. 16.18 EPI, SI, and efficiency rate

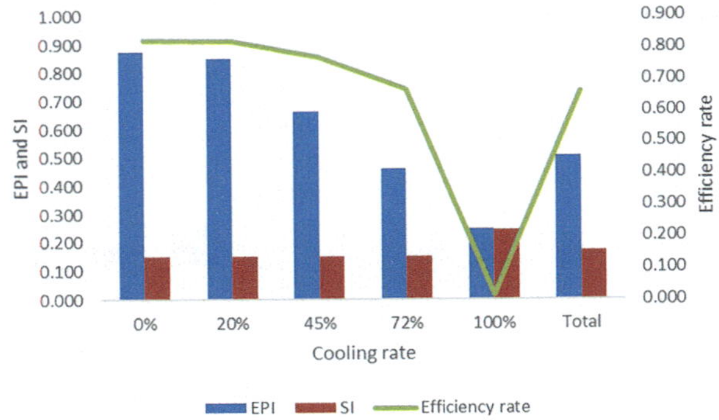

Fig. 16.19 Free-cooling application CO_2 emission savings

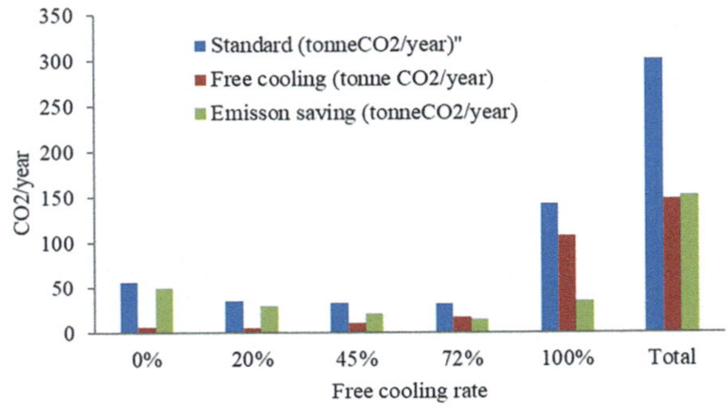

consumption, was calculated to be 300.92 tonsCO_2/year. By implementing the Fre-cooling application, the total CO_2 emission equivalent decreased to 148.32 tonsCO_2/year, resulting in an annual savings of 152.59 tonsCO_2/year, which represents a 50.71% reduction for such systems.

16.5 Conclusion

In this section, mold cooling through the use of free cooling is considered as an example process for industrial processes, and environmental impact management is examined, taking into account energy efficiency and environmental impact assessment for process cooling. In the study, the free-cooling potential was examined together with the mechanical system, based on the climate data of Bursa, which was taken as a reference. According to this:

1. While the exergy efficiency of the system is 12.5% in the mechanical system, the total efficiency with free-cooling support increases to 49.3%.
2. While the IP rate of the system is 76.51% for the mechanical system, it is increased to a potential of 25.72% with the free-cooling contribution of the system.

3. Choosing free cooling shows a 42% saving in environmental impact compared to the mechanical system. This rate shows an environmental pollution impact of 12.3% for SI. However, when the possible improvement potential is taken into account, it indicates a 19.8% improvement in efficiency.
4. According to the IPCC approach, choosing free cooling compared to mechanical system preference shows a 50.7% CO_2 emission saving.

All of these analyses demonstrate the benefits of natural cooling, particularly in situations where cooling preferences are influenced by climatic data. In fact, the application of free cooling is an effective engineering solution for cooling applications that have many functional differences in industrial settings. The study can be supported by life cycle analyses, and more analytical results related to sustainability can be produced.

References

Acül, H. (2007). Energy-saving free (natural) cooling systems in cold water air conditioning and process cooling applications. In *VII. National Plumbing Engineering Congress—TESKON 2007, Izmir (TR).*

Acül, H. (2009). Kuru Soğutuculu Doğal Soğutma Uygulamaları ile Binalarda Enerji Verimliliği. In *IX. Ulusal Tesisat Mühendisliği Kongresi, İzmir, Türkiye, 2009, Friterm A.Ş Teknik Dokumanları ve Uygulamaları*. Retrieved from http://www.friterm.com

ASHRAE. (2000). *Handbook 2000 systems and Equipment, Chapter 36, Chapter 38*. ASHRAE.

Bulut, H., Buyukalaca, O., & Yılmaz, T. (2001). "Bin weather data for Turkey", Department of Mechanical Engineering/University of Cukurova. *Applied Energy, 70*, 135–155.

Cansevdi, B. (2014). The importance of natural cooling and matters to be considered for its more effective use. In *XI. International Plumbing Technologies Symposium, Istanbul, 08–10 May* (pp. 290–297).

Cengel, Y. A., & Boles, M. A. (2006). *Thermodynamics: An engineering approach* (5th ed.). McGraw-Hill Series in Mechanical Engineering. ISBN-10: 0072884959.

Climaveneta. (2014). *Free cooling*. Teknik Broşürü. Retrieved from http://www.climaveneta.it

De Saulles, T. (2004). *BSRIA guide: Free cooling systems*. BSRIA.

DIN. (2019). *Deutsches Institut für Normung e. V. DIN SPEC 15240; Energy performance of buildings—Ventilation for buildings—Inspection of air-conditioning systems. 27.015; 91.140.30 (SPEC 15240)* (p. 53). Beuth Verlag GmbH.

EU (2001) Integrated Pollution Prevention and Control (IPPC) Source Document on the Application of Best Available Techniques in the Field of Industrial Cooling Systems, European Union (TR); https://webdosya.csb.gov.tr/db/ippc/icerikbelge/icerikbelge1124.pdf

EU (2019) Clean energy for all Europeans package, European Commission (Visited in March 27, 2025), https://wayback.archive-it.org/12090/20241209144917/https://energy.ec.europa.eu/topics/energy-strategy/cleanenergy-all-europeans-package_en

Eurovent. (1998). *A BREF-document for cooling systems*. Report number: 61350027. Coopers & Lybrand Umwelt, the European Committee of Air handling and Refrigeration Equipment Manufacturers.

Eurovent. (2016). *Performance efficiency standard for evaporative cooling equipment: Eurovent 9/12*. Eurovent. Retrieved from https://www.eurovent.eu/sites/default/files/Eurovent%20REC%209-12%20-%20Performance%20Efficiency%20Standard%20for%20Evaporative%20Cooling%20Equipment%20-%202016_0.pdf

Gerke, B. F., McNeil, M. A., & Tu, T. (2017). The International Database of Efficient Appliances (IDEA): A new tool to support appliance energy-efficiency deployment. *Applied Energy, 205*, 453–464.

Hincke, P., & Hainbach, C. (2018). Komponenten und Bauteile: Rückkühler. In IKET GmbH (Ed.), *Pohlmann Taschenbuch der Kältetechnik: Grundlagen, Anwendungen, Arbeitstabellen und Vorschriften* (22nd ed., pp. 324–335). VDE Verlag GmbH. ISBN 9783800741496.

ICS. (2024). *Technical catalog of company*. Retrieved February 15, 2024, www.industrialcooling.co.uk

Kakı, E. (2014). Free cooling and adiabatic cooling integrated chillers. In *XI. International Plumbing Technologies Symposium, Istanbul, 08–10 May* (pp. 270–286).

Liu, J., Su, L., Dong, K., & Liu, X. (2022). Critical temperature of free cooling using indirect opening cooling tower in data centres: A review. *International Journal of Ambient Energy, 43*, 1339–1344.

Monfet, D., & Zmeureanu, R. (2013). Calibration of a central cooling plant model using manufacturer's data and measured input parameters and comparison with measured performance. *Journal of Building Performance Simulation, 6*, 141–155.

Qi, X., Liu, Y., Guo, Q., Yu, S., & Yu, J. (2016). Performance prediction of a shower cooling tower using wavelet neural network. *Applied Thermal Engineering, 108*, 475–485.

Schlei-Peters, I., Kurle, D., Wichmann, M. G., Thiede, S., Herrmann, C., & Spengler, T. S. (2015). Assessing combined water-energy-efficiency measures in the automotive industry. *Procedia CIRP, 29*, 50–55.

Schlei-Peters, I., Wichmann, M. G., Matthes, I.-G., Gundlach, F.-W., & Spengler, T. S. (2017). Integrated material flow analysis and process modeling to increase energy and water efficiency of industrial cooling water systems. *Journal of Industrial Ecology, 22*, 41–54.

Schulze, C., Thiede, S., Thiede, B., Kurle, D., Blume, S., & Herrmann, C. (2019). Cooling tower management in manufacturing companies: A cyber-physical system approach. *Journal of Cleaner Production, 211*, 428–441.

Sogut, M. Z., Coşkun, S., Yamankaradeniz, R., & Karakoç, T. H. (2014). Free-cooling applications and energy efficiency in process cooling liquids. In *2nd National Air Conditioning Cooling Training Symposium and Exhibition 23–25 October, Balıkesir (TR)*.

Sogut, M. Z. (2021) New approach for assessment of environmental effects based on entropy optimization of jet engine. Energy, Volume 234, 121250, 2021, ISSN 0360-5442, https://doi.org/10.1016/j.energy.2021.121250

Táboas, F., & Vázquez, F. (2021). Pressure drops and energy consumption model of low-scale closed circuit cooling towers. *Processes, 9*, 974.

Tan, F. L., & Fok, S. C. (2008). Methodology on sizing and selecting thermoelectric cooler from different TEC manufacturers in cooling system design. *Energy Conversion and Management, 49*, 1715–1723.

TPC. (2024). *Company technical catalog*. Retrieved from http://www.totalprocesscooling.co.uk

Wang, J.-G., Shieh, S.-S., Jang, S.-S., & Wu, C.-W. (2013). Discrete model-based operation of cooling tower based on statistical analysis. *Energy Conversion and Management, 73*, 226–233.

Wenzel, P. M., Mühlen, M., & Radgen, P. (2023). Free cooling for saving energy: Technical market analysis of dry, wet, and hybrid cooling based on manufacturer data. *Energies, 16*(9), 3661. https://doi.org/10.3390/en16093661

Zhang, H., Shao, S., Xu, H., Zou, H., & Tian, C. (2014). Free cooling of data centers: A review. *Renewable and Sustainable Energy Reviews, 35*, 171–182.

Chapter 17
Assessment of Thermo-Economic Performance with Sustainable Heat Management in Trigeneration Processes

Özay Kas and M. Ziya Sogut

NomenclatureAbbreviations and Symbols

C	Cost
CCHP	Trigeneration
CHP	Cogeneration
COP	Coefficient of performance
E	Fuel
GT	Gas turbine
H	Heating
HVAC	Heating, ventilating, and air conditioning
ICE	Internal combustion engine
LiBr	Lithium bromide
MOSS	Modified single-stage LiBr-water CCHP flow diagram
PES	Primary energy saving
Q	Cooling
ST	Steam turbine
W	Electricity
η	Efficiency
Θ	Sustainable heat management index

Subscripts and Superscripts

a	Air

Ö. Kas (✉)
Graduate School of Science and Engineering, Piri Reis University, Istanbul, Turkey
e-mail: ozaykas@alumni.pru.edu.tr

M. Z. Sogut
Maritime Faculty, Piri Reis University, Istanbul, Turkey

M. Z. Sogut, M. Koray (eds.), *Energy Rationality and Management for Decarbonization*, Green Energy and Technology,
https://doi.org/10.1007/978-3-031-68497-5_17

abs	Absorption chiller
b	Hot water boiler
ch	Conventional chiller
chp	Cogeneration
e	Electricity
f	Facility/integrated structure
g	Grid
h	Heating
ng	Natural gas
q	Cooling
tot	Total
ul	Unavoidable losses

17.1 Introduction

Nowadays, decreasing energy consumption and costs (C) in high-level, multi-energy buildings has become not only an individual demand but also an important pressure tool in fight against global climate change. In this context, internal combustion engines (ICEs), which are evaluated in low carbon technologies, have started to be used as effective technologies in the solution of multiple energy needs in residential and commercial buildings. Cogeneration (CHP) systems that utilize heat energy together with electrical energy and trigeneration (CCHP) systems, in which cooling (Q) energy is added in these productions, are widely used as effective system preference for energy saving (Nesheim & Ertesva, 2007; Boschiero, 2014; Martínez-lera & Ballester, 2010; Wu et al., 2019). CCHP systems are versatile systems for efficient energy management. Energy consumption control, efficient planning of load distributions, and technology management of system components are seen as prominent. Especially for CCHP systems, it can be seen in studies that will improve system efficiency (η) (Cho et al., 2014; Santo, 2012). However, the economic and environmental effectiveness of CCHP technologies can be assessed by many criteria. Firstly, it is important that the effectiveness of such technologies is designed in accordance with the requirements defined for the structures in which it is used. Especially nowadays, design difficulties or performance variability may be the main reasons with investment costs for the low prevalence of these applications (Nesheim & Ertesva, 2007). Energy demand in buildings is directly dependent on the structural characteristics and the intended use. Environmental parameters should be considered as the basic criteria, especially in system capacity calculations. As a matter of fact, in field studies, it was observed that especially cold climate conditions directly affect CCHP system performance (Boschiero, 2014).

CCHP systems can be defined as basically fossil-fueled complex structures. Such system analysis generally depends on thermodynamic principles. Heat output processes together with electricity (W) generation can be evaluated as a criterion of system efficiency. Thus, heating (H) and cooling are also simultaneous demands besides electricity in integrated structures, and especially single-effect lithium bromide (LiBr) type absorption cooling technology is preferred in such applications (Ghafurian & Niazmand, 2018; Ahmadi et al., 2012) Thus, the heat energy obtained from the cogeneration unit(s) can be served to both cooling and heating systems at the same time with the flow chart formed in series. While heating efficiency is related to efficient utilization of the heat source in thermic processes, the coefficient of performance (COP) value of absorption cooling units depends on the temperature of the heat energy supplied to its generator circuit (Arcuri et al., 2007). Thus, the higher the inlet temperature, the greater the cooling energy obtained. Energy consumption values in integrated structures are not constant, but change according to activity, conditions and occupancy change even during the day (Martínez-lera & Ballester, 2010). Determining the design capacity and operation strategy of a CCHP system for heating, ventilating, and air-conditioning (HVAC) systems in settlements is a difficult process, because there are complex interactions between efficiency and economic parameters (Wu et al., 2019). Because of the high volatility in integrated structures' heat demand, big amount of simultaneous input data is required for reliable design evaluation (Cho et al., 2014). A considerable amount of literature has been published on this topic (Rezvan et al., 2013; Li et al., 2014; Fan et al., 2017; Rafsanjani, 2016; Wang et al., 2017).

Sometimes the heat energy obtained from the cogeneration system is prioritized in the heating system, which results in more economic savings depending on fuel (E) prices (Agarwal et al., 2020; Pantaleo et al., 2017). However, it is inevitable that this prioritization of the heating system in the flowchart will reduce the performance coefficient of the absorption chiller (Manu & Chandrashekar, 2016; Pandya et al., 2017). In particular, considering system performance with demand management in terms of thermal efficiency and costs will provide significant achievements. In this context, prioritization of waste heat obtained from cogeneration units depending on energy demand is examined from a thermo-economic point of view and then the decision processes are determined. In addition, a flowchart has been proposed to ensure that cooling performance does not degrade for a heating priority operation. Thus, this study is intended to contribute to the efficient use of trigeneration systems.

17.2 Heat Demand Control of the Trigeneration Processes

While power technologies have developed solutions within the scope of heat power generation, which has increased in recent years, a new perspective has emerged in which heating, cooling and power generation are combined solutions that take into account the sustainability of energy. In particular, the demand for these three heat demands is increasing in buildings that use trigeneration solutions with a cycle feature. However, the manageability of the waste heat generated in such combined power systems has prompted a re-evaluation, particularly in relation to cooling and absorption processes, as key components in energy solutions. Indeed, the absorber preferences directly impact the performance of cycles developed with different solution pairs. In order to manage sustainability in system performance, the manageability of this cycle, together with the waste heat potential, can be regarded as an optimisation problem. In such processes, absorption constitutes a thermal cycle that is initiated by heat. There is no considerable mechanical energy exchange, but only thermal energy with its surroundings. Absorption chillers typically evolve from a single-stage or double-stage system, originating from the point of heat transfer. The heat source that is generated for the absorber can be either direct or indirect. The performance of absorber-cooler pairs is categorised by optimisable solutions, namely LiBr-water or water-ammonia. Basically, the COP value of the absorption chiller increases when the inlet temperature of the generator increases (Modi et al., 2017; Siddiqui, 2017). On the other hand, there is a direct relation between temperatures of an absorption chiller's components, generator, absorber, evaporator, and condenser for optimum COP value and operation conditions. It is obvious that the minimum temperature of the generator decreases with evaporator temperature and increases with absorber and condenser temperature (Sanaye et al., 2016) (Fig. 17.1).

It is possible to use two different absorption chillers to achieve better performance values: the first one utilizes heat from jacket water circuit of an ICE as a single-effect LiBr-water hot water type, and the second one utilizes heat from high-temperature exhaust of an ICE as a double-effect LiBr-water exhaust type (Yang et al., 2017) (Fig. 17.2).

There are several technologies of CCHP systems: internal combustion engines (ICE), gas turbines (GT), steam turbines (STs), fuel cells, etc. for CHP units, and single-effect hot water, steam, exhaust types, double-effect steam, exhaust, etc. for absorption chillers. Although the absorption chillers utilizing high-temperature steam and exhaust media have higher COP values than the hot water type, they are not capable of serving the systems in integrated structures where hot water is used for heating and domestic hot water consumption and chilled water for HVAC systems simultaneously (ASHRAE, 2015).

There are also possibilities such as thermal energy storage to improve CCHP systems' efficiency (Pina et al., 2017, 2018; Facci et al., 2014). However, while thermal energy storage systems require sizable area and volume, it is hard to find proper areas in integrated structure buildings, especially if they are located in the center of cities where spaces are significantly limited and precious. In this context,

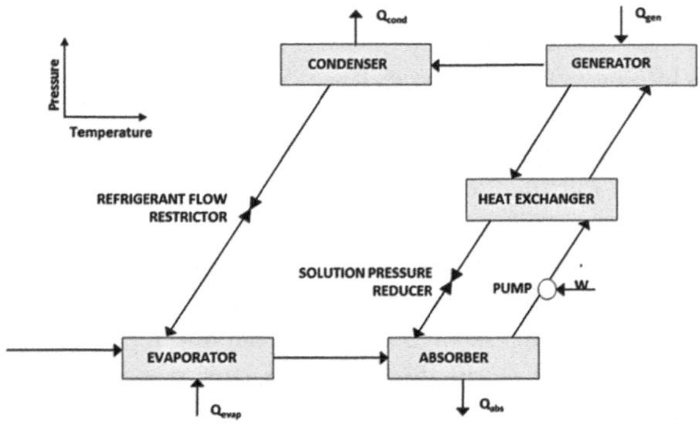

Fig. 17.1 Single-stage LiBr-water absorption refrigeration cycle (Santo, 2012)

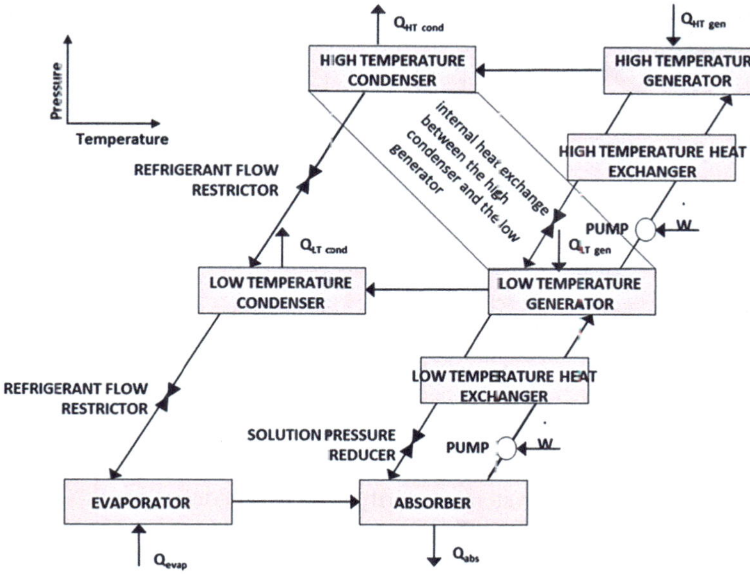

Fig. 17.2 Double-stage LiBr-water absorption refrigeration cycle (Santo, 2012)

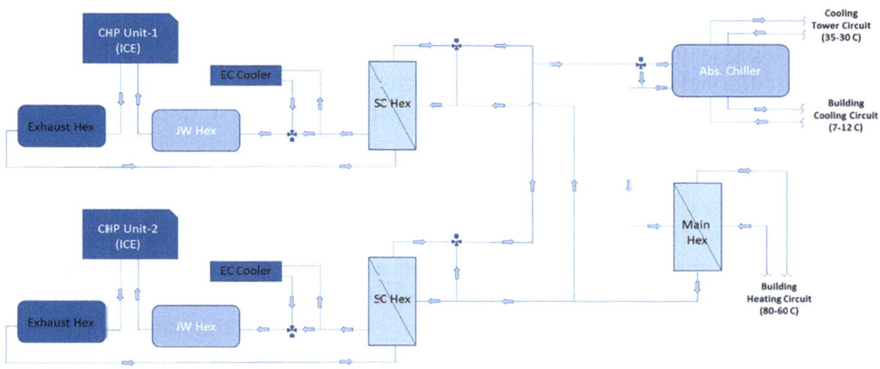

Fig. 17.3 Conventional single-stage LiBr-water CCHP flow diagram

a heat demand control approach is developed for these processes and demand management providing effective energy efficiency in system operation. Therefore, it is obvious that heat demand management is also considerable in such buildings.

Figure 17.3 shows a conventional single-stage LiBr-water CCHP flow diagram where the heat supplied from CHP unit(s) is utilized at the cooling and heating system simultaneously. However, this diagram allows to build the system cooling system primarily. Otherwise, it will be inevitable to decrease the COP value of absorption chiller. In the aim to overcome this disadvantage, modified single-stage

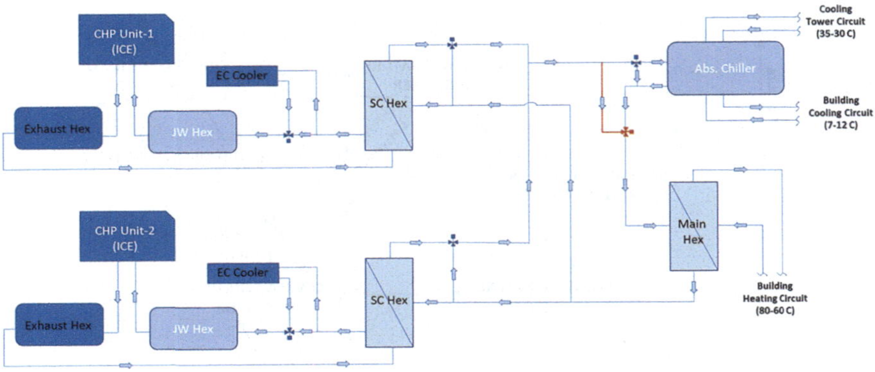

Fig. 17.4 Modified single-stage LiBr-water CCHP flow diagram (MOSS)

LiBr-water CCHP flow diagram (MOSS) has been developed, where it is possible to build the system heating system primarily without any decrease at COP value of absorption chiller shown in Fig. 17.4.

17.3 Heat Demand Management

Heat demand is primarily a service that has different purposes in all integrated structures. In traditional approaches, covering such needs is provided by direct or indirect system components, and domestic structures originated from fossil fuels. However, especially in integrated structures, capacity approaches are not a direct demand analysis but rather a choice according to determined capacities. This indicates high energy consumption and costs due to distributed consumer structures. As a matter of fact, field studies correspond to approximately 68% of fossil fuel resource consumption, depending on different supply sources of heat demand.

These results are important in terms of reducing fossil fuel consumption, which is the main cause of carbonization pollution, especially for the management and control of heat demand. Especially in the building sector, campuses with a potential of 22% will provide a manageable structure with system preferences. Today, conventional resources and different energy usage of energy demands have been eliminated, especially with a low-carbon approach. Along with sustainable resource management, structural needs management is also valuable in these systems. In this context, reducing primary fuel consumption in system preferences where low-carbon technologies are handled together is a result of demand management. Demand management has an approach methodology as given in Fig. 17.5.

Technology management in particular is a key indicator that directly affects system performance. However, this system approach should be taken with an integrated approach within an engineering discipline. In this respect, this study has revealed an approach in which heat demand management is evaluated together with

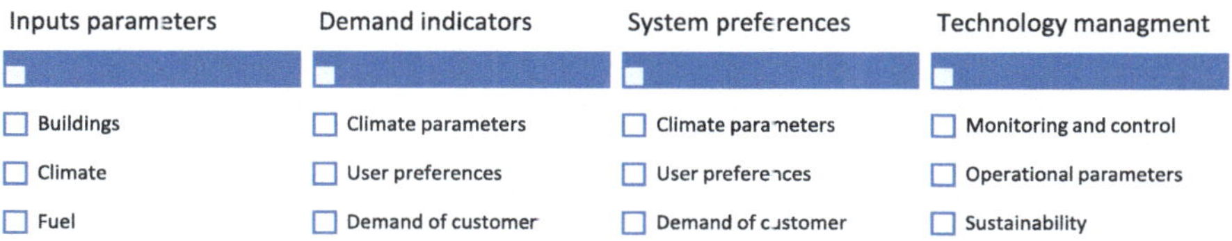

Fig. 17.5 Heat demand control and management approach

the system preference in a direct system choice. The developed study showed that direction preference in heat control is an important energy and cost saving in system preference and application.

17.4 Thermo-Economic Analysis

Energy systems in integrated structures are a thermodynamic preference in direct engineering approach. All thermal processes are interacted with the environmental conditions in which they are located. In this context, energy requirement is evaluated in a continuous flow control during an energy conversion process (Shelar et al., 2016). Trigeneration applications in integrated structures are the transformation of electricity, heating, and cooling energies obtained instantly with primary fuels. The system scheme and thermodynamic equations used for the approach proposed in this study are presented below.

Figure 17.6 represents a simplified diagram of an integrated structure with a trigeneration system.

Total electricity demand of the integrated structure is the sum of the facility itself and the conventional chiller, where it is supplied from cogeneration and grid network, is expressed as follows:

$$W_{\mathrm{f}} + W_{\mathrm{ch}} = W_{\mathrm{chp}} + W_{\mathrm{g}} \tag{17.1}$$

Energy balance of the cogeneration unit is expressed in the following equation (Frangopoulos, 2017):

$$E_{\mathrm{chp}} + E_{\mathrm{a}} = W_{\mathrm{chp}} + H_{\mathrm{chp}} + H_{\mathrm{ul}} \tag{17.2}$$

Heat supplied from the cogeneration unit can be utilized in a cooling or heating system, or both at the same time (Frangopoulos, 2017).

$$H_{\mathrm{chp}} = H_{\mathrm{abs}} + H_{\mathrm{h}} \tag{17.3}$$

Total heating demand of the integrated structure is supplied from cogeneration and hot water boiler, and expressed as follows:

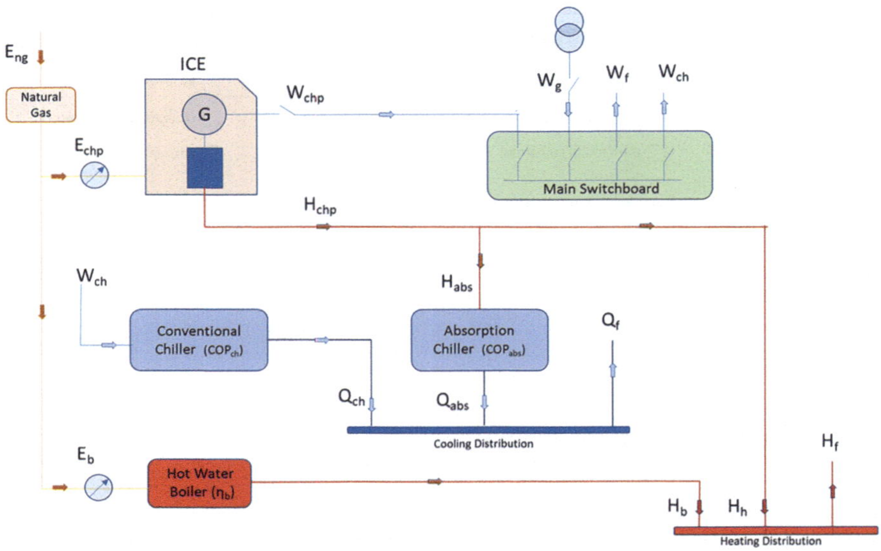

Fig. 17.6 Simplified diagram of an integrated structure with trigeneration system

$$H_f = H_h + E_b.\eta_b \tag{17.4}$$

Total cooling demand of the integrated structure is supplied from absorption chiller and conventional hot water boiler, and expressed as follows:

$$Q_f = (H_{abs} \cdot \text{COP}_{abs}) + (W_{ch} \cdot \text{COP}_{ch}) \tag{17.5}$$

If the heat supplied from CHP unit is utilized in the ***cooling system primarily*** as shown Eq. (17.5), then the heating demand of the integrated structure is supplied from the CHP unit and the hot water boiler, which is expressed as follows:

$$H_f = (H_{chp} - H_{abs}) + E_b.\eta_b \tag{17.6}$$

If $Q_f > H_{abs} \cdot \text{COP}_{abs}$ which means the cooling supplied by the CHP unit for the cooling system is not enough for the integrated structure's cooling demand. Then there will be no remaining heat for the heating system by the CHP unit; $(H_{chp} - H_{abs}) = 0$. All heating demand should be supplied by a hot water boiler: $H_f = E_b.\ \eta_b$.

If the heat supplied from CHP unit is utilized at ***heating system primarily*** as shown Eq. (17.4), then cooling demand of the integrated structure is supplied by the absorption chiller and conventional chiller, which is expressed as follows:

$$Q_f = (H_{chp} - H_h)\text{COP}_{abs} + (W_{ch} \cdot \text{COP}_{ch}) \tag{17.7}$$

If $H_f > H_h$, it means the heat supplied by the CHP unit for the heating system is not enough for the integrated structure's heating demand. Then there will be no remaining heat for the cooling system by the CHP unit; $(H_{chp} - H_h) = 0$. All cooling demand should be supplied by conventional chiller; $Q_f = W_{ch} \cdot COP_{ch}$.

Economic saving of the heat supplied by the CHP unit for the heating system is expressed as follows:

$$C_{CHP_h} = \frac{H_{chp} \cdot C_{ng}}{\eta_b} \tag{17.8}$$

Similar to Eq. (17.8), the economic saving of the heat supplied by the CHP unit for the cooling system is expressed as follows:

$$C_{CHP_q} = \frac{H_{chp} \cdot COP_{abs} \cdot C_e}{COP_{ch}} \tag{17.9}$$

If $C_{chp_h} > C_{chp_q}$, it means an economic saving of the heat supplied by the CHP unit for the heating system is more than for the cooling system. Then the heat shall be utilized in the ***heating system primarily***.

$$\text{Thus, if } \frac{C_{ng}}{\eta_b} > \frac{COP_{abs} \cdot C_e}{COP_{ch}} \text{ Heating System Primarily} \tag{17.10}$$

$$\text{If } \frac{C_{ng}}{\eta_b} < \frac{COP_{abs} \cdot C_e}{COP_{ch}} \text{ Cooling System Primarily}$$

In an aim to help facility managers decide to operate their trigeneration systems utilize waste heat for heating or cooling system primarily, the following "**Sustainable Heat Management Index**" can be used:

$$\theta = \frac{\frac{C_{ng}}{\eta_b}}{\frac{COP_{abs} \cdot C_e}{COP_{ch}}} \tag{17.11}$$

If $\boldsymbol{\theta} > \mathbf{1}$, then the heat supplied by CHP unit shall be utilized for **heating** system primarily.

If $\boldsymbol{\theta} < \mathbf{1}$, then the heat supplied by CHP unit shall be utilized for **cooling** system primarily.

17.5 Results and Discussion

Real field data belongs to an integrated structure which is a monolithic hospital building on 16,500 m^2 of land, having a 98,000 m^2 indoor area. Construction is made of reinforced concrete, including six floors above ground level and five basement floors underground. Exterior of the building is covered by composite aluminum panels, double-glazed windows. The integrated structure serves 7 days and 24 h without any interruption.

The structure is located in Üsküdar province of Istanbul. All climate conditions and parameters of Istanbul are applicable to this complex. Winter design temperature is −3 °C, summertime design dry bulb temperature is +35 °C (wet bulb: +24 °C).

The primary source of electricity is the national grid system connected at high voltage (34.5 kV) via 3 × 2500 kVA transformers. The secondary source is 4 × 1600 kVA diesel generators for emergency and 1.200 kVA UPS power for backup. The primary source for heating demand is the national natural gas network via 3 × 2.000 kWth hot water boilers and 20 tons/h domestic hot water. The primary source for cooling demand is electricity via 3 × 2.500 kWth water-cooled centrifugal chillers. The energy demand distribution of the facility is shown in Fig. 17.7, where annual electricity demand (excluding chillers' consumption) is 16.910 MWh/year, cooling demand is 19.095 MWh/year, heating demand is 8.111 MWh/year and domestic hot water demand is 3.910 MWh/year.

The COP value of the conventional chiller is 6.0. The annual electricity demand of the chiller is 3.182 MWh/year. Therefore, annual electricity demand (including chillers' consumption) is 20.092 MWh/year.

The active use of air-conditioning processes in the integrated structure emphasizes electricity consumption. In addition, demand for heating and cooling in air-conditioning plants has a potential of approximately 17% and 40% respectively in direct conventional load. The share of consumption distribution according to

Fig. 17.7 Energy demand distribution

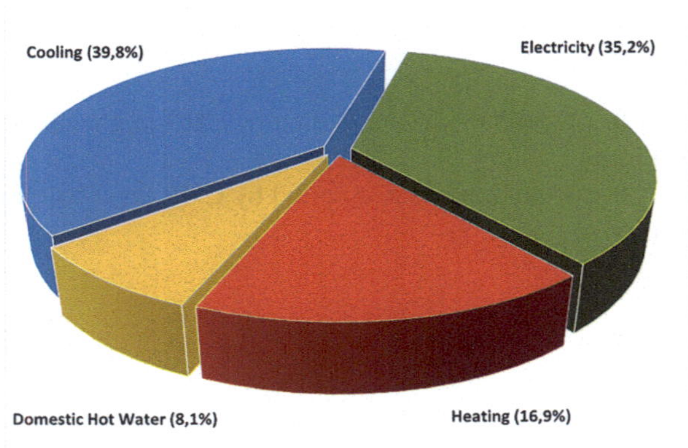

resources in energy management is also examined. According to the review, the electricity, cooling, heating, and domestic hot water demand distribution monthly changes are shown in Fig. 17.8.

Seasonal effects are remarkable in energy consumption. Heat demand, which has 17% of the total energy, stands out, especially for the winter months. Additionally, it is also a resource used in the system for the months of transition, April, May, and October. However, especially the monthly demand for electricity consumption is over 1.300 MWh/month, excluding December. The requirement for cooling in terms of heat management in the enterprise is also evaluated. In this context, monthly distributions were taken together with the total heat requirement. Total heat and cooling demand distribution monthly change is shown in Fig. 17.9.

Controlling seasonal transitions in heat demand management and thermal management of air-conditioning processes should be evaluated in a structure that needs attention. Especially considering the cooling requirements in special conditions, it is seen that cooling is required every month. This has revealed the necessity for a continuous control system in relation to heat management. The cooling demand required depending on the monthly consumption in the enterprise is 19.095 MWh/year and it has a potential of 39.8% of total energy demand.

The trigeneration processes designed for the enterprise have been developed as a model that is based on primary energy consumption, primarily electricity. The basic approach in the enterprise is planned to obtain the heat (heating, cooling, and domestic hot water) required by the facility together with the electricity and waste heat system, and the system diagram is given in Fig. 17.7. Accordingly, the facility has a trigeneration plant composed of two pcs. of internal combustion engines, two pcs. of exhaust heat exchangers and one pc. of single-effect LiBr hot water absorption chiller. Basic parameters of the trigeneration plant are shown in Table 17.1.

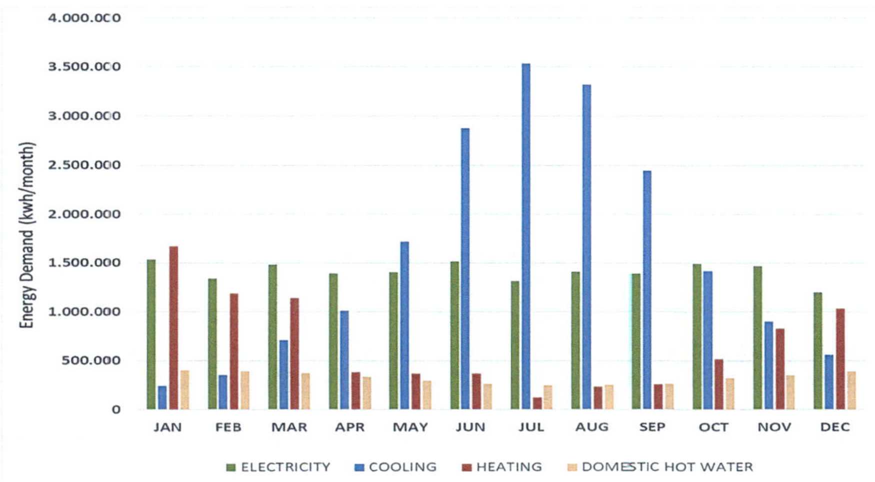

Fig. 17.8 Energy demand distribution monthly change

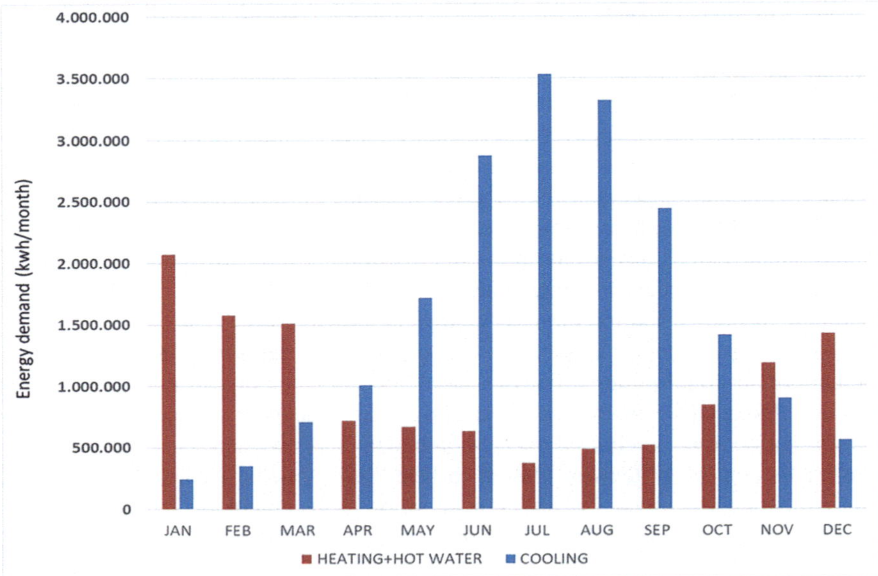

Fig. 17.9 Cooling and heat demand distribution monthly change

Table 17.1 Basic parameters of CCHP plant

Description	Unit	Data	Description	Unit	Data
ICE electrical output	kW	1.200	Exh. *T* at ICE outlet	°C	414
ICE heat output (95–75 °C)	kW	1.189	Exh. *T* at Exh.Exch. outlet	°C	120
ICE fuel input	kW	2.748	Exh. gas amount	kg/h	6.481
ICE electrical efficiency	%	43.6	Exh. heat exch. capacity	kW	577
ICE heat efficiency	%	43.3	Abs. chiller cooling capacity	kW	1.800
ICE total efficiency	%	86.9	Abs. chiller COP value	–	0.75

The effectiveness of trigeneration processes is primarily aimed at optimum covering energy demand. In this context, resource priority and coverage rates are valuable in system preferences. The CCHP plant realized 15.686 MWh of electricity (78.4% of total electricity demand), 9.292 MWh heating (75.1% of total heating demand), and 6.389 MWh of cooling energy (31.6% of total cooling demand) for the facility while the system is operated in heating priority mode, as shown in Fig. 17.4. CCHP plant's total annual operation period is 8.385 h where the total electricity generation capacity utilization rate is 77.9%. During the same period, total generated waste heat was 18.560 MWh/year and 96% of it utilized by heating and/or cooling systems.

Energy distribution monthly change by the CCHP plant is shown in Fig. 17.10 and energy demand coverage rates monthly change by CCHP plant is shown in Fig. 17.11.

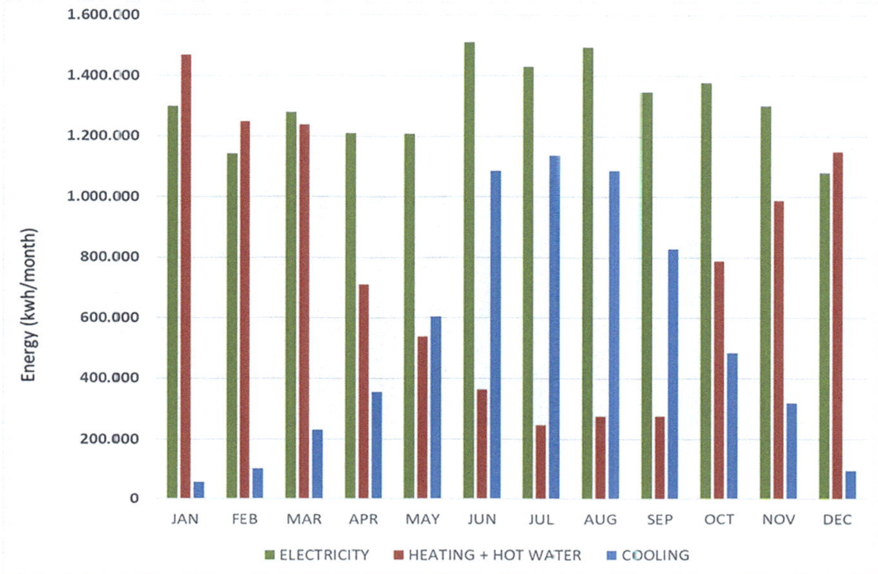

Fig. 17.10 Energy distribution by trigeneration plant monthly change

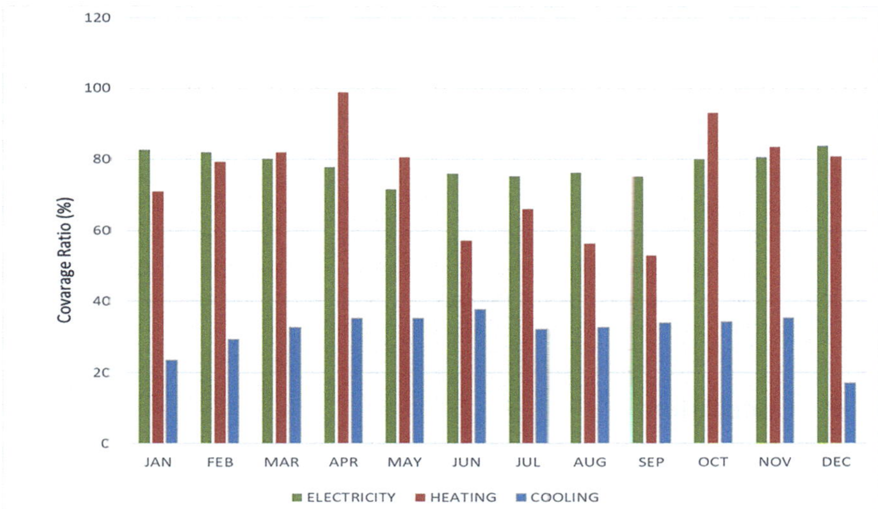

Fig. 17.11 Energy demand coverage rates by CCHP plant monthly change

The primary energy demand of the facility was 57.859 MWh/year with reference to 0.525 thermal power plant's electrical efficiency, 14% transmission and distribution loss, and 0.90 hot water boiler efficiency before CCHP plant's installation.

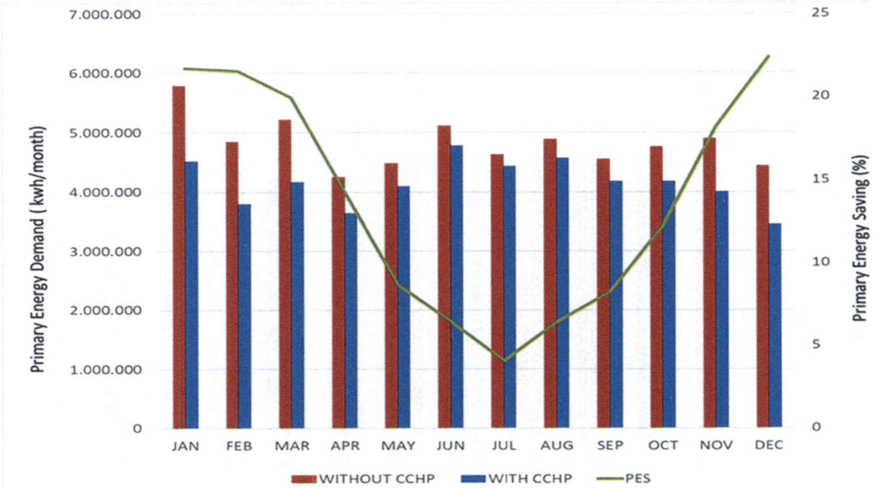

Fig. 17.12 Primary energy distribution and savings monthly change

Primary energy demand is reduced to 49.853 MWh after CCHP plant's installation, which provides 13.67% primary energy saving (PES). Primary energy distribution of monthly change and primary energy saving (PES) are shown in Fig. 17.12.

Trigeneration plant provides 410.930 USD/year cost saving with utilized heat for heating and cooling system in heating priority mode, where the share for heating is 71.1% and 28.9% for cooling. If the plant was operated in cooling priority mode, then cost savings would be 342.400 USD/year, where the share for heating is 49.4% and 50.6% for cooling. Thus, the modified system developed depending on demand management provides approximately 22.4% extra saving from the utilized heat for heating and cooling.

Cost savings comparison of heating and cooling priority modes is shown in Fig. 17.13. It is obvious that cost saving difference is significant in the spring and autumn seasons where heating and cooling loads are relatively near to each other.

17.6 Conclusion

This study has investigated the efficiency analysis to be provided in the system according to the preference of heating and cooling demands in a trigeneration system carried out according to thermo-economic principles. The study has shown that there is a direct relation between thermo-economic parameters, energy cost, and efficiencies.

Conventional trigeneration system and modified system were examined comparatively based on demand management optimization on real field study data. In the analysis, it was seen that the modified system developed depending on demand

Fig. 17.13 Cost saving comparison of heating and cooling priority modes

management provides approximately 22.4% extra savings from the utilized heat for heating and cooling.

Additionally, a new index named "Sustainable Heat Management Index" is introduced to help facility managers decide to operate their trigeneration systems utilize waste heat for heating or cooling priority.

References

Agarwal, S., Arora, A., & Arora, B. B. (2020). Energy and exergy analysis of vapor compression—Triple effect absorption cascade refrigeration system. *Engineering Science and Technology, an International Journal, 23*, 625–641.

Ahmadi, P., Dincer, I., & Rosen, M. A. (2012). Exergo-environmental analysis of an integrated organic Rankine cycle for trigeneration. *Energy Conversion and Management, 64*, 447–453.

Arcuri, P., Florio, G., & Fragiacomo, P. A. (2007). A mixed integer programming model for optimal design of trigeneration in a hospital complex. *Energy, 32*, 1430–1447.

ASHRAE. (2015). *Combined heat and power design guide*. ASHRAE.

Boschiero, D. (2014). An energy and exergy analysis of a high-efficiency engine trigeneration system for a hospital: A case study methodology based on annual energy demand profiles. *Energy and Buildings, 76*, 185–198.

Cho, H., Smith, A. D., & Mago, P. (2014). Combined cooling, heating and power: A review of performance improvement and optimization. *Applied Energy, 136*, 168–185.

Facci, A. L., Andreassi, L., Ubertini, S., & Sciubba, E. (2014). Analysis of the influence of thermal energy storage on the optimal management of a trigeneration plant. *Energy Procedia, 45*, 1295–1304.

Fan, C., Xiao, F., & Zhao, Y. (2017). A short-term building cooling load prediction method using deep learning algorithms. *Applied Energy, 195*, 222–233.

Frangopoulos, C. A. (2017). *Cogeneration-technologies, optimisation and implementation, Chapter 7*. The Institute of Engineering and Technology.

Ghafurian, M. M., & Niazmand, H. (2018). New approach for estimating the cooling capacity of the absorption and compression chillers in a trigeneration system. Nouvelle approche pour estimer la puissance frigorifique des refroidisseurs à absorption et à compression dans un système de trigénération. *International Journal of Refrigeration, 86*, 89–106.

Li, L., Mu, H., Gao, W., & Li, M. (2014). Optimization and analysis of CCHP system based on energy loads coupling of residential and office buildings. *Applied Energy, 136*, 206–216.

Manu, S., & Chandrashekar, T. K. (2016). A simulation study on performance evaluation of single-stage LiBr–H_2O vapor absorption heat pump for chip cooling. *International Journal of Sustainable Built Environment, 5*, 370–386.

Martínez-lera, S., & Ballester, J. (2010). A novel method for the design of CHCP (combined heat, cooling and power) systems for buildings. *Energy, 35*(7), 2972–2984.

Modi, B., Mudgal, A., & Patel, B. (2017). Energy and exergy investigation of small capacity single effect lithium bromide absorption refrigeration system. *Energy Procedia, 109*, 203–210.

Nesheim, S. J., & Ertesva, I. S. (2007). Efficiencies and indicators defined to promote combined heat and power. *Energy Conversion and Management, 48*, 1004–1015.

Pandya, B., Patel, J., & Mudgal, A. (2017). Thermodynamic evaluation of generator temperature in LiBr–water absorption system for optimal performance. *Energy Procedia, 109*, 270–278.

Pantaleo, A. M., Camporeale, S. M., Markides, C. N., Mugnozza, G. S., & Shah, N. (2017). Energy performance and thermo-economic assessment of a microturbine-based dual-fuel gas-biomass trigeneration system. *Energy Procedia, 105*, 764–772.

Pina, E. A., Lozano, M. A., & Serra, L. M. (2017). Optimal operation and marginal costs in simple trigeneration systems including thermal energy storage. *Energy, 135*, 788–798.

Pina, E. A., Lozano, M. A., & Serra, L. M. (2018). Thermoeconomic cost allocation in simple trigeneration systems including thermal energy storage. *Energy, 153*, 170–184.

Rafsanjani, H. N. (2016). Linking building energy-load variations with occupants' energy-use behaviors in commercial buildings: Non-intrusive occupant load monitoring (NIOLM). *Procedia Engineering, 145*, 532–539.

Rezvan, A. T., Gharneh, N. S., & Gharehpetian, G. B. (2013). Optimization of distributed generation capacities in buildings under uncertainty in load demand. *Energy and Buildings, 57*, 58–64.

Sanaye, S., Mustafa, M., & Tavakoli, F. (2016). Applying relative net present or relative net future worth benefit and exergy efficiency for optimum selection of a natural gas engine based CCHP system for a hotel building. *Journal of Natural Gas Science and Engineering, 34*, 305–317.

Santo, D. B. E. (2012). Energy and exergy efficiency of a building internal combustion engine trigeneration system under two different operational strategies. *Energy and Buildings, 53*, 28–38.

Shelar, M. N., Bagade, S. D., & Kulkarni, G. N. (2016). Energy and Exergy analysis of diesel engine powered trigeneration systems. *Energy Procedia, 90*, 27–37.

Siddiqui, M. A. (2017). Energy and exergy analyses for optimization of the operating temperatures in double effect absorption cycle. *Energy Procedia, 109*, 211–218.

Wang, F., Zhou, L., Ren, H., & Liu, X. (2017). Search improvement process-chaotic optimization-particle swarm optimization-elite retention strategy and improved combined cooling-heating-power strategy based two-time scale multi-objective optimization model for stand-alone microgrid operation. *Energies, 10*, 12.

Wu, D., Han. Z., Liu, Z., & Zhang, H. (2019). Study on configuration optimization and economic feasibility analysis for combined cooling, heating and power system. *Energy Conversion and Management, 190*, 91–104.
Yang, G., Zheng, C. Y., & Zhai, X. Q. (2017). Influence analysis of building energy demands on the optimal design and performance of CCHP system by using statistical analysis. *Energy and Buildings, 153*, 297–316.

Chapter 18
Pneumatic Propulsion of Vehicles

Abdul Hai Alami, Mohamad Ayoub, Adnan Alashkar, Fatima Alshamsi, Ayman Mdallal, Siren Khuri, and Selçuk Sarıkoç

18.1 Introduction

One of the most transformational technological advancements in civilization is the ability to transport goods and people reliably from one location to the other. The vehicles used during the Agricultural Revolution were powered by strong animals such as horses or oxen, depending on the distance to be covered or the tugging power, respectively. Although the propulsion mechanism shifted during the Industrial Revolution into steam-powered portable power plants, some terms describing vehicular performance parameters, such as horsepower remained faithful

A. H. Alami (✉) · M. Ayoub
Sustainable and Renewable Energy Engineering Department, University of Sharjah, Sharjah, United Arab Emirates

Sustainable Energy & Power Systems Research Centre, RISE, University of Sharjah, Sharjah, United Arab Emirates
e-mail: aalalami@sharjah.ac.ae

A. Alashkar
PhD Program in Materials Science and Engineering, American University of Sharjah, Sharjah, United Arab Emirates

F. Alshamsi
Sustainable and Renewable Energy Engineering Department, University of Sharjah, Sharjah, United Arab Emirates

A. Mdallal · S. Khuri
Sustainable Energy & Power Systems Research Centre, RISE, University of Sharjah, Sharjah, United Arab Emirates

S. Sarıkoç
Faculty of Engineering, Mechanical Engineering Department, Branch of Energy, Amasya University, Amasya, Turkey

M. Z. Sogut, M. Koray (eds.), *Energy Rationality and Management for Decarbonization*, Green Energy and Technology,
https://doi.org/10.1007/978-3-031-68497-5_18

to this era. In the field of environmentally sustainable transportation, alternative methods to replace conventional propulsion technologies are being sought. This shift is crucial in areas where the availability of conventional fuel has substantial ecological and socioeconomic consequences. An innovative method involves utilizing pneumatic propulsion systems that are operated by compressed air energy storage (CAES). This system signifies a fundamental change from traditional engines, with the goal of minimizing the environmental effect of transportation by reducing carbon emissions, noise pollution, and waste that the environment and nearby urban areas produce. There are various potential advantages to implementing such systems, including the promotion of cleaner technology adoption in residential and transportation sectors, as well as the encouragement of safer and more sustainable transportation-related activities.

This chapter highlights the position of pneumatic propulsion in the era where the world is departing from conventional propulsion technologies into all-electric vehicles on land, sea, and air. Moreover, this chapter focuses on the technological potential of pneumatic propulsion and discusses user sectors that best fit the application of pneumatically propelled vehicles.

18.2 Conventional Propulsion

18.2.1 Internal Combustion Engines

The current global transport infrastructure relies heavily on internal combustion engines for land, air, and sea vehicles. These engines are mainly fueled by gasoline and diesel. Conventional propulsion via internal combustion engines has played an essential role in shaping the way we move and connect across various terrains (Kalghatgi, 2018).

Vehicles with gasoline internal combustion engine (ICE) typically exhibit a wheel-to-wheel efficiency in the range of 11–27%. These engines are commonly used to fuel light vehicles and are characterized by their relatively cleaner emission metrics compared to their counterparts. Diesel ICE, on the other hand, exhibits an efficiency range of 25–37% and is known for its powerful torque and fuel efficiency. Vehicles operating on compressed natural gas are not widely utilized due to their relatively low efficiency range of 12–22% (Leach et al., 2020; Albatayneh et al., 2020).

The aviation industry owes much of its success to the development and refinement of ICEs. Piston engines, turbo engines, and Ramjet engines are all specialized forms of internal combustion engines specifically designed for conventional propulsion-based air vehicles. Both commercial and military aviation sectors rely extremely on these engines, showcasing the versatility and power of conventional propulsion in the air domain (Zhang et al., 2022).

Diesel ICEs have dominated the maritime transportation scene over the nineteenth and twentieth centuries. The transition from steam engines to diesel engines marked a revolutionary shift in maritime propulsion, with diesel engines currently holding the highest share of the market due to their higher efficiency, accessibility of spare parts, and well-established operational procedures (Inal et al., 2022).

While conventional propulsion has served as a solid base for various modes of transportation, it bears a lot of issues that drove the need for finding suitable alternatives to propulsion means for these sectors. Concerns about the environmental impact, fossil fuels consumption, and its sustainability have led to an interest in looking for alternative propulsion systems. Electric and novel hybrid technology-based systems are gaining traction as possible alternatives that promise reduced emissions and a more sustainable future for transportation (Lešnik et al., 2018).

18.2.2 Jet Engines

The aviation industry relies heavily on jet engines as a propulsion mechanism for aerial vehicles as they are known for their remarkable efficiency and power generation (Alves et al., 2020). The two most important types of jet engines, turboprop engines, and turbofan engines, have played a huge role in how the current aviation sector performs.

Turboprop engines are known for their inherent limitation in generating high power. These engines consume a significant portion of their energy for the propeller rotation and perform better at low speeds (Cengiz & Utlu, 2019). Designing a high-capacity turboprop aircraft is a complex procedure. The pursuit of an aircraft tailored for modern regional transport applications hinges on a set of meticulously defined topic level aircraft requirements. These parameters serve as the guiding principles for the design process, emphasizing the need to enhance passenger capacity, reduce specific fuel consumption, and minimize emissions (Nicolosi et al., 2021).

Turbofan engines, also known as "fanjets," are made of a set of fans and turbines combinations. The distinctive feature setting them apart lies in their by-pass ratio, also known as the airflow ratio. A lower airflow ratio is more favorable for civil aircrafts due to their performance on longer flights (Cengiz & Utlu, 2019). Their ability to provide high thrust at high altitudes, combined with reduced fuel consumption, makes them more desirable for airlines striving to balance performance and economic viability. The most essential key parameters to keep in mind when designing a turbofan engine are thermal efficiency, exergy, economics, and environmental impact. Tests on turbofan engine performance under varying speeds and altitudes are essential as they will highly affect thermal efficiency and exergy at different operational points (Akdeniz et al., 2022).

18.2.3 Nuclear Propulsion

Nuclear-based propulsion has become an attractive option for marine and submarine applications because of its benefits, such as high energy density, extended range, and compact size. Nuclear-powered propulsion, unlike traditional fuel-based propulsion systems, provides the distinct ability to produce a continuous stream of energy without requiring regular refilling. This makes them especially well-suited for submarines that remain submerged for long durations. Moreover, its compact and light design is well-suited for the limited area often seen in marine and submarine applications.

Ships equipped with nuclear propulsion systems utilize the heat created by nuclear fission to boil water to generate steam, which in turn drives the ship's turbines and propellers (Drosińska-Komor et al., 2022). Power and speed are improved, range is increased, and dependency on fossil fuels is decreased compared to conventional propulsion systems (Bayraktar & Pamik, 2023). When it comes to navy vessels, nuclear reactors not only provide propulsion, but they may also provide power for the ship's systems, cutting down on the need for backup generators and improving the vessel's efficiency (Bayraktar & Pamik, 2023). To extract nuclear power from nuclear reactors in order to drive nuclear-based propulsion systems, a Rankine cycle is utilized. The Rankine cycle (Dincer & Bicer, 2020) is the standard system for steam generation in thermal power generation including nuclear power (Liu et al., 2012). It employs a series of multiple connected stages to convert thermal energy into mechanical power. Water initially evaporates when heated to its boiling point in the boiler, absorbing heat from the nuclear reactor and converting it into steam. This steam then serves as the working fluid for the cycle. After evaporation, the steam passes through a process called superheating, where its temperature is raised above the boiling point in the superheater. This increases the efficiency of the cycle by raising the specific enthalpy of the steam. In the next phase, the superheated steam is supplied to the turbine, where it undergoes expansion over a series of blades. This process converts thermal energy into mechanical energy, which is then delivered to the propeller, driving the vessel forward. Finally, the procedure concludes in condensation, when the depleted steam, currently at a reduced temperature and pressure, reaches the condenser. Within the condenser, the steam transfers its residual thermal energy to a secondary fluid, commonly saltwater, resulting in the steam undergoing condensation and reverting to its liquid state. Subsequently, the condensed water is returned to the boiler, therefore completing the cyclic process. An investigation of three possible power cycles was carried out as a result of the discovery of alternatives to the traditional steam Rankine power cycle. The condensing supercritical CO_2 power cycle is characterized by three turbines and two compressors, and the Recompression supercritical CO_2 power cycle has one turbine and two compressors and the compact steam Rankine cycle. In contrast to the conventional steam Rankine cycle, the compact steam Rankine cycle has a high-speed turbine generator in addition to low-pressure and high-pressure turbines. This allows the cycle to maintain the same level of efficiency as its predecessor (Shirvan et al., 2016; Shirvan & Kazimi, 2016).

Nuclear power plants have become the fundamental source of propulsion for several types of vessels. Submarines rely on this vital backbone to achieve prolonged underwater endurance and possess strategic deterrent capabilities. Furthermore, nuclear-powered icebreakers have a vital function in guaranteeing accessible shipping routes in areas covered with ice, therefore promoting trade and transportation under challenging conditions. Nuclear power reactors are used on aircraft carriers to fulfill the significant energy requirements for operating both the carriers and their aircraft, hence enhancing their efficiency as remarkable naval forces (Yu, 2020).

18.2.4 Electric Propulsion

Electric propulsion-based vehicles are transforming the transportation industry by providing a more environmentally friendly and efficient alternative to conventional fuel-based vehicles. These systems utilize electric motors to propel vehicles forward, harnessing the efficiency and environmental friendliness of electricity compared to gasoline.

The electric motor is the main component of electric propulsion. It transforms electrical energy into mechanical energy, which is then used to directly power the vehicle. The conversion process is based on the fundamental principle of electromagnetism, which states that an electric current passing through a conductor produces a magnetic field. The interaction between this magnetic field and another magnetic field generates a force that induces motion in the conductor. In an electric motor, the conductor is usually in the form of a wire copper coil, while the magnetic field is produced by either permanent magnets or electromagnets. When an electric current passes through the coil, it produces an independent magnetic field. The interaction between the magnetic field generated by the coil and the stationary magnetic field results in a rotational force exerted on the coil. Subsequently, this rotational motion is transmitted to the vehicle's propulsion system, causing it to accelerate (Masuzawa et al., 2018; Kim, 2017).

There are four main types of electrically propelled vehicles: battery electric vehicles (BEVs), hybrid electric vehicles (HEVs), fuel cell electric vehicles (FCEVs), and plug-in hybrid electric vehicles (PHEVs) (Bilaţiu et al., 2019). BEVs completely rely on electricity stored in a battery pack, resulting in no emissions; however, it necessitates frequent charging. The vehicle's range on a single charge can vary between approximately 100 and 250 km, depending on factors such as road conditions, driver behavior, climate conditions, vehicle configuration, and battery age. HEVs integrate an electric motor and an internal combustion engine (ICE), resulting in an increased driving distance while maintaining fuel efficiency. In low-speed situations, the electric motor is used to decrease fuel consumption and greenhouse gas emissions. The hybrid vehicle shifts to its ICE when the electric motor cannot keep up with the demands of higher speeds. Furthermore, the vehicle's performance can be enhanced by operating the two engines simultaneously (Mustafi, 2022). On the other hand, in PHEV, the electric motor serves as the primary source of propulsion, with the ICE serving to either charge the battery pack as needed or guarantee propulsion in situations where the electric motor is unable to do so. Finally, FCEVs use fuel cells to generate electricity from hydrogen and oxygen, which powers the electric motor of the vehicle. Due to the fact that their only byproduct is water, FCEVs are more environmentally friendly than traditional vehicles. In addition, they have a fast hydrogen tank refill time that is on level with traditional gas-powered cars, making them a practical substitute (Bilaţiu et al., 2019). With the continuous advancement of battery technology, which allows for longer driving distances and quicker charging, electric propulsion is on the edge of becoming the dominant force in the automotive industry.

18.3 Impact of Conventional Propulsion Technologies

18.3.1 Economic Impact

Due to the volatility of fossil fuels, countries that are highly dependent on these conventional propulsion systems suffer from economic instability either directly or indirectly. An example of macroeconomic variables that are negatively affected as a result of these rapid and unpredictable changes in fuel cost include real wages and employment, selling prices, basic inflation, profits, and investment (Qahtan et al., 2022). In addition, the negative health and environmental impacts associated with the consumption of conventional propulsion systems are correlated with significant social costs that remain at a higher value than the retail price. The large magnitude of such externalities in the United States holds an average value of \$0.14–\$0.35/kWh as a result of fossil fuel usage which corresponds to the loss of \$361.7–886.5 billion annually, representing 2.5–6.0% of the national GDP (Machol & Rizk, 2013).

18.3.2 Technological Impact

Regular maintenance such as lubrication, alignment checks, and wear inspections are paramount for conventional systems, where such tasks are linked to high cost and consumption of arduous time. Despite efforts in design and optimization, there remains the issue of noise and vibrations generated at significant levels for these conventional systems that have adverse effects on the safety and comfort of the crew (Ammar et al., 2023).

18.3.3 Environmental Impact

Conventional propulsion systems are distinguished as one of the factors that have a significant effect on current global emissions due to the considerable amounts of greenhouse gas (GHG) emissions and air pollutants mainly released from the combustion reactions taking place in their engines. The atmosphere is constantly diffused with pollutants such as carbon monoxide (CO), hydrocarbons (HC), particulate matter (PM), and nitrogen oxide (NOx) that are released as a result of the inadequate combustion of the fuel used (Kumar et al., 2021). Depending on the type of propulsion system, other pollutants are additionally released into the atmosphere disrupting the ecological status. An example of this can be observed in ships where it is correlated with emissions consisting of transitional and alkali earth metals along with their soluble or insoluble chemical forms (Blasco et al., 2014). These common emissions are generated not only from the combustion in the usage phase but also throughout the entire lifecycle of these propulsion systems, allowing significant negative effects on both humans and ecosystems. A wide array of diseases that are not limited to premature mortality, heart attacks, and asthma has been subjected to the human population with greater threatened impact experienced by the population located in big cities and urban areas that are known for being exposed to such conventional propulsion means that utilize oil derivatives (Jurić & Županović, 2012). From an environmental perspective, the effect of the accumulated concentrations of these emissions has led to the increase of global warming, acidification, and photochemical ozone formation potential (Li et al., 2013). In addition, land and water degradation has been a consequence arising from the various processes involved in the consumption of fossil fuels as an energy source for these propulsion systems such as exploration, extraction, processing, and distribution.

18.4 Pneumatically Propelled Vehicles

Pneumatically propelled vehicles present a novel and environmentally friendly approach to transportation. These vehicles depend on the principles of pneumatics through utilizing compressed air to apply thrust and create motion. This technology is sought after and is seen as a promising alternative to traditional combustion engines, or newly integrated electric vehicles. One of their main advantages is the lack of direct emissions during operation. Pneumatic vehicles are propelled by air that is compressed and stored in tanks, where this compressed air is then discharged in controlled amounts into the surrounding environment to drive the vehicle. This concept not only provides an environmentally friendly solution to other propulsion counterparts but is also a simpler mechanism to control in comparison to ICE or electric vehicles. This concept of propulsion can be applied to various forms of transportation that include land, aerial, and maritime vehicles.

The environmental impact of pneumatic propulsion (or the lack of it) is one of its unique selling points. Since pneumatically propelled vehicles rely solely on air, which is both environmentally inert and sustainable, their contribution comes down to a net zero impact, all while attenuating air quality issues. A prominent challenge in the context of pneumatic propulsion, however, is energy storage and efficiency. Current pneumatic transportation means may implement propulsion mechanisms that come short in comparison to chemical fuel alternatives (that bear higher energy densities), which impose limitations on both the range and power density adaptability of these vehicles.

Further advances in pneumatic-based propulsion may fill gaps and address issues related to both vehicle design, in terms of weight and aerodynamics, as well as pneumatic propulsion subsystems. For example, optimizing air compression techniques and utilizing lightweight materials for compressed air storage tanks, will lead to an increase in the energy density of the propulsion system. Furthermore, integrating pneumatically propelled vehicles can lead to a much bigger role in transportation in achieving net zero-emission goals that are set by fossil fuel-leading countries, as well as countries that have set a countdown for halting the manufacturing of ICE vehicles. Moreover, advances can be made in terms of the acoustics in pneumatic propulsion systems to make them quieter and more appealing to customers and decision-makers alike.

18.4.1 Land Pneumatically Propelled Vehicles

Land vehicles that use pneumatic propulsion typically rely on compressed air that is integrated with a pneumatic powertrain, as well as pneumatic engine configurations. Moreover, heat management can be also utilized through techniques such as phase change material heat exchangers for extracting cooling effects from the expanding air for a maximum utilization of the propulsion system and its side effects (Karaca et al., 2022; Alami et al., 2022). The use of pneumatic propulsion in land vehicles can be done at various scales ranging from small-scale scooters to hybrid transit bus applications.

An example of a small-scale pneumatically driven vehicle is a scooter developed by Alami et al. (2020). The developed scooter has a compressed air configuration of up to five cylinders and pressures up to 6 bar. The compressed air is discharged into an air turbine that eventually drives an electric generator/motor module. Tests on the effects of operating pressure and compressed air volume yielded that increasing the pressure results in an increase of voltage and subsequently power output. Whereas an increase in volume only resulted in an increase in energy output (similar power output but for longer periods of time). Compared to a battery-based configuration for propulsion, the proposed compressed air system has a lower charging time (3.5 min vs. 8 h), a higher lifetime, and a lower environmental impact. In terms of operating time and distance, an operating pressure of 6.3 bar can support up to 180 m with a discharge time of 187 s and a speed of 3.47 km/h, whereas an operating pressure of 5 bar can support the same distance with a discharge time of 286 s and a speed of 2.27 km/h. Moreover, according to Daveler et al. (2017), pneumatic technology has the upper hand compared to battery alternatives due to its lower weight, lower operating cost, as well as increased user safety. A powered mobility device integrating rotary piston air motors was developed as a competitive alternative to a battery-based counterpart. The developed prototype was able to achieve 800 m using a 1.44 L air tank. On the other hand, to achieve a widespread and normalized adoption of such vehicles, a proper infrastructure to refuel and resupply these vehicles is as important as optimizing the propulsion mechanism in question. For example, Hung et al. (2016) developed an air supply station for pneumatically propelled scooters, which consists of a high-power air compressor, low-pressure cylinder, and a boost cylinder. The test scooter integrated a 1 kW four-blade air vane motor that was mounted near its rear wheel, along with mechanical couplings between the air vane motor shaft and the rear wheels. A maximum pressure of 10 kg/cm^2 was obtained from the developed supply station, and a maximum scooter velocity of ~34 km/h was achieved. The scooter and its supply station can be seen in Fig. 18.1.

On a larger scale, integrating pneumatic motors into cars presents a compelling prospect for advancing automotive technology toward sustainability and efficiency. The lightweight design facilitates seamless integration into various car models, potentially enhancing overall fuel efficiency and minimizing environmental impact. Pneumatic motors also exhibit the advantage of instantaneous torque, promoting improved acceleration and responsiveness in driving. Alami et al. (2021) carried out a technical feasibility of a pneumatically driven vehicle. The motor consisted of three 5-bar cylinders that are designed to discharge at tandem. Figure 18.2 shows the discharge estimation of the various cylinder combinations and pressures (Alami et al., 2021). The anticipated discharge times range up to 2 h (for the discharge of all three cylinders at 5 bar), contrasting with just over an hour when a single cylinder is discharged with a 5-bar initial discharge. The primary challenge in this mode of operation lies in the gradual decline in the quality of the discharged air. To address this, the concept of operating the system sequentially without fully depleting the

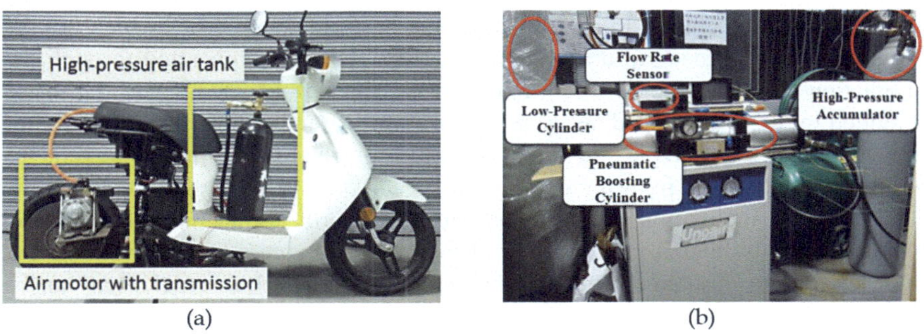

Fig. 18.1 (**a**) Pneumatically propelled scooter. (**b**) Supply station. (Hung et al., 2016—with Permission (Permission No: 5687430368370))

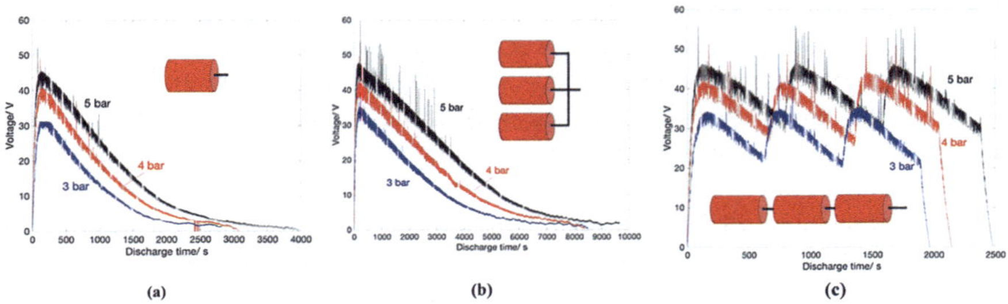

Fig. 18.2 Discharge estimation for (**a**) single cylinder, (**b**) all cylinders, and (**c**) sequential discharge. (Alami et al., 2021—with permission (Permission No: 5687430628111))

cylinders is introduced. Despite a decrease in discharge time (approximately 40 min), the energy quality remains high, comparable to the starting voltage for a single cylinder but sustained over the entire discharge period. Moreover, velocities in the range of 11 and 18 km/h were recorded for drive ratios of 1:1 and 1:0.75. Okonkwo et al. (2023) investigated the performance of a pneumatically driven vehicle to assess its potential applications. The propulsion system displayed a force of 1865 N, a torque of 93 Nm, and an efficiency of 78%. Pneumatic propulsion can also be employed as part of a hybrid engine. For instance, the addition of a pneumatic motor increased the torque by 20 N m and the power by 3 kW (Fang et al., 2021).

18.4.2 Aerial Pneumatically Propelled Vehicles

Aerial pneumatic vehicles include airplanes/drones that are lightweight and have the ability to provide storage space for compressed air. There are various components for a pneumatically driven aerial vehicle, which include the fuselage, a propeller, a motor, a valve for compressed air discharge, and pressurized tanks. Moreover, there are various physical parameters to be considered in the context of aerial pneumatic vehicles to achieve desirable performance metrics which are the thrust force, lift force, and drag force. Lift force, which is generated through the interaction and interface between a solid body and a fluid, along with the thrust provided by the compressed air system, allows liftoff and maintains the motion of aerial vehicles. Drag on the other hand is a force opposing the motion of air vehicles while airborne, which in turn reduces their speed. Drag is highly dependent on the shape and the inclination of an air vehicle. It is worth noting that weight force, which is directly related to the weight of the body and components of the vehicle, imposes various limitations on the amount of compressed air (fuel) that can be stored on board. A developed pneumatically driven drone is shown in Fig. 18.3, utilizing a Polyethylene terephthalate (PET)-based plastic body (2.28 L), which acts as the compressed air storage tank as well, with a mass of 0.230 kg and a wingspan

Fig. 18.3 Prototype of pneumatic drone midflight. (Author's original work)

of 1 m, a polylactic acid (PLA) 3D printed 2-blade propeller with a diameter of 25 cm, and a pneumatic engine. An air velocity of 18.125 m/s and a thrust force of 0.5706 N were generated.

Relatively larger vehicles and equipment have also been included in the discussion of pneumatically propelled aerial vehicles. For instance, Rehman et al. (2019), developed a pneumatic catapult for launching aerial drones. This catapult is able to launch 50–100 kg drones with speeds up to 50 m/s. The system comprises of a pneumatic power pack, rail, pullets, and an aircraft shuttle. Launch pressures for pneumatic drones were correlated with the mass of drones, for example, a launch pressure of 6 bar was required to launch a 25 kg drone, whereas a pressure of 15 bar was required to launch a 60 kg drone. Along with pneumatic propulsion, pneumatic systems in general can be used to accommodate different tasks in terms of air vehicles. For example, pneumatic-mechanical systems have been developed utilizing the pneumatic part as an actuation source. Iversen et al. (2021) developed a drone that utilizes pneumatics to deploy and retrieve sensors to power lines. The main components of this drone include a pressure tank, control valves, and the sensor unit to be deployed. It was highlighted that the ease and flexibility of compressed gas (air) allow for cheap and versatile actuation methods that facilitate automation in unmanned aerial vehicles.

18.4.3 Maritime Pneumatically Propelled Vehicles

Maritime transport constitutes a vital component of the global transportation network and plays a large role in driving significant commercial and economic activities worldwide (Fratila et al., 2021). With over 90% of global trade by quantity and more than 70% by value being facilitated through maritime transportation, it stands as a crucial and efficient alternative compared to air and land transport (Scarbrough

Ricardo et al., 2022). However, conventional fossil fuel propulsion systems utilized in maritime transport pose substantial challenges, particularly in terms of their detrimental environmental impact. These systems contribute significantly to marine pollution, releasing pollutants such as sulfur dioxide, nitrogen oxides, and particulate matter into the air and water, leading to degraded air quality along coastlines and adverse effects on marine ecosystems (Andersson et al., 2016). Furthermore, the carbon dioxide emissions exacerbate ocean acidification, contribute to climate change, and result in rising sea levels and unpredictable weather patterns, presenting considerable challenges to both maritime operations and coastal communities, where a 5% increase in emissions from the global shipping industry has been reported in 2021 (UNCTAD, 2022). The reliance on finite fossil fuel reserves also exposes the maritime industry to supply chain vulnerabilities and price fluctuations. To address these pressing issues, ongoing research emphasizes a transition toward cleaner alternatives, emphasizing advanced propulsion technologies and sustainable energy sources.

Pneumatic propulsion holds significant promise for maritime transport, offering a range of advantages tailored to the unique requirements of the maritime industry. Compressed air energy storage can be harnessed to propel ships, providing an environmentally friendly alternative to traditional marine propulsion systems. The zero-emission characteristic of compressed air aligns with the growing emphasis on sustainable shipping practices, contributing to reduced environmental impact and compliance with stringent emission regulations. The regenerative braking capabilities of pneumatic systems become particularly valuable in maritime applications, allowing vessels to recover and store energy during maneuvers and braking, thereby enhancing overall operational efficiency. The inherent safety features of compressed air, including the absence of flammable fuels, can enhance safety standards in maritime transport. While challenges such as energy density and infrastructure development need consideration, the potential for a cleaner, quieter, and more sustainable maritime transportation solution makes pneumatic propulsion a compelling avenue for exploration and innovation in the shipping industry. Pneumatic propulsion in maritime transportation involves integrating compressed air energy storage systems into vessels. High-pressure tanks securely store compressed air on board, with compressors converting mechanical energy into compressed air for subsequent use. Pneumatic motors or turbines then harness this stored energy to propel the ship, with a control system managing the release of compressed air for optimized efficiency. This technology can be part of hybrid propulsion systems, offering flexibility and extending the vessel's range, while the absence of emissions aligns with global efforts to reduce the environmental impact of maritime activities.

Small boats and ferries can benefit greatly from the utilization of pneumatic propulsion systems especially due to the low power that is required to drive them. For instance, a small ferry utilizing compressed air demonstrated satisfactory and cost-effective performance, delivering a work output of 1.014 kW and a torque of 11 N m at a pressure of 9 bar (Xu et al., 2021a). Moreover, the effectiveness of a pneumatic compressed air system, considering different vane angles and casing sizes was investigated (Ramasubramanian et al., 2021). The experimental findings indicated an average operational duration of 7 h at a pressure of 3 atm. Pneumatic motors can also play a role in hybrid propulsion systems. For instance, a supplementary pneumatic motor exhibited an energy efficiency of 62% and contributed a power output of 0.4 kW (Xu et al., 2021b). Another study was designed to uncover potential energy savings in a compressed air system (Dere & Deniz, 2021). Through pressure optimization, there is a notable 47.3% reduction in daily power requirements, a 58.2% decrease in compressed air unit costs, and a consequential annual reduction of 18.4 and 57.4 tons in fuel consumption and CO_2 emissions, respectively.

The application of pneumatic systems, particularly utilizing pneumatic motors, finds diverse and valuable use in maritime settings. On ships, pneumatic motors are aptly employed to power various appliances such as deck machinery including winches and cranes, contributing to efficient cargo handling and vessel operations. Additionally, pneumatic motors can drive ventilation systems, ensuring optimal air circulation throughout different compartments of the ship. In emergency scenarios, these motors prove crucial for powering life raft propulsion systems, offering a reliable backup in case of engine failure. Furthermore, the use of pneumatic systems in life rafts extends beyond propulsion, encompassing rapid inflation mechanisms to facilitate the swift deployment of life-saving equipment. The safety attributes of pneumatic systems, with their reduced risk of fire or explosion, make them particularly suitable for maritime applications. The compact and lightweight design of pneumatic motors is advantageous in environments where space and weight considerations are paramount. However, the successful application of these systems necessitates careful consideration of energy storage, efficiency optimization, and regular maintenance to ensure reliable and effective performance in challenging maritime conditions. Burciu et al. (2017) developed and tested an innovative pneumatic life raft. The raft consisted of various inflated compartments such as the pneumatic floor, canopy, and two ballast. The two-layer pneumatic floor includes a gravitational drainage system, while the canopy includes a lock for protection and deformation resistance. Computational fluid dynamics (CFD) have also been conducted to study the raft towing performance and compare it to the experimental results. Figure 18.4 shows the CFD and experimental life raft towing for velocities of 0.7 and 1.5 m/s (Burciu et al., 2017). The results show this novel pneumatic motor-based life raft is a great alternative to conventional life rafts. Another study investigated the aerodynamic performance of a pneumatic-based life raft (Jachowski et al., 2021). The study concluded the dependence of the aerodynamic drag on the angle of the wind raft and confirmed the adequate performance of the raft even with slight deformation. Pneumatic motors have also been applied to reduce vibrational stress on ship machines. For instance, the utilization of a pneumatic-based tuner reduced the torsional vibration of a piston machine from 15 to 5 N m (Grega et al., 2023). Another study utilized a pneumatic tuner to continuously tune a ship's propulsion system (Homišin et al., 2021). This innovative solution can replace the conventional employment of shaft coupling and other mechanical components, in result saving on operational cost and storage space.

A maritime pneumatic propulsion system has been developed by the authors of this work as seen in Fig. 18.5, consisting of liquefied petroleum gas (LPG) tanks, a

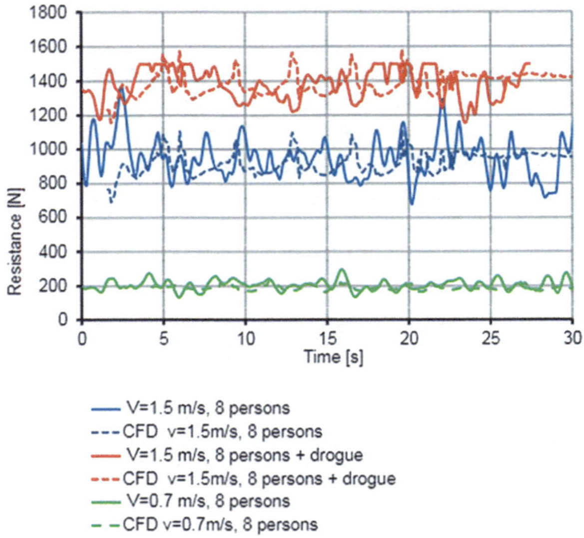

Fig. 18.4 Experimental and CFD model results for the life raft towing. (Burciu et al., 2017—open access)

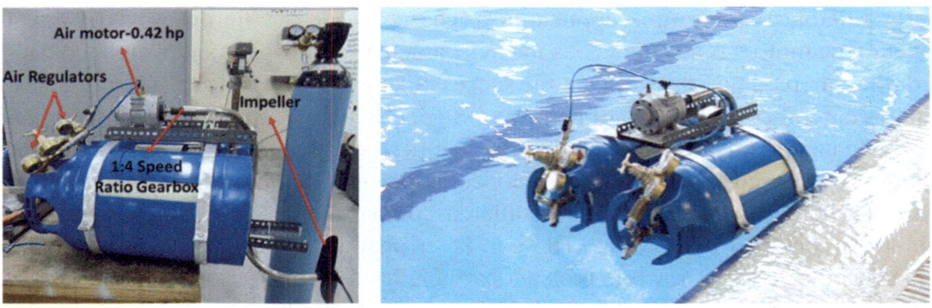

Fig. 18.5 Maritime pneumatic propulsion system. (Author's original work)

Table 18.1 Experimental setup testing results

Metric	Value
Travel distance	120 m
Operation time	200 s
Push force	185 N
System weight	27 kg

Author's original work

gearbox, an impeller, an air motor, and regulators. As seen in Table 18.1, the preliminary findings of this study are promising and show that CAES is feasible for small-scale marine propulsion. These conclusions include evaluations of the system weight as a whole, generated thrust, travel distance, and operating time. However, additional investigation is required to expand such an approach for wider use. It is anticipated that comparative assessments of diesel, electric, and CAES propulsion systems will be the main focus of future research.

18.5 Conclusion

In conclusion, the technical potential of pneumatic propulsion as a viable alternative to conventional propulsion methods was highlighted. The economic, technical, and environmental impact of conventional propulsion was discussed to offer a baseline for comparison. The utilization of compressed air in vehicles for pneumatic propulsion can be used across a wide spectrum of vehicles at various scales, such as pneumatic scooters, drones, cars, and maritime ferries. Moreover, there are two primary methods of coupling pneumatic propulsion systems in these vehicles, which are direct transmission from air motors to propellers or wheels via a gearbox, and indirect transmission through the combination of air motors and generators to power electric motors. Both implementation tracks show potential, that has been proven experimentally, in enhancing the efficiency and environmental friendliness of vehicle operation. These vehicles and corresponding pneumatic subsystems allow for the potential deviation from conventional propulsion system. Despite limitations in operating pressure and storage volume, pneumatic propulsion shows significant advantages when applied to vehicles operating on fixed routes. These routes allow for predictable operational and design parameters, including the sizing of compressed air tanks, operating pressures, and the efficient coupling with propellers and wheels.

References

Akdeniz, H. Y., Balli, O., & Caliskan, H. (2022). Energy, exergy, economic, environmental, energy based economic, exergoeconomic and enviroeconomic (7E) analyses of a jet fueled turbofan type of aircraft engine. *Fuel, 322*, 124165. https://doi.org/10.1016/J.FUEL.2022.124165

Alami, A. H., Alblooki, N., Tawalbeh, M., Almerashi, Z., Alkharousi, A., & Salameh, T. (2020). Compressed air for light vehicle propulsion. In *2020 Advances in Science and Engineering Technology International Conferences (ASET)* (pp. 1–6). IEEE. https://doi.org/10.1109/ASET48392.2020.9118278

Alami, A. H., Hawili, A. A., Fadel, M. I., Zwayyed, F., Barbarji, T., & Ghommem, M. (2021). Technical feasibility of a pneumatically driven vehicle. *Science of the Total Environment, 757*, 143937. https://doi.org/10.1016/j.scitotenv.2020.143937

Alami, A. H., et al. (2022). Cooling potential for hot climates by utilizing thermal management of compressed air energy storage systems. *Scientific Reports, 12*(1), 22066. https://doi.org/10.1038/s41598-022-26666-1

Albatayneh, A., Assaf, M. N., Alterman, D., & Jaradat, M. (2020). Comparison of the overall energy efficiency for internal combustion engine vehicles and electric vehicles. *Environmental and Climate Technologies, 24*(1), 669–680. https://doi.org/10.2478/rtuect-2020-0041

Alves, P., Silvestre, M., & Gamboa, P. (2020). Aircraft propellers—Is there a future? *Energies, 13*(16), 4157. https://doi.org/10.3390/EN13164157

Ammar, N. R., Almas, M. A., & Nahas, Q. H. (2023). Overview of the main benefits and challenges of different ship propulsion systems. *International Journal of Multidisciplinary and Current Research, 11*, 3. https://doi.org/10.14741/ijmcr/v.11.3.3

Andersson, K., Baldi, F., Brynolf, S., Lindgren, J. F., Granhag, L., & Svensson, E. (2016). *Shipping and the environment* (pp. 3–27). Springer. https://doi.org/10.1007/978-3-662-49045-7_1

Bayraktar, M., & Pamik, M. (2023). Nuclear power utilization as a future alternative energy on icebreakers. *Nuclear Engineering and Technology, 55*(2), 580–586. https://doi.org/10.1016/J.NET.2022.10.033

Bilaţiu, C. A., Cosman, S. I., Marţiş, R. A., Marţiş, C. S., & Morariu, S. (2019). Identification and evaluation of electric and hybrid vehicles propulsion systems. In *2019 Electric Vehicles International Conference, EV 2019*. https://doi.org/10.1109/EV.2019.8892965

Blasco, J., Durán-Grados, V., Hampel, M., & Moreno-Gutiérrez, J. (2014). Towards an integrated environmental risk assessment of emissions from ships' propulsion systems. *Environment International, 66*, 44–47. https://doi.org/10.1016/J.ENVINT.2014.01.014

Burciu, Z., Abramowicz-Gerigk, T., Jachowski, J., Kornacka, E., & Wawrzusiszyn, M. (2017). Experimental and numerical investigation of towing resistance of the innovative pneumatic life raft. *Polish Maritime Research, 24*(2), 40–47. https://doi.org/10.1515/pomr-2017-0048

Cengiz, A., & Utlu, Z. (2019). Analysis of different engine types in aircrafts with exegetic approach. *International Journal of Engineering and Natural Sciences, 2*(2), 1–6. Retrieved December 9, 2023, from https://dergipark.org.tr/en/pub/ijens/issue/62959/1005985

Daveler, B., Wang, H., Gebrosky, B., Grindle, G. G., Schneider, U., & Cooper, R. A. (2017). Integration of pneumatic technology in powered mobility devices. *Topics in Spinal Cord Injury Rehabilitation, 23*(2), 120–130. https://doi.org/10.1310/sci2302-120

Dere, C., & Deniz, C. (2021). Energy efficiency based operation of compressed air system on ships to reduce fuel consumption and CO_2 emission. *International Journal of Maritime Engineering, 161*(A2), 1088. https://doi.org/10.5750/ijme.v161iA2.1088

Dincer, I., & Bicer, Y. (2020). Integration of conventional energy systems for multigeneration. In *Integrated energy systems for multigeneration* (pp. 143–221). https://doi.org/10.1016/B978-0-12-809943-8.00004-2

Drosińska-Komor, M., Głuch, J., Breńkacz, Ł., & Ziółkowski, P. (2022). On the use of selected 4th generation nuclear reactors in marine power plants. *Polish Maritime Research, 29*(113), 76–84. https://doi.org/10.2478/pomr-2022-0008

Fang, Y., et al. (2021). Study of a hybrid pneumatic-combustion engine under steady-state and transient conditions for transport application. *International Journal of Engine Research, 22*(2), 528–539. https://doi.org/10.1177/1468087419860322

Fratila, A., Gavril, I. A., Nita, S. C., & Hrebenciuc, A. (2021). The importance of maritime transport for economic growth in the European Union: A panel data analysis. *Sustainability, 13*(14), 7961. https://doi.org/10.3390/SU13147961

Grega, R., Krajnak, J., Žuľová, L., Kačír, M., Kaššay, P., & Urbanský, M. (2023). Innovative solution of torsional vibration reduction by application of pneumatic tuner in shipping piston devices. *Journal of Marine Science and Engineering, 11*(2), 261. https://doi.org/10.3390/jmse11020261

Homišin, J., et al. (2021). Continuous tuning of ship propulsion system by means of pneumatic tuner of torsional oscillation. *International Journal of Maritime Engineering, 158*(A3), 992. https://doi.org/10.5750/ijme.v158iA3.992

Hung, Y.-H., Chen, J.-H., Wu, C.-H., & Chen, S.-Y. (2016). System design and mechatronics of an air supply station for air-powered scooters. *Computers & Electrical Engineering, 54*, 185–194. https://doi.org/10.1016/j.compeleceng.2015.12.016

Inal, O. B., Charpentier, J. F., & Deniz, C. (2022). Hybrid power and propulsion systems for ships: Current status and future challenges. *Renewable and Sustainable Energy Reviews, 156*, 111965. https://doi.org/10.1016/J.RSER.2021.111965

Iversen, N., Kramberger, A., Schofield, O. B., & Ebeid, E. (2021). Pneumatic-mechanical systems in UAVs: Autonomous power line sensor unit deployment. In *2021 IEEE International*

Conference on Robotics and Automation (ICRA) (pp. 548–554). IEEE. https://doi.org/10.1109/ICRA48506.2021.9560946

Jachowski, J., Książkiewicz, E., & Szwoch, I. (2021). Determination of the aerodynamic drag of pneumatic life rafts as a factor for increasing the reliability of rescue operations. *Polish Maritime Research, 28*(3), 128–136. https://doi.org/10.2478/pomr-2021-0040

Jurić, V., & Županović, D. (2012). View of ecological impacts of diesel engine emissions. *Promet—Traffic&Transportation, 24*(2), 151–160. Retrieved December 3, 2023, from https://traffic.fpz.hr/index.php/PROMTT/article/view/287/193

Kalghatgi, G. (2018). Is it really the end of internal combustion engines and petroleum in transport? *Applied Energy, 225*, 965–974. https://doi.org/10.1016/J.APENERGY.2018.05.076

Karaca, A. E., Dincer, I., & Nitefor, M. (2022). Analysis of a newly developed hybrid pneumatic powertrain configuration for transit bus applications. *Energy, 248*, 123557. https://doi.org/10.1016/j.energy.2022.123557

Kim, S.-H. (2017). Control of direct current motors. In *Electric motor control* (pp. 39–93). Elsevier. https://doi.org/10.1016/B978-0-12-812138-2.00002-7

Kumar, P. G., Lekhana, P., Tejaswi, M., & Chandrakala, S. (2021). Effects of vehicular emissions on the urban environment—a state of the art. *Materials Today Proceedings, 45*, 6314–6320. https://doi.org/10.1016/J.MATPR.2020.10.739

Leach, F., Kalghatgi, G., Stone, R., & Miles, P. (2020). The scope for improving the efficiency and environmental impact of internal combustion engines. *Transportation Engineering, 1*, 100005. https://doi.org/10.1016/J.TRENG.2020.100005

Lešnik, L., Kegl, B., Torres-Jiménez, E., & Cruz-Peragón, F. (2018). DSC used to follow the evolution of W/O emulsions versus time on ground and in space in the ISS (International Space Station). *Oil & Gas Science and Technology—Revue d'IFP Energies Nouvelles*. https://doi.org/10.2516/ogst/2020051

Li, T., Liu, Z. C., Zhang, H. C., & Jiang, Q. H. (2013). Environmental emissions and energy consumptions assessment of a diesel engine from the life cycle perspective. *Journal of Cleaner Production, 53*, 7–12. https://doi.org/10.1016/J.JCLEPRO.2013.04.034

Liu, B., Rivière, P., Coquelet, C., Gicquel, R., & David, F. (2012). Investigation of a two stage Rankine cycle for electric power plants. *Applied Energy, 100*, 285–294. https://doi.org/10.1016/J.APENERGY.2012.05.044

Machol, B., & Rizk, S. (2013). Economic value of U.S. fossil fuel electricity health impacts. *Environment International, 52*, 75–80. https://doi.org/10.1016/J.ENVINT.2012.03.003

Masuzawa, T., Osa, M., & Mapley, M. (2018). Motor design and impeller suspension. In *Mechanical circulatory and respiratory support* (pp. 335–377). Academic. https://doi.org/10.1016/B978-0-12-810491-0.00011-4

Mustafi, N. N. (2022). An overview of hybrid electric vehicle technology. In *Energy, environment, and sustainability* (pp. 73–102). Springer. https://doi.org/10.1007/978-981-16-8717-4_5/COVER

Nicolosi, F., Corcione, S., Trifari, V., & De Marco, A. (2021). Design and optimization of a large turboprop aircraft. *Aerospace, 8*(5), 132. https://doi.org/10.3390/AEROSPACE8050132

Okonkwo, B. U., Osuagwu, M. C., Chiabuotu, C. C., & Aladum, K. C. (2023). Design analysis of a pneumatic vehicle. *Journal of Basic and Applied Research International, 29*(2), 1–15. https://doi.org/10.56557/jobari/2023/v29i28250

Qahtan, A. S. A., Xu, H., Abdo, A. B., & Abdulsalam, A. (2022). Asymmetric impacts of disaggregated energy consumption and oil price fluctuations on the MENA net oil-exporting and importing economies. *Environmental Science and Pollution Research, 29*(37), 55830–55844. https://doi.org/10.1007/S11356-022-19558-1/TABLES/10

Ramasubramanian, S., Chandrasekaran, M., Baskar, S., Dowhithamaran, A., & Ravin. (2021). Design and development of pneumatic compressed air vehicle. *Materials Today Proceedings, 37*, 690–693. https://doi.org/10.1016/j.matpr.2020.05.637

Rehman, H., Siddiqui, B., Kumar, C. R., & Iqbal, N. (2019). Design optimization, manufacturing and testing of pneumatic catapult for UAV. In *Sixth International Conference on*

Aerospace Science & Engineering (ICASE), Islamabad, Pakistan. https://doi.org/10.1109/ICASE48783.2019

T. Scarbrough Ricardo et al., "Technological, operational and energy pathways for maritime transport to reduce emissions towards 2050," , 2022.

Shirvan, K., & Kazimi, M. (2016). Superheated water-cooled small modular underwater reactor concept. *Nuclear Engineering and Technology, 48*(6), 1338–1348. https://doi.org/10.1016/J.NET.2016.06.003

Shirvan, K., Ballinger, R., Buongiorno, J., Forsberg, C., Kazimi, M., & Todreas, N. (2016). Technology selection for offshore underwater small modular reactors. *Nuclear Engineering and Technology, 48*(6), 1303–1314. https://doi.org/10.1016/J.NET.2016.06.002

UNCTAD. (2022). *Review of maritime transport*. Retrieved February 25, 2023, from https://unctad.org/rmt2022

Xu, Y., et al. (2021a). Experimental research on universal characteristics and output performance of pneumatic motor for compressed air vehicle. *Journal of Energy Storage, 41*, 102943. https://doi.org/10.1016/j.est.2021.102943

Xu, Y., et al. (2021b). Experimental investigation of pneumatic motor for transport application. *Renewable Energy, 179*, 517–527. https://doi.org/10.1016/j.renene.2021.07.072

Yu, J. (2020). *Marine nuclear power technology* (pp. 1–472). Springer Singapore. https://doi.org/10.1007/978-981-15-2894-1/COVER

Zhang, B., Song, Z., Zhao, F., & Liu, C. (2022). Overview of propulsion systems for unmanned aerial vehicles. *Energies, 15*(2), 455. https://doi.org/10.3390/EN15020455

Chapter 19
Navigating Toward a Greener Future: An Analysis of Emission Reduction in Shipping

Igor Gritsuk, Andrii Golovan, Iryna Honcharuk, and Rima Mickiene

Nomenclature

CH_4	Methane
CO	Carbon monoxide
CO_2	Carbon dioxide
SO_X	Sulfur oxides
NO_X	Nitrogen oxides

19.1 Introduction

Shipping has a long history of contributing significantly to trade and economic growth around the world, and in 2021 it accounted for 80% of global trade traffic. Global maritime traffic is expected to increase by about 3% per year from 2021 to 2030 due to economic growth and the promotion of maritime transport to mitigate climate change. Shipping is also a source of pollution, accounting for 15% of total nitrogen oxides (NO_X), 13% of sulfur oxides (SO_X), and 3% of carbon dioxide (CO_2) emissions worldwide. CO_2 emissions contribute to global warming, and SO_X and NO_X gases rapidly turn into acidic substances that affect the chemical composition of soil and surface water, with consequences for organisms that live in acidified water.

I. Gritsuk (✉)
Kherson State Maritime Academy, Kherson, Ukraine

A. Golovan · I. Honcharuk
Odesa National Maritime University, Odesa, Ukraine

R. Mickiene
Lithuanian Maritime Academy, Klaipeda, Lithuania
e-mail: r.mickiene@lajm.lt

M. Z. Sogut, M. Koray (eds.), *Energy Rationality and Management for Decarbonization*, Green Energy and Technology,
https://doi.org/10.1007/978-3-031-68497-5_19

In the context of global efforts to decarbonize and reduce the impact of climate change, there is a critical need to develop and implement effective strategies to reduce these emissions. However, existing research and policy initiatives often do not emphasize a comprehensive approach that includes technological, economic, and governance aspects.

This problem becomes even more urgent given the expected growth of maritime transportation in the near future. Therefore, there is a need to develop integrated solutions that would ensure energy rationalization and decarbonization in the shipping sector.

19.1.1 Literature Review in Introduction

Currently, a number of publications are devoted to the study of air pollution prevention during ship operation (Wang et al., 2022; Silveira et al., 2022; Mandal et al., 2021; Herdzik, 2021). An analysis of these publications shows that the issues of determining the optimal method of preventing air pollution during ship operation have not been given due attention. Given the forecasted increase in global maritime traffic by approximately 3% per year from 2021 to 2030 due to economic growth and the promotion of maritime transport, this forecast creates the need to analyze and apply effective means of preventing air pollution.

19.1.2 Purpose of the Article in Introduction

The purpose of this article is to conduct a comprehensive analysis of the impact of shipping on atmospheric pollution and to identify effective strategies for energy rationalization and decarbonization in this area. The authors of the article aim to:

1. To assess the current state of CO_2, NO_X, and SO_X emissions in maritime transport and their impact on climate change and ecosystems.
2. To consider modern technological innovations that can contribute to reducing emissions, including alternative fuels and exhaust gas cleaning systems.
3. Analyze existing international regulations and policies governing this issue and suggest ways to optimize them for effective management.
4. Identify opportunities for integrated approaches that combine technological, managerial, and political aspects to achieve sustainable development.
5. To propose recommendations for future research and practical measures in the field of energy rationalization and decarbonization in shipping.

The article seeks to contribute to the development of effective ways of energy rationalization and decarbonization in maritime transport.

19.2 Methods

It is impossible to overestimate the role of the ocean as an important criterion for climate stabilization and sustaining human life and well-being, as well as a resource that needs to be protected and maintained. However, various assessments of the state of the world's oceans show that the biological resources of coastal marine areas continue to degrade (UNEP, 2018), and with the expected increase in human population to 9.7 billion by 2050 (UNCTAD, 2019), the negative impact on the marine environment will increase significantly.

In the context of sustainable maritime transport development, in addition to measures to minimize ocean pollution from shore-based sources, special actions are taken to reduce pollution from ships, protect the atmosphere by introducing new limits on sulfur emissions, manage ballast water, control the transportation of hazardous and harmful substances, and preserve coastal and marine areas, including beyond the limits of national jurisdiction (Golovan et al., 2020).

Ships are a significant contributor to air pollution. Along the main shipping routes, thick clouds of exhaust gases stretch for hundreds of kilometers and are clearly visible from airplanes and satellites (UNDRR, 2015). Ship exhaust gases contain:

- Carbon dioxide
- Nitrogen oxides
- Sulfur oxides
- Fine dust and volatile organic compounds

The reasons why ships are significant air pollutants are as follows (Fig. 19.1):

Firstly, this is due to the fact that ship engines are very powerful and consume a huge amount of fuel.

Secondly, this fuel is fuel oil, a cheap fuel with a high sulfur content, which is essentially oil refinery waste.

Thirdly, unlike cars or power plants, marine vessels are not sufficiently equipped with exhaust gas cleaning systems, as this problem has not been given due attention until now.

Nitrogen oxides and sulfur dioxide are concentrated primarily in the immediate vicinity of emission sources and accumulate along shipping routes (Fig. 19.2). This means that their environmental impact is significantly localized.

Fine dust, on the other hand, behaves differently. "Along the entire coastal strip up to three hundred kilometers wide, particulate matter is the main factor in the negative impact of ship exhaust on the environment. They pollute the atmosphere over much larger areas than all other ingredients" (Golovan et al., 2018).

Since January 1, 2020, the International Maritime Organization has banned the use of fuels with a high sulfur content.

According to the new regulations, ships are prohibited from burning any marine fuel with a sulfur content of more than 0.5% without additional exhaust gas cleaning

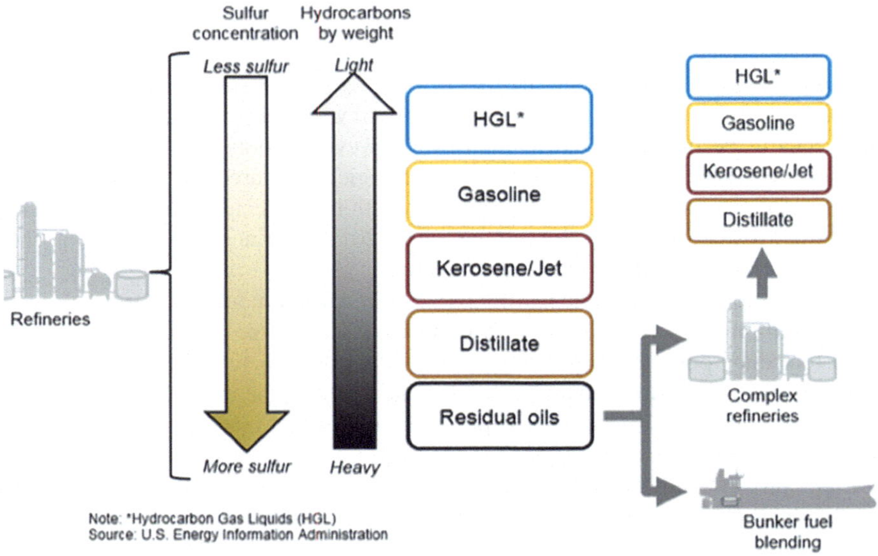

Fig. 19.1 Causes of air pollution from ships

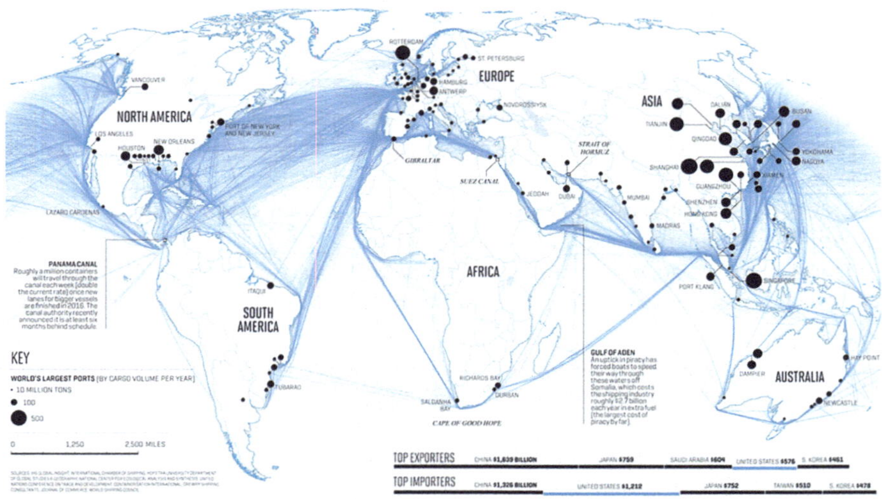

Fig. 19.2 Increased concentrations of nitrogen oxides and sulfur dioxide along shipping routes

(Golovan et al., 2019). International legal framework governing the prevention of air pollution from ships:

- International Convention for the Prevention of Pollution from Ships (MARPOL).

- Amendments to MARPOL Annex VI (resolution MEPC.203(62))—Energy efficiency.
- Amendments to MARPOL Annex VI (resolution MEPC.278(70))—Data Collection System for Fuel Oil Consumption by Ships.
- Guidelines for the development of a ship energy efficiency management plan (SEEMP) for 2022 (resolution MEPC.346(78)).
- Guidelines 2022 on the administrative verification of data on ship fuel oil consumption and operational carbon intensity (resolution MEPC.348(78)).
- Guidelines 2022 on the development and management of the International Maritime Organization (IMO) Fuel Oil Consumption Database (resolution MEPC.349(78)).
- MEPC Circular on the submission of data to the IMO data collection system on fuel consumption by ships from a State not party to MARPOL Annex VI (MEPC.1/Circ.871/Rev.1).
- Sample format for confirmation of compliance in accordance with regulation 5.4.5 of MARPOL Annex VI (MEPC.1/Circ.876).

There are four main ways to comply with the new standards and reduce sulfur oxide emissions (Fig. 19.3):

1. Switch to low-sulfur fuel. This transition will not require large financial investments from shipowners and will take only a few days to change the oil and adjust the engine to run on low-sulfur fuel.
2. Switch to alternative low-sulfur fuels, such as liquefied natural gas (LNG). Liquefied natural gas is a type of fuel that does not contain sulfur. LNG ships reduce nitrogen oxide emissions by 90% and sulfur oxide emissions by almost 100%.

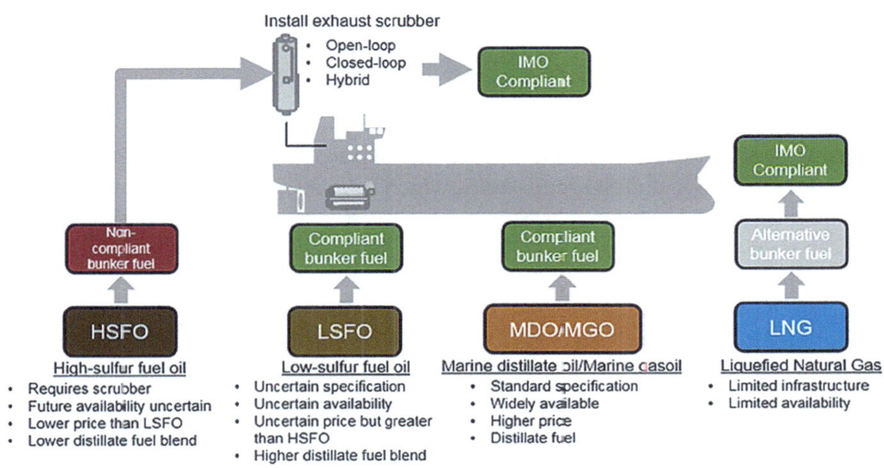

Fig. 19.3 Ways to reduce emissions of sulfur oxides and fine dust

3. Installation of exhaust gas cleaning systems, also known as scrubbers. Scrubbers are designed to remove sulfur oxides from the exhaust gases of a ship's engine, generator, or boiler.
4. Reduced speed. According to the independent research center Delft, if ship speeds are reduced by 30%, emissions will be reduced by two and a half gigatons between 2021 and 2030.

The ways to reduce carbon dioxide emissions are divided into two categories: structural and operational. Structural measures include the use of new structural materials, alternative energy sources (sun, wind), optimization of hull contours, and the introduction of new ship coatings. This category also includes an air lubrication system for the underwater hull, hybrid power plants, improvements to the main engine, auxiliary systems, and heat recovery systems. Operational measures include route planning based on time, fuel consumption, and hydrometeorological conditions, increasing the efficiency of the propulsion system, optimizing the vessel's differential and draft, and reducing speed. The most effective operational measures were route planning taking into account the above parameters, increasing the efficiency of the propulsion system, and reducing speed.

In the context of reducing nitrogen oxide emissions, there are also design and operational methods. Design measures include increasing the fuel injection pressure, optimizing the shape of the combustion chamber to obtain a homogeneous mixture, recycling part of the exhaust gases, using an emulsion of water and fuel, using exhaust gas after treatment, and switching to natural gas engines. Operational measures include limiting the use of main engines, using multifunctional fuel additives, fuel pretreatment, reducing the maximum cycle temperature, optimizing the fuel injection advance angle, and changing the parameters of the inflated air. The most effective operational measures are limiting the use of main engines, using multifunctional fuel additives, and fuel pretreatment.

19.3 Results and Discussion

This article analyzes in detail the impact of shipping on atmospheric pollution, in particular CO_2, NO_X, and SO_X emissions, which have a significant impact on global climate change and marine ecosystems. The effectiveness of various technologies that can reduce emissions, such as alternative fuels and exhaust gas treatment systems, is highlighted, and their potential for significant pollution reduction is identified. International regulations and policies are also reviewed, shortcomings in regulations are identified, and ways to improve them are suggested. The article provides specific strategies for decarbonization and energy rationalization in the shipping sector, including the transition to alternative fuels and the introduction of energy-efficient technologies. It highlights the need for an integrated approach that combines technological, managerial, and political aspects to achieve sustainable development in shipping and provides a number of recommendations for further research and practical measures aimed at energy rationalization and decarbonization.

Overall, the results of the paper confirm that effective energy management and decarbonization strategies can significantly reduce emissions in the shipping sector.

19.4 Conclusion

In the context of the current problem of air pollution from shipping, this article makes an important contribution to the field of energy rationalization and management for decarbonization. The article reviews the key aspects that affect CO_2, NO_X, and SO_X emissions in maritime transport and their impact on climate change and ecosystems.

1. The article emphasizes the need to optimize energy systems on ships as a key factor in reducing emissions.
2. Modern technologies that can reduce emissions, including alternative fuels and exhaust gas treatment systems, are considered.
3. The article analyzes the current international norms and recommends strategies for their effective implementation and continuous updating.
4. It highlights the importance of the transition from fossil fuels to a more sustainable energy mix in the context of decarbonization.
5. Emphasizes the need for an integrated approach that combines technological, managerial, and political aspects to achieve sustainable development.
6. Calls for further research to focus on the development of effective methods for energy rationalization and decarbonization in the shipping sector.

Thus, the problems of preventing air pollution during ship operation should be addressed in a comprehensive manner, taking into account that the most effective, less costly, and not requiring significant modernization of the ship itself are the following methods of preventing air pollution: exhaust gas purification, recycling, fuel pretreatment, and the use of water-fuel emulsions. Organizational measures include reducing the speed and restricting the use of main engines in port areas. The vessel is usually towed to the berth and taken to the outer roadstead by tugs. This article is an important step toward rational energy use and decarbonization in maritime transport.

References

Golovan, A., Rudenko, S., Gritsuk, I., Shakhov, A., Vychuzhanin, V., Mateichyk, V., Kononova, O., Kuric, I., Saga, M., Zenkin, E. Y., & E. (2018). Improving the process of vehicle units diagnosis by applying harmonic analysis to the processing of discrete signals. *SAE Technical Paper Series*. https://doi.org/10.4271/2018-01-1774

Golovan, A., Gritsuk, I., Popeliuk, V., Sherstyuk, O., Honcharuk, I., Symonenko, R., Saravas, V., Volodarets, M., Ahieiev, M., Pohorletskyi, D., & Khudiakov, I. (2019). Features of mathematical modeling in the problems of determining the power of a turbocharged engine according to the characteristics of the turbocharger. *SAE International Journal of Engines, 13*(1), 5–16. https://doi.org/10.4271/03-13-01-0001

Golovan, A., Gritsuk, I., Kurtsev, M., Ischuka, O., & Vrublevskyi, R. (2020). Aspects of remote monitoring of the transport vessel under operating conditions. *ICTE in Transportation and Logistics, 2019*, 295–301. https://doi.org/10.1007/978-3-030-39688-6_37

Herdzik, J. (2021). Decarbonization of marine fuels—The future of shipping. *Energies, 14*(14), 4311. https://doi.org/10.3390/en14144311

Mandal, A., Biswas, J., Farooqui, Z., & Roychowdhury, S. (2021). A detailed perspective of marine emissions and their environmental impact in a representative Indian port. *Atmospheric Pollution Research, 12*(10), 101194. https://doi.org/10.1016/j.apr.2021.101194

Silveira, R. S., Corrêa, S. M., & Neto, N. M. (2022). Possible influence of shipping emissions on metals in size-segregated particulate matter in Guanabara Bay (Rio de Janeiro, Brazil). *Environmental Monitoring and Assessment, 194*(11), 828. https://doi.org/10.1007/s10661-022-10517-7

UNCTAD. (2019). *Handbook of statistics 2019*. Retrieved December 10, 2019, https://unctad.org/webflyer/handbook-statistics-2019

UNDRR. (2015). *Sendai framework for disaster risk Reduction 2015–2030*. UNDRR. Retrieved from https://www.undrr.org/publication/sendai-framework-disaster-risk-reduction-2015-2030

UNEP—UN Environment Programme. (2018). *Emissions gap report 2018*. UNEP—UN Environment Programme. Retrieved from https://www.unep.org/resources/emissions-gap-report-2018

Wang, G., Zhu, Z., Liu, Z., Liu, X., Kong, F., Nie, L., Gao, W., Zhao, N., & Lang, J. (2022). Ozone pollution in the plate and logistics capital of China: Insight into the formation, source apportionment, and regional transport. *Environmental Pollution, 313*, 120144. https://doi.org/10.1016/j.envpol.2022.120144

Chapter 20
Modification of Low-Speed Marine Diesel Engine for Slow Steaming Operation

Dimitar Popov

Nomenclature

bhp	Brake horsepower
CMCR	Normal service output
CO_2	Carbon dioxide
HFO	Heavy fuel oil
IMO	International Maritime Organization
MCR	Maximum continuous rate
MEPC	Marine Environment Protection Committee
NOx	Nitrogen oxides
PM	Particular matters
SFOC	Specific fuel oil consumption
SOx	Sulfur oxides
T/C	Turbocharger
TEU	Twenty-foot equivalent unit

20.1 Introduction

The 12-cylinder Sulzer RTA96C engine (Fig. 20.1) with a maximum continuous rate (MCR) of 61,350 kW (83,440 bhp) was introduced in December 1994 and quickly gained popularity, powering many of the largest container liners in the world. For container ships, class post-Panamax 6500 TEU (Fig. 20.2), the average design ship speed used is 22.7 knots. The design ship speed used for several years

D. Popov (✉)
NVNA, Varna, Bulgaria

Faculty of Engineering, Nikola Vaptsarov Naval Academy, Varna, Bulgaria
e-mail: d.popov@nvna.eu

M. Z. Sogut, M. Koray (eds.), *Energy Rationality and Management for Decarbonization*, Green Energy and Technology,
https://doi.org/10.1007/978-3-031-68497-5_20

Fig. 20.1 SULZER RTA96C engine maintenance

Fig. 20.2 Container vessel 6500 TEU

has been relatively high because of the relatively low fuel price (Wartsila, 2018; MAN Diesel & Turbo, 2013). However, because of the considerable increase in fuel prices and MARPOL 73/78 air emission restrictions, the design ship speed was reduced and the applied ship speed in service was reduced. Today, controlling NOx emissions in compliance with Annex VI of the MARPOL73/78 convention is standard for marine diesel engines. Currently, fleet average ship speeds vary as the shipping sector adapts to changing market circumstances. In recent years, speeds were reduced in response to an increasing supply of ships, a slower increase in demand for maritime transport and rising fuel prices. Within most speed ranges, fuel consumption and consequently emissions of carbon dioxide (CO_2) and sulfur oxides (SOx) are reduced in line with energy consumption. Only at very low speeds, the amount of fuel needed to provide a unit of output energy may increase somewhat, although this can be prevented by derating the engine. Many shipping companies have had an experience with slow steaming in recent years. Even at very low engine loads, they have encountered only a few problems and these problems could be solved by small changes to engine design. Hence, it appears that there are very few technical constraints to slow steaming mode (Hristov et al., 2018; Hristov & Ivanov, 2019).

As a rule of thumb, engine power output is a third power function of speed. Hence, when a ship reduces its speed by 10%, its engine power is reduced by 27%. Because it takes longer to sail a given distance at a lower speed, the energy required for a voyage is reduced by 19% (a quadratic function). Within most speed ranges, fuel consumption and consequently emissions of carbon dioxide and sulfur oxides are reduced in line with the energy consumption. Only at very low speeds, the amount of fuel needed to provide a unit of output energy may increase somewhat, although this can be prevented by derating the engine (Kalgora & Christian, 2016; Faber, 2012).

The speed restriction, respectively engine power should be expressed in average speed over ground so that it can be monitored and verified. Mandatory slow steaming is legally feasible either under a global agreement or unilaterally as a condition of entry to a port. 'Speed Reduction—the key to the fast and efficient reduction of greenhouse gas emissions from ships' (MEPC 61/5/10). The proposal contained references to evidence showing that speed reduction can result in considerable reductions in fuel use and CO_2 emissions (Agrawal et al., 2010).

The main aim of globally regulated slow steaming is to reduce CO_2 emissions from shipping. Sulzer RTA-C engines achieve the limits of International Maritime Organization (IMO) regulations through a few minor design changes. The listed engine can operate in compliance with Regulation 13 of Annex VI of MARPOL 73/78, which limits IMO-NOx emissions, by cutting out one of the three sets of turbochargers (T/C). Originally equipped with three sets of ABB TPL85B12 turbochargers, the engine's specific fuel oil consumption (SFOC) at the normal service output (CMCR) 52,150 kW at 92.5 rpm and reference conditions according to ISO 3046-1 is 166.1 g/kWh, using diesel fuel oil with a net calorific value of 42.7 MJ/kg.

20.2 Experimental Methodology

SFOC is a measure of the fuel efficiency of a marine diesel engine, particularly 2-stroke engines. It quantifies the amount of fuel (typically in grams or kilograms) that an engine consumes per unit of power output, usually expressed in units like kilowatts or brake horsepower (bhp). SFOC is essentially a measure of an engine's efficiency in converting fuel into useful work, such as mechanical power or electricity. Lower SFOC values indicate greater fuel efficiency because the engine is using less fuel to generate a given amount of power. SFOC can vary significantly depending on the load on the engine. Engines are typically most efficient at or near their design load, and SFOC tends to increase at lighter loads. Therefore, it is important to specify the load condition when discussing SFOC.

The following sections present the function of SFOC depending on effective engine power, conducted on the 12-cylinder Sulzer RTA96C main engine of a post-Panamax Class container ship. SFOC is calculated as grams per kilowatt hour. Engine output is recorded during sea passages by shaft horsepower meter in kilowatts. The data collection time is divided into two periods.

In the first period, the data was collected during the ship sea-trial test, when the main engine uses all three turbochargers. The selected engine specifications are provided in Table 20.1.

The main engine Sulzer 12RTA96C during the sea-trial test used heavy fuel oil (HFO) meeting ISO 8217 with properties shown in Table 20.2.

In the second period, the data were collected in July 2022 during various sea passages with the engine modified so that one of the three turbocharges was cut out. During voyages, the main engine burned HFO type RMG 380 meeting ISO 8217 specifications.

The fuel specifications are provided in Table 20.3.

Regression analysis is used on the collected data as a statistical method to determine the relationship between a dependent variable (represented by "*x*") and one or

Table 20.1 Selected data for the main propulsion engine

Ship type	Container vessel, 6500 TEU
Engine type	12RTA96C
Built in	2003, DU-SULZER
MCR	61,350 kW
Rated speed	97.7 rpm
Bore × stroke	960 mm × 2500 mm
Turbocharger type	3 sets TPL85-B12

Table 20.2 Properties of the fuel used in the vessel

HFO properties	Method	RMG 380
Density @ 15 °C	ISO 12185	0.9841 kg/L
Viscosity @ 50 °C	ISO 3104	115 mm^2/s (cSt)
Low caloric value	Calc.	40.19 MJ/kg

Table 20.3 HFO specification

HFO properties	Method	RMG 380 VLS
Density @ 15 °C	ISO 12185	0.9538 kg/L
Viscosity @ 50 °C	ISO 3104	119 mm^2/s (cSt)
Sulfur	ISO 8754	0.42 %m/m
Net specific energy	Calc.	41.55 MJ/kg
Gross specific energy	Calc.	44.00 MJ/kg

more independent variables (represented by "y"). In our case, the dependent variable is the main engine's effective power (output at kW), while the independent variable is SFOC (g/kWh). R-squared and the Pearson correlation coefficient (*R*) are commonly used to assess the goodness of fit of the model. A higher *R*-squared value indicates a better fit of the model. The Pearson correlation coefficient measures the strength and direction of the linear relationship between the dependent variable and independent variables. It ranges from −1 to 1, where values closer to −1 or 1 indicate a stronger linear relationship, while values close to 0 indicate a weak or no linear relationship. The sign of the correlation coefficient indicates the direction of the relationship. A positive correlation means that as SFOC increases, the main engine's effective power also tends to increase, and vice versa. A negative correlation means the variables move in opposite directions. To assess the appropriateness of modifying the main engines, we can consider both the *R*-squared value and the correlation coefficient (*R*). If the *R*-squared value is high and the correlation coefficient is close to 1 or −1, it suggests a strong linear relationship between SFOC and the main engine's effective power. This would indicate that modifying the main engines (changing from three T/C to two T/C) significantly impacts the engine's effective power in a consistent and predictable manner. By comparing the graphs of the obtained equations for the operation of the main engine with three T/C and the modified operation with two T/C, we can visually analyze how the effective power of the engine changes for the same values of SFOC. If the graphs show a similar trend and are highly correlated, it provides further evidence supporting the appropriateness of the modification. In summary, the criteria to use for assessing the appropriateness of modifying the main engines include a high *R*-squared value indicating a good fit of the regression model and a correlation coefficient (*R*) close to 1 or −1, suggesting a strong linear relationship. Additionally, visual analysis of the graphs can provide insights into the consistency of the relationship between SFOC and the main engine's effective power.

By analyzing the above results, we are to confirm or reject this method of modifying the main engines as appropriate.

20.3 Determining the Relationship Between SFOC and Effective Engine Power in Different Operating Modes

20.3.1 SFOC and Engine Output During Sea-Trial Test with Three Turbochargers in Operation

Sea-trial test was performed on different engine loads—50%, 75%, 80%, 85%, and 100% of MCR. All data are collected, and the results are shown in Table 20.4.

Figure 20.3 shows the relationship between the SFOC and the main engine output.

Table 20.4 Sea-trial test records

Engine load (% of MCR)	Engine power output (kW)	Engine speed (rpm)	SFOC (g/kWh)
48.9	30,000	86.6	207.25
74.8	45,900	96.8	189.99
78.6	48,200	98.9	188.67
79.1	48,500	99.2	190.42
85.2	52,300	100	190.14
97.0	59,500	105	195.95

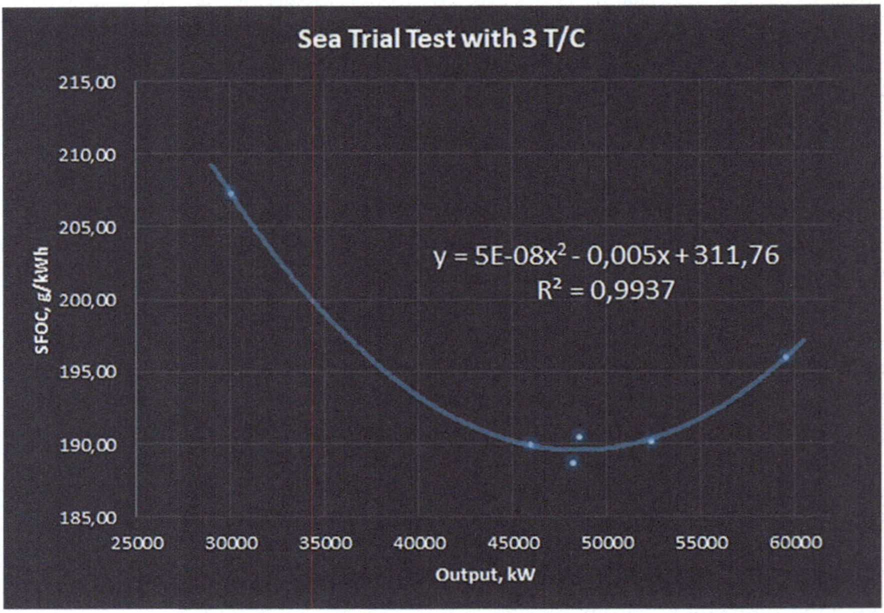

Fig. 20.3 SFOC during sea-trial tests

Using the data from Table 20.4, we plot the points in the coordinate system SFOC/output. Using regression analysis on the collected data, we found the quadratic equation (20.1) that expresses the relationships between the dependent variable "*x*" and the independent variable "*y*". The minimum value of the function *y* (20.1) is obtained at the vertex of the parabola (Fig. 20.3). The *y*-coordinate of the vertex is given by Eq. (20.2), and the corresponding *x*-coordinate is found by substituting that value of *y* into the equation. Using Eq. (20.1), the minimum value of the function "*y*" is calculated as shown in Eq. (20.2):

$$y = 5e - 8x^2 - 0.005x + 311.76 \quad (20.1)$$

$$\mathrm{y}_{\min} = \mathrm{c} - \mathrm{b}^2 / 4\mathrm{a} \quad (20.2)$$

where from Eq. (20.1):

$c = 311.76$

$b = 0.005$

$a = 5e - 8$

By applying the coefficients, a, b, c from Eq. (20.1) to Eq. (20.2), for $y$min is obtained:

$$y_{\min} = 311.76 - (0.005)^2 / 4 * 5 * 10^{-8} \tag{20.3}$$

$$y \min = 186.76 \mathrm{g/kWh} \tag{20.4}$$

By applying $y$min as SFOC 186.76 g/kWh to Eq. (20.1), the new quadratic equation (20.5) and (20.6) are:

$$5e - 8x^2 - 0.005x + 311.76 = 186.76 \tag{20.5}$$

$$5e - 8x^2 - 0.005x + 125 = 0 \tag{20.6}$$

By applying the coefficients, $a = 5e - 8$, $b = -0.005$ from Eq. (20.6) to Eq. (20.7) for $x$min is obtained:

$$x \min = -b/2a \tag{20.7}$$

$$x \min = 0.005 / 2(5e - 8) \tag{20.8}$$

$$x \min = 50{,}000 \mathrm{kW} \tag{20.9}$$

The minimum SFOC 186.76 g/kWh corresponding to engine output 50,000 kW is proportional to 81.5% of MCR.

20.3.2 SFOC and Engine Output During Sea Passage with One of the Three T/C Cutout and Slow Steaming Mode

Main engines SULZER RTA96C type are normally designed for optimum performance in a range from 75% up to 90% of MCR. Therefore, some countermeasures are required to ensure optimum performance and optimum combustion during slow steaming mode of operation below 50% of contracted MCR.

During the sea passage, main engine parameters data were collected and processed under variable engine load. The results are shown at Table 20.5.

Figure 20.4 shows the relation between the SFOC and the engine output during the slow steaming mode and one of the three T/C cutouts.

Analyzing the relationship between the dependent variable x and independent variable y, it was observed that there is a polynomial relationship Eq. (20.10). Using

Table 20.5 Sea passage records

Engine load (% of MCR)	Engine power output (kW)	Engine speed (rpm)	SFOC (g/kWh)
8.1	5000	42.8	227.48
8.5	5200	44.6	228.67
9.4	5740	46.4	228.45
10.8	6600	47	223.89
12.8	7840	51.5	218.81
13.0	7995	53	218.10
29.6	18,187	67	193.82
33.0	20,250	67.8	190.79
33.5	20,550	68.1	192.25
32.8	20,150	70	191.27
36.8	22,550	72	189.67
40.0	24,560	75	190.6

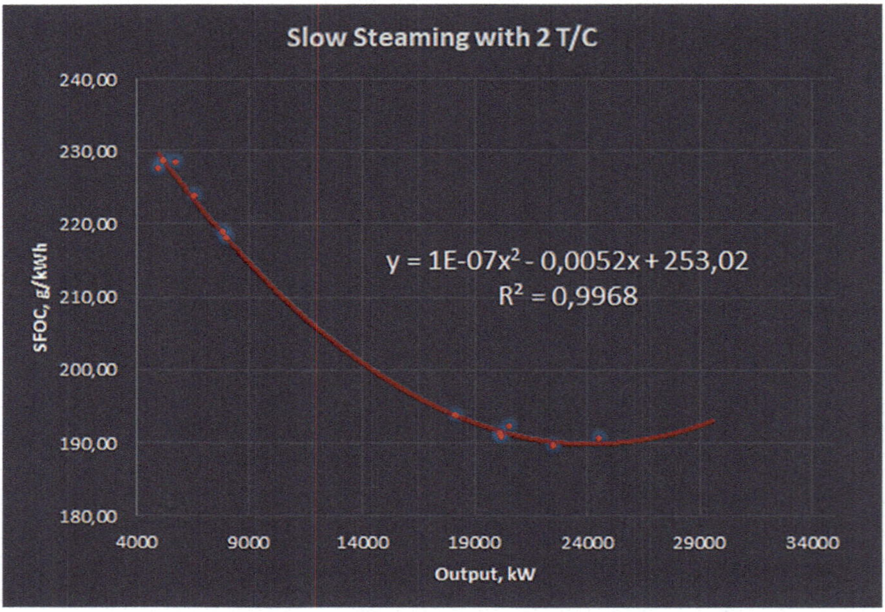

Fig. 20.4 SFOC during sea passage with two turbochargers in operation

Eq. (20.10), the minimum value of the function *ymin* (20.11) is calculated as follows:

$$y = 1e - 07x^2 - 0.0052x + 253.02 \quad (20.10)$$

$$y\min = c - b^2 / 4a \quad (20.11)$$

where from Eq. (20.10):

$c = 253.02$

$b = 0.0052$

$a = 1e - 07$

By applying the coefficients, a, b, c from Eq. (20.10) to Eq. (20.11), for $y$min is obtained (20.12):

$$y\min = 253.02 - 0.00522/4(1e-7) \quad (20.12)$$

$$y\min = 185.42\text{g}/\text{kWh} \quad (20.13)$$

By applying $y$min as SFOC 185.42 g/kWh to Eq. (20.10), the new quadratic equations (20.14) and (20.15) are:

$$1e - 7x^2 - 0.0052x + 253.02 = 185.42 \quad (20.14)$$

$$1e - 7x^2 - 0.0052x + 67.6 = 0 \quad (20.15)$$

By applying the coefficients, a, b, from Eq. (20.15) to Eq. (20.16) for $x$min is obtained (20.18):

$$x\min = -b/2a \quad (20.16)$$

$$x\min = 0.0052/2(1e-7) \quad (20.17)$$

$$x\min = 26{,}000\text{kW} \quad (20.18)$$

The minimum value determined for SFOC ($y$min) is 185.42 g/kWh, which corresponds to an engine output of 26,000 kW and it is proportional to 42.4% of MCR.

20.3.3 Comparing the Results Obtained

In the first case Sect. 20.3.1, R-squared is equal to 0.9937, so Pearson correlation coefficient R is equal to $|0.9968|$. These results indicate a strong correlation between the SFOC and effective power. In the second case Sect. 20.3.2, R-squared is equal to 0.9968, so Pearson correlation coefficient R is equal to $|0.9984|$. These results also indicate a strong correlation between the SFOC and effective power. Therefore, we can conclude that the data collected in both cases are highly dependent, with coefficients close to 1, and we can continue to find the mathematical relationships between x and y.

The obtained Eqs. (20.1) and (20.10) indicate that the relationship between SFOC and the effective power is nonlinear and forms a parabolic curve. The coefficients "a" of x^2 determine the steepness of the curve. The second equation (20.10) has a higher coefficient "a" of x^2 ($1e - 7$) compared to the first equation (20.1) where the coefficient "*a*" of x^2 is $(5e - 8)$. This means that the second equation will result in a steeper curve compared to the first equation.

The coefficient "*b*" of *x* is (−0.0052) in the second equation is slightly larger in magnitude compared to the coefficient "*b*" of *x* is (−0.005) in the first equation. This suggests that the slope of the curve in the second equation will be slightly steeper than the slope in the first equation.

The constant terms "*c*" is (253.02 and 311.76) represent the *y*-intercepts of the curves when "*x*" is zero. The second equation has a higher constant term, indicating that its curve will be shifted upward compared to the first equation.

In summary, the second equation (20.10) results in a steeper and narrower quadratic curve that is shifted upward compared to the first equation (20.1). The differences in the coefficients and constants lead to variations in the shape, position, and orientation of the curves. The results obtained are shown in Fig. 20.5.

20.4 Conclusion

DU-Sulzer RTA-C engines have excellent performance, combustion conditions, and a wide range of engine loads in operation. The specifications of a turbocharger are normally set to obtain the optimum engine performance under normal operation

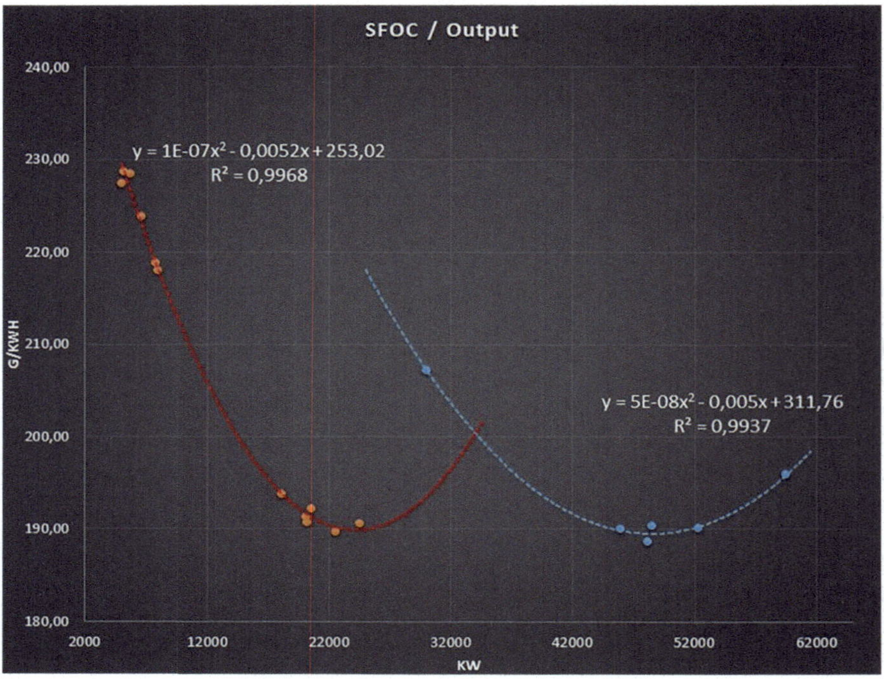

Fig. 20.5 Comparing the results

load between 80% and 90% of MCR. Reduction of the number of turbochargers or modification of the turbochargers improves engine performance for the planned slow steaming operation between 10% and 50% of MCR.

The method and data analysis shown can be applied to other engine types to determine the effectiveness and feasibility of applying the turbine modification (cut-out method) for slow steaming of operation.

As a continuation of the positive effect of the turbocharger cut-out method, Wartsila offers a kit for cutting off, through which an operating mode optimization of the work is achieved in slow steaming mode. The approach is applicable to engines with two or more turbochargers. The device is automated and connected to the engine management system. If necessary to reduce engine load, one turbocharger unit cuts out. Thus, the remaining turbochargers operate at higher efficiency and supply enough amount of scavenging air with appropriate pressure. The efficiency increase is a result of a relative increase of the exhaust gas pressure, compared with the case with a full set of turbochargers on low load mode. The flow velocity of the gases passing through the turbines remaining in operation will be higher. Modes of engine load vary from 10% to 100%.

MAN Diesel & Turbo for its engines, also offers an option with a cut-off turbocharger, having improved technical solutions for individual elements of the engines, during the use of which their technical deterioration is avoided at slow steaming mode. These are fuel nozzles with zero sack volume (slide fuel valves) and an electronically controlled cylinder lubrication system—alpha lubricator (Hristov, 2016).

In the case of an engine with a single turbocharger, operating mode below 50% of MCR can be achieved by a system with variable turbine geometry, which makes the control system of the engine more complicated and sophisticated.

Acknowledgments The collection and analysis of the data presented are due to the author's extensive experience as a chief engineer on board various merchant ships. Daily problems and challenges on board the ship require adequate and quick measures. It is the author's professionalism and responsibility that are the basis of the in-depth analysis of the problem and its presentation to the respected audience.

References

Agrawal, H., Welch, W. A., Henningsen, S., Miller, J. W., & Cocker, D. R. (2010). Emissions from main propulsion engine on container ship at sea. *Journal of Geophysical Research, 115*(D23), D23205.

Faber J. (2012). *Regulated slow steaming in maritime transport an assessment of options, costs and benefits*. Retrieved from www.cedelft.eu

Hristov, D. (2016). Features of the technical operation of the main engines at low load modes. In *Doctoral Scientific Conference, Varna* (pp. 61–75).

Hristov, D., & Ivanov, I. (2019). Study of the impact of technical condition of the turbocharger unit on nitrogen oxides emissions of engine running on natural gas with pre-chamber ignition and lean fuel-air mixture. In *Doctoral Scientific Conference, Varna* (pp. 30–35). Nikola Vaptsarov Naval Academy.

Hristov, D., Ivanov, I., & Bakalov, I. (2018). Turbocharger technical condition influence on the carbon dioxide emissions on natural gas fueled engine. In *Varna: First International Marine Engineering Conference* (pp. 187–191).
Kalgora, B., & Christian, T. M. (2016). The financial and economic crisis, its impacts on the shipping industry, lessons to learn: The container-ships market analysis. *Open Journal of Social Sciences, 04*(01), 38–44.
MAN Diesel & Turbo. (2013). *Propulsion trends in container vessels two-stroke engines*. MAN Diesel & Turbo.
Wartsila Corporation. (2018). *SULZER RTA-C Technology review 2 3*. FIN-00531 Helsinki. Retrieved July 18, 2023 from www.wartsila.com

Chapter 21
Sustainability-Based Assessment of Aircraft Pre-conditioned Air Supplying Systems in Green Transition

Enver Yalcin and T. Yiğit Gencel

21.1 Introduction

As the demand for air transport increases, emissions related to the airport and aircraft also increase. In this context, this situation forces relevant stakeholders to take measures to minimize environmental impacts from airports and aircraft. Expanded regulatory programs and global climate protection initiatives are being developed, which require the implementation of state-of-the-art and clean technologies in the aviation industry, as well as the need to change operational practices to reduce emissions.

Air conditioning is required for all aircraft, whether during flight or on the ground. In this context, the question and the answer of how to solve this inevitable need arise. In today's world where energy efficient and environmentally friendly system solutions come to the fore, it is expected that the system analysis in flight period and parking position will be different from each other.

All aircraft are actually equipped to provide cabin comfort conditions in both hot and cold conditions. The aircraft uses an onboard auxiliary power unit (APU) for these applications. When an aircraft approaches a boarding bridge or takes a parking position, before the main engines shut down, the APU is turned on. In this context, APU takes over to meet the aircraft's power demand and cooling or heating needs. Also, auxiliary bleed air, which is used as pneumatic starters in the main engines of

E. Yalcin (✉)
Mechanical Engineering Department, Engineering Faculty of Balikesir University, Balikesir, Turkey
e-mail: eyalcin@balikesir.edu.tr

T. Y. Gencel
R&D Department, Gesbey Enerji Türbini Kule Üretim San. ve Tic. A.Ş., Balikesir, Turkey
e-mail: talatyigit.gencel@gri.com.es

M. Z. Sogut, M. Koray (eds.), *Energy Rationality and Management for Decarbonization*, Green Energy and Technology,
https://doi.org/10.1007/978-3-031-68497-5_21

the aircraft, is produced by APU. This bleed air is drawn from the compressor section of the engine, depending on the pneumatic pressure. This is because the bleed air is routed to the main engines, especially in the initial commissioning of the main engines of large aircraft. From this moment on, the APU will continue to operate until all service and passenger reception are completed and the main engines are started. This means a working time of 1–2 h depending on the body size of the aircraft.

APU is driven by a turbine engine which is fed from the main kerosene fuel tanks used during the flight period. Traditionally, the APU continues to run to circulate conditioned air into parked aircraft. The APU usage on the parked aircraft is an uneconomical choice having high cost and low efficiency as well as having a devastating effect on the environment because of the very low efficient turbine of 4%. For example, when the APU operates for a narrow-body aircraft, it consumes approximately 125 L/h of jet fuel, while a large-body jumbo jet consumes about 550 L/h of jet fuel to which must be added the cost of maintenance for APU (Adelte, 2019; Sikorski, 2010).

In such a situation, alternative solutions are needed to replace the APU. First of all, it is essential to set up a 400 Hz power supply and a system to prepare and supply the conditioned air externally required by the aircraft. For this purpose, in practice, various air-conditioning units such as aircraft cooling system (ACU), pre-conditioned air (PCA), direct expansion refrigeration system (DX), air handling unit (AHU), centralized air system (CAS), point of use (POU), and diesel air-conditioner (DAC) are used. In this study, only PCA systems for cooling purposes are considered as an alternative to APU (Bullerdick, 2012).

External conditioning of the air to be supplied to the aircraft forms the basis of these applications. In this context, besides the different temperatures at different times of the day, the variability of these constraints regionally is the main problem encountered in practice. In this respect, it is very important that the system has the flexibility to adapt to variable load conditions.

In sizing the systems, ambient conditions, demanded power types, and desired cabin conditions (temperature and relative humidity) have to be considered. The aforementioned temperature and relative humidity indicators directly affect the system capacity as well as the number of people onboard. If fresh air is not circulated to the inside, the cabin temperature will increase to 50–55 °C within a very short time. After 5 min, the concentration of CO_2 increases and then cabin air almost becomes breathless. On the other hand, these parameters determine the differential according to the outside conditions during the day.

For the correct use of the energy, the system efficiency can be increased on the cooling systems via the equipment (electronic expansion valve + inverter + soft starter, etc.) and field applications such as floating condensation and evaporation in the application areas, time-dependent shifting evaporation, floating condensation, and glowing optimization (use of electronic expansion valve) speed control (use of inverter) in compressors, use of frequency inverter in three-phase fan motors or dimmer driver card in single-phase fan motors to change the load-dependent speed of fans, and use of electronically controlled (EC) motors in fans. Recently, the use of phase change material (PCM) has also been included as an alternative solution approach.

Latent energy storage systems are frequently used because of the energy saving and high system efficiency it provides in the heating and cooling sector. Because of the lower volume per unit of stored energy requirement and the narrower temperature range that occurs during phase-shifted heat transfer, latent energy storage systems have attracted the attention of many researchers, and many studies have been conducted. Besides, classification, properties, application areas of phase change materials, and general literature studies on geometry and mathematical modeling are remarkable (Abhat, 1983; Dutil et al., 2011).

In this study, energy consumption values were calculated by reference to the external weather conditions of Istanbul Airport, which is located between the northern hemisphere and among the 40–42 parallels for three cases.

21.2 Method

21.2.1 System Description

In the scenario, when the aircraft is in the parking position and the APU is inoperative, the AHU operates independently of the aircraft and the outside air is pre-conditioned.

Four different cases were handled within the scenario.

Case 1: Instantaneous operation of single-stage vapor compression refrigeration system with 100% fresh air.
Case 2: Application of floating condensation with constant temperature differences with reference to the outside temperatures during the day.
Case 3: Storage of coldness at night by using the phase change materials.
Case 4: Air conditioning via APU.

21.2.2 Boundary Conditions

Analyses were made on the basis of the first law of thermodynamics. To take into consideration in the analysis, the following assumptions were made for the calculations;

- In the summer design condition having an outside air temperature of 35 °C, the air cycle return temperature is set to +20 °C. With the air inlet temperature being −5 °C, the air handling cycle operates at a temperature difference of 25 °C.
- At full load operation, the rate of airflow is 2 kg/s for wide/narrow-body aircraft.

- In AHU circuit, R404A refrigerant was used as working fluid at −15 °C evaporation temperature.
- In air-cooled condensers, the difference between the ambient temperature at which the condenser is located and the coolant liquid condensation temperature at the condenser is taken as 15 K as it is stated in EN 327 “Heat exchangers. Forced convection air-cooled coolant condensers. Test procedures for establishing performance” and Eurovent 7/C/005 test standards (EN 327:2014, 2014).
- As mentioned above, in a floating condensation application, the temperature differences between the refrigerant condensation temperature and outside temperature at the condenser is taken as 15 K.
- The refrigeration system is operated under steady-state conditions.
- The heat transfer rate across from the compressor shell to the environment was neglected.
- Fouling in the condenser and freezing in the evaporator are neglected.
- The pressure drops are neglected.
- In the analysis, kinetic energy and potential energy are neglected.
- Subcooling and superheating of 5 °C were applied in the heat pump (primary circuit).
- Changes in atmospheric air properties were neglected.

21.2.3 Governing Equations

The heat pump system operates according to the single-stage vapor compression refrigeration cycle. Subcooling and superheating values were taken as equal (Fig. 21.1).

The heat transfer rate in the evaporator can be calculated via the following equation;

The heat transfer rates and work of the main core elements of a refrigeration system are defined by the following equations (Çengel, 2011).

$$\dot{Q}_{\mathrm{e}} = m_{\mathrm{ref}}\left(h_1 - h_4\right) \tag{21.1}$$

The rate of heat rejection from the refrigerant to the environment in the condenser can be determined by using the following equation:

$$\dot{Q}_{\mathrm{c}} = m_{\mathrm{ref}}\left(h_2 - h_3\right) \tag{21.2}$$

where:

Q—heat load, W
m_{ref}—mass flow rate of refrigerant, kg/s
h—enthalpy, kJ/kg

The compressor work is defined from the following equation:

Fig. 21.1 The pressure-enthalpy (*P*-*h*) diagram of heat pump system

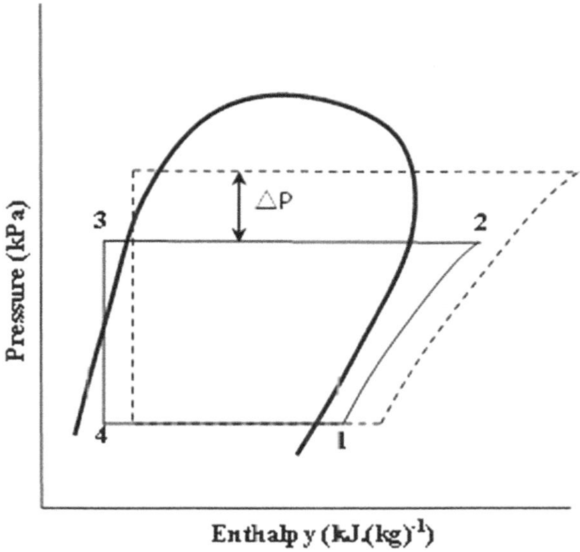

$$\dot{W}_{\text{cmp}} = m_{\text{ref}}\left(h_2 - h_1\right) / r_{\text{v}} \tag{21.3}$$

where:

W_{cmp}—compressor work, W.

Volumetric efficiency of compressor is as follows (Wang et al., 2007)

$$\eta_{\text{v}} = 0.851 - 0.041\frac{P_{\text{c}}}{P_{\text{e}}} \tag{21.4}$$

where P_{c} and P_{e} are condensing and evaporating pressures, respectively.

The performance of refrigeration system is expressed by coefficient of performance (COP), which is determined as follows:

$$\text{COP} = \frac{Q_{\text{e}}}{W_{\text{cmp}}} \tag{21.5}$$

The COP expression does not include fan energy consumptions of evaporator and condenser because their energy requirements are significantly lower than the energy consumption of the compressor.

21.3 Results and Discussion

In this scenario, four different cases are examined and the results are presented below. Case 4 uses only jet fuel; in other words, APU stands for operational status, and is compared only in terms of energy consumption. The time zone in which the outdoor temperature is higher for the reference Istanbul is July and August, between 14 and 16 h. The most unfavorable conditions for AHU, which is used to meet the ventilation requirement, are the mentioned time zones. In the meantime, cooling system performance is also low.

In this context, the graph showing the change in condensation temperatures for the conventional direct expansion DX cooling system, surface condensing cooling system, and phase change material (PCM) added cooling system is shown in Fig. 21.2.

Since a constant condensate pressure is set in the DX system application, condensate temperature and pressure are maintained at unnecessarily high values at both high and low outside temperatures, thus reducing system efficiency. By means of an inverter, the condenser fan speed is applied to the outside air temperature by applying variable condensing pressure to maintain a constant temperature difference of 15 K, thus achieving a more efficient system than the conventional DX system. Between 00.00 and 06.00 h, the PCM added system is on, it is off during the remaining time interval.

Both the compressor volumetric efficiency and the COP are adversely affected by the increase in condensation temperature. In addition, the mass flow rate of the refrigerant in the system increases. A 5 K increase in condensation temperature results in an increase of 15–16% in COP and an increase of approximately 20% in compressor energy consumption. However, compared to conventional applications, COP and power consumption can be improved by using PCM.

Especially when the PCM is applied together with the floating condensate process, the COP value of the system can be improved by up to 30%. COP variation in PCM and non-PCM applications and conventional DX applications is shown in

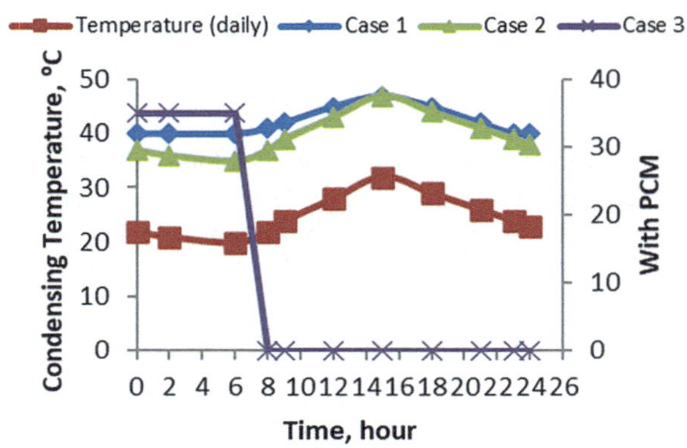

Fig. 21.2 Condensing temperature changings

Fig. 21.3 COP values

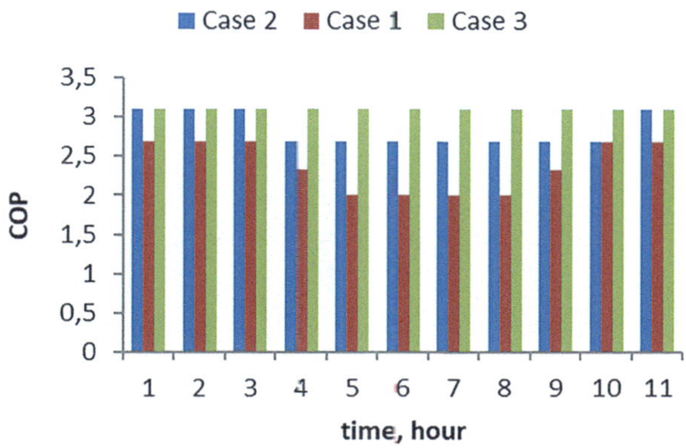

Fig. 21.3. The graph shows the cold energy storage via PCM and the values of the AHU system in the off state are the same as those in the on position. In fact, the system is off for the remaining 18 h except for 6 h.

Regarding energy consumption values, it can be seen that all AHU applications stand out incomparably compared to APU usage conditions. The use of central AHU with PCM instead of APU will save up to 95% of energy (Fig. 21.4).

21.4 Conclusion

When APU systems and PCA systems are compared, no matter which PCA system, they will stand out as systems with higher energy efficiency and lower emission levels.

Taking advantage of passive cooling applications as much as possible will always be the best choice for a parked aircraft.

Centralized PCA practices at all airports will make a significant contribution to the development of policies for future carbon footprint reduction.

Kerosene is needed when the plane is in the air. It should not be used if there are other sources of energy on the ground.

Perhaps the initial investment cost may be high, but with the adoption of a life of 10–15 years, it will pay off soon.

In the context of the energy needs of PCA systems being met by the use of vertical axis wind turbines that can be integrated into the grid system, it will become an environmentally friendly system with zero carbon emissions, such as a self-managing closed circuit system.

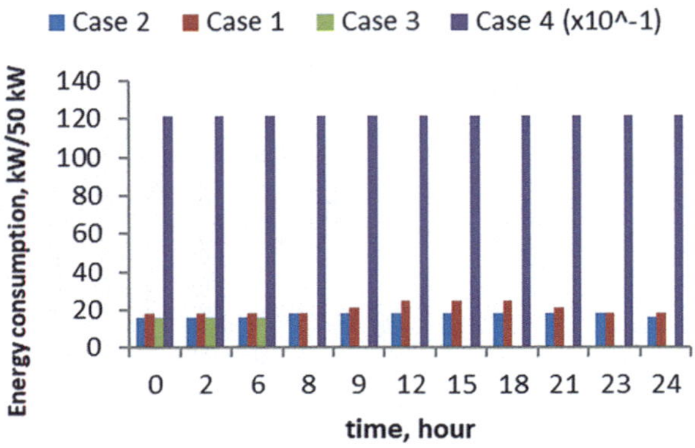

Fig. 21.4 Energy consumptions of AHUs

References

Abhat, A. (1983). Low temperature latent heat thermal energy storage: Heat storage materials. *Solar Energy, 30*, 313–332.

Adelte. (2019). *AAFF_ZEPHIR inverter brochure*. Retrieved from https://www.adelte.com/downloads/ADELTE_preconditioned_air_units_airports_ZEPHIR.pdf

Bullerdick, B. (2012). *The basics of preconditioned air*. Retrieved from https://www.aviationpros.com/gse/gpus-pcas-power-carts-accessories/article/10704019/the-basics-of-preconditioned-air

Çengel, Y. A. (2011). *Heat and mass transfer*. Güven Bilimsel. (in Turkish).

Dutil, Y., Rousse, D. R., Salah, N. B., Lassue, S., & Zalewski, L. (2011). A review on phase-change materials: Mathematical modeling and simulations. *Renewable and Sustainable Energy Reviews, 15*(1), 112–130.

EN 327:2014. (2014). Heat exchangers—Forced convection air cooled refrigerant condensers—Test procedures for establishing performance.

Sikorski, E. (2010). Air-conditioning of parked aircraft by ground-based equipment. In *International Refrigeration and Air Conditioning Conference. Paper 1086*. Retrieved from http://docs.lib.purdue.edu/iracc/1086

Wang, F., Maidment, G., Missenden, J., & Tozer, R. (2007). The novel use of phase change materials in refrigeration plant. Part 2: Dynamic simulation model for the combined system. *Applied Thermal Engineering, 27*, 2902–2910.

Chapter 22
Can Green Energy Transition Be Achieved Despite Investment Disputes?

Ceyda Süral Efeçınar and Başak Başoğlu

22.1 Energy Charter Treaty

Foreign investments in the field of energy are unique with their large sizes and long periods. To put it briefly, the investor shall contribute a significant amount of investment for a lengthy period of time. Additionally, natural resources of energy are within the scope of the sovereignty of states which can determine its own energy policies unilaterally. Therefore, there is a tension between the expectations of the investors and the sovereign power of the state when energy investments are at stake (Baltag, 2012, 3). Consequently, the investors do not wish to take their disputes to national courts to avoid unfavorable procedures, nationalistic or hostile approaches and they prefer a neutral, international, and flexible mechanism, namely, international arbitration (Vorburger & Petti, 2018, 1278). As explained below, the Energy Charter Treaty (ECT) is one of the instruments which provide significant protection for investors and an opportunity to refer to arbitration.

By 1991, the communist regimes in Russia and throughout Europe collapsed leaving former communist states with economic failure, outdated technologies, and infrastructures, thus an emerging target for Western states where the energy consumptions are highly exceeding production. However, investment in these states also involved excessive bureaucracy and political unsteadiness. As a result of these circumstances, the idea of a multilateral international agreement in the field of energy investments appeared. Followed by the signature of the European Energy Charter at the end of 1991 by many European countries, Russia, and the United States, the Energy Charter Treaty came into force on April 16, 1998, which created

C. S. Efeçınar (✉) · B. Başoğlu
Piri Reis University, Istanbul, Turkey
e-mail: csural@pirireis.edu.tr; bbasoglu@pirireis.edu.tr

M. Z. Sogut, M. Koray (eds.), *Energy Rationality and Management for Decarbonization*, Green Energy and Technology,
https://doi.org/10.1007/978-3-031-68497-5_22

a multilateral regime for the protection and promotion of foreign investments in the energy field (Baltag, 2012, 7–9).

The Energy Charter Treaty includes various provisions on different aspects of international commerce, such as competition, transit, transfer of technology, and access to capital. Part III of the ECT titled "Investment Promotion and Protection" infers certain obligations on the states such as non-discrimination, fair and equitable treatment toward foreign investors; ban on direct or indirect expropriation. Article 26 of the ECT concerns the settlement of disputes between an investor and a contracting host state. According to Article 26, disputes between a contracting state and an investor of another contracting state relating to an investment, which concerns an alleged breach of an obligation of the state under Part III shall, if possible, be settled amicably. However, if an amicable settlement cannot be realized within 3 months, the investor may choose one of the following methods for dispute resolution: (a) litigation before the courts or administrative tribunals of the contracting state; or (b) any applicable, previously agreed dispute settlement procedure; or (c) arbitration before one of the following institutions (1) the International Centre for Settlement of Investment Disputes, established pursuant to the Convention on the Settlement of Investment Disputes between States and Nationals of other States if the Contracting State of the Investor and the Contracting State are both parties to the ICSID Convention; or (2) the Arbitration Institute of the Stockholm Chamber of Commerce. Another option for the investor is an ad hoc arbitration by a sole arbitrator or arbitration tribunal established under the Arbitration Rules of the United Nations Commission on International Trade Law.

According to ECT Statistics of December 1, 2023, since 2001, 162 investor-state arbitration cases concerned energy sources. As a result of 44 concluded cases where a breach of the ECT is found and damages are awarded, total damages awarded in cases relating to fossil fuels are approximately 41 billion Euros; and in cases relating to renewables, approximately 23 billion Euros. These two energy sources constitute the main subject matter of the cases as 93% of the cases arise out of them. The most breached obligation of the states is fair and equitable treatment with 25.2%; and expropriation follows with 16.6% (https://www.energychartertreaty.org/cases/statistics/).

22.1.1 Indirect Expropriation

Expropriation is the act of the government taking privately owned property. Although mostly the states secure the right to own private property, they also reserve the power to take it away for a public purpose. According to Article 13/1 of the ECT, "*Investments of Investors of a Contracting Party in the Area of any other Contracting Party shall not be nationalized, expropriated or subjected to a measure or measures having effect equivalent to nationalization or expropriation (hereinafter referred to as "Expropriation") except where such Expropriation is: (a) for a purpose which is in the public interest; (b) not discriminatory; (c) carried out under due process of law; and (d) accompanied by the payment of prompt, adequate and effective compensation.*"

This provision outlines the circumstances under which expropriation is considered to be perfectly legitimate. While the law recognizes the right to expropriate, states often avoid direct expropriation due to the negative impact it may have on their investment climate. Therefore, states seek alternative methods to interfere in order to erode ownership rights, without physically taking or transferring the title (indirect expropriation). Determining whether an expropriation or merely a policy choice has occurred is one of the most challenging issues for tribunals in claims of indirect expropriation. According to the Art. 13/1 of the ECT, if an expropriation has taken place, the states are obligated to provide prompt, adequate, and effective compensation. Otherwise, the investor will bear the economic consequences of the state's policy choice.

22.1.2 Fair and Equitable Treatment

According to Article 10/1 of the ECT, "*Each Contracting Party shall, in accordance with the provisions of this Treaty, encourage and create stable, equitable, favourable and transparent conditions for Investors of other Contracting Parties to make Investments in its Area. Such conditions shall include a commitment to accord at all times to Investments of Investors of other Contracting Parties fair and equitable treatment. Such Investments shall also enjoy the most constant protection and security and no Contracting Party shall in any way impair by unreasonable or discriminatory measures their management, maintenance, use, enjoyment or disposal. In no case shall such Investments be accorded treatment less favourable than that required by international law, including treaty obligations. Each Contracting Party shall observe any obligations it has entered into with an Investor or an Investment of an Investor of any other Contracting Party.*"

This provision has been understood to encompass an obligation to act transparently; refrain from arbitrary or discriminatory measures; and not frustrate the investor's legitimate expectations of the legal framework in a way to adversely affect its investment. The criterion of "legitimate expectations" is accepted to cover all these aspects and therefore forms the dominant and core element of fair and equitable treatment obligation of the states. A state cannot give rise to legitimate expectations of a stable legal framework of investment incentives; and later abruptly hinder or frustrate such expectations (Noilhac, 2020, 22–23; Faccio, 2020, 4; Biggs, 2021, 101).

22.2 Disputes Concerning States' Decisions to Prohibit Certain Energy Sources

In the disputes over the state's decision to ban certain energy sources, investors argue that such decisions constitute indirect expropriation and tend to question the adequacy of compensation.

22.2.1 *Coal Phase-Out: Cases Against the Netherlands*

In 2019, the Netherlands decided to phase-out coal in electricity generation by 2030 and passed the Law on the Prohibition of Using Coal in Electricity Production (Wet verbod op kolen bij elektriciteitsproductie, Staatsblad), following the outcome of the Urgenda case, in which the Hague Court of Appeals ruled that the Dutch government had acted unlawfully and violated its duty of care under Articles 2 and 8 of the European Court of Human Rights (ECHR) by failing to take the necessary measures to reduce greenhouse gas emissions by at least 25% by the end of 2020, as committed to in the Paris Agreement. The phase-out decision, effective until the end of 2029, with compensation schemes under certain conditions, led to a lawsuit by RWE, German investor. In 2021, RWE challenged the adequacy of the compensation before the ICSID (RWE AG and RWE Eemshaven Holding II BV v. Kingdom of the Netherlands, ICSID Case No. ARB/21/4). RWE argued that the Law Prohibiting Coal violates the Netherlands' obligations under Articles 13 and 10 of the ECT. This is because the Netherlands had acted unfairly, inequitably, discriminatory, and unreasonably while indirectly expropriating RWE's investments without providing compensation, complying with the obligations set forth in their investment agreements, or providing adequate protection and security. The case is still pending and its outcome would definitely affect the future phase-out decisions.

In 2007, Uniper, another German investor, had agreed to build a coal-fired power plant in the Netherlands and the plant was opened in 2016 with an expected life of 40 years to cover the costs amounting to 1.6 billion Euros which is impacted by the coal phase-out decision in 2019. Uniper filed an ICSID claim against the Dutch government; however, it withdrew the claim in 2022 following a bail-out package agreed upon with the German government upon the nationalization of Uniper due to the financial difficulties Uniper faced as a result of the energy price crisis and the war in Ukraine. (Uniper SE, Uniper Benelux Holding B.V. and Uniper Benelux N.V. v. Kingdom of the Netherlands, ICSID Case No. ARB/21/22 (Order of Discontinuance 17 March 2023)).

22.2.2 *Nuclear Phase-Out: Vattenfall v. Germany (II)*

In 2011, following the Fukushima accident in Japan, Germany decided to discontinue the use of nuclear power and adopted the 13th Amendment to the Atomic Energy Act (Dreizehntes Gesetz zur Änderung des Atomgesetzes) which established decommissioning dates for nuclear power plants in Germany. In 2012, Vattenfall, a Swedish investor, brought the issue before the German Constitutional Court (Bundesverfassungsgericht) claiming that its rights to the distribution of residual energy and property rights secured by the German Constitution (Grundgesetz) were affected by this decision. In 2016, the Court ruled that the 13th Amendment was partially unconstitutional because it violated Vattenfall's property rights. This was due to the fact that the electricity output allowances allocated to each power plant could not be utilized by the fixed decommissioning dates and the 13th Amendment did not provide compensation schemes for the investments. As a result, the Court ordered the German Parliament to draft new provisions and thus the 18th Amendment to the Atomic Energy Act (Achtzehntes Gesetz zur Änderung des Atomgesetzes) was finally enacted in 2021.

Vattenfall did not challenge the claim of indirect expropriation in ICSID but before the German Constitutional Court. Nevertheless, Vattenfall also initiated an ICSID case against Germany, not to challenge the policy choice, but to claim compensation of 4.7 billion Euros (Vattenfall AB and others v. Germany, ICSID Case No. ARB/12/12, Decision 31 August 2018). The case was later discontinued as the parties agreed on a Settlement Agreement, with Germany agreeing to pay 1.4 billion Euros of compensation.

22.3 Disputes Concerning Roll-Back of Renewable Energy Incentives

The core of disputes concerning the roll-back of renewable energy incentives, dealt with below, is to assess whether the investor's legitimate expectations are infringed by the states' actions to amend the legal framework applicable to renewable energy investments and relied on by the investors.

22.3.1 *Spanish Renewables Saga*

It was at the end of 1997 when Spain first made a distinction between traditional electricity generation plants and those producing electricity from renewable energy sources where the latter was entitled to a premium higher than the wholesale market price (Law 54/1997 of 27 November 1997). This was followed by Royal Decree ('RD') 436/2004 which brought an additional option to premium payment; to obtain a higher tariff for electricity produced. RD 661/2007 established new tariffs and premiums for facilities incorporating renewable energy. These incentives attracted many foreign investors; however, 2008 came with an economic crisis. Therefore, Spain had to amend its legislation. With RD 1565/2010, Spain started to limit the advantages offered to renewable energy providers, and other limitations followed until 2013. In 2013, with RDL 9/2013, the distinction provided to the plants producing electricity from renewable energy sources has been abolished so that they are treated equally with the traditional ones (Noilhac, 2020, 24–26).

As a result of roll-back of the incentives, Spain faced a large number of arbitral proceedings which are called as "Spanish renewables saga".

In all these cases, the claimants, renewable energy investors, based their claims on Article 10/1 of the Energy Charter Treaty.

First of these proceedings concerned the legislative amendments before 2013. In the first case, *Charanne*, the 2010 amendments had been challenged where the tribunal held that the essential features of the legislative framework were not suddenly and unexpectedly eliminated (Charanne B. V. & Construction Investments S.A.R.L. v the Kingdom of Spain, SCC Arb No 062/2012 (Award 21 January 2016)). The 2013/2014 amendments were first challenged in *Isolux* along with the 2010 amendments. In this case, the investment had been realized in 2012, so the tribunal considered that after the 2010 amendments, the 2013/2014 modifications did not frustrate the legitimate expectations of the investors (Isolux Infrastructure Netherlands B.V. v Spain, SCC Arb No 2013/153 (Award 17 July 2016)).

In the following decisions concerning the 2013 amendment, the complete abolishment of favorable treatment to renewable energy producers, the tribunals concluded that Article 10/1 of the ECT is breached by Spain. Only in *Stadtwerke*, the tribunal held that there is no commitment contractually assumed by a state to freeze its legislation in favor of an investor (Stadtwerke München GmbH, RWE Innogy GmbH and others v Kingdom of Spain, ICSID Case No ARB 15/1 (Award 2 December 2019)). However, in none of the other cases, the tribunals required a strict contractual commitment by the state not to change its legislation. Although there is no such clear commitment, the tribunals decided that the legitimate expectations of foreign investors had been breached. In *Eiser*, the tribunal decided that the investors had a legitimate expectation that the original investment regime for renewable energy would be preserved. In the wording of the tribunal, "*regulatory regimes cannot be radically altered as applied to existing investments in ways that deprive investors who invested in reliance on those regimes of their investment's value*" (Eiser Infrastructure Limited and Energía Solar Luxembourg S.à.r.l. v Kingdom of Spain, ICSID Case No ARB 13/36 (Award 4 May 2017)). Therefore, in this case and in the following cases, the tribunals took into account the economic impact of legislative amendments on the investors, the deprivation of the value of the investment, and dispossession of a significant part of their anticipated revenues, to assess whether their legitimate expectations had been violated (Novenergia II—Energy & Environment (SCA) (Grand Duchy of Luxembourg), SICAR v The Kingdom of Spain, SCC Arb No 2015/063 (Award 15 February 2018); Foresight Luxembourg Solar and al. v Kingdom of Spain, SCC Arb No 2015/150 (Final Award 14 November

2018); REEF Infrastructure (G.P.) Limited and RREEF Pan-European Infrastructure Two Lux S.àr.l. v Kingdom of Spain, ICSID Case No ARB 13/30 (Decision 30 November 2018); SolEs Badajoz GmbH v Kingdom of Spain, ICSID Case No ARB 15/38 (Award 31 July 2019); Hydro Energy 1 S.àr.l. and Hydroxana Sweden AB v Kingdom of Spain, ICSID Case No ARB 15/42 (Decision 9 March 2020)). In *9REN*, it is clearly stated that a guarantee of certain premiums or tariffs for at least 25 years provided by Spain in itself constitutes a legitimate expectation, bearing in mind the very large costs of renewable energy projects (9REN Holding S.à.r.l. v Kingdom of Spain, ICSID Case No ARB 15/15 (Award 31 May 2019)). In *Antin*, the tribunal also set forth that the abolishment of advantages provided to renewable energy producers is arbitrary and the rate of return provided by the new regime depends on governmental discretion (Antin Infrastructure Services Luxembourg S.à.r.l and Antin Energia Termosolar B.V v Kingdom of Spain, ICSID Case No ARB 13/31 (Award 15 June 2018)). Further, in *PV Investors*, the tribunal accepted that the tariffs and incentives cannot stay unchanged due to economic and technological variations in the renewable energy sector, but the complete abolishment of special tariffs and incentives is unreasonable (The PV Investors v Kingdom of Spain, PCA Case No 2012-14 (Final Award 28 February 2020)).

In order to tackle this flow of cases, Spain adopted Royal Decree-Law 17/2019 promising to pay a reasonable return of 7.398% for the next two remuneration periods to all renewable energy producers that had benefited from the regulatory regime before 2013/2014. However, investors shall guarantee not to pursue arbitral claims, withdraw from pending arbitrations, and/or waive their rights to collect their compensation granted according to the finalized arbitral awards (Ilie & Cercel, 2021, 63).

22.3.2 Cases Against Italy, Czech Republic, and Romania

It was with the Legislative Decree 387/2003 that the Italian National Energy Agency adopted an incentive framework for solar energy producers from whom the Agency would purchase electricity at a price above the average market price. Each investor, who is found eligible by the Agency as a result of reaching certain production targets, would be granted a certain right of tariffs which will remain constant for 20 years. Subsequently, between 2011 and 2014, Italy brought certain limitations to the incentive regime, which in the end, left investors with different options all leading to the reduction of the amount of tariffs initially granted to the investor (Faccio, 2020, 7–8).

There are eight cases against Italy, in five of which Italy prevailed; but in the other three, it had to pay compensation for breaching the fair and equitable treatment principle. In *Blusun* (Blusun S.A., Jean-Pierre Lecorcier and M. Stein v Italian Republic, ICSID Case No ARB 14/3 (Award 27 December 2016)), *Belenergia* (Belenergia S.A. v Italian Republic, ICSID Case No ARB 15/40 (Award 6 August 2019)), *Sunreserve* (Sunreserve Luxco Holdings S.a.r.l., Sunreserve Luxco Holdings II S.a.r.l., Sunreserve Luxco Holdings III S.a.r.l, SCC Arb No 2016/32 (Award 25

March 2020)), *Eskosol* (Eskosol S.p.A. in liquidazione v. Italian Republic, ICSID Case No. ARB/15/50, Award, 4 September 2020) and *Silver Ridge* (Silver Ridge Power BV v. Italian Republic, ICSID Case No. ARB/15/37 Award, 26 February 2021) after finding that there was no specific commitment by Italy, the tribunals decided that the actions of Italy are not grossly unreasonable, unfair, radical or arbitrary. On the contrary, the amendments are foreseeable and proportional to the public purpose to be achieved and within the state's right to regulate. Further, in these cases, the effect of the regulatory changes on the economic value of the investment was taken into consideration. In none of these cases, the economic loss attributed by the investors was found to be sufficient. In *Sunreserve*, for example, the decline of 7% in the enterprise value of the investor's subsidiary companies and a total decline of 19% in the value of the investment was not deemed to deprive the investor of fair remuneration. However, in *CEF Energia* (CEF Energia B.V v The Italian Republic, SCC Arb No 2015/158 (Award 16 January 2019) and *Greentech* (Greentech Energy Systems A/S, NovEnergia II Energy & Environment (SCA) SICAR, and NovEnergia II Italian Portfolio SA v The Italian Republic, SCC Arb No 2015/095 (Award 23 December 2018)) and *ESPF* (ESPF Beteiligungs GmbH, ESPF Nr. 2 Austria Beteiligungs GmbH and InfraClass Energie 5 GmbH & Co. KG v. Italian Republic, ICSID Case No. ARB/16/5, Award, 14 September 2020)), the investors were one of those which had reached the necessary production target to obtain a 20-year constant tariff guarantee, therefore, in these cases the tribunals decided that there was a specific commitment by Italy and it is under the duty to pay compensation.

On 31 December 2014, Italy notified its withdrawal from the ECT. The withdrawal took effect on 1 January 2016 according to Article 47/2 of the ECT where it is prescribed that any withdrawal shall take effect upon the expiry of 1 year after the date of the receipt of the notification by the Depositary.

The Czech Republic first introduced its support package for renewable energy in 2005 with Act 180 on the Support of Electricity Generation from Renewable Energy Sources. The incentive was to enjoy a purchasing price fixed at a certain level guaranteeing a 7% annual return on eligible investments for 15 years. In 2010, relevant legislation had been amended, and moreover in 2011, a new solar levy of 26% was presented on profits obtained from renewable energy plants. Six different arbitration proceedings had been initiated against the Czech Republic in all of which the state prevailed because the tribunals concluded that the legislative amendments made to the incentives legislation did not forbid the investors from achieving the 7% annual rate of return and the full remuneration during 15 years as predicted (Ilie & Cercel, 2021, 66–67).

It was in 2004 when the first green certificate system was introduced in Romania for renewable energy producers. After the 2008 financial crisis, however, the initial system has been gradually amended between 2008 and 2018 to the detriment of investors (Ilie & Cercel, 2021, 56–60). The first case against Romania is *LSG Building Solutions and others v. Romania* (ICSID Case No. ARB/18/19) where the tribunal concluded that Romania is liable for violation of Art. 10(1) of the ECT, "*on account of Romania's decision to alter the essential characteristics of the Green Certificate support scheme, on which Claimants had relied when making their investments, and to unreasonably impair the enjoyment of Claimants' investments*" (Decision on Jurisdiction, Liability and Principles of Reparation, 11 July 2022). There are two other pending cases before ICSID arbitration: *Fin.Doc S.r.l. and others v. Romania* (ICSID Case No. ARB/20/35), *EP Wind Project (Rom) Six Ltd. v. Romania* (ICSID Case No. ARB/20/15).

22.4 Results and Discussion

In the cases concerning the phase-out decisions, the investors argue that these decisions constitute a violation of Article 13 of the ECT, because the states do not pay adequate compensation to the investors. Apart from the *RWE v Netherlands* case, these cases were resolved with a Settlement Agreement. Nevertheless, the case law shows that in case of phase-out decisions, such policy choices undoubtedly constitute indirect expropriation, and therefore such decisions shall be accompanied by adequate compensation schemes for the investors which will create a massive economic burden on those states.

In the cases concerning roll-back of renewable energy incentives, the infringement of Article 10 of the ECT, the right of investors to a fair and equitable treatment, had been at stake. As the core element of this right is to protect and uphold the legitimate expectations of the investor; the tribunals assessed whether such an expectation has arisen and has been demolished in the specific cases. The legitimate expectation can arise from a general legislative framework or specific commitments by the state. Of course, if there are specific commitments given to the investors, the area of maneuver of the state is tighter and responsibility is greater. The form of the specific commitment may differ, it may be implied from the legislation itself, or may be directed to the specific investor in a correspondence; or take place in a public announcement or press release (Biggs, 2021, 115–119). If the source of legitimate expectations is the general legislative framework, then the investor shall be entitled to be protected against unreasonable, disproportionate, arbitrary, and radical changes made by the state which also cause devaluation of the investment.

As a result of a review of relevant case law, it can be concluded that although foreign investors cannot be completely exempt from any legislative changes concerning the regime of their investment; if the regime that they rely upon is radically amended in a way to deprive them of the economic advantages and revenues that they had prospected; then there is a breach of the ECT, so, the state is under an obligation of compensation. Additionally, if there was a specific commitment by the state to freeze the legislative framework as promised to the investors, then there is absolutely a violation of fair and equitable treatment obligation by the state.

22.5 Conclusion

In July 2023, United Nations Secretary-General António Guterres declared that the era of global warming was over and that an "era of global boiling" had begun, calling on the world's governments to take urgent actions to limit temperature rises and avoid harmful consequences of climate change. This requires the states to take their commitments under the Paris Agreement more seriously and urgently change their energy policies, such as phasing out from certain energy sources or promoting renewable energy. However, as explained above, phasing out from certain energy sources is considered to be indirect expropriation. As a result, the states that choose to phase out should pay high amounts of compensation to investors affected by such decisions. The global financial crisis also affects those states to avoid paying compensation. This arises the question of whether the green energy transition could be achieved due to the high compensation claims in the investment disputes.

The essence of the fair and equitable treatment obligation of the states is to ensure a balance between foreign investors' confidence on the legislation governing their long-term investments and the host states' sovereign right to adapt its legislation to new necessities and policies (Zannoni, 2020). In other words, the aim of the ECT is to realize a balance between the sovereign rights of the state over its energy resources and the establishment of an atmosphere favorable to the influx of investments (The PV Investors v Kingdom of Spain, PCA Case No 2012-14 (Final Award 28 February 2020)). Thus, the state must take into account the effects on investors in its decision-making process (RWE Innogy GmbH and RWE Innogy Aersa S.A.U. v Kingdom of Spain, ICSID Case No ARB 14/34 (Decision on Jurisdiction, Liability, and Certain Issues on Quantum 30 December 2019)). The state is not free to unreasonably and radically alter the legislative regime applied to existing investments in a way that severely deprives investors of their investment's value (Biggs, 2021, 110). Therefore, one aspect of ECT and investor-state arbitrations initiated against states is the promotion and protection of investments in the renewable energy sector (Garcia-Perrote & Wisniewski, 2020, 249). However, on the other hand, possible consequences of such arbitrations may discourage states to grant incentives for the renewable energy sector if they feel pressure or limitation on their sovereign power. In other words, the states may refrain from granting substantial advantages only not to bind themselves for a certain period of time which may, in the end, hinder green transition.

As a matter of fact, although international investment in renewable energy has tripled since 2015, it has slowed down in 2022. Furthermore, the growth is only valid for developed countries, whereas, no international investment project on renewables has been introduced in more than 30 developing countries. The costs of renewable energy technologies are rising due to high inflation rates, so, the energy transition will require higher amounts of investment in renewable energy generation, energy efficiency, and energy infrastructure (United Nations Conference on Trade and Development UNCTAD World Investment Report 2023). The provisions of the ECT shall be adapted to safeguard low-carbon energy sources or to impose restraints on old-style energy investments. However, although there are modernization efforts (https://www.energychartertreaty.org/modernisation-of-the-treaty/), it is difficult to rely on a fundamental change in the near future (Garcia-Perrote & Wisniewski, 2020, 270). The proposal of the European Union to exclude fossil fuel investment protection never materialized (European Parliament Briefing PE 754.632 December 2023).

Under the current legal regime, the bilateral or multilateral international investment agreements, especially the ECT, shall be interpreted in a manner to promote and protect such investments. This is referred to as the "Climate Change Element" that needs to be taken into account by the arbitral tribunals in deciding investor-state disputes (Burstein, 2020, 116–125). Indeed, several arbitral institutions, led by the International Chamber of Commerce, formed working groups to study these issues (ICC Publication on Resolving Climate Change Related Disputes through Arbitration and ADR).

On the other hand, there will be lessons learned by the legislators from these cases. First, inflexible promotional incentive regimes will be abandoned for those that are adaptable to technological and economic fluctuations. Second, the assurances and guarantees given to the investors will be elaborated more carefully without leaving room for any ambiguity. And lastly, instead of costly investor-state arbitration as a dispute settlement mechanism, the states will give preference to amicable settlement methods such as direct negotiation or mediation (Burstein, 2020, 115–116).

It must be noted, however, that one impact of investor-state arbitrations is the withdrawal of some European states such as Italy, France, Germany, and Poland from the ECT. What is even more significant is the decision of the European Union to withdraw. The European Commission proposed withdrawal by the EU as "*it considers the Treaty to be no longer compatible with the EU's climate goals under the European Green Deal and the Paris Agreement, predominantly due to concerns over continued fossil fuel investments*" (European Parliament Briefing PE 754.632 December 2023). According to ECT Article 47, the withdrawal takes effect at the end of 1 year following the notification of withdrawal for future investments; however, continues to apply for 20 years for existing investments.

References

Baltag, C. (2012). *The Energy Charter Treaty: The notion of investor*. Kluwer Law International.

Biggs, J. (2021). The scope of investors' legitimate expectations under the FET standard in the European renewable energy cases. *ICSID Review—Foreign Investment Law Journal, 36*(1), 99–128.

Burstein, B. D. (2020). Green investment disputes: The interaction between investment arbitration and the climate change agenda. *Revista Brasileira de Arbitragem, XVII*(68), 95–125.

Faccio, S. (2020). The assessment of the FET standard between legitimate expectations and economic impact in the Italian solar energy investment case law. *Questions of International Law, 71*, 3–20.

Garcia-Perrote, G., & Wisniewski, E. (2020). Time for a clean Energy Charter Treaty? Examining the role of ISDS in promoting and protecting renewable energy investments. In A. M. Anderson & B. Beaumont (Eds.), *The investor-state dispute settlement system: Reform, replace or status quo?* (pp. 247–270). Kluwer Law International.

Ilie, L., & Cercel, L. (2021). Romania's recent renewable energy disputes: Similar issues, same outcomes? *Revista Româna de Arbitraj, 15*(3), 52–75.

Noilhac, A. (2020). Renewable energy investment cases against Spain and the quest for regulatory consistency. *Questions of International Law, 71*, 21–39.

UNCTAD. (2023). *World Investment Report 2023*. UNCTAD.

Vorburger, S., & Petti, A. M. (2018). Arbitrating energy disputes. In M. Arroyo (Ed.), *Arbitration in Switzerland: The practitioner's guide* (pp. 1277–1319). Kluwer Law International.

Zannoni, D. (2020). The legitimate expectation of regulatory stability under the Energy Charter Treaty. *Leiden Journal of International Law, 33*(2), 1–16.

Chapter 23
Investigation of Energy Efficiency Practices in Buildings and Determination of Roadmap for Zero-Energy Buildings Target

Umit Unver

23.1 Introduction

The polar ice caps melt because of the global warming. The melting of glaciers causes changes in the major currents of the oceans, such as the North Equatorial Current and the Gulf Stream, etc. Changes in major currents cause changes in air currents, unprecedented climatic events, and various disasters. Authorities state that greenhouse gases are the main responsibles for global warming, which causes this significant climate change.

The whole world is trying to develop a common strategy to reduce greenhouse gas emissions and keep global warming below 2 °C. The most important part of the action plan is to shift to renewable energy by reducing the use of fossil fuels, one of which is the leading cause of greenhouse gas emissions. The most useful method to get over this problem is the energy efficiency (EE) studies.

EE studies are divided into three major sectors. These are namely: industry, transportation, and building sectors. Among these three sectors, the transportation sector has its own characteristics and tendencies. The automotive industry is investing at R&D studies at a satisfactory rate to develop new technologies and transportation vehicles toward the goals of more EE and less greenhouse gas emissions.

Since the industry is profit-oriented and any application that reduces costs increases profitability, EE studies are welcomed by the industry sector. In other words, EE studies in industry continued and went deeper without any resistance from the sector. A significant amount of knowledge has been gained as a result of EE studies in industry carried out between 1973 and 2023. A period of 50 years became quite a history in the field. The sector is encouraged and guided on EE with guiding publications such as "Best Available Technologies" published by the EU.

U. Unver (✉)
Mechanical Engineering Department, Yalova University, Yalova, Turkey

M. Z. Sogut, M. Koray (eds.), *Energy Rationality and Management for Decarbonization*, Green Energy and Technology,
https://doi.org/10.1007/978-3-031-68497-5_23

Energy studies, evaluation of studies, and development of application strategies in EE studies in industry have become routine today. Many applications have been experienced so far.

However, the situation is slightly different in the building sector. Constructors, owners, and users of buildings are often not the same people. Therefore, deciding on EE practices in the building sector requires more support than in the industrial sector. However, EE efforts can be monitored the most clearly and can be directed most easily in the building sector since it is controlled by local governments.

Additionally, ownership and use of buildings are more specific in law. The people who are the addressees of all laws and legislation are more clear. Therefore, legislation regarding buildings has an important place in the fight against global warming.

23.1.1 Developments and Legislation Regarding EE in Buildings

EE studies in the world started after the oil crisis in 1973. Fawkes (2016) divided this developmental process into six stages. These phases are respectively called "energy conservation phase" (1973–1981), "energy management phase" (1981–1993), "energy procurement phase" (1993–2000), "carbon reduction phase" (2000–2010), and "energy efficiency phase" (2010–2020) and the one we are in today is called "efficiency as a resource phase" (2020–2030).

The US pioneered one of the earliest EE studies. Besides governmental efforts, Private enterprises were established in the field. The Alliance Commission on National EE Policy is one of these organizations. According to the Alliance, 41% of US energy consumption takes place in buildings. The Alliance states that most of the energy consumption, which increased by 48% between 1980 and 2009, was due to the increase in the use of home electronics, as well as the increase in the total floor area in buildings and the average square meter per house, as well as the demand for other energy services. Informing that, thanks to the published legislation and the implemented program, energy consumption per unit floor area has decreased by 11% in residential buildings and 21% in commercial buildings since 1980.

The first publication on EE in buildings in the USA was ASHRAE Standard 90.1, Energy Savings in New Building Design (1975). Then, the first building EE law was adopted in 1976 (ALLIANCE, 2013).

The first legislation published by the European Union was the EE legislation in buildings proposed in 2001 (European Union, 2001), later published as Directive 2002/65/EC (European Parliament and the Council of the European Union, 2003). This legislation was developed in 2008 (European Union, 2008) and took its final form in 2010 and was published as the Energy Performance of Buildings Directive (European Union, 2010). In this directive, each country evaluates buildings according to its own climatic conditions, more use of renewable energy sources and the use of passive methods in building heating and cooling was recommended. In addition, nearly zero-energy buildings (nZEB) were defined and it was decided that all new buildings to be built in member countries until December 31, 2020 should be nZEBs.

23.1.2 *Latest Situation Today (Why Do We Need a Zero-Energy Building)*

An addition was made in 2018 to the directive published by the EU in 2010. The annex states that member states should determine their targets and local indicators and monitor them (European Commission, 2018). The revised directive was submitted to the commission in December 2021 to add the European Green Deal to this legislation in 2021 (European Commission, 2021).

23.2 Introduction to Zero-Energy Buildings (ZEB)

23.2.1 *Introduction and Description of ZEB*

Today, the highest energy performance in EE in buildings is defined for passive buildings. In general, buildings are classified according to the amount of energy consumption as follows: standard energy (E < 65 kW h/m^2/year), low energy (E < 40–50 kWh/m^2/year), ultra-low energy (E < 20–30 kWh/m^2/year), and passive houses (<15 kWh/m^2/year) (Ionescu et al., 2015).

The application that takes this definition one step further is zero-energy building applications. Nowadays, different definitions of ZEB are proposed according to site, energy source, cost, and emission (Sartori et al., 2012; Taherahmadi et al., 2021). However, a simple definition of net ZEB given by Alajmi et al. (2020) is as follows: "an energy-efficient building where, on a source energy basis, the actual annual delivered energy is less than or equal to the on-site renewable exported energy." More detailed descriptions can be found in the literature (Torcellini et al., 2010; D'Agostino & Mazzarella, 2019).

However, the truth is that regardless of the definition, there are two important points of the zero-energy building concept: (1) minimum energy consumption and (2) the production of energy is as much as consumption from renewable energy sources (Karanfil & Unver, 2022; Hermelink et al., 2013).

23.2.2 *Key Features of ZEB*

There are five important components to ensure maximum EE in ZEB. These are the building components designed to minimize energy consumption. Since building heating and cooling systems are responsible for 80% of the energy consumed in buildings, most of them are basically defined to minimize the energy required for heating and cooling. Namely:

1. Super Insulation: It is one of the most important components of EE in buildings since a well-insulated building will reduce the energy required both for heating and for cooling (Zecevic et al., 2020).
2. Super Tightness: One of the reasons for significant energy loss in buildings is leaks. In ZEB, air tightness should be ensured in order to reduce energy consumption for heating and cooling by preventing leaks from windows, doors, and other openings. For this reason, the air tightness in ZEBs is required to be less than 0.6 vol/h (Nematchoua et al., 2021).
3. Thermal Recovery: One of the important features of the ZEBs is that the energy of the air-conditioned air is disposed of after being recovered. Thus, it is possible to benefit from the energy of the conditioned indoor air and reduce energy consumption not only in the heating season but also in the cooling season (Li et al., 2022).
4. Design without Thermal Bridge: It is desired to prevent energy losses by designing building columns and beams in a way that does not create a thermal bridge (Shehadi, 2020).
5. Sun Shades: By using suitable sunshades, the entrance of direct sunlight is prevented in summer. The shadings should be designed to provide passive heating by letting the sun's rays into the space in winter.
6. Energy Production from Renewable Energy Resources: Photovoltaic systems, which are very popular these days, can be used for renewable energy production, as well as solar thermal systems, wind energy, geothermal energy, biogas, or biomass energy (Kasaeian & Sarrafha, 2020).

As you can see, zero-energy building components are not actually a concept full of unknowns. But the concept is not very well known either. ZEB is not widely accepted and preferred. Because of this, local governments need to be more active to encourage the implementation of ZEBs. In order to achieve ZEB target, countries must guide and encourage people by publishing legislations.

However, since zero-energy building practice is not very common, and conversion of a large amount of building stock in a short time may cause problems. Therefore, it is recommended to switch to ZEB with a well-designed strategy.

In the following sections of the chapter, a suitable road map for the transition to ZEB is presented.

23.2.3 A Route to Achieve Zero-Energy Built Environment Target

ZEB are structures designed and constructed to use renewable resources, energy-efficient technologies and advanced building design strategies to achieve zero net energy consumption. ZEB are designed to produce as much energy as they consume on an annual basis (Lohwanitchai & Jareemit, 2021). Therefore, they are extremely ecology-friendly structures. The cost of living in a ZEB is considered to be low. However, it is not common. There are valid reasons why such advantageous and useful buildings are preferred so little. Within the scope of this study, first, the obstacles will be presented, and then a proposal of a route to achieve the ZEB target will be given.

23.2.3.1 Obstacles in the Construction of ZEB

There are several obstacles to building ZEBs, namely:

1. Actors are different and independent from each other: One of the most important obstacles to achieve ZEB is that the people who own, use, and construct the building are different people. Since the priorities of each of the different actors are also different, this difference triggers other obstacles
2. High upfront costs: Constructing ZEB often requires significant investments in energy-efficient technologies and in renewable energy systems, leading to higher construction costs compared to conventional buildings (Lohwanitchai & Jareemit, 2021).
3. Lack of awareness and understanding: Many stakeholders, including building owners, developers, architects, and contractors, may not fully understand the concept of ZEB or the benefits of ZEB (Souaid et al., 2021).
4. Limited availability of qualified professionals: The design, construction, and maintenance of ZEB require specific knowledge and skills that may be limited in the current market (Nematchoua et al., 2021). Policy and regulatory barriers: Existing policies and regulations may not adequately support the construction of ZEB and hinder their widespread adoption.
5. Resistance to change: There may be resistance to adopting new construction methods and technologies, especially if they are perceived as unfamiliar or risky by sector professionals (Laustsen, 2008).
6. Limited access to financing: Securing adequate financing for zero-energy building projects can be challenging as financial institutions are hesitant to invest in relatively new and innovative construction practices (Souaid et al., 2021).

23.2.3.2 Strategic Action Plan for ZEB

In order for a country to fully transition to ZEB, despite all the obstacles listed above, a correct strategy must be developed and move toward the goal. Within this study, we have gathered the ten-step strategic action plan that is recommended to achieve the ZEB target is presented below.

Step 1: Defining Goals and Targets: The first step in achieving the ZEB target should be to determine the destination and the target (Crawley et al., 2009). Building energy performance: means the amount of energy required to meet the energy demand associated with the typical use of a building, including the energy used for heating, cooling, ventilation, hot water, and lighting, calculated or measured for different types of buildings. Clear targets should be set for the energy consumption of buildings, the energy production to meet this consumption, and energy performance (Marszal et al., 2011). Goals and objectives should be specific (Sartori et al., 2012), measurable, achievable (Economidou et al., 2020), relevant and time-bound. These goals should include four key features such as energy saving, sustainable development, ecology, and greening throughout the life cycle of the building (Kolokotsa et al., 2011).

Step 2: Performing Energy Audits and Assessments: One of the main difficulties in assessing the energy performance of buildings is that building energy data is often limited data obtained from a single electricity and/or gas meter (Alajmi et al., 2020). From most of the existing buildings, limited data can be obtained. However, in order to achieve the ZEB goal, it is very important to carry out detailed energy audits, collect data, and evaluate the results (Jones & Bogus, 2010). For detailed inspection, it is especially recommended to measure the temperature, relative humidity, and other physical parameters in the spaces. Thus, the causes of energy inefficiencies can be determined, energy consumption targets can be set, and renewable energy production targets can be determined. Thus, the size of the investment required for renewable energy production is revealed (Wang et al., 2010; Mathur & Muthukumaraswamy, 2018). Also, it must be noted that the more EE is achieved the less the renewable investment is!

Next, conduct a comprehensive feasibility analysis for potential investments and formulate a policy by assessing factors such as employment generation and market size. It is advisable to commission building audits and measurements from a dedicated and specialized institute. This institute could be a private entity, a university, or an organization exclusively focused on conducting building assessments. Even if it is a government institution, it is preferable for it to maintain an autonomous structure. This ensures that it remains unaffected by daily policy changes and promotes standardized evaluations.

Moreover, it is highly advisable to periodically reevaluate building measurements and monitor performance in both new constructions and renovated structures. This ongoing assessment, conducted after the project completion, serves as a crucial step to gather valuable data. If the effectiveness of implemented measures diminishes over time, monitoring becomes instrumental in preventing the recurrence of similar errors in future projects. Consistent monitoring and assessment contribute to continuous improvement.

Step 3: Developing an EE Plan: An EE plan should be developed by making a detailed analysis with the information obtained from the audits (Mills, 2011). The plan should be developed locally, taking into account the local building type, local architecture, and local materials. The developed plan should include region-specific measures, technologies, and strategies.

These strategies may include improving insulation, improving heating, ventilation, and air conditioning (HVAC) systems, installing energy-efficient lighting and appliances, and integrating renewable energy sources (Yandri et al., 2020). Which renewable energy sources will be used and how much will be used must be in the master plan. If necessary, deciding on new investments such as new sealing, insulation, and energy conversion should be taken into consideration within this plan. The use of local products should be prioritized in planning.

Step 4: Implementation of Energy Efficiency Measures: It is crucial to put into action the measures and strategies outlined in the EE plan. This may involve upgrading equipment and systems, as well as installing renewable energy systems (Li et al., 2022). The plan should also include improvements in insulation, optimization of HVAC systems, integration of smart building technologies for enhanced control and monitoring, and the installation of renewable energy systems such as solar panels, geothermal energy systems, or wind turbines (Zheng et al., 2017; Tejero-González et al., 2016).

Step 5: Monitoring and Evaluating Performance: It is important to regularly monitor and evaluate the energy performance of the building with correct indicators in order to check whether any measures implemented reach the target and their effectiveness (Bilardo et al., 2024). For this, it is recommended to collect and analyze data on energy consumption, renewable energy production, indoor comfort levels, and overall building performance (Yandri et al., 2020).

Within the scope of this study, there is a potential to develop software for continuous monitoring. The performance monitoring and evaluation process will uncover projects and improvements that require attention over time. Therefore, through regular and rigorous EE monitoring, the extent to which the set targets are achieved can be compared (Perlova et al., 2015).

The continuous monitoring outlined in this step can be conducted by the institution mentioned in step 2. This monitoring is automated through specialized software, with performance indicators tailored to the building's location (Bilardo et al., 2024). Any abrupt changes or deviations are promptly reported to relevant parties, and necessary precautions are implemented. Moreover, over time, the effectiveness of energy efficiency (EE) can be enhanced through experimental data and standardized applications.

Step 6: Ensuring Continuous Building Performance Improvement: Implementing the Plan-Do-Check-Action (PDCA) cycle is crucial for continuous improvement, utilizing monitoring and measurement data carried out by local governments or private institutes. After each action, targets for enhancing energy performance should be reassessed, and the institution should monitor the achievement of new targets.

For a country or region to transition completely to ZEB, there must be continuity in the monitoring system, improvement processes, renewable energy production, and optimization (European Commission, 2021; Sanz et al., 2015).

To maintain continuous improvement, after minimizing energy requirements for heating and cooling (Cabeza & Chàfer, 2020), it is recommended to investigate the energy consumption of lighting and electrical appliances. The responsibility lies with management to formulate appropriate policies for such research and to facilitate collaboration among project stakeholders (Shen et al., 2016).

Step 7: Raising Awareness and Ensuring Stakeholder Participation: To achieve the zero-energy building goal globally or locally, it is necessary to provide plenty of allies. It is very difficult or impossible to achieve a goal that is not adopted by governments, governments or the public. It is not possible for a project that is resisted by everyone to succeed. Therefore, it is recommended to ensure that both the public and the administrators embrace and adopt it.

Since the ZEB goal is about the sustainability of the world, all parties are actually expected to be allies of this project. In reality, only if the parties are informed and trained will they embrace the goal of ZEB. Therefore, educational campaigns, workshops, and training sessions should be organized to create a zero-energy building understanding in society and increase support (Evans et al., 2017). Implementing a comprehensive education and engagement program involving building owners, occupants, architects, engineers, contractors, and policy makers will help develop a collaborative environment for the successful realization of ZEB. By engaging and educating building occupants, building occupants should be educated on energy-saving practices, efficient resource use, and the benefits of ZEB. Thus, users' preferences begin to be questioned and shift to more energy-efficient alternatives.

Step 8: Collaborate with Research Institutions and Industry Experts: Building partnerships with industry experts, researchers, and organizations is crucial to staying up to date on the latest technologies, best practices, and policy changes in the field of ZEB. Thus, a catalog can be created that guides researchers in implementing innovative solutions and overcoming challenges (Deng et al., 2014; Belussi et al., 2019).

In the building industry, since the same people construct, own, inspect, and use the building, it is difficult to include everyone in the project at once. Therefore, collaboration with professionals and stakeholders from all walks of life can provide significant contributions. If possible, it is recommended to recruit stakeholders from architects, engineers, contractors, energy consultants, government agencies, and financial institutions. With the increase of stakeholders, a synergy space can be created to exchange knowledge, share best practices, and collaborate on research and development initiatives related to zero-energy building construction (Roudi, 2015). Thus, it becomes possible for stakeholders to come together from the first stage of the project and benefit from the experience and knowledge of all parties to make an integrated design (Erebor et al., 2021).

Step 9: Scale Up: It is important to document and share results, lessons learned, and best practices at the end of a successful implementation for each region covered. In this way, it may be possible for other people to see these models and adopt them, get inspired, and take them as an example (Cabeza & Chàfer, 2020). In this way, the spread effect can be increased.

Thanks to cooperation with stakeholders and scale-up, the acceptance and preference of zero-energy building transformation projects in society will increase exponentially. Such projects can go beyond prestige and become an unnamed secret necessity in society.

Step 10: Advocating Supportive Policies and Incentives: zero-energy building transformation is an important and large project and requires abundant stakeholder participation (Brozovsky et al., 2021). This project is associated with the sustainability of the world all over the world. Therefore, it stands out as a project that concerns all humanity. Government support and incentives need to be published so that this situation does not create the perception in individuals that "I cannot individually finance this very important project for humanity." In this way, individuals can be honored to be a part of such an important project and provide voluntary support to the project.

Additionally, support and incentives will encourage the building owner to become a stakeholder in the project (Liu et al., 2019). This type of incentive method would be particularly convincing for building owners who are not users of the buildings (Zhang et al., 2021).

23.3 Evaluation

Although ZEB have come to the fore frequently in recent decades, they are still unknown to the general public. It is extremely important to conduct numerous pilot applications to increase awareness in order to set an example for both industry professionals and users. For private or individual users, attempting an unknown application is perceived as embarking on an adventure. No one wants to index their savings to a question mark. Therefore, it is especially recommended that the public side increase the number of zero-energy building examples in every location by conducting pilot projects.

In order to verify and certify the performance of ZEB, it is important to increase and disseminate certificates such as green building or net zero-energy building certificates (Efe et al., 2022). In addition, information on developing technologies, innovations, and industry trends should be shared by organizing national/international meetings on the subject (Taherahmadi et al., 2021).

23.4 The Future of ZEB

The future of ZEB has great potential to transform our built environment and reduce our dependence on fossil fuels (Shehadi, 2020). It will be important to continue research and advancement in renewable energy technologies, building materials, and design strategies to further increase the adoption of ZEB (Kasaeian & Sarrafha, 2020). Policy decisions and global collaborations will also play an important role in establishing a supporting regulatory framework and international standards for ZEB (Blumberga et al., 2018). To successfully achieve ZEB on a larger scale, a collaboration between architects, engineers, builders, policy makers, and stakeholders is essential. Together, these stakeholders can work to develop innovative solutions, share knowledge and resources, and overcome the challenges associated with zero-energy construction (Camarasa et al., 2019). By following this systematic approach, we can pave the way for a future where ZEB becomes the norm, not the exception, leading to significant reductions in greenhouse gas emissions and a more sustainable built environment.

23.5 Conclusion: Toward Sustainable Construction

Achieving the goal of constructing ZEB requires a holistic and systematic approach that includes research and advancement in technology, supporting policy decisions, collaboration between stakeholders, and continuous monitoring and optimization (Jia et al., 2022). This approach will not only help reduce greenhouse gas emissions but also contribute to the development of a more sustainable and resilient built environment, benefiting both current and future generations. Consequently, achieving the goal of constructing ZEB requires a holistic and systematic approach that includes technological research and advancement, supporting policy decisions, collaboration between stakeholders, and continuous monitoring and optimization (Frappe-Seneclauze & Kniewasser, 2015).

References

Alajmi, A., Short, A., Ferguson, J., Vander Poel, K., & Griffin, C. (2020). Detailed energy efficiency strategies for converting an existing office building to NZEB: A case study in the Pacific Northwest. *Energy Efficiency, 13*(6), 1089–1104. https://doi.org/10.1007/s12053-020-09861-9

ALLIANCE. (2013). *The history of energy efficiency*. Retrieved from https://www.ase.org/sites/ase.org/files/resources/Mediabrowser/ee_commission_history_report_2-1-13.pdf

Belussi, L., et al. (2019). A review of performance of zero energy buildings and energy efficiency solutions. *Journal of Building Engineering, 25*(March), 100772. https://doi.org/10.1016/j.jobe.2019.100772

Bilardo, M., Kämpf, J. H., & Fabrizio, E. (2024). From Zero Energy to Zero Power Buildings: A new paradigm for a sustainable transition of the building stock. *Sustainable Cities and Society, 101*(January), 105136. https://doi.org/10.1016/j.scs.2023.105136

Blumberga, A., Cilinskis, E., Gravelsins, A., Svarckopfa, A., & Blumberga, D. (2018). Analysis of regulatory instruments promoting building energy efficiency. *Energy Procedia, 147*, 258–267. https://doi.org/10.1016/j.egypro.2018.07.090

Brozovsky, J., Gustavsen, A., & Gaitani, N. (2021). Zero emission neighbourhoods and positive energy districts—A state-of-the-art review. *Sustainable Cities and Society, 72*(May), 103013. https://doi.org/10.1016/j.scs.2021.103013

Cabeza, L. F., & Chàfer, M. (2020). Technological options and strategies towards zero energy buildings contributing to climate change mitigation: A systematic review. *Energy and Buildings, 219*, 110009. https://doi.org/10.1016/j.enbuild.2020.110009

Camarasa, C., Nägeli, C., Ostermeyer, Y., Klippel, M., & Botzler, S. (2019). Diffusion of energy efficiency technologies in European residential buildings: A bibliometric analysis. *Energy and Buildings, 202*, 109339. https://doi.org/10.1016/j.enbuild.2019.109339

Crawley, D., Pless, S., & Torcellini, P. (2009). Getting to net zero energy buildings. *ASHRAE Journal*, 1–12. Retrieved from http://www.google.com/url?sa=t&source=web&ct=res&cd=2&url=http://www.stanford.edu/group/peec/cgi-bin/docs/home/events/2009/public_discussions/

presentation_Selkowitz.pdf&ei=DlJGSvimPIHaNaCF3aTB&usg=AFQjCNGQgaTzt2zJXfRxbcdGcd2lE-J6WQ

D'Agostino, D., & Mazzarella, L. (2019). What is a Nearly zero energy building? Overview, implementation and comparison of definitions. *Journal of Building Engineering, 21*, 200–212. https://doi.org/10.1016/j.jobe.2018.10.019

Deng, S., Wang, R. Z., & Dai, Y. J. (2014). How to evaluate performance of net zero energy building—A literature research. *Energy, 71*(2014), 1–16. https://doi.org/10.1016/j.energy.2014.05.007

Economidou, M., Todeschi, V., Bertoldi, P., D'Agostino, D., Zangheri, P., & Castellazzi, L. (2020). Review of 50 years of EU energy efficiency policies for buildings. *Energy and Buildings, 225*, 110322. https://doi.org/10.1016/j.enbuild.2020.110322

Efe, O., Özdemir, R., Işik, S., Durmuş, İ., & Ünver, Ü. (2022). Assessment of the Yalova University Engineering Faculty Building using the B.E.S.T. green building certification system. *International Journal of Sustainable Energy, 41*(11), 1759–1777. https://doi.org/10.1080/14786451.2022.2109027

Erebor, E. M., Ibem, E. O., Ezema, I. C., & Sholanke, A. B. (2021). Energy efficiency design strategies in office buildings: A literature review. *International Conference on Energy and Sustainable Environment, 665*(1), 012025. https://doi.org/10.1088/1755-1315/665/1/012025

European Commission. (2018). *Amending Directive 2010/31/EU on the energy performance of buildings and Directive 2012/27/EU on energy efficiency* (Vol. L156, pp. 75–91).

European Commission. (2021). *Proposal for a Directive of the European Parliament and of the Council on the energy performance of buildings (recast). COM (2021) 802* (Vol. 0426).

European Parliament and the Council of the European Union. (2003). *Directive 2002/65/EC of the European Parliament and of the Council.*

European Union. (2001). *Energy policy: Energy performance of the buildings of the Union, energy saves and efficiency 2001/0098(COD)* (Vol. 2001/0098).

European Union. (2008). *Energy performance of buildings*. Recast Repealing 2008/0223(COD).

European Union. (2010). *On the energy performance of buildings Directive 2010/31/EU.*

Evans, M., Roshchanka, V., & Graham, P. (2017). An international survey of building energy codes and their implementation. *Journal of Cleaner Production, 158*, 382–389. https://doi.org/10.1016/j.jclepro.2017.01.007

Fawkes, S. (2016). *A brief history of energy*. Retrieved from https://www.onlyelevenpercent.com/, https://www.onlyelevenpercent.com/a-brief-history-of-energy-efficiency/

Frappe-Seneclauze, T.-P., & Kniewasser, M. (2015). *Tom-Pierre Frappé-Sénéclauze. The path to 'net-zero energy' buildings in BC.* Pembina Institute.

Hermelink, A., et al. (2013). *Towards nearly zero-energy buildings: Definition of common principles under the EPBD.* European Commission.

Ionescu, C., Baracu, T., Vlad, G. E., Necula, H., & Badea, A. (2015). The historical evolution of the energy efficient buildings. *Renewable and Sustainable Energy Reviews, 49*, 243–253. https://doi.org/10.1016/j.rser.2015.04.062

Jia, J., Huang, Z., Deng, J., Hu, F., & Li, L. (2022). Government performance evaluation in the context of carbon neutrality: Energy-saving of new residential building projects. *Sustainability (Switzerland), 14*(3), 1274. https://doi.org/10.3390/su14031274

Jones, B., & Bogus, S. M. (2010). Decision process for energy efficient building retrofits: The owner's perspective. *Journal of Green Building, 5*(3), 131–146. https://doi.org/10.3992/jgb.5.3.131

Karanfil, G., & Unver, U. (2022). ZERO ENERGY BUILDINGS: At a glance. In *2022 7th International Conference on Smart and Sustainable Technologies, SpliTech 2022* (pp. 1–4). https://doi.org/10.23919/SpliTech55088.2022.9854301

Kasaeian, A., & Sarrafha, H. (2020). *Solar energy systems: An approach to zero energy buildings.* Elsevier.

Kolokotsa, D., Rovas, D., Kosmatopoulos, E., & Kalaitzakis, K. (2011). A roadmap towards intelligent net zero- and positive-energy buildings. *Solar Energy, 85*(12), 3067–3084. https://doi.org/10.1016/j.solener.2010.09.001

Laustsen, J. (2008). *Energy efficiency requirements in building codes, energy efficiency policies for new buildings*. Retrieved from http://www.iea.org/g8/2008/Building_Codes.pdf
Li, Z., Zou, Y., Tian, M., & Ying, Y. (2022). Research on optimization of climate responsive indoor space design in residential buildings. *Buildings, 12*(1), 59. https://doi.org/10.3390/buildings12010059
Liu, Z., Zhou, Q., Tian, Z., Jie He, B., & Jin, G. (2019). A comprehensive analysis on definitions, development, and policies of nearly zero energy buildings in China. *Renewable and Sustainable Energy Reviews, 114*(July), 109314. https://doi.org/10.1016/j.rser.2019.109314
Lohwanitchai, K., & Jareemit, D. (2021). Modeling energy efficiency performance and cost-benefit analysis achieving net-zero energy building design: Case studies of three representative offices in Thailand. *Sustainability (Switzerland), 13*(9), 5201. https://doi.org/10.3390/su13095201
Marszal, A. J., et al. (2011). Zero energy building—A review of definitions and calculation methodologies. *Energy and Buildings, 43*(4), 971–979. https://doi.org/10.1016/j.enbuild.2010.12.022
Mathur, A., & Muthukumaraswamy, S. A. (2018). On the enhancement of building energy conservation using energy auditing. In *Proceedings of the International Conference on Intelligent Sustainable Systems, ICISS 2017* (pp. 720–724). https://doi.org/10.1109/ISS1.2017.8389267
Mills, E. (2011). Building commissioning: A golden opportunity for reducing energy costs and greenhouse gas emissions in the United States. *Energy Efficiency, 4*(2), 145–173. https://doi.org/10.1007/s12053-011-9116-8
Nematchoua, M. K., Orosa, J. A., Ricciardi, P., Obonyo, E., Sambatra, E. J. R., & Reiter, S. (2021). Transition to zero energy and low carbon emission in residential buildings located in tropical and temperate climates. *Energies, 14*(14). https://doi.org/10.3390/en14144253
Perlova, E., Platonova, M., Gorshkov, A., & Rakova, X. (2015). Concept project of zero energy building. *Procedia Engineering, 100*(January), 1505–1514. https://doi.org/10.1016/j.proeng.2015.01.522
Roudi, F. (2015). *Understanding net zero energy building concept through precedents from different climate zones*. Eastern Mediterranean University.
Sanz, R., Corredera, Á., Hernández, J. L., Samiengo, J., Vicente, J. M., & Bujedo, L. A. (2015). Towards the integration of monitoring systems to support the evaluation of nearly Zero Energy Buildings through Key Performance Indicators. In *2015 IEEE Workshop on Environmental, Energy, and Structural Monitoring Systems, EESMS 2015—Proceedings* (pp. 90–95). https://doi.org/10.1109/EESMS.2015.7175858
Sartori, I., Napolitano, A., & Voss, K. (2012). Net zero energy buildings: A consistent definition framework. *Energy and Buildings, 48*, 220–232. https://doi.org/10.1016/j.enbuild.2012.01.032
Shehadi, M. (2020). Net-zero energy buildings: Principles and applications. In *Zero-energy buildings—New approaches and technologies* (pp. 1–16).
Shen, L., He, B., Jiao, L., Song, X., & Zhang, X. (2016). Research on the development of main policy instruments for improving building energy-efficiency. *Journal of Cleaner Production, 112*(2016), 1789–1803. https://doi.org/10.1016/j.jclepro.2015.06.108
Souaid, C., van der Heijden, H., & Elsinga, M. (2021). Institutional barriers to near zero-energy housing: A context specific approach. *Sustainability (Switzerland), 13*(13), 7135. https://doi.org/10.3390/su13137135
Taherahmadi, J., Noorollahi, Y., & Panahi, M. (2021). Toward comprehensive zero energy building definitions: A literature review and recommendations. *International Journal of Sustainable Energy, 40*(2), 120–148. https://doi.org/10.1080/14786451.2020.1796664
Tejero-González, A., Andrés-Chicote, M., García-Ibáñez, P., Velasco-Gómez, E., & Rey-Martínez, F. J. (2016). Assessing the applicability of passive cooling and heating techniques through climate factors: An overview. *Renewable and Sustainable Energy Reviews, 65*, 727–742. https://doi.org/10.1016/j.rser.2016.06.077
Torcellini, P., Pless, S., Lobato, C., & Hootman, T. (2010). Main street net-zero energy buildings: The zero energy method in concept and practice. In *Proceedings of the ASME 2010 4th International Conference on Energy Sustainability ES2010* (pp. 1–9).

Wang, X., Huang, C., & Cao, W. (2010). Energy audit of building: A case study of a commercial building in Shanghai. *Asia-Pacific Power and Energy Engineering Conference, 50478113*, 2005–2008. https://doi.org/10.1109/APPEEC.2010.5448746

Yandri, E., et al. (2020). Implementation of walk-through audits for designing energy management system: A first step towards an efficient campus. *IOP Conference Series: Earth and Environmental Science, 490*(1), 012005. https://doi.org/10.1088/1755-1315/490/1/012005

Zecevic, N., Dzaferovic, E., & Husika, A. (2020). Defining criteria for the selection of measures to increase energy efficiency in public buildings. *IOP Conference Series: Materials Science and Engineering, 960*(3), 032052. https://doi.org/10.1088/1757-899X/960/3/032052

Zhang, S., Fu, Y., Yang, X., & Xu, W. (2021). Assessment of mid-to-long term energy saving impacts of nearly zero energy building incentive policies in cold region of China. *Energy and Buildings, 241*, 110938. https://doi.org/10.1016/j.enbuild.2021.110938

Zheng, D. L., Yu, L. J., & Tan, H. W. (2017). Design and optimization of zero-energy-consumption based solar energy residential building systems. *IOP Conference Series: Earth and Environmental Science, 93*(1), 012055. https://doi.org/10.1088/1755-1315/93/1/012055

Chapter 24
Energy Efficiency in the Road Transportation Sector

A. Murat Köseoğlu

24.1 Concept of Energy Efficiency

Energy efficiency is a concept that means using less energy to do a job, thus reducing energy waste, reducing energy costs, and reducing environmental pollution (Energy Star, 2023). It is one of the most efficient and cost-effective ways to combat climate change, reduce energy costs for consumers, and increase the competitiveness of businesses (EERE (a), 2022). Energy efficiency can be achieved in a variety of ways, including the use of energy-efficient products, homes, and buildings, as well as the implementation of energy-efficient practices and behaviors (EIA, 2022; Constellation, 2022).

Energy efficiency is a concept related to energy saving that involves using less energy by adjusting behaviors and habits. Energy saving and efficiency may be related but have different definitions in the energy literature. While energy saving involves using less energy by adjusting behaviors and habits, energy efficiency involves using technology that requires less energy to perform the same function (Constellation, 2022).

Energy efficiency provides different benefits such as cost savings, improved efficiency, increased productivity, and job creation. It also helps reduce greenhouse gas emissions and other air pollution, combat the problem of climate change, and protect human health and the environment (Energy Star, 2023; EERE (a), 2022).

Examples of energy-saving products are energy-saving light bulbs, which use less energy to produce the same amount of light, or thermal insulation or double-glazed windows, which increase the overall efficiency and comfort of a home or building (Energy Star, 2023). With energy-efficient homes, less energy is used to

A. M. Köseoğlu (✉)
Istanbul Gedik University, Istanbul, Turkey
e-mail: murat.koseoglu@gedik.edu.tr

M. Z. Sogut, M. Koray (eds.), *Energy Rationality and Management for Decarbonization*, Green Energy and Technology,
https://doi.org/10.1007/978-3-031-68497-5_24

heat and cool buildings and power electronic devices, while energy-efficient manufacturing facilities use less energy to produce products (EERE (a), 2022). Despite the benefits of energy efficiency in the transportation sector, electric vehicles have challenges such as limited availability, high upfront costs, limited range and charging infrastructure, maintenance and service services, and lack of awareness and training (EESI, 2021).

In many countries around the world, energy efficiency standards are established that encourage or require electricity companies operating in their countries to reduce their customers' electricity and natural gas usage in targeted amounts and according to a defined timetable or program. Energy efficiency standards are like renewable energy standards in many ways. Like renewable energy standards, energy efficiency standards vary in requirements and timing across countries. Energy efficiency standards policies may have separate reduction targets for electricity sales, electricity demand, and consumption. In many cases, utilities need to save energy through demand-side management programs. Some countries provide financial incentives, such as tax credits or rebates, to consumers and businesses for purchasing energy-efficient appliances (EIA, 2022).

Decarbonizing and affordability of energy are two notable qualities that will become even more important in the effort to reduce greenhouse gas emissions by 50% by 2030 under the Paris Agreement. As the built environment and transportation sector become less dependent on fossil fuels over time, investments in electricity generation will need to keep pace with increasing demand. Renewable energy, which is increasingly cheaper to install than polluting sources, will continue to account for the majority of new generations added to the grid (Bresette, 2023).

24.2 The Importance and Potential of Energy Efficiency in Road Transportation

Sustainable transportation refers to low- and zero-emission, energy-efficient, and affordable transportation modes, as well as electric and alternative fuel vehicles. Benefits of sustainable transportation; cost savings on fuel and vehicles; reduced carbon emissions from burning fossil fuels, resulting in less air pollution; job creation through increased vehicle and battery production; and better accessibility to reliable, affordable transportation options (En-ROADS, 2020; EERE (b), 2022).

Energy efficiency in road transportation is very important in reducing energy consumption, greenhouse gas emissions, and air pollution. The main points that reveal the importance and potential of energy efficiency in road transportation are as follows:

- Energy efficiency in road transport can be achieved in different ways, including the use of energy-efficient vehicles, adopting improved driving habits, and implementing energy-efficient practices and policies (EEA, 2022).

- Energy-efficient vehicles, such as hybrid and electric cars, use less energy to travel the same distance as conventional internal combustion engine (ICE) vehicles, reducing fuel consumption and emissions (NRC, 2010).
- Improved driving habits, such as reducing idling time, maintaining proper tire pressure, and avoiding aggressive driving, can also increase energy efficiency and reduce fuel consumption (AAR, 2019).
- Policies and practices that promote energy efficiency in road transport include investing in public transportation, building bicycle paths, and creating high-density pedestrian-friendly urban areas (EEA, 2022).
- Improving energy efficiency in road transport can lead to increased productivity and employment, as well as significant cost savings for consumers and businesses (NRC, 2010).
- Energy efficiency in road transport is also necessary to reduce dependence on fossil fuels, increase energy security, and mitigate the effects of climate change (En-ROADS, 2020).
- The potential for energy efficiency in road transport is very large. There are many improvement opportunities in this area, including the use of alternative fuels, the development of more efficient vehicles, and the implementation of policies and practices that encourage energy efficiency (AAR, 2019).

In today's competitive world, the energy spent on transportation is very important for the economy. Limited access to affordable energy, especially in developing countries, causes negativities in socioeconomic growth (Altinay & Karagol, 2004). Population growth, infrastructure development, and trade activities increase the diversity of energy consumption (Halicioglu, 2009; Zhang & Lin, 2012). Therefore, energy supply and consumption are some of the basic activities that reflect the economic development of a country.

In many countries, energy-intensive transport plays a key role in people's access to socioeconomic opportunities. The intensity of transportation can turn the environment into an unhealthy environment and a greenhouse gas producer. This situation makes greenhouse gas emissions resulting from the transportation sector one of the serious environmental problems. Therefore, the negative impact of transportation energy must be addressed to achieve higher energy efficiency (Zhang et al., 2011).

As a result, energy efficiency in road transportation is of great importance in reducing energy consumption, greenhouse gas emissions, and air pollution. This can be achieved in a variety of ways, including using energy-efficient vehicles, improving driving habits, and implementing energy-efficient policies and practices. The potential for energy efficiency in road transport is significant and there is much room for improvement (Akbar et al., 2020).

24.3 Current Status of Energy Consumption in the Sector and Future Expectations

The current situation of energy consumption in the road transport sector in terms of energy efficiency is that it significantly affects global energy use and greenhouse gas emissions (NRC, 2010; US EIA 2016,). Road transportation accounts for 75% of the energy used in this area and when evaluated from this perspective, it has the greatest potential in terms of energy efficiency (SEforALL, 2022).

However, there have been no major developments in the efficiency improvement of the energy intensity used in road transportation since the early 1970s. Fuel efficiency in new vehicles has improved for all modes, but differences in vehicle fleets (larger cars with more powerful and heavier payloads) and in-car usage (underutilization of capacity) have absorbed much of the positive impact in most countries (Akbar et al., 2020).

Prospects for energy efficiency in road transport are positive and there are many improvement opportunities. A study by the International Council on Clean Transportation suggests that it is possible to increase energy efficiency in trucks by 35% by 2035 (SEforALL, 2022).

The European Union has set a target of reducing greenhouse gas emissions from transportation by 60% by 2050, and these developments will require significant improvements in energy efficiency (Akbar et al., 2020). Using alternative fuels, developing more efficient vehicles, and implementing policies and practices that encourage energy efficiency are some of the ways to achieve these goals (EIA, 2016).

Given the significant impact of road transportation energy on the economy and the negative impact on the environment, numerous studies have focused on transport efficiency in terms of CO_2 emissions (Liu et al., 2018).

The effects of road transport energy on the environment, society, and the country's economy are quite large, therefore it has become the focus of attention of many researchers. When the CO_2 emission patterns in the transportation sector of the seven largest carbon-emitting economies were examined, it was found that China, the USA, Russia, Canada, India, and Brazil increased their emission rates between 2000 and 2015, while this rate decreased in Japan (Solaymani, 2019).

The energy consumption and emissions of China's transportation until 2050 were evaluated. As a result of the study, it was concluded that by 2050, energy consumption may reach values that are very difficult to predict, and this may cause significant health and economic losses (Liu et al., 2018). The change in CO_2 emissions in the Tunisian transportation industry was analyzed. The main effects of energy efficiency solutions on reducing CO_2 emission rates in fuels are shown (Talbi, 2017).

In a study estimating the direct energy impact of road passenger transportation in China, it was found that as the mileage increases, the direct negative impacts in different regions also increase proportionally (Zhang et al., 2015). In another study, it was stated that compared to developed economies, Japan stands out with its exemplary energy efficiency structure rather than energy intensity and that the Japanese transportation industry can be used as a good example (Lipscy & Schipper, 2013).

The road transport sector contributes significantly to global energy use and greenhouse gas emissions. Although the energy intensity of road transport has not improved much since the early 1970s, there are many opportunities for improvement. Future prospects for energy efficiency in road transport are promising, with opportunities for significant improvement through the use of alternative fuels, the development of more efficient vehicles, and policies and practices that promote energy efficiency.

24.4 Environmental Impacts and Economic Benefits of Implementing Energy Efficiency Strategies

It has been found that economic impact and carbon intensity have a strong positive correlation. Therefore, it is necessary to specifically focus on cost reduction and environmental protection in terms of reducing greenhouse gas emissions (Amicarelli et al., 2019).

A study in China found that an increase in human-based technology transfer will increase CO_2 emissions, while a decrease in infrastructure-based technology transfer will reduce CO_2 emissions. It has also been found that innovation and infrastructure development through increased absorptive capacity can help reduce China's carbon intensity (Mariyakhan et al., 2020).

A study conducted in the United States found that suburban commercial centers had average CO_2 emissions of 6.94% and 26.92%, higher than urban commercial centers. It has been determined that CO_2 emissions in wholesale centers are almost three times less than in urban trade centers (Li et al., 2016).

On the other hand, it has been suggested to conduct information campaigns to raise awareness about the importance of energy use for sustainable economic development in Romania (Brătucu et al., 2019). Another study found that more than 80% of energy consumption comes from fuel, and demand increases over time (Al-Mofleh et al., 2010).

In another study, the emission level resulting from energy consumption was discussed and the impact models in Malaysia were analyzed. This study found that the total oil emissions consumed by transportation are almost equal to the net amount of oil imported (Ji & Chen, 2006).

Developing and implementing different and effective strategies to increase energy efficiency will provide both environmental and economic benefits. Some of the environmental impacts that will occur with the implementation of these strategies are as follows:

- Energy efficiency strategies contribute significantly to reducing greenhouse gas emissions that contribute to climate change (Energy Sage, 2022).

- Energy efficiency strategies help reduce air pollution, which has negative effects on human health and the environment (EERE (c), 2022).
- Energy efficiency strategies support reducing water use, which is important for conserving water and preserving natural ecosystems (EPE, 2023).
- The economic benefits of implementing energy efficiency strategies include:
  - Energy efficiency strategies help reduce energy costs for consumers and businesses, resulting in significant cost savings (Energy5 (a), 2023).
  - Implementing energy efficiency strategies creates jobs in many industries, including manufacturing, construction, and engineering (Energy5 (b), 2022).
  - Energy efficiency strategies help businesses become more competitive by reducing energy costs and increasing operational efficiency (EPA, 2023).
  - Energy efficiency strategies contribute to economic growth by reducing energy costs, creating jobs, and increasing productivity (ECLA, 2021).

In this context, implementing energy efficiency strategies provides significant environmental and economic benefits, including reducing greenhouse gas emissions, cost savings, job creation, and increased competitiveness. These benefits highlight the importance of promoting and implementing energy efficiency strategies in the road transport sector to achieve a sustainable future.

24.5 Various Examples of Implementing Energy Efficiency Strategies

In a globalized world, key actors in different industries must constantly innovate to develop products and services to address the short-, medium-, and long-term impacts of their activities on the environment (GFEI, 2011). The transportation sector is a major consumer of energy, accounting for 19% of global energy consumption. The sector is expected to be responsible for 97% of the increase in global oil consumption between 2013 and 2030 (IEA (a), 2012).

The impacts of an oil-dominated transport sector on energy consumption and greenhouse gas emissions indicate that reducing fuel use in the sector should be one of the highest priorities for all countries. Because the transportation sector is responsible for a quarter of energy-related greenhouse gas emissions. Unless radical measures are taken to curb them, transport emissions are expected to rise from 7.7 gigatons to nearly gigatons by 2050 (UNCSD, 2011).

Transport is the largest source of energy-related greenhouse gas emissions in 45% of countries and the second largest source in the remainder. Consequently, transportation must be considered a key element in any effective solution (ITF, 2016).

The 2030 Agenda for Sustainable Development, approved by the United Nations in 2015, proposed the sustainable development goals (SDGs) for global development by 2030. 423 (1) The agenda sets out medium and long-term goals and objectives, focusing on all-round development. 423 Agenda that emphasizes the importance of having long-term investment plans or road maps that can address today's infrastructure gaps. 423 Also, to prepare countries to cope with future challenges and opportunities by taking full advantage of them (United Nations, 2016).

There are different energy efficiency strategies that can be implemented in the road transportation sector to reduce energy consumption and energy costs. Some of these strategies are:

- Use of energy-efficient vehicles: The use of energy-efficient vehicles such as hybrid and electric cars reduces fuel consumption and emissions (EERE (c), 2022).
- Improved driving habits: Small changes in driving behavior, such as reducing idling time, maintaining proper tire pressure, avoiding aggressive driving, improving energy efficiency, and reducing fuel consumption (Thacker et al., 2018).
- Optimum route planning: Optimum route planning helps reduce fuel consumption and emissions by minimizing distance traveled and avoiding traffic congestion (Energy5 (a), 2023).
- Use of alternative fuels: Using alternative fuels such as biodiesel and compressed natural gas reduces emissions and increases energy efficiency (Energy Renovables, 2019).
- Aerodynamic design: Using aerodynamic design features such as aerodynamic bodies and low rolling resistance tires increases fuel efficiency and reduces emissions (Energy5 (a), 2023).
- Efficient logistics operations: Efficient logistics planning, such as optimizing delivery routes and reducing empty distances, increases energy efficiency and reduces costs (EERE (c), 2022).
- Intelligent transportation systems: Using intelligent transportation systems, such as traffic management systems and vehicle-to-vehicle communication, increases energy efficiency and reduces emissions (EERE (d), 2022).

Implementation of the mentioned energy efficiency strategies in the road transportation sector. In addition to reducing greenhouse gas emissions, providing cost savings, creating employment, and increasing competitiveness, it can also provide significant environmental and economic benefits (United Nations, 2021).

24.6 The Relationship Between Energy Efficiency in Road Transportation and Sustainability Goals

Many sustainable development goals are linked to sustainable transport and thus to energy efficiency, through specific targets and indicators. Other sustainable development goals are also indirectly linked through the enabling role of sustainable transport. For example, it is estimated that sustainable and resilient infrastructure, including transportation, will be effective in achieving 92% of all sustainable development goals. In this context, the connections between energy efficiency and sustainability targets through transportation are stated below (Thacker et al., 2018).

- Sustainable Development Goal 1 End Poverty Everywhere and In All Forms: Improved transportation networks and digital communications can provide access to services by delivering vital products and information to facilities such as schools or hospitals in rural or remote areas.
- Sustainable Development Goal 2 End Hunger, Ensure Food Security, Improve Nutrition, and Promote Sustainable Agriculture: More efficient transportation connecting producers and consumers will increase the sustainability of the food system and increase the diversity of food options available to consumers.
- Sustainable Development Goal 3 Ensure Healthy Lives and Promote the Well-Being of All Ages: Where access to facilities is not possible, transport networks, education, and public awareness in rural or remote communities are vital for the provision of health services, including prevention.
- Sustainable Development Goal 4 Providing Inclusive, Equal, Quality Education, and Promoting Lifelong Learning Opportunities for All: The availability of convenient transportation facilities may enable students to progress to higher education in another city or region.
- Sustainable Development Goal 5 Achieving Gender Equality, Empowering All Women and Daughters: Improved transportation enables the delivery and administration of sexual and reproductive health supplies and services to all communities.
- Sustainable Development Goal 6 Achieving Water and Sanitation for All and Their Sustainable Management: The use of transportation networks and digital communications infrastructure provides access and capabilities to advance this goal (EPA (a), 2014).
- Sustainable Development Goal 7 Providing Access to Cheap, Reliable, Sustainable, and Modern Energy for All: This goal emphasizes the significant interdependence between the energy and transportation sectors. Technological trends, such as increasing fuel efficiency in vehicles and electrification of transportation, can both reduce energy demand and reduce fossil fuel emissions on a large scale if they are made more widely available.
- Sustainable Development Goal 8 Supporting Sustainable, Inclusive, Economic Growth, Full and Productive Employment, and Human Business for All: Secure and accessible energy sources, transportation networks to connect producers and consumers, infrastructure systems that require digital technology and communication. It plays an important role in increasing national economic growth and productivity, to increase added value and ensure efficiency at all stages of the value chain (Estache, 2006; Estache & Garsous, 2012).
- Sustainable Development Goal 10 Reducing Inequality Within and Among Countries: Due to traditional gender roles, domestic responsibilities, including fuel and water collection, can limit the economic and educational opportunities of women and girls in some regions. This disadvantage can be eliminated through the provision of modern energy or water supply infrastructure, or alternatively

through efficient transport that allows these services to be delivered directly to communities and households.
- Sustainable Development Goal 11 Making Cities and Human Settlements Inclusive, Secure, Resilient, and Sustainable: Using increased digital technology, transportation systems can be designed to reduce traffic accidents, while public transportation and sanitation facilities can be expanded to provide safer and more accessible options for all people. Particular attention is needed in cities to reduce or eliminate the negative health effects on air quality of emissions from transportation sectors (IEA (b), 2016).
- Sustainable Development Goal 12 Ensuring Sustainable Consumption and Production Models: Ensuring adequate cooling throughout the production process, increasing the efficiency of food transportation, and using digital technology to improve production and consumption choices can help limit these losses. Increasingly replacing face-to-face meetings with online, telephone or video communication could contribute to reducing traffic congestion and transport emissions (Parfitt et al., 2010).
- Sustainable Development Goal 17 Strengthening Implementation Tools and Revitalizing Global Partnership for Sustainable Development: The targets for this Goal can also be strongly supported by transport and digital communication networks that can bring together actors and leverage global partnerships for the most effective implementation of sustainable development policies.

Similarly, achieving other SDGs and related targets could facilitate the construction of sustainable transport systems and improve access to them. These can be achieved, for example, by increasing household incomes, improving data use capacities and policy coherence, as well as deploying innovative technology solutions to reduce climate and other environmental impacts. The links between sustainable transport and different sustainable development goals and targets, in addition to interdependencies between countries and stakeholders, must be taken into account in the planning, investment, and implementation actions of multiple stakeholders to accelerate progress toward achieving the sustainable development goals and climate targets (IDS, 2015).

24.7 Potential Barriers to Achieving Energy Efficiency in the Road Transportation Sector

There are some potential barriers to achieving energy efficiency in the road transport sector. These are some of the obstacles:

- Many roads, bridges, and railways were built decades ago and lack the latest advances in energy efficiency technologies, resulting in increased energy consumption and emissions (Energy5 (c), 2023; NRC, 2010).

- Implementing energy-efficient technologies and practices requires significant investments, which can be an obstacle for many organizations (Estache & Garsous, 2012).
- A lack of supportive policies and regulations can hinder the adoption of energy-efficient technologies and practices (Lah, 2015; Roeth et al., 2013).
- Resistance to change is another significant obstacle to achieving energy efficiency in the road transport sector (Rehmatulla & Smith, 2015).
- Private investments in energy efficiency are higher in countries/sectors where public investments in infrastructure are lacking (Lah, 2015).
- Energy efficiency measures can have long payback periods, which can be a barrier for many organizations (Estache & Garsous, 2012).
- The high upfront costs of energy-efficient vehicles can be a barrier for consumers with low purchasing power (Lah, 2015).
- The limited availability of alternative fuels such as electricity and hydrogen can be a barrier to achieving energy efficiency in the road transport sector (ECLA, 2021).
- The limited financing structure of governments often prevents infrastructure projects in energy-saving transportation systems investments that require significant financial resources (Rehmatulla & Smith, 2015).
- Inefficient and inadequate planning in traffic and infrastructure projects leads to suboptimal transportation networks. The resulting bottlenecks, traffic congestion, and excessive travel distances lead to unnecessary energy consumption in both urban and rural areas (Energy5 (c), 2023).

To overcome these obstacles, innovative solutions and policies must be implemented, such as investing in public transportation, building bicycle paths, creating high-density pedestrian-friendly urban areas, and providing incentives and subsidies for the adoption of energy-efficient technologies and practices (Roeth et al., 2013).

24.8 Duties for Policymakers

Policymakers play a significant role in promoting energy efficiency in road transport. They support energy efficiency for the various benefits it provides. These benefits include reducing energy waste and saving costs for governments to achieve decarbonization goals, improving energy reliability and resilience, reducing air pollution, improving public health, and promoting economic development (NGA, 2021).

They can implement a variety of measures to encourage the adoption of energy-efficient practices and technologies. Policymakers can also create regulations and policies aimed at reducing emissions, lowering energy consumption, and encouraging innovation in the road transport sector. Some of the duties of policymakers regarding energy efficiency in road transport can be summarized as follows:

- Policymakers can develop and implement policies and regulations that promote energy efficiency in road transport, such as fuel efficiency standards, carbon pricing, and incentives for the adoption of energy-efficient vehicles (EERE (e), 2022; Energy5 (d), 2023; Karlo, 2023; UNECE, 2015).
- Policymakers can invest in sustainable transport infrastructure, such as public transport, bicycle paths, and pedestrian-friendly urban areas, to promote energy efficiency and reduce emissions (NGA, 2021).
- Policymakers can encourage the adoption of alternative fuels such as biodiesel and compressed natural gas to reduce emissions and increase energy efficiency (UNECE, 2015).
- Policymakers can encourage energy-efficient behaviors, such as reducing idling time and maintaining appropriate tire pressure, through education and awareness campaigns (EERE (e), 2022).
- Policymakers can provide financing and incentives for the adoption of energy-efficient technologies and practices, such as tax credits and grants (IEA (c), 2020).
- By encouraging innovation and the development of sustainable solutions, policymakers can allocate funds to finance research and development in energy-efficient technologies (Thacker et al., 2018).

Policymakers can increase public awareness of the benefits of using energy-efficient systems through public awareness campaigns that promote advantages in terms of cost savings, reduced environmental impact, and improved air quality. The energy efficiency measures taken by policymakers in the United States and the resulting emissions reductions are ambitious and vital to achieving state and federal decarbonization goals. Sixteen Governors and D.C. have established and enacted goals for 100% clean energy or zero-carbon electricity generation (Cimino & Rackley, 2020).

The Governor of Massachusetts signed new climate legislation into law, and the Governor of Louisiana issued an executive order in 2020 setting a goal of making Louisiana carbon neutral by 2050 and joined the US Climate Alliance in May 2021 (LOG, 2021). The Biden-Harris Administration has also set decarbonization targets and prioritized energy efficiency with executive directives that put "the United States on a path to achieve net-zero economy-wide emissions by no later than 2050" and identify investments in energy efficiency and renewable energy (MOG, 2021). The recently proposed federal infrastructure plan aims to achieve 100% carbon-free electricity by 2035 and calls for the adoption of a unified federal clean energy and energy efficiency standard (WH, 2021).

As a result, policymakers play an important role in promoting energy efficiency in road transport. In this context, they can implement various measures to encourage the adoption of energy-efficient practices and technologies, invest in sustainable transportation infrastructure, encourage the adoption of alternative fuels, encourage energy-efficient behavior, and provide financing and incentives. These activities are necessary to achieve sustainability targets and reduce greenhouse gas emissions in the sector, especially by ensuring energy efficiency in road transportation.

24.9 Results and Discussion

The size and density of the road transportation sector, which plays a fundamental role in the life of societies and is the lifeblood of economic activities, causes it to be an important source of energy consumption (EPA (b), 2023). The fact that the sector has great potential for energy efficiency is due to the widespread use of fossil fuels in road transportation. The sector is of critical importance in achieving sustainability goals. At the same time, energy efficiency in road transportation offers great advantages both environmentally and economically. In this context, the study examined energy efficiency in the road transportation sector, which is an important research and application area (Energy Star, 2023).

Energy efficiency means using less energy to do a job, reducing energy waste, reducing energy costs, and reducing environmental pollution. Properly implemented energy efficiency policies are one of the most efficient and cost-effective ways to combat climate change, reduce energy costs for consumers, and increase the competitiveness of businesses (EERE (a), 2022). In addition to the use of houses and buildings, energy efficiency can be achieved in different ways such as developing energy-efficient practices and behaviors and using energy-efficient products (Energy Sage, 2022).

The energy efficiency potential is quite high in road transportation. The use of alternative fuels, the development of more efficient vehicles, and the implementation of policies and practices that encourage energy efficiency reveal many improvement opportunities in this field (En-ROADS, 2020). Taking advantage of these opportunities is very important for the economy, considering the energy spent on transportation in today's competitive environment (AAR, 2019).

Lack of access to affordable energy causes various negativities in socioeconomic growth, especially when developing countries are evaluated (Halicioglu, 2009). Population growth, infrastructure development, and trade activities increase energy consumption (NRC, 2010). Therefore, some of the most important activities that reflect the economic development of a country are energy supply and consumption (Zhang & Lin, 2012).

24.10 Conclusion

In terms of its impact on global energy use and greenhouse gas emissions, the road transportation sector significantly affects energy consumption in terms of energy efficiency (NRC, 2010). Road transportation accounts for 75% of the energy used in this area and when evaluated from this perspective, it has great potential in terms of energy efficiency. Solid fuels dominate the road transport sector. Therefore, in order to reduce the sector's impacts on energy consumption and greenhouse gas emissions, reducing fuel use should be one of the highest priorities for all countries (EET, 2022), because the road transport sector is responsible for a quarter of energy-related greenhouse gas emissions (EIA, 2016). Transport emissions are expected to approximately double by 2050 unless radical measures are taken to curb the dominance of solid fuels in the road transport sector (IEA (a), 2012).

The majority of sustainable development goals are linked to sustainable transport (and therefore to energy efficiency) through specific targets and indicators. Other sustainable development goals are also indirectly linked through the facilitating role of sustainable transport (Energy5 (c), 2023). It is estimated that sustainable and resilient infrastructure, including transport, will be effective in achieving 92% of all sustainable development goals. There are actions to be taken and steps to be taken to overcome potential obstacles to achieving energy efficiency in the road transport sector. These activities are innovative solutions and policies such as investing in public transportation, building bicycle paths, creating high-density pedestrian-friendly urban areas, and providing incentives and subsidies for the adoption of energy-efficient technologies and practices (Roeth et al., 2013; ECLA, 2021).

References

AAR (Association of American Railroads). (2019). *The economic impact of America's freight railroads*. Retrieved from https://www.aar.org/wp-content/uploads/2018/05/AAR-Economic-Impact-US-Freight-Railroads.pdf

Akbar, U., Popp, J., Khan, H., Khan, M. A., & Oláh, J. (2020). Energy efficiency in transportation along with the belt and road countries. *Energies*. https://doi.org/10.3390/en13102607

Al-Mofleh, A., Taib, S., & Salah, W. A. (2010). Malaysian energy demand and emissions from the transportation sector. *Transport, 25*, 448–453. Retrieved from https://www.researchgate.net/publication/49598857_Malaysian_energy_demand_and_emissions_from_the_transportation_sector

Altinay, G., & Karagol, E. (2004). Structural break, unit root, and the causality between energy consumption and GDP in Turkey. *Energy Economics, 26*, 985–994. Retrieved from https://www.sciencedirect.com/science/article/abs/pii/S0140988304000593

Amicarelli, V., Christian, B., Giovanni, L., & Gallucci, T. (2019). Energy efficiency policies in non-residential buildings: The case of the University of Bari Aldo Moro. *Amfiteatru Economic, 21*, 845–860. Retrieved from https://econpapers.repec.org/article/aesamfeco/v_3a21_3ay_3a2019_3ai_3as13_3ap_3a845.htm

Brătucu, G., Constantin, C. P., Chițu, I. B., Grădinaru, E., & Dovleac, L. (2019). Approaching the bioeconomy in terms of increasing the energy efficiency of households in Romania. *Amfiteatru Economic, 21*, 90–104. Retrieved from https://ideas.repec.org/a/aes/amfeco/v21y2019i50p90.html

Bresette, D. (2023). *Driven to be more energy efficient, distribution transformers are more than meets the eye*. Retrieved from https://www.eesi.org/articles/view/driven-to-be-more-energy-efficient-distribution-transformers-are-more-than-meets-the-eye

Cimino, E., & Rackley, J. (2020). *Governors leading on energy transitions: An overview of state energy goals*. NGA. Retrieved from https://www.nga.org/center/publications/governors-leading-energy-transitions/

Constellation. (2022). *Energy conservation vs. energy efficiency: What is the difference?* Retrieved from https://www.constellation.com/energy-101/what-is-energy-conservation.html

ECLA (The Economic Commission for Latin America). (2021). *Freight transport by road: Tools and strategies for energy efficiency and sustainability*. Retrieved from https://www.cepal.org

EEA (European Environment Agency). (2022). *Energy efficiency*. Retrieved from https://www.eea.europa.eu/publications/ENVISSUENo12/page027.html
EERE (a) (Office of Energy Efficiency and Renewable Energy). (2022). *Energy efficiency: Buildings and industry*. Retrieved from https://www.energy.gov/eere/energy-efficiency-buildings-and-industry
EERE (b) (Office of Energy Efficiency and Renewable Energy). (2022). *Sustainable transportation and fuels*. Retrieved from https://www.energy.gov/eere/sustainable-transportation-and-fuels
EERE (c) (Office of Energy Efficiency and Renewable Energy). (2022). *Environmental impacts of clean energy*. Retrieved from https://www.energy.gov/eere/environmental-impacts-clean-energy
EERE (d), Office of Energy Efficiency and Renewable Energy. (2022). *Sustainable steps boosting energy efficiency in the transportation sector*. Retrieved from https://www.energy.gov/eere/environmental-impacts-clean-energy
EERE (e), Office of Energy Efficiency and Renewable Energy. (2022). *Transportation system efficiency*. Retrieved from https://afdc.energy.gov/conserve/system_efficiency.html
EESI (Environmental and Energy Study Institute). (2021). *Energy efficiency*. Retrieved from https://www.eesi.org/topics/energy-efficiency/description
EET (Energy Efficiency in Transport). (2022). Retrieved from https://www.seforall.org/energy-efficiency-for-sustainable-development/energy-and-transport
EIA (a) (Energy Information Administration). (2022). *Energy efficiency and conservation*. Retrieved from https://www.eia.gov/energyexplained/use-of-energy/efficiency-and-conservation.php
EIA (b) (Energy Information Administration). (2016). *Transportation sector energy consumption*. International Energy Outlook 2016. Retrieved from https://www.eia.gov/outlooks/ieo/pdf/transportation.pdf
Energy Renovables. (2019). *12 Strategies to step up global energy efficiency*. Retrieved from https://www.renewableenergymagazine.com/energy_saving/12-strategies-to-step-up-global-energy-20191205
Energy Sage. (2022). *The environmental benefits of energy efficiency*. Retrieved from https://www.energysage.com/energy-efficiency/environmental-impact-of-ee/
Energy Star U.S. Environmental Protection Agency. (2023). *Energy efficiency*. Retrieved from https://www.energystar.gov/about/how_energy_star_protects_environment/energy_efficiency
Energy5 (a). (2023). *The economic impact of energy efficiency measures*. Retrieved from https://energy5.com/the-economic-impact-of-energy-efficiency-measures
Energy5 (b). (2022). *Sustainable steps boosting energy efficiency in the transportation sector*. Retrieved from https://energy5.com/sustainable-steps-boosting-energy-efficiency-in-the-transportation-sector
Energy5 (c). (2023). *Addressing barriers to energy efficiency in the transportation-infrastructure*. Retrieved from https://energy5.com/addressing-barriers-to-energy-efficiency-in-the-transportation-infrastructure
Energy5 (d) (2023). *Role of government policy in promoting energy efficiency in rail*. Retrieved from https://energy5.com/addressing-barriers-to-energy-efficiency-in-the-transportation-infrastructure
En-ROADS. (2020). *Transport–energy efficiency*. Retrieved from https://docs.climateinteractive.org/projects/enroads/en/latest/guide/transport_ee.html
EPE U.S. Environmental Protection Agency (2023) , Local Energy Efficiency Benefits and Opportunities, https://www.epa.gov/statelocalenergy/local-energy-efficiency-benefits-and-opportunities
EPA (a) (Environmental Protection Agency). (2014). *Promoting technology innovation for clean and safe water*. Retrieved from https://www.epa.gov/sites/default/files/2014-04/documents/clean_water_blueprint_final.pdf
EPA (b) (Environmental Protection Agency). (2023). *Local energy efficiency benefits and opportunities*. Retrieved from https://www.epa.gov/statelocalenergy/local-energy-efficiency-benefits-and-opportunities

Estache, A. (2006). *Infrastructure: A survey of recent and upcoming issues*. World Bank. Retrieved from https://documents.worldbank.org/en/publication/documents-reports/documentdetail/519111468337245410/infrastructure-and-development-a-survey-of-recent-and-upcoming-issues

Estache, A., & Garsous, B. (2012). *The impact of infrastructure on growth in developing countries*. IFC Economics Notes. Retrieved April 2012, from https://www.researchgate.net/publication/322569814_The_Impact_of_Infrastructure_on_Growth_in_Developing_Countries_Dynamic_Panel_Data_Analysis

GFEI (Global Fuel Economy Initiative). (2011). *50by50 prospects and progress—Executive summary*. Retrieved from www.globalfueleconomy.org/Documents/Publications/executive-summary.pdf

Halicioglu, F. (2009). An econometric study of CO_2 emissions, energy consumption, income, and foreign trade in Turkey. *Energy Policy, 37*, 1156–1164. Retrieved from https://www.sciencedirect.com/science/article/abs/pii/S0301421508007027

IDS (Institute for Development Studies). (2015). *Knowledge sharing and development in a digital age*. Institute for Development Studies Policy Briefing Issue 87, Brighton. Retrieved from https://www.itoca.org/articles/1575831111PB87_AGID258_KnowledgeShare_Online.pdf

IEA (a) (International Energy Agency). (2012). *Energy technology perspectives 2012: Pathways to a clean energy system*. Retrieved from https://iea.blob.core.windows.net/assets/7136f3eb-4394-47fd-9106-c478283fcf7f/ETP2012_free.pdf

IEA (b) (International Energy Agency). (2016). *World energy outlook*. Special Report: Energy and air pollution. International Energy Agency. Retrieved from https://www.iea.org/reports/energy-and-air-pollution

IEA (c) (International Energy Agency). (2020). *Energy efficiency and economic stimulus*. Retrieved from https://www.iea.org/articles/energy-efficiency-and-economic-stimulus

ITF (International Transport Forum-OECD). (2016). *International Transport Forum Report*. Retrieved from https://www.itf-oecd.org/

Ji, X., & Chen, G. Q. (2006). Exergy analysis of energy utilization in the transportation sector in China. *Energy Policy, 34*, 1709–1719. Retrieved from https://www.sciencedirect.com/science/article/pii/S0301421505000480

Karlo, H. (2023). Identifying policy areas for the transition of the transportation sector. *Energy Policy, 178*, 113591. Retrieved from https://www.sciencedirect.com/science/article/pii/S0301421523001763

Lah, O. (2015). The barriers to low-carbon land-transport and policies to overcome them. *European Transport Research Review, 7*, 5. https://doi.org/10.1007/s12544-014-0151-3

Li, J., Lo, K., Zhang, P., & Guo, M. (2016). Consumer travel behaviours and transport carbon emissions: A comparative study of commercial Canters in Shenyang, China. *Energies, 9*, 765. Retrieved from https://www.mdpi.com/1996-1073/9/10/765

Lipscy, P. Y., & Schipper, L. (2013). Energy efficiency in the Japanese Transport Sector. *Energy Policy, 56*, 248–258. Retrieved from https://www.sciencedirect.com/science/article/abs/pii/S0301421512010981

Liu, L., Wang, K., Wang, S., Zhang, R., & Tang, X. (2018). Assessing energy consumption, CO_2 and pollutant emissions and health benefits from China's transport sector through 2050. *Energy Policy, 116*, 382–396. Retrieved from https://www.sciencedirect.com/science/article/abs/pii/S0301421518300958

LOG (Louisiana Office of the Governor). (2021). *Gov. Edwards joins bipartisan U.S. Climate Alliance, commits to addressing climate change at the state level*. State of Louisiana. Retrieved from https://gov.louisiana.gov/index.cfm/newsroom/detail/3107

Mariyakhan, K., Mohamued, E. A., Khan, M. A., Popp, J., & Oláh, J. (2020). Does the level of absorptive capacity matter for carbon intensity? Evidence from the USA and China. *Energies, 13*, 407. Retrieved from https://www.researchgate.net/publication/338580835_Does_the_Level_of_Absorptive_Capacity_Matter_for_Carbon_Intensity_Evidence_from_the_USA_and_China

MOG (Massachusetts Office of the Governor). (2021). *Governor Baker signs climate legislation to reduce greenhouse gas emissions, protect environmental justice communities*. Commonwealth of Massachusetts. Retrieved from https://www.mass.gov/news/governor-baker-signs-climate-legislation-to-reducegreenhouse-gas-emissions-protect-environmental-justice-communities

NGA (National Governors Association). (2021). *State energy efficiency policy in a new era: A toolkit for governors*. Retrieved from https://www.nga.org/wp-content/uploads/2021/10/Energy-Efficiency-Toolkit.pdf

NRC (National Research Council). (2010). *Real prospects for energy efficiency in the United States*. The National Academies Press. https://doi.org/10.17226/12621

Parfitt, J., Barthel, M., & Macnaughton, S. (2010). Food waste within food supply chains: Quantification and potential for change to 2050. *Philosophical Transactions of the Royal Society B: Biological Sciences, 365*(1554), 3065–3081. Retrieved from https://www.researchgate.net/publication/45659717_Food_Waste_within_Food_Supply_Chains_Quantification_and_Potential_for_Change_to_2050

Rehmatulla, N., & Smith, T. (2015). Barriers to energy efficiency in shipping: A triangulated approach to investigate the principal agent problem. *Energy Policy, 84*, 44–57. https://doi.org/10.1016/j.enpol.2015.04.019

Roeth, M., Kircher, D., Smith, J., & Swim, R. (2013). *Barriers to the increased adoption of fuel efficiency technologies in the North American on-road freight sector*. International Council on Clean Transportation (ICCT). Retrieved from https://theicct.org/publication/barriers-to-adoption-of-fuel-efficiency-technologies-in-the-north-american-on-road-freight-sector/

SEforALL. (2022). *Energy efficiency in transport*. Retrieved from https://www.seforall.org/energy-efficiency-for-sustainable-development/energy-and-transport

Solaymani, S. (2019). CO_2 emissions patterns in 7 top carbon emitter economies: The case of transport sector. *Energy, 168*, 989–1001. Retrieved from https://www.sciencedirect.com/science/article/abs/pii/S0360544218323569

Talbi, B. (2017). CO_2 emissions reduction in road transport sector in Tunisia. *Renewable and Sustainable Energy Reviews, 69*, 232–238. Retrieved from https://www.sciencedirect.com/science/article/abs/pii/S1364032116309583

Thacker, S., Adshead, D., Morgan, G., Crosskey, S., Bajpai, A., Ceppi, P., Hall, J. W., & O'Regan, N. (2018). *Infrastructure: Underpinning sustainable development*. UNOPS. Retrieved from https://www.itrc.org.uk/itrcpublications/infrastructure-underpinning-sustainable-development-2/

US EIA Energy Information Administration (2016), Transportation Sector Energy Consumption, International Energy Outlook 2016, https://www.eia.gov/outlooks/ieo/pdf/transportation.pdf

UNCSD (United Nations Commission on Sustainable Development). (2011). *Global overview of fuel efficiency and motor vehicle emission standard: Policy options and perspectives for international cooperation*. UNCSD. Retrieved from https://www.un.org/esa/dsd/resources/res_pdfs/csd-19/Background-paper3-transport.pdf

UNECE (United Nations Economic Commission for Europe). (2015). *United Nations best policy practices for promoting energy efficiency, New York & Geneva*. Retrieved from https://unece.org/DAM/energy/se/pdfs/geee/pub/ECE_Best_Practices_in_EE_publication.pdf

United Nations. (2016). *The sustainable development goals report 2016*. Retrieved from https://unstats.un.org/sdgs/report/2016/the%20sustainable%20development%20goals%20report%202016.pdf

United Nations. (2021). Sustainable transport, sustainable development. *Interagency Report for Second Global Sustainable Transport Conference*. Retrieved from https://sdgs.un.org/sites/default/files/20210/Transportation%20Report%202021_FullReport_Digital.pdf

White House. (2021). *Fact sheet: The American jobs plan*. The White House. Retrieved from https://www.whitehouse.gov/briefing-room/statementsreleases/2021/03/31/fact-sheet-the-american-jobs-plan/

Zhang, C., & Lin, Y. (2012). Panel estimation for urbanization, energy consumption and CO_2 emissions: A regional analysis in China. *Energy Policy, 49*, 488–498. Retrieved from https://www.sciencedirect.com/science/article/abs/pii/S0301421512005587

Zhang, M., Li, G., Mu, H. L., & Ning, Y. D. (2011). Energy and exergy efficiencies in the Chinese transportation sector, 1980–2009. *Energy, 36*, 770–776. Retrieved from https://www.sciencedirect.com/science/article/abs/pii/S0360544210007346

Zhang, Y. J., Peng, H. R., Liu, Z., & Tan, W. (2015). Energy rebound effect for road passenger transport in China: A dynamic panel quantile regression approach. *Energy Policy, 87*, 303–313. Retrieved from https://www.sciencedirect.com/science/article/abs/pii/S0301421515301087

Chapter 25
Driving Sustainable Energy Efficiency in Industry: An Investigative Approach

M. Ziya Sogut

Nomenclature

CO_2	Carbon dioxide
CUSUM	Cumulative sum
EMS	Energy management system
EPIs	Energy performance indicators
HDD	Heating Degree Days
ISO	International Organization for Standardization
UNDP	United Nations Development Programme

25.1 Introduction

Global climate change concerns, intensified since the 1990s, necessitate a drastic reduction in fossil fuel reliance. The industrial sector, a significant energy consumer, is crucial to this transition. The COVID-19 pandemic's impact on sectors, including reduced production and increased energy costs, highlighted the critical need for dynamic energy management strategies. Balancing economic sustainability with the imperative for fossil fuel reduction presents a complex challenge for businesses (Moehle et al., 2008). The associated costs of transitioning to alternative energy sources are substantial, posing significant investment challenges. Despite these difficulties, national energy policies must consider not only supply and demand balance but also sustainable growth. Studies like Pusnik et al.'s (2016) energy efficiency index for Slovenia's industrial sector demonstrate the value of targeted approaches. Successfully mitigating fossil fuel use in industry requires aligning national goals

M. Z. Sogut (✉)
Maritime Faculty, Piri Reiss University, Istanbul, Turkey

M. Z. Sogut, M. Koray (eds.), *Energy Rationality and Management for Decarbonization*, Green Energy and Technology,
https://doi.org/10.1007/978-3-031-68497-5_25

with global targets, predicting future energy consumption patterns, and prioritizing energy efficiency for all stakeholders.

Global energy security concerns and national savings targets are increasingly focusing on reducing energy consumption across sectors. The COVID-19 pandemic, while causing a significant 4% contraction in global energy demand in 2020, has since been followed by a partial recovery, with a 0.5% increase in 2021 compared to pre-crisis levels (Orhan, 2021). This resurgence in demand, especially for fossil fuels (estimated at 85% of total consumption), points to a continuation of high fossil fuel reliance despite the crisis. While 2020 witnessed a 14% contraction, the 2021 increase represents a 3% rise from the pre-crisis baseline. The ongoing effects of the Russian crisis have further complicated fossil fuel supply dynamics. Looking ahead, projections suggest continued, albeit moderate, growth in industrial energy consumption, with a projected 1.2% increase in the industrial sector's energy use by 2050. Furthermore, non-OECD countries are forecast to exhibit a much sharper rise in energy demand, potentially tripling the industrial sector's growth rate by 2040 (IEA, 2021a). These trends underscore the critical need for proactive energy management strategies within the industrial sector, effectively prioritizing both energy efficiency and consumption patterns as part of comprehensive national strategies (IEA, 2021b). Effective energy management is a structured approach to reducing energy costs and environmental impacts while maintaining service quality and output levels (Onaygil, 2015). Its importance in modern industrial sectors cannot be overstated; it necessitates a robust institutional framework and dynamic operational procedures. Efficient energy management, driving up output quality, fosters competition and simultaneously mitigates environmental concerns. This is further supported by extensive research on the subject (Fahad et al., 2017; Schulze et al., 2016; Mirzakhani et al., 2017). Growing social pressure to curb fossil fuel consumption and address climate change further emphasizes the need for proactive energy management strategies. The emergence of standards like ISO 50001 for energy management, building on the successes of ISO 9001 for quality and ISO 14000 for environmental sustainability, signals a crucial cultural shift (Mohamad et al., 2014). This shift demands a change in business practices, emphasizing energy efficiency and responsible consumption within organizational structures. Implementing effective energy management programs and leveraging existing energy management infrastructure has become a significant challenge for many businesses.

Existing literature offers various perspectives on the structural needs of energy management within industries. Studies like that of Ates and Durakbasa (2012) explore process issues and sectoral energy intensities in energy management applications. Other research, such as Finnerty et al.'s (2017), has developed novel energy management techniques, proposing a methodology based on four key criteria for optimal performance. This research highlights the importance of tailored approaches for different sectors. Various energy management systems and applications, including monitoring and supervisory control and data acquisition (SCADA) systems, are also documented. Models like EMMM 50001, ISO 50001-based Energy Management Maturity Model, and EUMMM offer frameworks for evaluating and improving energy management practices (Bojana & Filipović, 2016; EnergyCAP,

2015; Mineeva, 2016). However, these models often fail to account for the unique management cultures prevalent within specific companies, particularly those with multiple production or service units. Handling such diverse operational aspects within a single, holistic framework requires a robust, sustainable management program.

The global shift towards sustainable development requires a significant transformation in industrial practices, particularly about energy consumption. While industrial processes are often a critical component of economic prosperity, they frequently exhibit high energy intensities, which contribute significantly to greenhouse gas emissions and resource depletion (Bojana & Filipović, 2016). This requires a proactive approach to energy efficiency, which will encourage industries to pursue a more sustainable future. The objective of this study is to develop and implement strategies that will facilitate the achievement of sustainable energy efficiency in industrial contexts. The extant literature underscores the significance of integrated energy management systems (e.g., ISO 50001) in attaining energy efficiency gains. However, these systems frequently lack the requisite dynamic adaptability and responsiveness to address industrial operations' complex and often unpredictable nature. This research addresses this gap by proposing a novel energy management model (Uysal & Sogut, 2017) that incorporates dynamic monitoring, real-time data analysis, and adaptive control strategies, aiming to achieve optimal energy efficiency and operational sustainability within industrial enterprises. This introduction will outline the motivations for this study, highlighting the challenges and opportunities presented by the current state of industrial energy efficiency, and detail the proposed methodology for developing and evaluating the novel energy management model.

25.2 Navigating Industrial Energy Transitions

Industrial energy transitions present a complex interplay of technological advancements, regulatory pressures, economic incentives, and societal expectations. In order to successfully navigate this transition, it is necessary to adopt a multifaceted approach that encompasses investment in renewable energy sources, the development and adoption of energy-efficient technologies, the implementation of robust energy management systems, and the creation of supportive policies and regulations. Furthermore, fostering collaboration between industry, government, and research institutions is critical to accelerate innovation, facilitate knowledge sharing, and ultimately achieve a sustainable and resilient industrial sector.

The industrial sector, a significant energy consumer (approximately 29% of global energy use), also employs a substantial portion of the global workforce (around 23%). Energy is a vital input for a wide range of industrial processes, from manufacturing and production to logistics. Moreover, corporate sustainability is increasingly recognizing the importance of effectively managing both energy and environmental factor. This has led to energy efficiency becoming a crucial national and sectoral priority, particularly for energy-intensive industries (Lei, 2024). The need for sustainable energy practices extends beyond economic considerations, demanding institutional frameworks to address environmental concerns. Research by the United Nations Development Programme (UNDP) highlights the massive energy consumption of key industrial sectors: chemical and petrochemical (33%), iron and steel (17%), cement (9%), and pulp and paper (5%). These four sectors alone account for 64% of total industrial energy consumption, demonstrating the critical role energy plays in product costs and overall industrial sustainability (UNDP, 2016; EIA, 2016). European Union studies on industrial energy efficiency provide valuable insights into consumption patterns. Research by Yeen and Ravi (Brian et al., 2024), for example, indicates that eight industrial sectors consumed approximately 98% of the EU's total energy, representing roughly 25% of the overall consumption (Tsemekidi Tzeiranaki et al., 2020). Foresight analysis suggests that consumption in iron and steel, and the chemical sectors will likely continue to rise (Fig. 25.1).

Global industrial energy demand is exhibiting a pronounced upward trend. Using 2012 as a baseline, OECD countries' energy consumption increased by roughly 16%, rising from 73 quadrillion BTUs to 83 quadrillion BTUs. In contrast, non-OECD countries are projected to see a significantly steeper rise, potentially reaching 225 quadrillion BTUs, representing an increase of approximately 51% (IEA, 2021a). This stark difference highlights the distinct energy-intensive characteristics of industrial sectors in non-OECD nations, while OECD countries exhibit a more moderate yet notable upward trajectory. Industrial energy consumption is largely reliant on fossil fuels like natural gas, petroleum products, coal, and petcoke. This reliance, particularly for countries or companies heavily dependent on imported energy, creates vulnerabilities in energy security and corporate sustainability. The

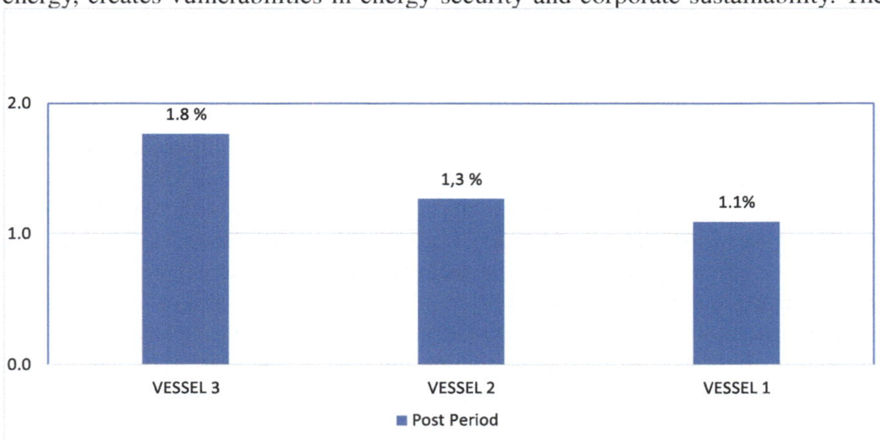

Fig. 25.1 Final energy consumption of industrial sectors in EU28 (Tsemekidi Tzeiranaki et al., 2020)

ongoing Russian crisis has further exposed these vulnerabilities, impacting energy supply and national strategies (Priyadarshi et al., 2023). To address these challenges, a shift towards resource sustainability in national energy policies, including increasing the use of domestic energy sources, is essential. Simultaneously, implementing energy efficiency measures becomes critical to reducing overall energy demand and bolstering energy security.

The industrial sector, consuming roughly one-third of global final energy, is characterized by diverse and energy-intensive operations. Developing a robust energy management infrastructure focused on efficiency and cost reduction is crucial for energy-intensive sub-sectors. Numerous studies have explored this, particularly in identifying potential energy savings. These analyses often employ various institutional approaches to forecasting and quantifying potential savings, as detailed in Fig. 25.2.

National strategies for the industrial sector consistently prioritize energy management actions. Numerous strategic documents and action plans incorporate these priorities, aiming to achieve corporate and national targets. For instance, the Energy Efficiency Directive (2012/27EU) Annex VI outlines minimum criteria for energy audits as part of energy management systems (EU, 2012). The EU has also revised its 2030 targets, highlighting key pillars for national strategies, including energy efficiency, renewables, greenhouse gas emission reductions, interconnections, and research and innovation (EU, 2019). Many countries have adopted similar initiatives. Turkey, For example, Turkey's National Energy Efficiency Strategy emphasizes reducing energy losses and improving energy efficiency to decrease energy intensity as a key step. It is aimed for at least a 10% reduction in energy intensity across all sub-sectors by a specified date (Sogut & Karakoc, 2023). These measures demonstrate the significant potential for savings within the industrial sector but also underscore the ongoing challenge of effectively managing energy in manufacturing.

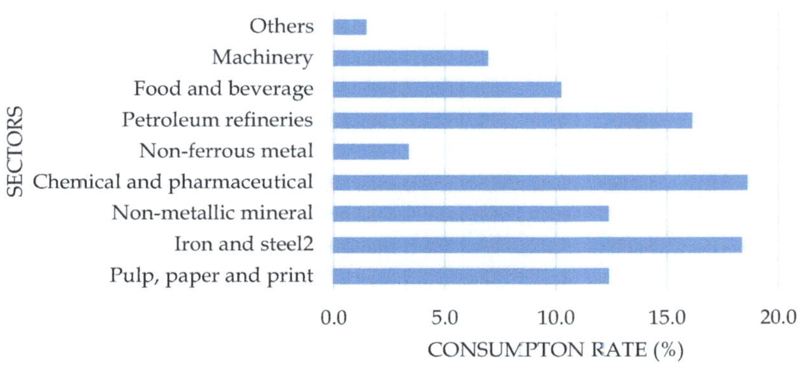

Fig. 25.2 Sectorial energy saving and energy intensity values (IEA, 2008)

25.2.1 Towards Sustainable Model of Energy Management

To limit global warming, global CO_2 emissions must be reduced by at least 30% below 1990 levels by 2050 (Fig. 25.2). Developed nations face a far steeper reduction target (approximately 80%) while developing and newly industrialized countries are permitted a maximum 30% increase. Without fundamental changes in energy systems, emissions in these nations are projected to double, or even quadruple, which necessitates a rapid shift in energy production and consumption towards renewable energy and increased efficiency (Zhu et al., 2023). Given the uncertainties surrounding climate system behavior, these emission reduction targets represent minimal requirements. Controlling energy consumption and establishing sustainable energy policies require a holistic approach, best modeled by a conceptual map defining a sustainable energy management structure (EU, 2012). Effective sustainable energy management necessitates the integration of all relevant components, which in turn dictates the formation of appropriate management elements. Energy management, at its core, is a strategy for efficient energy use without compromising product or service quality. A holistic energy management system, aligned with national strategies, is built on a management framework tailored to specific sector needs. The most widely recognized and effective model in this regard is the internationally recognized ISO 50001 Energy Management System (Yücel & Muhsin, 2016).

Different energy management models have been developed for industrial sectors, often adapting to specific production scales (Santi et al., 2024). These models frequently involve implementing measures without disrupting local production or service delivery. Various approaches exist, ranging from companies with ISO 50001 certification and robust SCADA monitoring systems, to those employing local metering for continuous monitoring at key points. Industrial entities typically base their energy management on their own production efficiency benchmarks and projected consumption, leading to the implementation of energy management systems without a fundamental shift to a truly efficient energy management model. Even companies certified under ISO 50001 sometimes fail to achieve continuous energy improvement. Therefore, effective energy management systems must possess a dynamic structure, encompassing various factors. The proposed Energy Management Model is designed to enhance the effectiveness of ISO 50001, as outlined in Fig. 25.3, before detailed evaluation.

The competition between energy supply and energy efficiency is a complex interplay of economic, environmental, and social factors. While expanding energy supply often emphasizes economic growth and affordability, focusing on energy efficiency prioritizes resource conservation, reduced environmental impact, and long-term sustainability. This competition highlights the need for a balanced approach that integrates both strategies to achieve economic prosperity without compromising environmental integrity or future energy security. Effective policy frameworks must incentivize both supply-side development and demand-side efficiency improvements to maximize overall energy resource utilization.

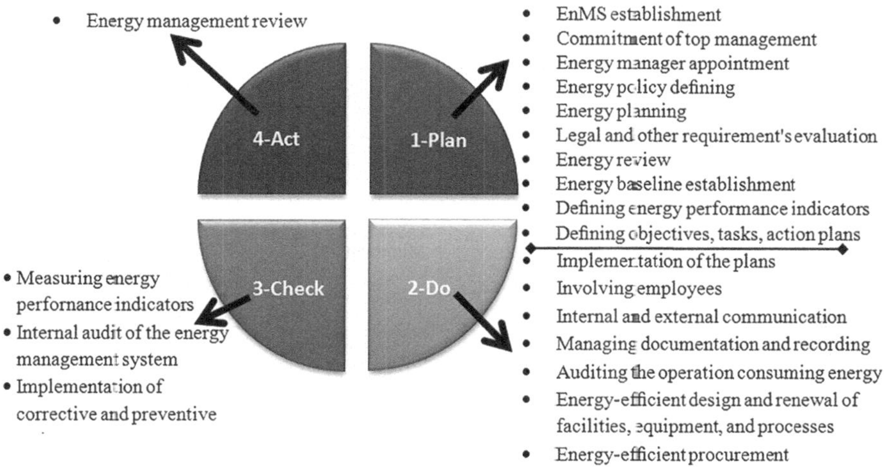

Fig. 25.3 The continuous improvement cycle of the ISO-50001 energy management system

Industrial energy management is a complex and multifaceted field, encompassing a dynamic interplay between economic, energy, and environmental considerations. From a business perspective, energy management is primarily regarded as the administration of a pivotal economic input, given that energy exerts a direct influence on production parameters. However, the growing societal awareness of environmental concerns has led industrial facilities to increasingly consider the mitigation or elimination of their environmental impact. In the present era, environmental sustainability constitutes a pivotal component of sustainable development. Consequently, energy management systems are obliged to contemplate the environmental ramifications of energy utilization in conjunction with economic efficiency. The interrelationships between these three factors delineate the scope of energy management, which can be directly correlated with the facility's energy consumption and management. It is evolving with the objective of ensuring the sustainability of energy management based on energy efficiency in conjunction with direct supply. This relationship directly points to an inverse correlation, as illustrated in Fig. 25.4.

In the literature, an architectural framework was developed to establish a foundation for energy management, enabling designers to treat energy as a fundamental design principle. The concept of managing energy as a critical resource is not novel. Energy architecture incorporates several ideas, including classifying energy as a resource, prioritizing requests, accounting for consumers, allocating resources based on constraints and leveraging feedback to guarantee quality of service. A three-component decomposition is used to manage scarce shared resources. Figure 25.5 shows three components: (1) a policy interface, (2) a resource monitoring and control mechanism, and (3) a management module. Section 25.2 describes these components' roles in the proposed architectural framework. This energy management architecture will address and prevent these issues.

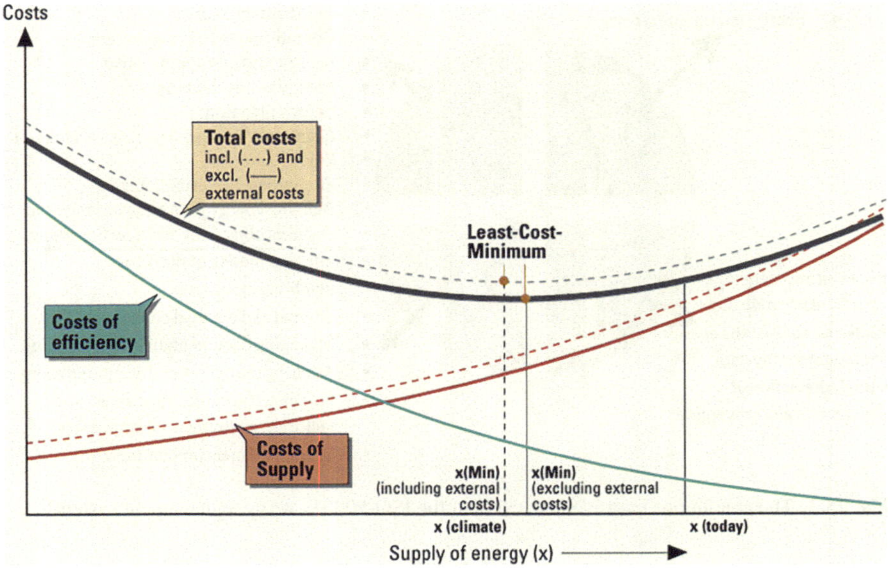

Fig. 25.4 Correlation of energy supply and energy efficiency (Gtz, 2004)

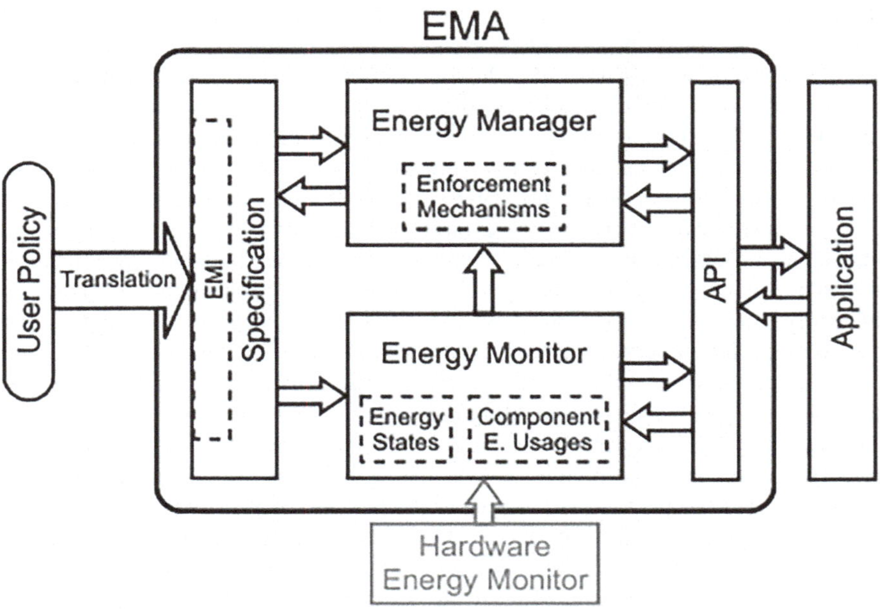

Fig. 25.5 Flow scheme of energy management architecture (Jiang et al., 2007)

Effective energy management requires establishing clear, direct, and indirect relationships and procedures among all system components. While ISO 50001 mandates this, energy management system implementations often fall short of desired outcomes due to fragmented or inadequately integrated systems, highlighting a fundamental flaw in the design. A robust, dynamic management system necessitates a thorough review and restructuring of the existing setup. Figure 25.6 depicts the proposed energy efficiency management model, outlining a structured approach to establishing such a system. Crucially, the energy management system must be built upon a firm corporate identity. ISO 50001, as a corporate model, defines improvement phases, but each defined improvement step must be supported by a specific, assigned action with defined responsibilities.

The proposed energy efficiency management model's implementation must be seamlessly integrated with the existing corporate structure. Therefore, a comprehensive assessment methodology for the energy management system has been developed, outlined as a flowchart in Fig. 25.6. Each stage of this model is further detailed with a specific action plan.

Step 1: Structuring and responsibility: Regardless of ISO 50001 certification status, establishing an energy management system (EMS) requires a clear management framework and defined responsibilities within the industrial organization. This framework's primary responsibility is securing top management's commitment to continuous improvement. This step necessitates:

- Appointing an energy management lead and forming a management team.
- Conducting a self-assessment and defining the scope and boundaries of the energy policy. The team must clearly define the energy policy for the organization.

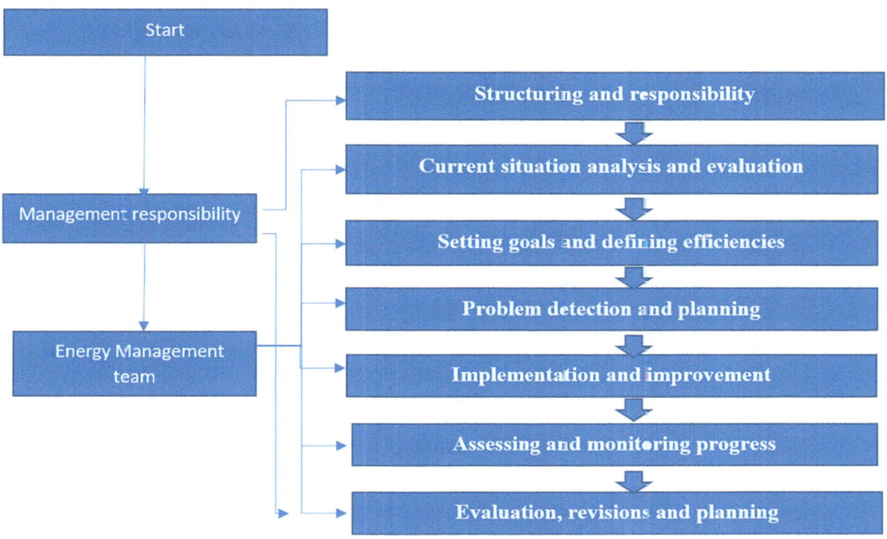

Fig. 25.6 Energy efficiency management model

- Defining the tasks and responsibilities of the designated team.
- Ensuring top management representation within the project team.
- Scheduling periodic meetings for status reviews and progress monitoring.
- This initial stage lays the groundwork for a successful EMS implementation by establishing a clear organizational structure, accountability, and a commitment to improvement at the highest levels.

Step 2: Current situation analysis and evaluation: Effective energy management hinges on thorough energy efficiency analysis and implementation. The energy unit must first identify key energy consumers, focusing on those accounting for over 80% of total energy consumption. Existing data potential within the organization regarding energy consumption and generation must be assessed and evaluated, and the underlying infrastructure scrutinized.

Based on the specific characteristics of the facility, energy performance indicators (EPIs) should be evaluated, possibly using regression or demand analysis. These analyses should reveal the IT infrastructure, data collection methods, and establish a clear baseline for potential energy consumption reductions. A preliminary study approach for energy management context is developed in Fig. 25.7. Energy consumption standards are defined by comparing current energy consumption with target levels and evaluating energy performance indices. Industrial

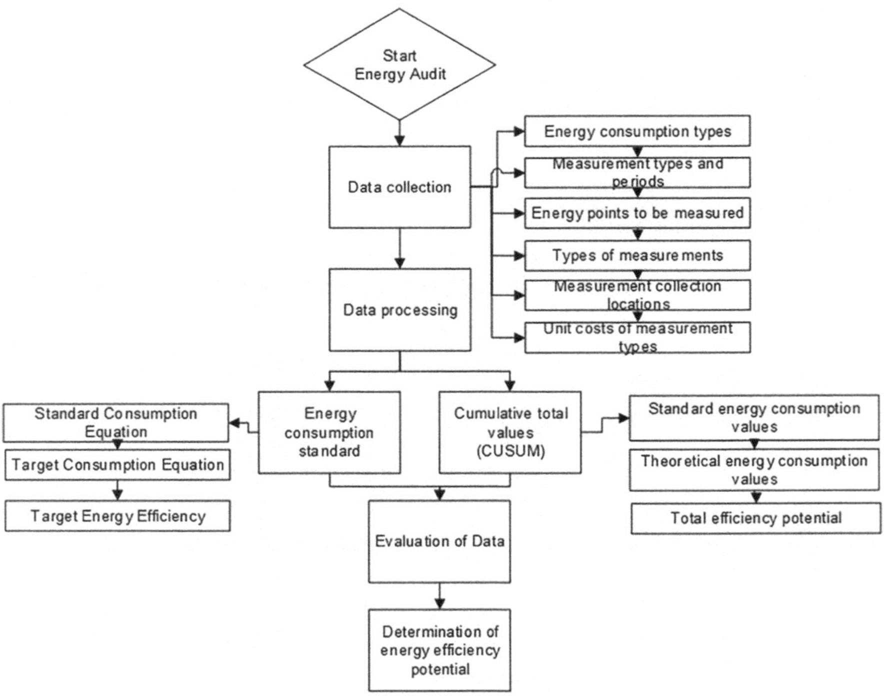

Fig. 25.7 Flow scheme of energy audit

production, however, is contingent on numerous variables. These variables can be categorized as either specific or controllable. Specific variables directly influence energy demand based on production levels, forming the basis of the equation describing the energy-product relationship. Controllable variables, in contrast, are factors outside of the production process itself, such as energy system operations, process controls, and maintenance standards; these are actively managed by the organization to minimize energy consumption. The standard equation, fundamentally, represents a linear relationship between energy requirements and specific variables, a key equation for energy efficiency analysis.

$$E = a + b(P) \tag{25.1}$$

In this linear equation, the constant "*a*" represents the coefficient "*b*", and "*P*" represents the specific variable defining production. However, a process might involve multiple variables (*x*, *y*, *z*), in which case the linear equation becomes;

$$E = a + b(x) + c(y) + d(z) \tag{25.2}$$

While three different linear equations can be used for energy assessments, the equation presented above is the fundamental model for directly evaluating the production-consumption relationship. Regression analysis using energy consumption data initially determines this standard equation. Then, considering consumption patterns associated with this equation, targets are set for the entire factory. These target-derived equations have the same form as the standard equation. Expected energy consumption under target conditions should be calculated and compared to the current situation. Another crucial parameter for evaluating energy performance is the cumulative sum (CUSUM) graph. This method uses the least squares method to define the potential of energy data and cumulatively aggregates these potentials. The CUSUM graph, or cumulative totals, identifies potential savings or trends in consumption related to targeted consumption within the system under analysis. Constructing the CUSUM graph involves calculating the cumulative sum of differences between actual energy consumption and target or projected energy consumption and identifies cumulative energy savings potential. The resulting graph visually displays performance changes. Negative regions or values indicate improvements in facility performance, while positive values reflect declining performance. To generate a CUSUM graph, the cumulative sum of the difference between actual and target energy consumption is calculated. This is expressed as:

$$\sum_{i=n}^{i=1} \dot{E}_{total} = \sum_{i=n}^{i=1} \dot{E}_{real} - \sum_{i=n}^{i=1} E_{expext} \tag{25.3}$$

where $\sum_{i=n}^{i=1} \dot{E}_{total}$ is the cumulative total savings potential, $\sum_{i=n}^{i=1} \dot{E}_{real}$ is the real energy consumption, $\sum_{i=n}^{i=1} E_{expext}$ refers to the expected consumption of the energy users. The primary goal of energy management in this facility is to identify the cumulative potential for savings. This is achieved by considering target consumption levels and their associated costs. The potential of the target energy consumption is referred to the following equation:

$$\sum_{i=n}^{i=1} \dot{E}_{cu} = \sum_{i=n}^{i=1} \dot{E}_{expect,d} + \sum_{i=n}^{i=1} E_{expext,d+1} \tag{25.4}$$

where $\sum_{i=n}^{i=1} \dot{E}_{cu}$ is the cumulative energy consumption, $\sum_{i=n}^{i=1} \dot{E}_{expext,d}$ is the monthly expected difference consumption energy, $\sum_{i=n}^{i=1} E_{expext,d+1}$ is the expected consumption difference of the next month.

Step 3: Setting goals and defining efficiencies: Following initial assessments or studies, target setting and performance evaluation activities should be implemented. All facilities implementing an energy management system must project improvement targets and potential savings for each critical energy use point within their operation's scope. These targets should be communicated across all relevant units, and an action plan infrastructure should be developed.

Step 4: Problem identification and planning: Based on target values derived from energy consumption analysis, problem areas within key energy users should be identified. Before developing action plans, these problems should be categorized by the energy management team into three groups: those requiring no investment, those requiring minimal investment, and those necessitating significant financial resources. Action plans should be prioritized based on this categorization, and each action should detail its estimated timeframe and implementation methodology.

Step 5: Implementation and improvement: The success of the energy management system (EMS) hinges on developing and implementing solutions based on identified efficiency potential. However, organizational commitment to these improvements will only materialize through demonstrable results. Therefore, the energy management team must prioritize and sequence action plans and their implementations. Effective solutions tailored to the operational parameters of the business will serve as significant motivators. To this end, comprehensive awareness training for the entire organization, task-specific process improvement training, and specialized expert training should be planned. The study should meticulously record the performance of implemented actions against targeted processes. This process also evaluates the effectiveness of the management system itself.

Step 6: Assess and monitor progress: Effective implementation of an energy management system (EMS) requires a well-structured management program, recognizing its importance as a significant achievement for participating businesses. Key success metrics include a robust communication plan, comprehensive training and awareness campaigns, identification of efficiency measures for consumption and capacity relationships, continuous monitoring and evaluation, and development of a knowledge management system for the EMS. This comprehensive monitoring process should encompass all aspects of the program, from procedures and daily actions to key energy users and energy performance indicators, creating a clear control framework.

Step 7: Evaluation, revisions, and planning: Sustained energy management system (EMS) performance and continuous improvement are achieved through periodic evaluations at the end of each cycle, scrutinizing implemented actions and their results, potentially using life-cycle analysis when necessary, and incorporating lessons learned from previous cycles and procedures into future improvements. Communicating successful outcomes to all stakeholders fosters awareness and supports EMS sustainability, requiring diverse methods of internal and external dissemination. A clear methodological framework, like the seven-step process outlined below, ensures the continuation of this cycle.

25.3 Case Study

This Energy Management Model is demonstrated using a rod component manufacturing facility specializing in automotive replacement parts. The initial analysis focused on energy infrastructure and available energy management tools. Without an EMS, the facility's energy consumption per product was approximately 0.87 kWh. With an annual production capacity of roughly 6.75 million units per year (based on a 3-year average), the plant's primary energy sources are electricity (approximately 70.12% of annual consumption) and natural gas (29.88%). Annual energy consumption averages 7.27 GWh, reflecting a production process requiring significant energy. The plant supported an energy management system program, and the program's implementation phases were subsequently evaluated using the proposed methodology. Initiating the proposed method involved establishing a dedicated energy management representative and forming an energy team, with clearly defined roles and responsibilities. This was followed by developing and disseminating a customized energy policy to foster a culture of energy efficiency throughout the organization. The newly formed team conducted a gap analysis of the energy management profile, using the existing structure as a benchmark. The results of this evaluation are presented in Fig. 25.8.

The effectiveness of the implemented program is significantly measured by the self-assessment process. Positive evaluations of the results by the company and management are crucial for program sustainability. These assessments, rated on a scale of 0 to 5, indicate the initial and final positions of various departments. For example, key energy users, initially scoring a 2, progressed to a widely understood framework by the end of the program, demonstrating increased awareness and understanding within the organization. This structured program, outlining the team's work, effectively established the energy management system. The second phase involved a current state analysis and evaluation, where the company established benchmarks for energy performance indicators. Using these benchmarks, crucial energy users underwent regression analysis, allowing for the development of a cumulative sum (CUSUM) graph. Regression analyses were conducted using a

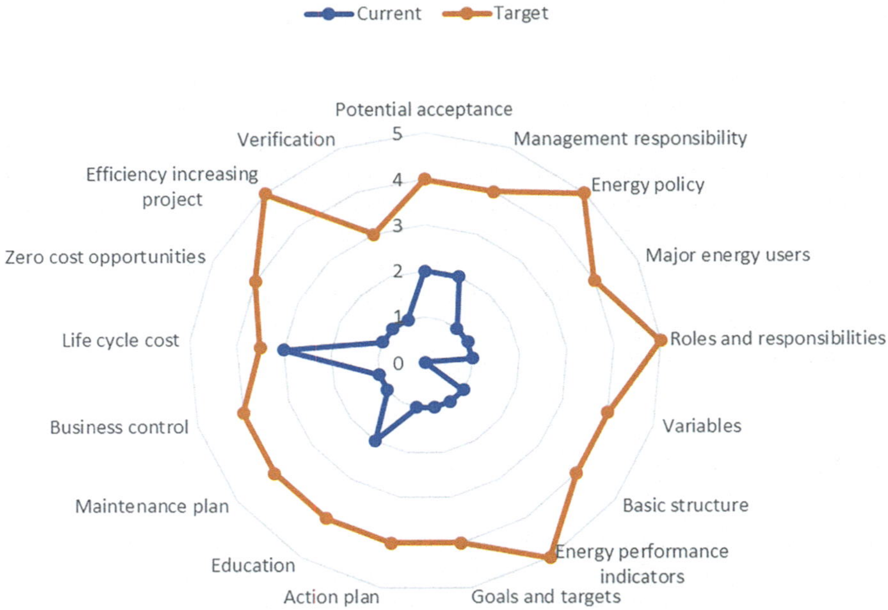

Fig. 25.8 Gap analysis

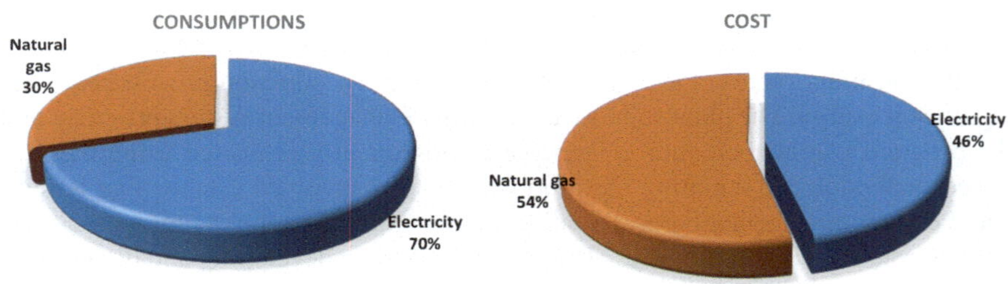

Fig. 25.9 Energy consumption distribution

reference year for the facility, resulting in a relationship equation ($y = a + bx$) between energy consumption and product output, also taking R^2 values into account. The facility's energy consumption distribution is illustrated in Fig. 25.9.

Analyzing 3 years of data, the facility's total energy consumption, approximately 718.85 TEP/year, was found to be roughly 70% attributable to electricity use. Examining the relationship between energy consumption and production based on 3 years of data revealed a non-linear correlation between these variables. Using an average annual production volume of 6,750,000 units and an average annual energy consumption of 8.63 GWh over the 3-year period, the changes in these values were assessed, and the results are presented in Fig. 25.10.

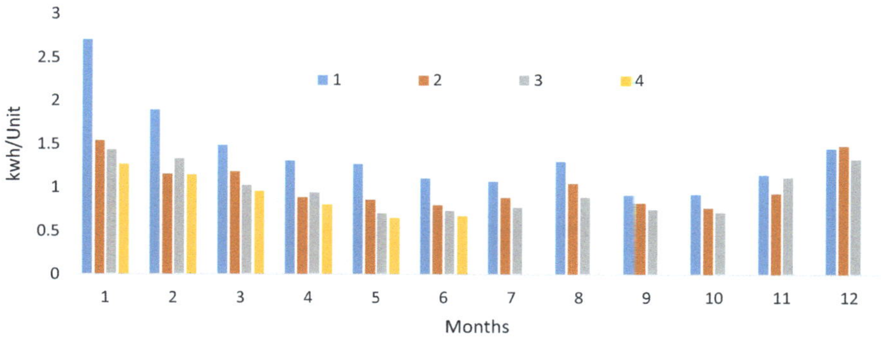

Fig. 25.10 Unit energy and production change rates

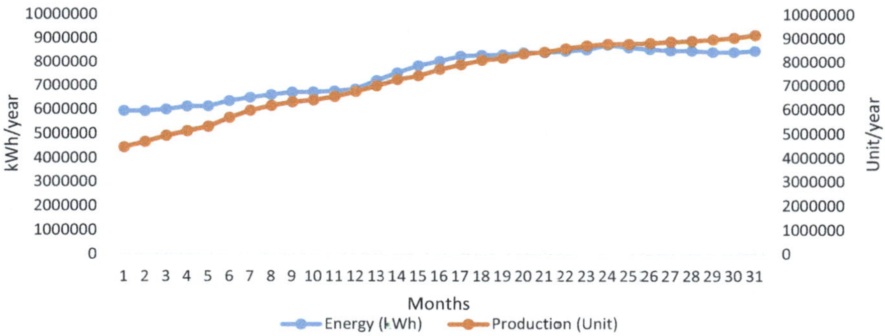

Fig. 25.11 Annual energy and production distribution

The distribution of energy consumption reveals the influence of non-production factors on the facility's energy usage. Furthermore, the annual patterns of consumption and production exhibit a similar trend, with a demonstrably increasing consumption rate over the years. Analysis of the annual production-consumption relationship based on the past 42 months' data, presented in Fig. 25.11, further illustrates this upward trend and highlights the non-linear correlation between energy consumption and production output.

The consumption patterns, while showing a proportional relationship to production, also reveal a persistent upward trend in energy consumption. Standard and target consumption relationships were considered, along with overall facility consumption trends. The analysis highlighted Heating Degree Days (HDD) as significant variables affecting energy use in conjunction with production. The correlation between variables and energy consumption within this framework produced an R^2 value of 93.31%. Using the past 18 months of data as a target year, the study investigated the parameters influencing energy consumption, using the methodology displayed in Fig. 25.12. This led to the identification of critical energy users and subsequent analysis of energy performance metrics and the cumulative sum

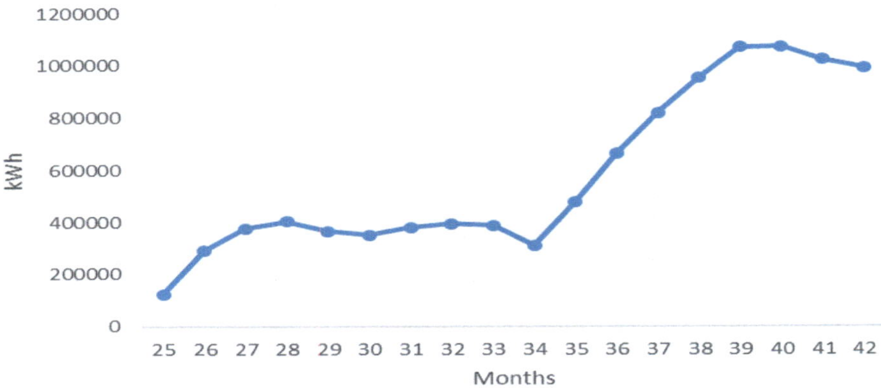

a. Standard

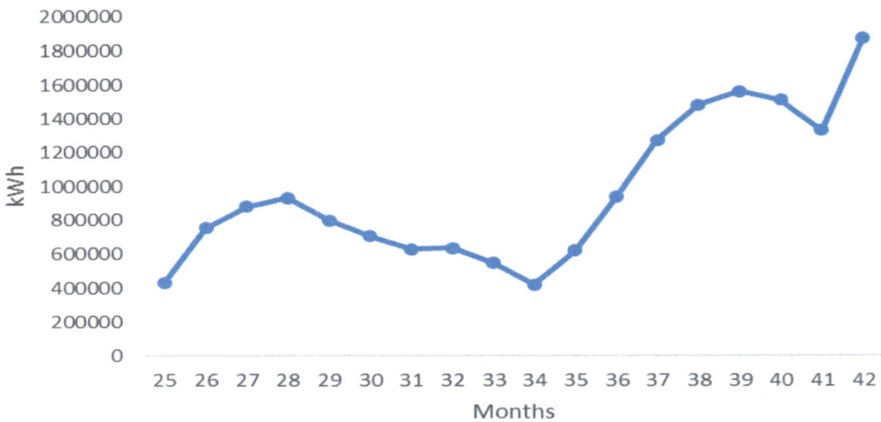

b. Target

Fig. 25.12 Production/energy consumption relationship CUSUM chart. (**a**) Standard. (**b**) Target

(CUSUM) graph. Results showed that while both calcination and compressors were substantial electrical consumers, calcination accounted for approximately 85% of the natural gas consumption, making it a key target area for energy optimization efforts. Consequently, the initial focus of the study was on these three systems.

The equation (Y = 195,387.7 + (0.471243 * Production) + (1139.4 * HDD)), derived for the base months shown in Fig. 25.12, represent the expected consumption as a function of production volume for each process. These equations were used to project consumption changes for the following year. From these projections, considering energy consumption intensity per process, the facility's target energy

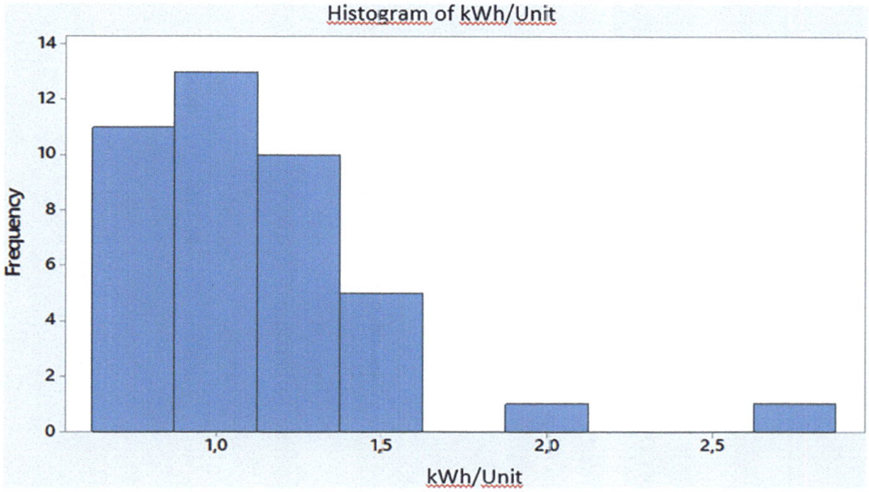

Fig. 25.13 Energy/production histogram

efficiency was determined to be 14.22%. However, another approach to evaluating efficiency was through consumption histograms. Figure 25.13 presents the facility's consumption histogram for per-unit production.

While the business's 4-year average consumption was 1115 kWh per unit, the target, based on the consumption histogram, was set at 0.875 kWh per unit. This distribution yielded a target efficiency of 10.07% for the enterprise.

Energy managers and process personnel should collaborate on achieving this target. The fourth step of energy management, which includes problem identification, and planning, is crucial. This phase involves categorizing potential solutions into those requiring no investment, low investment, and high investment, while meticulously examining operational parameters and production criteria. Examples of identified actions for this business are detailed in Table 25.1.

The energy management team, led by its designated representative, must implement the identified actions. These improvements can be realized in real time with this management model. Crucially, the fifth step involves implementing the improvements and evaluating their impact. The projected total improvement potential is 155.3 TEP per year, which will result in a CO_2 emissions reduction of 497 tons per year. This represents a significant achievement for the facility's overall savings targets. Finally, the facility's analysis indicated an average annual return on investment potential of 2.15 years.

The sixth and seventh steps of the methodology focus on ensuring the continuity of energy management and evaluating the process to inform future years. In this context, the facility's gains should be assessed separately by the energy management team and all top managers. Analysis revealed that the electricity consumption target was not met. Based on this finding, a plan for repeating the studies and generating new actions for critical energy users has been developed. Considering the

Table 25.1 Examples of actions by major energy users

		Annual saving potential				Amount of CO_2 reduction	Investment cost	Payback period	Implemtation
Actions	Energy	Energy	Unit	Ton/year	$/year	Ton/year	$/year	Year	Term
ENMS Program	Electricity						75.00		Short
	Natural gas								
Boiler Combustion Optimization	Natural gas	7059	m^3	5.8	7553	14	3750	0.50	Medium
Production Department Heat System Improvement	Natural gas	132,513	m^3	109.3	141,789	268	483,215	3.41	Long
Ventilation	Electricity		kWh				46,768		Long
Compressor Heat Recovery	Electricity	42,862	kWh	3.7	12,430	20	29,950	2.41	Long
Removal of Compressed Air Leaks	Electricity	316,800	kWh	27	91,872	146	30,000	0.33	Long
Compressed Air Production Automation	Electricity	64,576	kWh	6	18,727	30	15,000	0.80	Long
Fixture Replacement	Electricity	43,200	kWh	4	13,586	20	52,800	3.89	Long
Total	–	–	–	155.3	285,957	497	668,983	42,036	Medium

results and program sustainability, this facility has effectively established a management infrastructure that dynamically supports the implementation of an ISO 50001 Energy Management System.

25.4 Conclusion

This study developed an energy management model designed to facilitate a more dynamic application of the ISO 50001 Energy Management System in industry, exemplified through a case study of an automotive parts manufacturing facility. The proposed methodology's key elements are presented, along with guidance on establishing infrastructure for the structural implementation of ISO 50001 in industrial settings. The application within the automotive parts facility focused on creating a management framework and optimizing existing operations. Annual energy efficiency potential, calculated using CUSUM analysis, reached 14.12%, while the histogram-based approach yielded a potential of 10.07%. These calculations suggest annual savings of $285,957, with a projected average payback period of 2.15 years after considering potential investment costs. Critically, a significant portion of these savings can be achieved through non-investment actions, demonstrating the potential for reduced energy-related costs with effective management practices. Additionally, sustainable carbon management initiatives, using energy equivalents as a reference, projected total CO_2 savings of 497 tons per year.

This method offers an effective solution for industry's energy management needs, representing a viable management model with sustainable impact. The process should be viewed as a framework for optimizing energy systems in businesses, concurrently establishing the infrastructure necessary for a holistic and dynamic ISO 50001 Energy Management System implementation. Future priority areas for energy systems should focus on sustainable carbon management to drive low-cost, high-quality production, and the increased adoption of renewable energy sources.

References

Ates, S. A., & Durakbasa, N. M. (2012). Evaluation of corporate energy management practices of energy intensive industries in Turkey. *Energy, 45*(1), 81–91.

Bojana, J., & Filipović, J. (2016). ISO 50001 standard-based energy management maturity model—Proposal and validation in industry. *Journal of Cleaner Production, 112*(Pt 4), 2744–2755.

Brian, S., Hatem, M., Sekiguchi, H., & Hassan, H. (2024). New approach for zero energy industries via integrating energy management and renewable energy system: Enviro-economic investigation. *Energy and Buildings, 319*, 114573. https://doi.org/10.1016/j.enbuild.2024.114573

EIA. (2016). *International energy outlook 2016*. Section from U.S. Energy Information Administration. Retrieved November 15, 2024, from https://axelkra.us/wp-content/uploads/2023/01/industrial.pdf

EnergyCAP. (2015). *Energy Management Maturity Model*. Retrieved November 15, 2015, from http://info.energycap.com/hubfs/EM_maturityModel.pdf

EU. (2012). Directive 2012/27/EU of The European Parliament and of The Council. *Official Journal of the European Union*. Retrieved from https://eur-lex.europa.eu/legal-content/EN/TXT/PDF/?uri=CELEX:32012L0027&from=EN

EU. (2019). *National energy and climate plans (NECPs), the 28 draft NECPs in June 2019 (COM/2019/285)*. Retrieved from https://energy.ec.europa.eu/topics/energy-strategy/national-energy-and-climate-plans-necps_en

Fahad, M., AliNaqvi, S. A., Atir, M., Zubair, M., & Shehzad, M. M. (2017). Energy management in a manufacturing industry through layout design. *Procedia Manufacturing, 8*, 168–174.

Finnerty, N., Sterling, R., Coakley, D., Contreras, S., & Keane, M. M. (2017). Development of a Global Energy Management System for non-energy intensive multi-site industrial organisations: A methodology. *Energy, 136*, 16–31.

Gtz. (2004). The challenge: Affordable and sustainable energy services for all. In *Towards sustainable energy systems: Integrating renewable energy and energy efficiency is the key*. Wuppertal Institute for Climate Environment Energy in the Science Centre North Rhine-Westphalia.

IEA. (2008). *World energy outlook*. World Energy Agency. Retrieved from https://www.iea.org/media/weowebsite/2008-1994/WEO2008.pdf

IEA. (2021a). *Global energy review 2021. Assessing the effects of economic recoveries on global energy demand and CO_2 emissions in 2021*. International Energy Agency.

IEA. (2021b). *International energy outlook 2021*. International Energy Agency. Retrieved from https://www.eia.gov/outlooks/ieo/pdf/industrial.pdf

Jiang, X., Taneja, J., Jorge, O., Arsalan, T., Prabal, D., Jaein, J., David, C., Philip, L., & Scott, S. (2007). An architecture for energy management in wireless sensor networks. *ACM SIGBED Review, 4*(3), 31–36. https://doi.org/10.1145/1317103.1317109

Lei, M. (2024). Application of energy sustainability model based on optical sensing technology in intelligent warehousing performance management in green manufacturing industry. *Thermal Science and Engineering Progress, 54*, 102789. https://doi.org/10.1016/j.tsep.2024.102789

Mineeva, A. (2016). *Improvement of energy efficiency in a gold mining industry*. Master tezi, Lappeenranta University of Technology School of Business and Management Industrial Engineering and Management. Retrieved November 15, 2015, from https://core.ac.uk/download/pdf/39982660.pdf

Mirzakhani, M. A., Tahouni, N., & Panjeshahi, M. H. (2017). Energy benchmarking of cement industry, based on Process Integration concepts. *Energy, 130*(1), 382–391.

Moehle, N., Busseti, E., Boyd, S., & Wytock, M. (2008). *Dynamic energy management*. Standford University. Retrieved from https://web.stanford.edu/~boyd/papers/pdf/dyn_ener_man.pdf

Mohamad, F., Abdullah, N. H., Kamaruddin, N. K., & Mohammad, M. (2014). Implementation of ISO50001 energy management system. In *2014 International Symposium on Technology Management and Emerging Technologies, Bandung, Indonesia* (pp. 275–280). https://doi.org/10.1109/ISTMET.2014.6936518

Onaygil, S. (2015). *Sanayide Verimlilik, Enerji Yönetimi ve Tasarruf*. 5. Ulusal Verimlilik Kongresi, Ankara, 6–7 Ekim.

Orhan, A. (2021). Outlook of energy in the world and Turkey 2021, Turkish Informatics Association "human and technology" events 2. In *Workshop on Environment and Natural Resources in a Human Focus in a Digitalizing World, Sustainable Energy Policies, May 30*. Retrieved from https://enerji.mmo.org.tr/wp-content/uploads/2021/05/TBD-_%C3%87al%C4%B1%C5%9Ftay_27.04.2021_Sunum_OrhanAyta%C3%A7.pdf

Priyadarshi, R. C., Himani, M., Krishna, G. B., & Chaudhery, M. H. (2023). Severe deterioration in food-energy-ecosystem nexus due to ongoing Russia-Ukraine war: A critical review. *Science of the Total Environment, 902*, 166131. https://doi.org/10.1016/j.scitotenv.2023.166131

Pusnik, M., Al-Mansour, F., Susic, B., & Gubina, A. F. (2016). Gap analysis of industrial energy management systems in Slovenia. *Energy, 108*, 41–49.

Santi, B., Chanin, P., Chalermpan, F., Pailin, N., & Mohamed, A. H. (2024). Energy efficiency characteristics analysis for process diagnosis under anomaly using self-adaptive-based SHAP guided optimization. *Energy, 309*, 133074. https://doi.org/10.1016/j.energy.2024.133074

Schulze, M., Nehler, H., Ottosan, M., & Thollander, P. (2016). Energy management in industry—A systematic review of previous findings and an integrative conceptual framework. *Journal of Cleaner Production, 112*(5), 3692–3708.

Sogut, M.Z., & Karakoc, T. H. (2023). *Investigation of energy efficiency road map of industrial facility*. 15th International Green Energy Conference, IGEC2023, July 10–13, Glasgow, UK.

Tsemekidi Tzeiranaki, S., Bertoldi, P., Paci, D., Castellazzi, L., Ribeiro Serrenho, T., Economidou, M., & Zangheri, P. (2020). *Energy consumption and energy efficiency trends in the EU-28, 2000–2018. EUR 30328 EN*. Publications Office of the European Union, ISBN 978-92-76-21074-0, JRC120681. https://doi.org/10.2760/847849

UNDP. (2016). *Delivering sustainable energy in a changing climate strategy note on sustainable energy 2017–2021*. United Nations Development Programme. Retrieved November 15, 2024, from https://un-energy.org/wp-content/uploads/2017/01/UNDP-Energy-Strategy-2017-2021.pdf

Uysal, M. P., & Sogut, M. Z. (2017). An integrated research for architecture-based energy management in sustainable airports. *Energy, 140*(Part 2), 1387–1397. https://doi.org/10.1016/j.energy.2017.05.199

Yücel, M., & Muhsin, H. (2016). ISO 50001 based integrated energy management system and organization performance. *Journal of Advances in Technology and Engineering Research, 2*(3), 59–65. Retrieved from https://www.researchgate.net/publication/329099116_YAYIN_ESER_02_yucel_halis_ISO_50001/figures?lo=1

Zhu, W., Huang, B., Zhao, J., Chen, X., & Sun, C. (2023). Impacts on the embodied carbon emissions in China's building sector and its related energy-intensive industries from energy-saving technologies perspective: A dynamic CGE analysis. *Energy and Buildings, 287*, 112926. https://doi.org/10.1016/j.enbuild.2023.112926

Chapter 26
Walk-Through Audit Model for Enterprise's Decision Processes Regarding Energy Management

M. Ziya Sogut

Nomenclature

ANN	Artificial neural network
BEMS	Building Energy Management Systems
BRANN	Bayesian regularization ANN
CO_2	Carbon dioxide
NDCs	Nationally determined contributions

26.1 Introduction

Since the Paris Agreement's implementation, global efforts to address climate change have significantly increased. Signatory governments committed to reducing emissions through Nationally Determined Contributions (NDCs) to limit global warming to well below 2 °C above pre-industrial levels, ideally 1.5 °C, by 2050 (Kaled et al., 2025; EEA, 2016). This practically means substantial increases in renewable energy use, accelerated energy efficiency, scaled-up investment in clean technologies, rationalisation of fossil fuel subsidies, and reduced energy and resource intensity. Energy, with its global impact, emerges as a significant contributing factor to climate change, which is a prominent issue in the present era. As an indicator with a direct impact on the economy, energy is a priority for national strategies, particularly in developing countries where rising energy demands necessitate substantial investments (Ganda, 2019; Tiwari & Mohammed, 2024). The management of energy is currently situated within an institutional framework comprising intricate structures with direct economic, social, and environmental impacts. In essence, the objective of these three components is to facilitate the production of

M. Z. Sogut (✉)
Maritime Faculty, Piri Reiss University, Istanbul, Turkey

M. Z. Sogut, M. Koray (eds.), *Energy Rationality and Management for Decarbonization*, Green Energy and Technology,
https://doi.org/10.1007/978-3-031-68497-5_26

goods and services at the lowest possible cost while minimising environmental impact. As articulated by Capehart, Turner and Kennedy (2012), energy management is defined as 'the wise and effective use of energy to maximise profits (minimise costs) and improve competitive positions', encompassing action steps to achieve this fundamental objective. From a broader perspective, energy management has horizontal and vertical priorities (Tandon et al., 2012). These serve the larger structure in terms of energy and environmental sustainability, as well as important features in terms of cost and efficiency in corporate goals.

Currently, the predominant approach to decarbonisation and the advancement of energy and environmental sustainability hinges upon the cultivation of a managerial culture that places a premium on energy efficiency. Indeed, energy management represents a core capability that enables companies to enhance their productivity and improve the quality of their products or services through innovative approaches, including the utilisation of low-carbon or zero-carbon technologies, energy-efficient technologies, the application of new materials and new production processes that are sustainable or circular in transformation, and the integration of new technologies in operational processes (Bergougui, 2024). It is frequently regarded as an organisational framework.

The efficacy of energy management within organisational structures is contingent upon the existence of feasible processes. The determination of an institution's energy consumption behaviour, the development of potential target points for decision-making processes, and the provision of support for decision-making processes in accordance with improvement priorities are primarily associated with effective and comprehensive audit processes (Abdelaziz et al., 2024; Liu et al., 2024a, b). These processes are characterised by a systematic approach, particularly in the context of organisational decision-making. They entail a discipline of application. In terms of scope, energy auditing strives to achieve a balance between total energy inputs and their utilisation, with the objective of identifying all energy flows within a facility. It measures energy use according to its discrete functions (Amaral et al., 2021).

Energy management has functional characteristics that are pertinent to organisational structures. Energy management has a range of applications across different sectors, with the energy audits that support it fulfilling a variety of functions within this context. For example, Nakayama et al. (2024) developed a new method for optimising energy and water efficiency in high-performance buildings. Using data from Building Energy Management Systems (BEMS) and industry standards, the method profiles energy and water use, benchmarking against other buildings, identifies opportunities for carbon reduction, and analyses the cost-effectiveness of improvements. Applying this to a Tokyo building, the analysis found substantial potential for energy savings (10.1%) and CO_2 emissions reductions (8.9%) through cost-effective measures.

According to this study, key contributions include a structured energy audit tailored to BEMS data, a broader focus on water use, and quantifying the remaining efficiency potential even in high-performing buildings. Excell et al. (2024) stated that energy audit is a new automated energy audit tool that uses hourly electricity data and simulations to identify building energy efficiency issues remotely. They created a database of building energy profiles and matched unknown buildings to this database using machine learning, pinpointing potential inefficiencies. Testing on a large dataset of school buildings showed high accuracy (91–99%) for many factors, but less accuracy for control systems and boiler retrofits. Study of Adino et al. (2024) conducted an energy audit of Meta Abo brewery in Ethiopia to assess energy efficiency and greenhouse gas emissions. Using an indirect energy audit method and the greenhouse gas protocol, the researchers found significant energy losses from boiler inefficiencies, poor insulation, and steam leaks. High diesel generator consumption also contributed to substantial electricity costs. Greenhouse gas emissions increased from 2014 to 2018. Implementing recommended energy-saving measures, such as improving boiler efficiency and insulation, could save a substantial amount of energy (20.4 TJ annually) and reduce CO_2 emissions (455 tons annually) at a cost of approximately 8.5 million Ethiopian. Study of Marcos and Plangklang (2024) demonstrated the need for AI in energy audits by developing a MATLAB-based application to analyse energy consumption in a Philippine university building. Using a conventional energy audit for data collection, the researchers trained an artificial neural network (ANN) model to predict energy use. The study found that a Bayesian Regularization ANN (BRANN) performed better than the Levenberg–Marquardt algorithm in predicting voltage unbalance, current unbalance, lighting consumption, and plug loads, demonstrating improved accuracy in energy consumption estimations. The results highlight the potential of AI to improve the efficiency and accuracy of energy audits. The study of Kaab et al. (2024) assesses the energy efficiency and environmental impact of different barley irrigation methods (flood, sprinkler, and drip) throughout their life cycles. Using a life cycle assessment (LCA) framework, the research will collect data from a representative barley farm to analyse energy inputs (fossil fuels, electricity, etc.) and outputs (yield and energy content) at each production stage. Environmental impacts, including greenhouse gas emissions, water usage, soil erosion, and pesticide use, will also be quantified for each irrigation system. The results are intended to inform farmers, policymakers, and stakeholders about the sustainability of different irrigation choices for barley production. A significant challenge for many businesses is the optimisation of energy management. This is due to many factors, including fragmented decision-making processes, a lack of clear accountability, and disconnect between strategic objectives and operational energy practices. This frequently culminates in suboptimal energy efficiency, augmented operating expenses, and the inability to leverage substantial energy-saving prospects.

26.2 Energy Management

Energy management is an organisational process that is primarily considered in businesses with the objective of minimising energy-related costs and improving a company's competitive position. Energy management encompasses all operations linked to business processes, including services, product design, equipment, and product shipment (Yang et al., 2024). The specific scope of energy management will depend on the nature of the business in question. In addition to this, energy management includes opportunities to minimise and eliminate waste from the business. The primary objective of energy management is to guarantee the manageability of energy in buildings, production, and industry. This is achieved through a holistic systems approach, which optimises energy-related activities in accordance with the structural characteristics of the business in question. A considerable number of businesses establish continuous improvement objectives while integrating energy management into their Total Quality Management (TQM) strategy (Wen et al., 2021). This integration should encompass energy management practices with the objective of reducing costs. The principal objectives of energy management programmes are to enhance energy efficiency, curtail energy consumption, mitigate greenhouse gas emissions, and facilitate efficacious communication on energy matters.

The contemporary free trade structure is predicated on the imperative of profit or budgetary allocations for profit-oriented enterprises. Consequently, the viability of new activities is contingent upon their cost-effectiveness, which in turn is a function of their ability to justify implementation. Energy management has been demonstrated to be a cost-effective strategy, with studies indicating savings of 5–15% with minimal capital investment. Over time, savings of approximately 30% are typical, with some organisations achieving savings of up to 50–70%. These savings are primarily the result of improvements made in relation to operational frameworks. New, energy-efficient buildings have been demonstrated to operate on only 20% of the energy required by existing buildings, thereby resulting in an 80% reduction in energy use. Energy management represents one of the most effective programmes for increasing profits and reducing costs in the manufacturing, industrial, and commercial sectors.

Effective energy management can be considered a fundamental aspect of organisational structures that are based on the dual objectives of profitability and environmental sustainability. Energy management is essentially conceptualised as a cultural transformation with the objective of enhancing efficiency, and as a pivotal managerial instrument within operational frameworks.

The aforementioned steps, based on the manageability of energy, can be evaluated as follows: firstly, as an institution, the ability to move away from the threat of price increases and penalties; secondly, as a corporate culture, the optimisation of energy use behaviour from source to user; thirdly, the increase of efficient behaviour and awareness of consumption and costs; fourthly, the evaluation of performance structures and the proactive identification of problems through the real-time monitoring of consumption behaviour; fifthly, the management of energy purchase prices and the development of accurate consumption plans; and finally, the management of the demand response of the market due to production and service with energy-efficient consumption.

The field of energy management espouses a perspective that prioritises efficiency and sustainability across a range of contexts, from industrial buildings to service areas. The concept of energy efficiency is structured within a management framework that is itself managed and held to account by the entire corporate structure. This is achieved within a knowledge-based management culture.

Energy management primarily aims at management behaviour to improve the consumption behaviour of energy users, targeting a multitude of outputs to businesses and workplaces, from improved comfort and productivity to reduced operational capabilities, costs, and greenhouse gas emissions. As like in Fig. 26.1, an energy management system for organisational structures brings together several key components to develop a success story (Keith, 2011): Energy planning covering strategic energy use projections, energy optimisation focusing on the implementation of efficiency measures, monitoring and control systems providing proactive responses to potential energy inefficiencies, and reporting and billing tools providing formal information and a framework for tracking progress towards energy-saving targets.

Today, effective corporate energy management systems have achieved an effective and standardised framework with the ISO 50001 energy management system since 2011 (da Silva Gonçalves & dos Santos, 2019). Together with the corporate framework in energy management, this transition includes a comprehensive review of existing energy-related policies and procedures, a critical analysis of energy consumption patterns, and the development of measurable targets aligned with the organisation's overall strategic goals (da Silva Gonçalves & dos Santos, 2019). Most importantly, the ISO 50001 transition in energy management requires a cultural shift that encourages a proactive approach to energy efficiency that permeates all levels of the organisation. This includes training employees in energy-saving practices, integrating energy performance indicators into decision-making processes, and implementing robust monitoring and reporting mechanisms to track progress and identify areas for continuous improvement. A successful transition to this transformation in organisational structures not only ensures compliance with

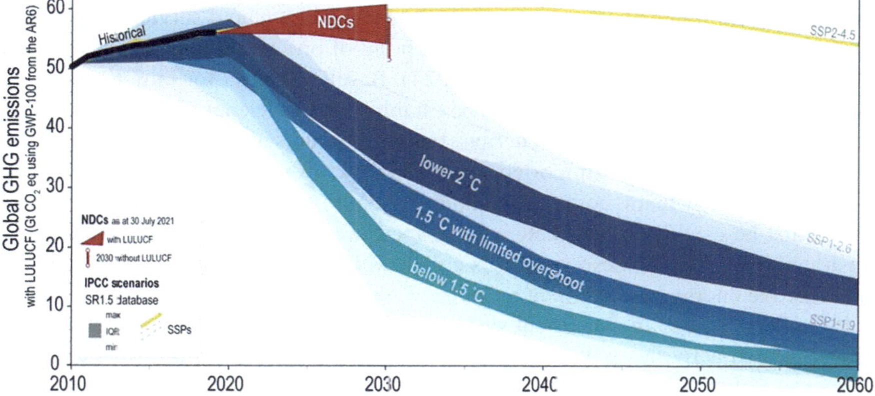

Fig. 26.1 Energy management framework (Keith, 2011)

the ISO 50001 standard, but also contributes to significant energy savings, reduced operational costs, and a more sustainable business model.

26.2.1 ISO 50001 Energy Management System

The ISO 50001 energy management system represents the latest and most advanced practices in energy management, building on existing national and local standards. The standard was developed through years of collaboration between energy management experts from more than 60 countries, facilitating the establishment of a robust energy performance framework. The standard enables organisations to identify and capitalise on opportunities to improve energy efficiency, to prioritise performance improvements, and to document the process in a rigorous manner. The ISO 50001 standard, published on 9 June 2011, was effectively driven by UNIDO demand and resulted in the establishment of an ISO energy management committee in 2008 (Dzene et al., 2015).

The ISO 50001 energy management system is designed to facilitate the implementation of processes aimed at understanding a facility's baseline energy consumption. The ISO 50001 energy management system facilitates the development of action plans and the setting of consumption reduction targets (Liu et al., 2024a, b). Furthermore, it is designed to align with other relevant standards, including ISO 9001 (Quality Management), ISO 14001 (Environmental Management), and BS EN 16001 (Energy Management Systems) (Jovanović & Filipović, 2016). This integrated approach ensures that energy management practices are aligned with wider corporate objectives and environmental sustainability, guaranteeing the delivery of a holistic strategy for improving energy efficiency.

While there is considerable overlap between ISO 50001 and TS EN 16001:2009, the former introduces significant revisions, primarily through its restructured numbering and more detailed clauses. Significant additions include the assessment of energy consumption and efficiency in equipment, products, and services, with consideration given to their useful life from the point of procurement to the design and refurbishment stages (Zhang & Zhang, 2024). Baseline values for initial and ongoing performance monitoring must also be established, while energy policies must be aligned with the procurement of energy-efficient products and services, and energy-efficient designs must be supported. Additionally, top management is required to establish an energy team, comprising management representatives, and clear procedures must be established for determining measurement needs and calibration. As illustrated in Fig. 26.2, these processes collectively result in a more comprehensive and effective energy management system.

ISO 50001 energy management system (EnMS) is a framework applicable to all energy-related activities within an organisation. It forces the organisation to fulfill its energy policy commitments, proactively improve its energy performance, and demonstrate compliance with the requirements of the international standard. Moreover, it serves as a tool for measuring energy use and consumption, enabling

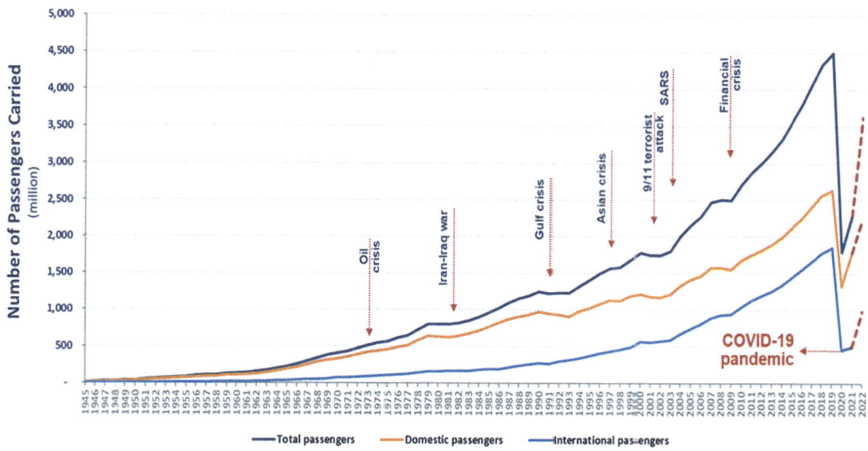

Fig. 26.2 Flow steps of the energy management system

the planning of energy-efficient practices through operational controls and adjustments in organisational behaviour.

In order to optimise energy performance, it is necessary to adopt a systematic approach. It is anticipated that the implementation of non-investment and cost-effective sustainable measures will be prioritised before the introduction of capital-intensive improvements. Indeed, as evidenced by studies such as that of Fiedler and Mircea's (2012) study anticipated and targeted the adoption of ISO 50001 standards, particularly in Europe and Romania, underscoring the potential for energy savings (Ranky, 2012). The governance framework of ISO 50001 enables the derivation of privileges, including the development of public subsidies. For example, in countries such as Germany, these subsidies may take the form of a reduction in electricity and gas taxes, as well as a reduction in fees linked to processes. The financial benefits associated with energy management, coupled with increased efficiency, provide a compelling rationale for prioritising these low-cost optimisation strategies before significant capital investments. The global adoption of ISO 50001 standards has been acknowledged in numerous applications. Examples of organisations that have adopted ISO 50001 include the Dahanu Thermal Power Plant in India, Schneider Electric, Porsche, Lamborghini, and Bad Eisenkappel in Europe, as well as several municipalities in Taiwan and China (Duglio, 2011; Ranky, 2012). In Malaysia, the most prominent EnMS certifiers include industry-leading companies such as Indah Water Konsortium Sdn Bhd, Kuala Lumpur Golf and Country Club Berhad, and Pahang Cement Sdn Bhd. This illustrates the growing recognition of the advantages of the ISO50001 framework and its potential for extensive implementation across a diverse range of sectors and geographical regions (Halbhavi et al., 2015).

The ISO 50001 standard primarily offers a solution methodology that is built upon existing organisational structures. This structured process directly addresses the current state of energy use through the implementation of audit procedures. This methodological approach is based on a documented energy audit methodology, which employs specific criteria for the assessment of energy consumption data. The energy audit process employs a data-driven approach to evaluate an organisation's performance, relating consumption patterns to actual measurements and thereby identifying opportunities for improvement. The objective of the audit process is to establish a baseline energy consumption profile and select relevant Energy Performance Indicators (EnPIs) associated with energy users. Additionally, it identifies the monitoring capability required to support effective future continuous improvement of the EnMS. To conduct the review, the organisation must first establish a list of equipment, identify different energy uses, and collect detailed energy consumption data for a specified period, which is typically a full year on a monthly basis.

26.3 Energy Audit Framework

Despite the multifaceted benefits of energy conservation, its implementation remains slow. Even in market economies where costs directly affect prices, industrial and other sectors often lag in adopting energy-saving investments. This is even more pronounced in developing nations. Several factors contribute to this delay: (a) a perceived lack of immediate need for improvement due to existing operational efficiency, (b) the complexity and potential disruptions associated with retrofitting equipment (requiring careful planning and potentially production downtime), (c) the frequent fragmentation of energy-saving measures into many small projects, (d) reduced financial resources due to recent economic hardship, and (e) a focus on increasing production over improving efficiency, often driven by a lack of prioritisation of energy conservation by top management.

Energy efficiency is a key driver for environmental performance improvements in industrial settings, offering a rapid and cost-effective route to enhancement Wang et al. (2025). Industries with significant energy inputs increasingly need to ensure consistent, high-quality energy supply at the lowest possible cost. The potential for increased operational costs due to energy consumption emphasises the crucial need for effective energy management Sujanto et al. (2024). Implementing efficient energy management, enabling more efficiency use of energy, and reducing associated expenses not only extends the lifespan of energy resources but also generates significant economic savings.

Implementing energy management or efficient energy use strategies in businesses involves a multi-stage process. This process typically includes:

1. Preliminary energy-saving assessments, encompassing energy audits (data collection, processing, and analysis).
2. Securing management and personnel support, including securing commitment and participation from relevant personnel.

3. Designing and implementing energy-saving adjustments, including leveraging market research and data analysis to optimise existing systems for maximum efficiency, and calculating the cost and return on investment of proposed energy-saving measures.
4. Ongoing monitoring and evaluation to track the effectiveness and impact of implemented strategies.

The evaluation of study assessments based on the development of energy efficiency strategies is conducted in conjunction with a flow framework. In particular, for models based on the energy management framework, the initial step required of enterprises is to ascertain the current situation. As illustrated in Fig. 26.3, the energy audit framework employs a flow methodology and is regarded as a potential means of defining the energy efficiency of the enterprise. The flow presented in this study addressed two fundamental constructs, which directly define the potential for energy efficiency.

An energy audit encompasses a wide range of energy studies, from a preliminary facility walkthrough to identify key problems to a detailed analysis of various energy efficiency options, considering their financial viability for investors.

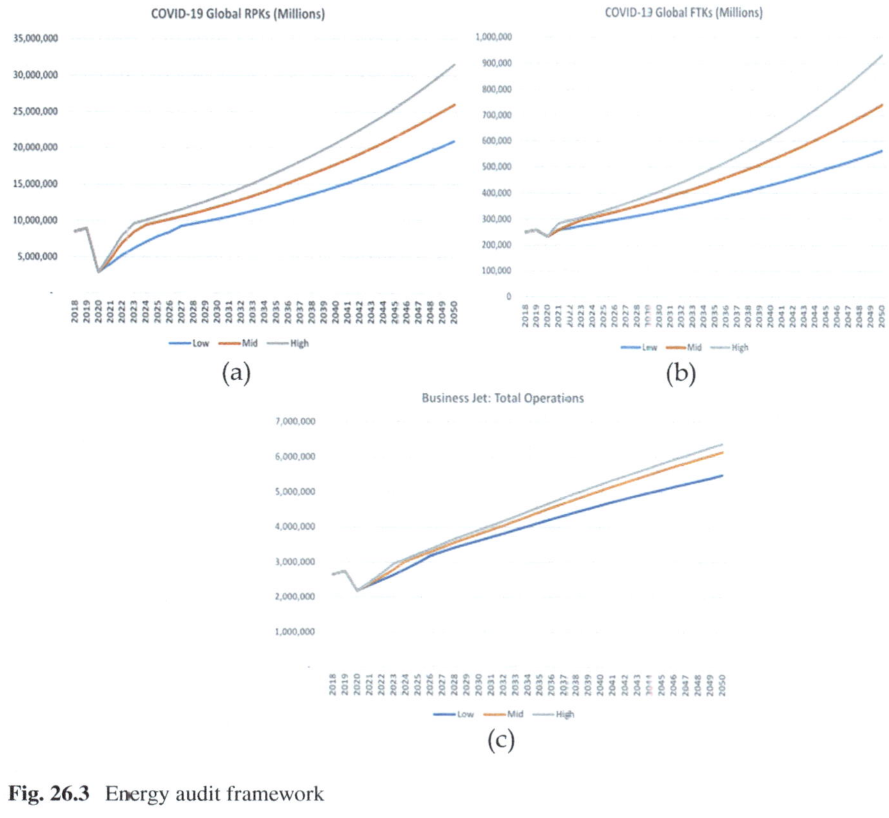

Fig. 26.3 Energy audit framework

Benchmarking: It involves comparing energy consumption data to reference values from similar facilities or simulation models to pinpoint areas of excessive or unacceptable operating costs.

Walk-through audit: A preliminary energy audit, sometimes called a simple, screening, or walk-through audit, is a quick and basic assessment. It involves minimal staff interviews, a review of utility bills and operational data, and a physical walk-through of the facility to identify obvious energy waste and inefficiencies.

Detailed/general energy audit: A general energy audit, also known as a mini-audit, detailed energy audit, or complete site energy audit, builds upon the preliminary audit by collecting more in-depth operational data and conducting a more thorough evaluation of potential energy conservation measures. This enhanced analysis may include advanced on-site measurements and sophisticated computer simulations to precisely assess the effectiveness of proposed energy retrofits.

Investment-grade audit: In corporate environments, energy infrastructure upgrades must contend with other capital investment needs. Both energy-related and non-energy projects are evaluated using the same financial criteria, typically prioritising projects with the highest projected return on investment (Sharma et al., 2021; Wikipedia, 2024a, b; Ramya & Femina, 2014).

This section details the crucial energy audit performed to facilitate a preliminary energy-saving assessment for an industrial facility. The audit meticulously collected, processed, and analysed the facility's energy and production data. Effective energy management within factories hinges on a robust energy audit. This involves compiling monthly production and consumption figures into detailed charts, which are then used to create comprehensive energy analysis graphs. These graphs, covering all aspects of the facility, visually compare energy consumption patterns to production output. This comparison is essential for identifying potential areas of energy inefficiency.

A crucial aspect of industrial energy management involves meticulously comparing energy consumption against production output. This comparative analysis is fundamental in identifying specific areas for potential energy savings, forming the basis for tailored energy policies. Sophisticated calculations are then employed to quantify these potential savings, consistently using standardised units such as tons for fuel, kilowatt-hours for electricity, and tons for production. The ultimate aim of the energy audit is to leverage accurate data and well-defined solutions to achieve maximum energy efficiency, ultimately optimising resource utilisation and minimising operational costs.

Effective energy audits begin with a robust data collection strategy. This initial phase isn't merely about gathering numbers; it's about meticulously defining the scope and methodology. Prioritising clarity, a comprehensive approach should first identify the specific types of energy being consumed (e.g. electricity, gas, and steam) and corresponding production outputs (e.g. units manufactured and volume produced). Secondly, precise identification of points of energy use and production is critical. This necessitates careful selection of measurement locations. Equally important is determining the appropriate measurement instruments and establishing regular, consistent measurement periods (ideally 10–20 data sets). The duration of the data collection phase must be long enough to capture relevant trends; for instance, weekly measurements for at least 10 weeks or monthly measurements for at least a year are recommended to provide a comprehensive overview of energy consumption patterns relative to production cycles.

26.4 Case Study

This study has conducted a comprehensive audit of energy consumption patterns in a manufacturing enterprise, with a particular focus on energy recovery. The audited enterprise utilises a considerable quantity of electrical energy in its production processes. An analysis of the energy users revealed a notable consumption of energy associated with motor-driven power systems and planned production areas. Additionally, the production process encompasses the heating of batteries through injection and the utilisation of two cooling towers for cooling purposes, which collectively highlight a substantial energy utilisation in these pivotal components. Consequently, the study concentrated on these areas to identify potential avenues for enhancing energy efficiency and establishing a robust foundation for an effective energy management system.

The audit further assessed the facility's current energy management practices, revealing gaps and inconsistencies in data collection, monitoring and reporting procedures. Moreover, areas requiring better control mechanisms were identified. Lack of documented procedures for energy consumption monitoring, a fragmented approach to data collection in different production units, and the absence of a centralised system for monitoring energy use were observed. These findings were analysed in a comprehensive and integrated energy efficiency assessment to improve data accuracy, streamline operations, and support continuous improvement efforts.

The review of the enterprise was limited to the existing production areas of the enterprise. However, even though the company has an investment plan for expansion, direct energy consumption behaviour was addressed and growth targets were not taken into account in the audit assessment. The walk-through assessment of the production processes is brief:

- Current situation analysis
- Identification of problem areas and solutions
- Action plans
- It is planned as an implementation procedures

26.4.1 Situation Analysis

This study has conducted a comprehensive audit of energy consumption patterns in a manufacturing enterprise, with a particular focus on energy recovery. The audited enterprise utilises a considerable quantity of electrical energy in its production processes. An analysis of the energy users revealed a notable consumption of energy associated with motor-driven power systems and planned production areas. Additionally, the production process encompasses the heating of batteries through injection and the utilisation of two cooling towers for cooling purposes, which collectively highlight a substantial energy utilisation in these pivotal components. Consequently, the study concentrated on these areas to identify potential avenues for enhancing energy efficiency and establishing a robust foundation for an effective energy management system (Tables 26.1 and 26.2).

In electric motors, measurements and analyses have been made for high capacities in both drive and non-drive motors. The operating performances of the motors

Table 26.1 Engine list of the manufacturing process

Motor location	Power (kW)	Current (A)	Driver Yes/No
Machine extruder motor	132	230	Yes
Machine extruder motor	160	276	Yes
AGROMER engine	200	348	No
Cone clamping motor	110	158	No
Crushing motor	110	205	No
Shereder shredder motor	132	230	Yes
Machine side feed motor	30	55	Yes
Machine output motor	75	133	Yes
Machine cutting motor	3.48	5.6	No
Fan motor	5.5	10.5	No
Machine side feeding motor	30	55	Yes
Machine cutting motor	3.5	6.7	No
Machine fan suction motor	5.5	9.9	No
Separator output motor	22	42.7	Yes
Pool bottom motor	3	6.24	No
Separator outlet	7.5	16	No
Bottom separator outlet	3	6.24	No
Bottom separator (sludge)	3	6.24	No
Separator outlet	22	42.7	No
Lower elevator motor 2×	2.2	4.8	No
Hydraulic motor	2.2	5	No
Oil motor	1.1	2.6	No
Hydraulic motor	7.5	16	No
Drum	3	6.48	No
Suction fan	55	96	No
Band	2.2	4	Yes

Table 26.2 Cooling towers

	Speed (m/s)	Flow rate (m³/h)	Engine (kW)	Inlet temp. (°C)	Exit temp. (°C)
1	0.75	9.55	4 + 2	35	29
2	1.59	20.43	5 + 2.2	17.6	16.8

Table 26.3 Agromer engine

Engine power (kW)	Drawn energy (kWh)	Hours of operation (h)	Avg power drawn (kWh)	Maximum power (kWh)	Minimum power (kWh)	PF
200	220.51	5600	192.47	383.83	121.15	0.89
Annual consumption and cost	(kWh)	$				
	157,046.4	9871.5				

Performance

	V	*I*	PF	kWh (measured)	kWh (theoretical)	Load factor	Annual savings (kWh/year)	Annual savings ($/year)	Efficiency
1	371.5	375.18	0.86	207.37	238.72	0.87			0.8790
2	359.4	208	0.81	104.75	128.03	0.82			
3	376.5	798	0.89	462.60	514.58	0.90			
			0.85	258.24	293.77	0.86	58,591.79	3682.91	

are exemplified and evaluated with efficiency performance analyses depending on the motors. Analyses for each motor are given separately in Tables 26.3, 26.4, 26.5, and 26.6.

In the evaluations, it was determined that the motor efficiencies remained quite low and especially the PF values remained well below 90% for all motors. The saving costs, especially depending on the efficiency ratio, remained quite high for the motors. In the analysis, an evaluation was also made for inverter motors. The studies carried out on this subject are given in Tables 26.7 and 26.8.

The performances of other non-inverter motors with different features in the enterprise were also examined holistically (Table 26.9).

These analyses are supported by an inventory study, and annual evaluations are made based on the engine efficiency class. The efficiency evaluations of the engines in the engine inventory were examined according to the charge between E1 and E4 classes, and their distributions are given in Fig. 26.4.

In the study, only the replacement of E1 motors with E4 shows an annual saving of 192.08 MWh. In the light of this data, the problem areas were re-examined.

Table 26.4 Cone clamping motor

Engine power (kW)	Drawn energy (kWh)	Hours of operation (h)	Avg power drawn (kWh)	Maximum power (kWh)	Minimum power (kWh)	PF
110	121.28	5600	30.693	43.095	20.228	0.74
Annual consumption and cost	(kWh)	$				
	50,7287.2	31,886.63				

Performance

	V	I	PF	kWh (measured)	kWh (theoretical)	Load factor	Annual savings (kWh/ year)	Annual savings ($/year)	Efficiency
1	375.978	63.64	0.75	30.88	40.98	0.75			0.7652
2	366.1	45	0.62	17.56	28.22	0.62			
3	380.5	96	0.83	52.39	62.56	0.84			
			0.73	33.61	43.92	0.74	82,503.96	36,301.74	

Table 26.5 Crushing motor

Engine power (kW)	Drawn energy (kWh)	Hours of operation (h)	Avg power drawn (kWh)	Maximum power (kWh)	Minimum power (kWh)	PF
110	121.28	5600	70.148	81.424	53.833	0.707
Annual consumption and cost	kWh	$				
	286,339.2	17,988.9				

Performance

	V	I	PF	kWh (measured)	kWh (theoretical)	Load factor	Annual savings (kWh/year)	Annual savings ($/year)	Efficiency
1	373.79	149.54	0.72	69.33	95.73	0.72			0.727
2	368.3	101	0.65	41.89	63.71	0.66			
3	378.3	208.5	0.75	102.75	135.09	0.76			
			0.71	71.33	98.18	0.71	101,754.60	6396.03	

26.4.2 Problem Areas

1. The company has been assessed for energy efficiency, and the following areas of concern have been identified.
2. There is no production plan in the ever-growing structure. This is an infrastructure problem in terms of direct future loads.
3. In the new investment plans, the company does not have a study in terms of motor performance.

Table 26.6 Shredder engine

Engine power (kW)	Drawn energy (kWh)	Hours of operation (h)	Avg power drawn (kWh)	Maximum power (kWh)	Minimum power (kWh)	PF
132	145.53	5600	28.608	66.851	12.27	0.45
Annual consumption and cost	kWh	$				
	65,4763.2	41,156.54				

Performance

	V	*I*	PF	kWh (measured)	kWh (theoretical)	Load factor	Annual savings (kWh/ year)	Annual savings ($/ year)	Efficiency
1	402.449	88.48	0.44	27.11	60.99	0.44			0.5852
2	369.9	52.5	0.25	8.47	33.26	0.25			
3	413.8	335	0.66	158.52	237.42	0.67			
			0.45	64.70	110.56	0.46	22,547.9	69,447.49	

Table 26.7 Machine extruder

Engine power (kW)	Drawn energy (kWh)	Hours of operation (h)	Avg power drawn (kWh)	Maximum power (kWh)	Minimum power (kWh)	PF	Load rate of change	Con. difference
132	145.53	5600	105.845	110.009	95.184	0.74	0.135	81581.77 (kWh)
Annual consumption and cost	kWh	$						5128 ($)
	222,236	13969.1						

Performance

	V	*I*	PF	kWh (measured)	kWh (theoretical)	Load factor	Annual savings (kWh/year)	Annual savings ($/year)	Efficiency
1	398.57	204.92	0.66	92.83	139.89	0.66			0.6556
2	337.82	146.5	0.52	44.69	84.76	0.53			
3	419.8	268.5	0.70	136.30	193.05	0.71			
			0.63	91.28	139.23	0.63	157,836.97	9921.18	

4. The current motor loads are quite high. These are structures that generate significant harmonics (electrical disturbances).
5. The preference for pulleys in motors reduces performance and will create significant pulley costs in the future.
6. The frequency operating ranges of the inverter motors are not determined in accordance with the activity. Problems have been observed with inverter setpoints.

Table 26.8 Machine extruder

Engine power (kW)	Drawn energy (kWh)	Hours of operation (h)	Avg power drawn (kWh)	Maximum power (kWh)	Minimum power (kWh)	PF	Load rate of change	Con. difference
160	176.41	5600	92.36	108.55	51.66	0.626	0.5241	34719.87
Annual consumption and cost	kWh	$						2182.39 $
	470,680	29,586						

Performance

	V	*I*	PF	kWh (measured)	kWh (theoretical)	Load factor	Annual savings (kWh/year)	Annual savings ($/year)	Efficiency
1	402.02	199.83	0.66	91.31	137.59	0.66			0.6542
2	381.3	125.5	0.52	43.21	81.96	0.53			
3	420.1	244	0.70	123.96	175.56	0.71			
			0.63	86.16	131.70	0.63	158,904.33	9988.27	

Table 26.9 Another motor without inverter

	Units	Real (kWh)	Theoretical (kWh)	Difference (kWh)
Power, std	kW	132	132	30
Engine speed	d/d	1500	1500	
Power, in	kW	176	146	
Efficiency		0.7	0.95	$
Cost of energy consumption	$/kWh	0.44	0.44	1.89
Cost of energy drawn from the grid in 1 h	$	11.06	9.18	
			Total year ($)	7542.86

7. High-power motors are operated without a driver at low power.
8. The cooling processes of the plant have low performance and poor installation infrastructure.
9. Especially in the large cooling tower, the temperature difference is almost 1 °C. This is a very poor value, and it was found that the pumps were running at unnecessarily high speeds.
10. Cooling is done with an open system. In addition to water and chemical costs, this situation creates a rust problem, especially on metal surfaces.
11. Approximately 51,000 kWh of electricity is used to cool hot water with an average capacity of 30 m^3/h and approximately 60 °C. This is an uncontrolled expenditure in terms of energy costs. This can be controlled by heat recovery.
12. High-speed flows have been observed at motor speeds, especially in Archimedean systems. If this is not necessary, it implies high energy costs.

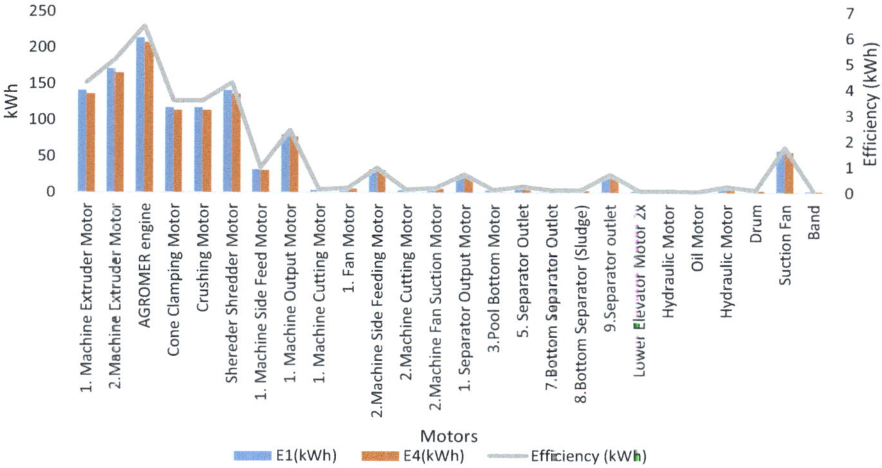

Fig. 26.4 Engine performance inventory

13. The company does not have an energy management plan. It is particularly important to define unit costs of production.
14. There is no infrastructure for monitoring energy consumption and costs.

26.4.3 Energy Efficiency Opportunities

1. Energy-efficient motor start and drive preference.
2. Driving preference on just four main motors can save up to 40% on average consumption costs. If the operating parameters are set correctly, the total savings in motor consumption can be as high as 65%.
3. Preferring pulleys in motors means losing up to 15% of energy under load. In this context, the preference for direct coupled or geared systems is important in applications.
4. High movement has been observed in motors due to high-power operation. This effect should be reduced in terms of processing capacity. To do this, the motor requirements should be re-evaluated in addition to the drive system preference. The manufacturer should be consulted during manufacture.
5. In the cooling process, the company should make a design according to the desired heat values in the screw cooling and water circuits in the production. However, a project study should first be carried out by evaluating the new lines to be installed. The collector line should be designed accordingly. Closed system dry cooler and chiller feeds should then be taken as required. Direct investment in large capacity is wrong.
6. A management infrastructure should be established to improve the decision support element for energy efficiency and management. The institutional structure

is particularly important in terms of maintenance and sustainability of the operation.

26.4.4 Action Plans

This study primarily includes the energy efficiency study of the business. In the study, action plans were evaluated with an energy management system understanding. In this context, energy efficiency opportunities in the business were presented as an action plan together with the performance enhancing project suggestions defined in the business.

26.4.4.1 Actions That Do Not Require Investment

Action 1	Energy-efficient purchasing
Background	It has been observed that energy efficiency is not taken as a basis in purchasing procedures in the business. Significant losses have been observed in business investments for this reason. Purchasing costs in many energy-consuming elements have a very low ratio to life cycle costs. Energy labels should be questioned, especially in engine preferences, in purchasing processes. Choosing a higher energy class will create a 12% saving
Objective	Creating an energy-efficient structure
Expectation	Improving efficiency for each component in energy-consuming processes

Action 2	Institutionalisation of energy efficiency
Background	Although it is far below institutional boundaries, there is a need for a corporate manager to improve energy efficiency. In this context, there is a need to develop the energy manager infrastructure of a staff to be assigned
Objective	An energy management infrastructure should be established by employing an energy manager (if possible, with an industrial energy manager certificate) in the business
Expectation	Reducing energy consumption in production with optimum operating parameters

Action 3	Efficiency of operating parameters
Background	Although it is far below institutional boundaries, there is a need for a corporate manager to improve energy efficiency. In this context, there is a need to develop the energy manager infrastructure of a staff to be assigned
Objective	Improving energy efficiency by optimising operating parameters
Expectation	Reducing energy consumption in production with optimum operating parameters

26.4.4.2 Actions Requiring Low Investment

Action 1	Drawing of infrastructure projects
Background	Especially the cooling process of the business. Recycling. Heat recovery. The growth potential of the business should be taken into consideration, and the project should be designed. Investment plans should be managed accordingly
Objective	Ensuring financial manageability in the future projections of the business and establishing energy efficiency at the project base
Expectation	The business: • Being able to define what will be done in the process • Being able to make energy-efficient product choices • Creating and controlling consumption targets • Creating a project-based infrastructure

Action 2	Installation and monitoring of filter meters
Background	The facility has different operating areas. In this context, the business consumption behaviours differ. In particular, filter meters and, if necessary, efficiency programs should be developed to determine unique consumption points
Objective	Controlling and managing energy consumption
Expectation	• Determine consumption densities in different processes in the business • Develop measures according to important consumption points • Optimise needs according to consumption habits

Action 3	Developing the energy management model
Background	There is a need for a management element that will support decision-making processes in the development process of the business. Especially considering the ever-increasing energy demand, a management model that can be monitored with an information system will provide significant gains
Objective	Developing an integrated energy management system
Expectation	• Determine the consumption density of businesses • Develop actions according to production and consumption data • Provide an energy-efficient infrastructure • Provide business optimisations according to consumption needs

26.4.4.3 Actions Requiring Investment

Action 1	Meeting the driver needs in engines
Background	Fluctuations of up to 80% in the company's motor consumption have been observed. In this context, especially in power motors, the driver preference should be selected and applied in accordance with the motor
Objective	Managing the need for low energy consumption in engines

Action 1	Meeting the driver needs in engines
Expectation	• Determine consumption fluctuations of motors • Define limits according to production and consumption data • Proper driver selection • Provide operational optimisations according to consumption needs

Action 2	Cooling system revision
Background	Procurement, installation, and commissioning of the system according to the project need created before making an investment for the cooling system of the business
Objective	Creating an energy-efficient, need-based, and sustainable system
Expectation	• Defining needs by creating project discoveries • Establishing the infrastructure and determining a sustainable model • Creating a closed system heat recovery system

Action 3	Energy efficiency in new production lines
Background	Fluctuations of up to 80% in the company's motor consumption have been observed. In this context, motor preferences for new investment systems should be reviewed accordingly. The new structure needs to be revised according to energy systems
Objective	Establishing a production infrastructure with low consumption
Expectation	• Determining consumption points in production • Defining limits according to production and consumption data • Choosing and designing appropriate systems • Ensuring business optimisations according to consumption needs

26.5 Conclusion

The present study has been conducted with the primary objective of examining energy efficiency and environmental sustainability in the context of industrial buildings. The study has established a framework for the prioritisation of energy management initiatives in the field, as well as developed a set of audit and assessment methodologies. The study has also yielded the development of a case study on the walk-through audit, which was conducted within the scope of the HAS project. The results obtained in the study are as follows:

a. Engine inventory targeting approximately 86% of the enterprise was created and energy efficiency potentials were evaluated according to the measurement data. The efficiency of motors ranged from 50% to 90%.
b. The energy efficiency assessment of the enterprise, as indicated by the motor energy class, exhibited a potential of 3.08% in total load.

c. Fourteen problem areas were identified in the plant, with six primarily being identified as energy opportunities.
d. The efficiency strategies for use in improving energy management in the study institutional infrastructure were also evaluated, with three strategies defined for each depending on the nature of the investment.

The walk-through audit is an essential component of the overall evaluation process, as it identifies the key issues that impact the effectiveness of action and decision-making processes within energy administrations. The formulation of corporate budget plans and potential energy efficiency measures constitutes a pivotal phase in thisprocess, which serves as the foundation for investment decision-making. Subsequent detailed survey processes are then initiated. This process is endorsed and provides a framework for enhancing energy efficiency within the enterprise. These types of efficiency studies are linked to detailed audits and are based on each energy user. should be supported by retrofits. In particular, the studies should include energy efficiency and investment feasibility.

References

Abdelaziz, A., Santos, V., Dias, M. S., & Mahmoud, A. N. (2024). A hybrid model of self-organizing map and deep learning with genetic algorithm for managing energy consumption in public buildings. *Journal of Cleaner Production, 434*, 140040. https://doi.org/10.1016/j.jclepro.2023.140040

Amaral, A. R., Rodrigues, E., Gaspar, A. R., & Gomes, Á. (2021). Lessons from unsuccessful energy and buildings sustainability actions in university campus operations. *Journal of Cleaner Production, 297*, 126665. https://doi.org/10.1016/j.jclepro.2021.126665

Bergougui, B. (2024). Algeria's pathway to COP28 and SDGs: Asymmetric impact of environmental technology, energy productivity, and material resource efficiency on environmental sustainability. *Energy Strategy Reviews, 55*, 101541. https://doi.org/10.1016/j.esr.2024.101541

Capehart, B. L., Turner, W. C., & Kennedy, W. J. (2012). *Guide to energy management* (7th ed.). The Fairmont Press.

da Silva Gonçalves, V. A., & dos Santos, F. J. M.-H. (2019). Energy management system ISO 50001:2011 and energy management for sustainable development. *Energy Policy, 133*, 110868. https://doi.org/10.1016/j.enpol.2019.07.004

Duglio, S. (2011). Energy management systems: From EN 16001 to ISO 50001. *Romanian Distribution Committee Magazine, 2*(4).

Dzene, I., Polikarpova, I., Zogla, L., & Rosa, M. (2015). Application of ISO 50001 for implementation of sustainable energy action plans. *Energy Procedia, 72*, 111–118. https://doi.org/10.1016/j.egypro.2015.06.016

EEA. (2016). *Circular economy in Europe—Developing the knowledge base*. EEA Report No 2/2016. European Environment Agency.

Excell, L. E., Andrews, A., & Jain, R. K. (2024). E-Audit: A "no-touch" energy audit that integrates machine learning and simulation. *Energy and Buildings, 317*, 114360, ISSN 0378-7788, https://doi.org/10.1016/j.enbuild.2024.114360

Fiedler, T., & Mircea, P. M. (2012). Energy management systems according to the ISO50001 standard—Challenges and benefits. In *International Conference on Applied and Theoretical Electricity (ICATE), Craiova*.

Ganda, F. (2019). The impact of innovation and technology investments on carbon emissions in selected organization for economic co-operation and development countries. *Journal of Cleaner Production, 217*, 469–483.

Halbhavi, S. B., Datar Vani, P., Kulkarni, S. G., Shambhavi, P., & Terani Pallavi, S. (2015). Energy auditing: A walk through survey of library building of institute to reduce the lighting cost. *International Journal of Innovative Research in Electrical, Electronics, Instrumentation and Control Engineerng, 3*(8), 52–54.

Jovanović, B., & Filipović, J. (2016). ISO 50001 standard-based energy management maturity model—Proposal and validation in industry. *Journal of Cleaner Production, 112*(Part 4), 2744–2755. https://doi.org/10.1016/j.jclepro.2015.10.023

Kaab, A., Khanali, M., Shadamanfar S., & Jalalvand, M. (2024). Assessment of energy audit and environmental impacts throughout the life cycle of barley production under different irrigation systems. *Environmental and Sustainability Indicators, 22*, 100357, ISSN 2665-9727, https://doi.org/10.1016/j.indic.2024.100357

Kaled, H. M., Maysoon, M. H., Mohammed, A., Erhart, R., Sameer, A., Aws, Z. S., Azizbek, K., Nouha, G. A., & Maha, B. (2025). Assessing climate strategies of major energy corporations and examining projections in relation to Paris Agreement objectives within the framework of sustainable energy. *Unconventional Resources, 5*, 100127. https://doi.org/10.1016/j.uncres.2024.100127

Keith, M. (2011). *Energy management and optimization*. ABB Westerville. Retrieved December 1, 2024, from https://library.e.abb.com/public/a85d6f39c1a0fe718525789d005bc295/energy%20management%20and%20optimization_paper.pdf

Liu, J., Rao, Z., Liu, G., & Liao, S. (2024a). Promotion of distributed energy system considering capacity allocation optimization and multi-stakeholder decision-making behaviors. *Journal of Cleaner Production, 437*, 140464. https://doi.org/10.1016/j.jclepro.2023.140464

Liu, J., Meng, X., & Wu, J. (2024b). Multi-stage cooperative planning among shared energy storage operator and multiple prosumers in regional integrated energy system considering long-term uncertainty. *Journal of Energy Storage, 103*(Part A), 114244. https://doi.org/10.1016/j.est.2024.114244

Marcos, F. L., & Plangklang, B. (2024). A high accurate user-friendly energy audit platform of a university building using ANN Bayesian regularization and Levenberg-Marquardt algorithm. *Energy Reports, 11*, Pages 2220–2235, ISSN 2352-4847, https://doi.org/10.1016/j.egyr.2024.01.062

Nakayama, S., Yan, W., & Fujita, A. (2024). Developing an energy audit methodology for assessing decarbonization potential in high performance buildings. *Energy Conversion and Management: X, 24*, 100765, ISSN 2590-1745, https://doi.org/10.1016/j.ecmx.2024.100765

Ramya, L. N., & Femina, M. A. (2014). Energy auditing—A walk-through survey. *International Journal of Advance Research in Electrical, Electronic and Instrumentation Engineering, 3*(Special issue 2), 406–411.

Ranky, P. G. (2012). Sustainable energy management and quality process models based on ISO50001:2011. In *The International Energy Management Standard, in IEEE International Symposium on Sustainable Systems and Technology (ISSST)*. IEEE.

Sharma, P., Salkuti, S. R., & Seong-Cheol, K. (2021). Energy audit: Types, scope, methodology and report structure. *Indonesian Journal of Electrical Engineering and Computer Science, 22*(1), 45–52. https://doi.org/10.11591/ijeecs.v22.i1.pp45-52

Sujanto, R. Y., Sarah, G. M. H., & Yeneneh, T. N. (2024). Enhancing environmental sustainability in a Circular Waste Bioeconomy: A hierarchical framework driven by operational efficiency and agro-energy management. *Cleaner and Circular Bioeconomy, 9*, 100115, ISSN 2772-8013, https://doi.org/10.1016/j.clcb.2024.100115

Tandon, A., Jitendra, K., & Pokharna, M. (2012). Energy audit and management. *International Journal of Latest Technology in Engineering Management & Applied Science (IJLTEMAS), 1*(3), 47–50.

Tiwari, S., & Mohammed, K. S. (2024). Unraveling the impacts of linear economy, circular economy, green energy and green patents on environmental sustainability: Empirical evidence from OECD countries. *Gondwana Research, 135*, 75–88. https://doi.org/10.1016/j.gr.2024.07.018
Wang, J., Wang, Y., Zhang, S., & Yi, B. (2025). Drivers of environmental externality reduction in China's electric power industry: A spatial-temporal analysis. *Journal of Environmental Management, 373*, 123612, ISSN 0301-4797, https://doi.org/10.1016/j.jenvman.2024.123612
Wen, X., Cao, H., Hon, B., Chen, E., & Hongcheng, L. (2021). Energy value mapping: A novel lean method to integrate energy efficiency into production management. *Energy, 217*, 119353. https://doi.org/10.1016/j.energy.2020.119353
Wikipedia. (2024a). Energy audit—"Type of energy audit". Retrieved November 29, 2014, from https://en.wikipedia.org/wiki/Energy_audit
Wikipedia. (2024b). PDCA ZIRKEL ENGL.png. Retrieved 1 December, 2024, from https://en.m.wikipedia.org/wiki/File:PDCA_ZIRKEL_ENGL.png
Yang, S., Ruyin, L., Hong, C., Meifen, W., & Qianwen, L. (2024). How do perceived organizational green climate and corporate energy-saving behaviors promote each other? An integrated evaluation model. *Environmental Impact Assessment Review, 105*, 107368. https://doi.org/10.1016/j.eiar.2023.107368
Zhang, Z., & Zhang, L. (2024). Government green procurement policies and corporate green total factor productivity. *Finance Research Letters, 65* 105556. https://doi.org/10.1016/j.frl.2024.105556

Chapter 27
Analyzing CO_2 Emission: A Case Study of Batumi State Maritime Academy

Guram Vadachkoria, Avtandil Tsitskishvili, and Givi Tsitskishvili

27.1 Introduction

As it is widely acknowledged, the world is currently encountering one of the most important and critical challenges: global climate change, which has been occurring for decades. More precisely, it implies a deleterious impact on the environmental and atmospheric conditions of the earth. As common instances, the consequences of climate change are notable in temperature, water scarcity, etc. Regardless the natural disasters, climate change takes place as a result of various external and internal factors, especially due to the human activities. According to the European Commission's explanation (European Commission, 2022), human activity is gradually influencing the earth's climate by adding enormous amounts of greenhouse gases (GHGs) to those occurring naturally in the atmosphere. 2011–2020 was the warmest decade recorded, with global average temperature reaching 1.1 °C above pre-industrial levels in 2019. "Human-induced global warming is presently increasing at a rate of 0.2°C per decade".

As previously stated, greenhouse gas emissions are a principal contributor to influencing the climate. Carbon dioxide, in particular, is the primary greenhouse gas that accounts for around three-quarters of all emissions. In accordance with NASA, the release of CO_2 in the atmosphere has increased 50% in past decades (NASA, n.d.), which can also be emitted from dissimilar and disadvantageous man-made activities. To mitigate the unfavorable results related to climate change, every organization and institution undertakes numerous initiatives and operations. The United Nations (UN) Framework Convention on Climate Change (UNFCCC) stated, that The Paris Agreement aims to prevent dangerous man-made interference with the

---

G. Vadachkoria (✉) · A. Tsitskishvili · G. Tsitskishvili
Batumi State Maritime Academy, Batumi, Georgia
e-mail: g.tsitskishvili@bsma.edu.ge

M. Z. Sogut, M. Koray (eds.), *Energy Rationality and Management for Decarbonization*, Green Energy and Technology,
https://doi.org/10.1007/978-3-031-68497-5_27

global climate system (Paris Agreement, 2015) and based on the UN's estimations to limit global warming to 1.5 °C, greenhouse gas emissions must begin falling by 7.6% (United Nations, 2019) each year starting in 2020. Apart from that the (IMO, n.d.) International Maritime Organization (IMO) sets goals to reduce to a minimum the GHG emissions from various types of ships. In June 2021, IMO adopted short-term measures to reduce the carbon intensity of all ships by 40% by 2030, compared to 2008.

As is recognized, Georgia is an officially admitted member of the UN and IMO. It is a signatory of the Paris Agreement (UNFCC, 2016), as well. Despite that membership, Georgia is included among the developing countries, which has a negative effect on the environment. To exemplify, considering the recently available data, the emission of CO_2 in Georgia has been approximately 8,610,414 tons since 2016 (Ministry of Environment Protection and Agriculture of Georgia, 2021).

The mentioned fact obligates us to minimize the level of emitted CO_2 and impede global warming, commencing even with incremental advances and improvements. Therefore, under the idea of "Think globally, act locally," authors believe analyzing CO_2 emissions from the local maritime education center Batumi State Maritime Academy (BSMA) is of utmost importance and purpose.

27.2 Materials and Methods

The detrimental influence on climate change from BSMA gradually becomes distinctive and vital. Thus, the principal purpose of the research is to determine to what extent BSMA impacts the global issue, more specifically, the total amount of emitted carbon dioxide in the environment.

The conducted survey required several procedures. Generally, the primary approach was quantitative research, which was implemented in order to acquire accurate energy consumption and emissions information. It was achieved by calculating and evaluating energy utilization and subsequent CO_2 emissions during academic work and education processes. During the dissimilar operations in the academy, electrical energy and natural gas are consumed, both of which make a significant contribution to emissions.

In the initial stages, participants gathered and analyzed the energy bills provided by the financial department, to obtain an annual energy consumption estimate for Batumi State Maritime Academy. Then, to further break down the information, the authors collected data about energy consumers on the premises. The research project was conducted throughout the academic year 2022–2023. However, despite the given statistics, we wanted to verify those numbers and proceeded with the exploration individually. After which, students observed and assessed each establishment of the academy, which particularly encompasses: the library, classrooms, personal cabinets, welding workplace, and machinery department. Moreover, researchers gained specific information and manuals about dissimilar types of equipment. Then, the CO_2 emission factor was applied to consume electricity and natural gas.

Respectively, participants generated a comprehensive and detailed overview of internal and outdoor energy consumption. Additionally, based on the period of consumption and the instructions for heating boilers, researchers estimated the approximate quantity of total natural gas usage, as natural gas is not consumed regularly.

27.3 Results and Discussion

The accumulated and evaluated information represented divergent electrical devices, systems, apparatus, and equipment. Especially assuming electrical energy, the academy produces carbon dioxide. CO_2 emissions from various types of machines, which are used for practical works, such as welding apparatus, vertical drills, phrasing lathen machine and abrasive wheels. Hence, the complete amount of consumption of those appliances equals 13,293 kWh and the emission of CO_2 reaches 1401,966 kg.

Furthermore, researchers distinguished light panels by their type and calculated the total number of them in the entire university. More precisely, outdoor illumination quantified 24.16277 kg of carbon dioxide. Regarding the indoor lightings, the amount of emission in study rooms was approximately 41.10192 kg of CO_2 and in the machinery compartment, the value exceeded 21.44448 kg. Besides the separately estimated numbers, the research resulted in a total amount of consumption of electric energy during educational processes and an academic year attained to the 665,900 kWh; consequently, the emitted carbon dioxide was equivalent to 67,921.8 kg, which is indicated in Fig. 27.1.

Concerning consumption of natural gas, depending on the provided information, 1 m^3 of CNG gas produced 10.55 kWh of power and the efficiency of heating boilers was 92%. Accordingly, based on the supplied manuals of central heating boilers the survey revealed that the whole emitted carbon dioxide through a year is about 127,155.9 kg, which showed the unfavorable condition of BSMA and in addition, 1 year mostly implies from 5 to 6 months (Table 27.1 and Fig. 27.2).

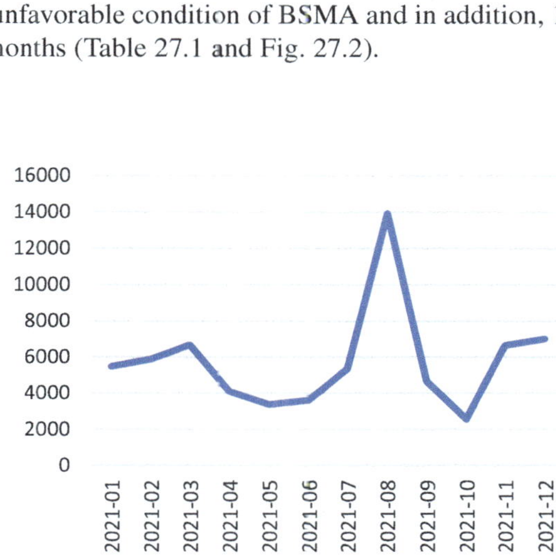

Fig. 27.1 The graph of consumed electric energy in 2022

Table 27.1 The parameters of central heating boilers and annual consumption

Power (KW)	1 m^3 CNG—produced energy (10.55 kW)	Hourly CNG consumption (m^3)	Hourly CO_2 emission (kg)	Yearly consumption (m^3)
349	10.55	33.1	62.85	44,658.8
174	10.55	16.5	31.34	22,265.4

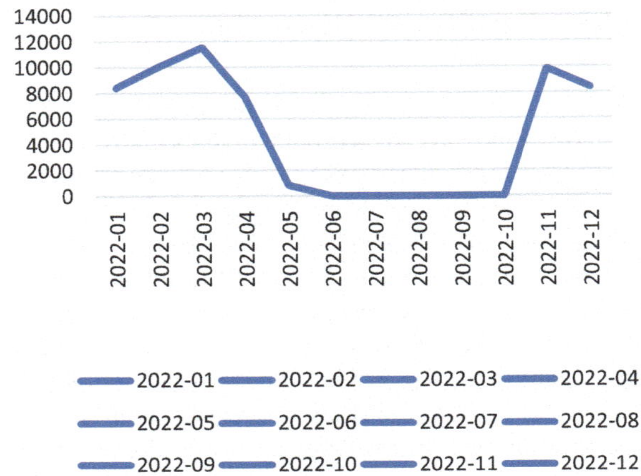

Fig. 27.2 The graph of consumed natural gas in 2022

27.4 Conclusion

The survey findings demonstrated that BSMA does not have an energy management system or structure to minimize the inauspicious and hazardous consequences regarding the adverse contribution in global warming. It should be mentioned that the academy requires contemporary approaches to reduce the negative impact on the climate and GHG emissions (UNFCC, 2016).

Irrespective of that the research group suggested alternative perspectives and strategies, in order to diminish the number of emissions and modify the issue. The contemporary technology gives the possibility to change certain electric devices with more efficient, systematized, and simultaneously energy-saving appliances. To illustrate, light bulbs with a power of 0.048 kW, which are commonly used in classrooms or in libraries, can be changed into LED panels with the function of a time relay. This unique function allows to turn off the electricity automatically after the prescribed time. In addition, the custom illumination on stairs and in the hallway can be replaced by sensor lights. As electricity for lighting accounts for approximately 15% of global power consumption and 5% of worldwide greenhouse gas (GHG) emissions, replacement by modern technology will lead to an improvement. The study can serve as a starting point for implementing CO_2 emissions reduction activities from maritime academy operations. Such as establishing an energy management plan and promulgating energy efficiency campaigns.

References

European Commission. (2022). *Causes of climate change*. European Commission—Climate Action. Retrieved from https://climate.ec.europa.eu/climate-change/causes-climate-change_en

International Maritime Organization. (n.d.). *Lowering containership emissions through just-in-time arrivals*. Green Voyage 2050. Retrieved from https://greenvoyage2050.imo.org/lowering-containership-emissions-through-just-in-time-arrivals/

Ministry of Environment Protection and Agriculture of Georgia. (2021). *Georgia's 2030 climate change strategy*. Retrieved from https://mepa.gov.ge/En/Files/ViewFile/50123

NASA. (n.d.). *Carbon dioxide—Climate change: Vital signs of the planet*. NASA. Retrieved from https://climate.nasa.gov/vital-signs/carbon-dioxide/

United Nations. (2019). *Why it matters—Goal 13: Climate action*. Retrieved from https://www.un.org/sustainabledevelopment/wp-content/uploads/2019/07/13_Why-It-Matters-2020.pdf

United Nations Framework Convention on Climate Change. (2016). *The Paris Agreement*. UNFCCC. Retrieved from https://unfccc.int/process-and-meetings/the-paris-agreement

Chapter 28
The Analysis of Climate Change Awareness and Environmental Attitudes at Batumi State Maritime Academy

Iakob Makharadze and Givi Tsitskishvili

28.1 Introduction

Due to its negative impact on human health and the environment, climate change is one of the most significant problems of the twenty-first century. According to the United Nations Framework Convention on Climate Change Annual Report 2021, climate change is a global phenomenon caused by the numerous activities of individuals, communities, and countries worldwide. Climate change is already affecting all parts of the planet in multiple ways, and the effects will worsen with further warming (UNFCCC, 2022).

One of the key aspects of climate change is greenhouse gas emissions. The Paris Agreement aims to strengthen the global response to climate change, and climate change education, public awareness, and public participation are enhanced under the agreement (Paris Agreement, 2015). Within the context of the problem, the International Maritime Organization also developed the greenhouse gas (GHG) strategy, with the goal of reducing carbon emissions from international shipping by 40% by 2030 (MEPC, 2018, IMO, 2021). Georgia, as a member of the IMO and as a signatory state to the Paris Agreement, the Kyoto Protocol, and a number of other climate change agreements, has a commitment to show progress in this field (MEPA, 2021).

Like many of the world's developing countries, climate change has a significant impact on the environment and human health in Georgia. It should be highlighted that climate change processes have intensified in Georgia, with a wide range of negative consequences. According to "Georgia's 2030 Climate Change Strategy (Mitigation)", the anticipated negative consequences will become much more

I. Makharadze (✉) · G. Tsitskishvili
Batumi State Maritime Academy, Batumi, Georgia
e-mail: G.tsistkishvili@bsma.edu.ge

M. Z. Sogut, M. Koray (eds.), *Energy Rationality and Management for Decarbonization*, Green Energy and Technology,
https://doi.org/10.1007/978-3-031-68497-5_28

pronounced in the future, and without the development of climate-resilient practices, climate change mitigation measures, and improvements to the country's preparedness and abilities, the most climate-sensitive sectors will become more vulnerable, and other negative effects of climate change will also rise (MEPA, 2021). Therefore, climate change awareness and its proper understanding among Georgians is critical since it leads to collective action for adaptation to the environment.

Educational institutions play a significant role in developing the knowledge, awareness, attitudes, and sustainable actions required to accomplish the objective of a sustainable Earth. They have an important role in shaping future generations, improving sustainability efforts, and accepting responsibility. Especially, tertiary education institutions have a strong potential to influence changes in environmental awareness and attitudes (Keleş, 2017).

Batumi State Maritime Academy (BSMA) is a tertiary educational institution that plays a key role in promoting environmental awareness and developing environmental attitudes. The analysis of employees' and students' awareness of climate change includes two levels of research, and the results can be generalized to Georgian educational institutions and used as a cornerstone for the development of the BSMA's environmental impact mitigation plan.

The objective of Batumi State Maritime Academy is to mitigate the impact of climate change, and the level of awareness of students and employees must be considered while adapting to climate change in the high educational institution. The aim of this article is to analyze the level of perception, awareness, and knowledge of BSMA employees and students regarding climate change and determine the association between climate change attitudes and environmental behavior. Batumi State Maritime Academy was selected as the research area, and appropriate quantitative methods were used to collect and analyze the relevant information using the questionnaires.

28.2 Materials and Methods

Climate change issues are becoming increasingly prominent on the BSMA agenda. The objective of the paper was to determine the current level of knowledge and awareness of climate change among students and employees at Batumi State Maritime Academy.

In order to collect data, questionnaires were distributed which were filled out by 70 employees and 302 students. This figure corresponds to 15% of students and personnel. The research was carried out in the 2022–2023 academic year. The objectives of the survey were clarified to the respondents at the beginning of the questionnaire.

This research used a three-part questionnaire. The first part examined the demographic characteristics of the students and employees. The second part measured climate change awareness, environmental consciousness, and behavior. The third part measured the level of awareness of green transport and waste management. Appropriate quantitative approaches have been used to analyze the data collected from the questionnaires. The study provides analyses of the questionnaire and suggests activities to address the issues revealed.

28.3 Results and Discussion

As indicated in Table 28.1, there is a significant gender imbalance among the students surveyed. More than 78% of responses are from male students, which is an equitable result given the gender balance of Maritime Academy students. In terms of employee gender distribution, the situation is completely reversed. Almost 63% of the employees are female.

As shown in Table 28.2, the largest part of the surveyed students, 94%, belongs to the age group of 15–24 years. The average age of students is 19.86, and the median age is 19. The majority of employees (70.1%) belong to the young and middle-aged groups. The average age of employees is 41.72, and the median age is 41.

Students from both educational levels participated in the study, with undergraduate students making up the majority. As shown in Table 28.3, 296 students answered

Table 28.1 Gender distribution of students and employees

Gender	Students	Percentage (%)	Employees	Percentage (%)
Female	59	19.5	44	62.9
Male	238	78.8	24	34.3
No answer	5	1.7	2	2.8
Total	302	100	70	100

Table 28.2 Age distribution of students and employees

Age	Students	Percentage (%)	Age	Employees	Percentage (%)
15–19	171	56.6	20–29	6	8.6
20–24	113	37.4	30–39	23	32.9
25–29	9	3	40–49	20	28.6
30–34	1	0.3	50–59	11	15.7
35–39	3	1	60–69	7	10
No answer	5	1.7	No answer	3	4.2
Total	302	100	Total	70	100

Table 28.3 Academic degree of students

Academic degree	Students	Percentage (%)
Bachelor's degree	285	96.3
Master's degree	11	3.7
Total	296	100

the question about their education level. Two hundred and eighty-five (96.3%) study at the bachelor's level, while 11 (3.7%) study at the master's level.

The number of students by faculty is shown in Fig. 28.1. Out of 298 students, 145 study at the Faculty of Navigation, 84 at the Faculty of Engineering, and 69 at the Faculty of Business and Management. According to Fig. 28.1, 67 employees answered the question about the position. Almost half of the employees, 33 (49.3%), hold administrative positions, while 17–17 work in academic positions or work in both.

As shown in Fig. 28.2, 46.7% of 296 students and 65.7% of 67 employees are aware or fully aware of the impacts of climate change. 16.9% of students and 9% of employees are unaware of climate change. The result indicated that employees have a higher level of awareness regarding the environmental impacts of climate change.

The majority of employees and students believe that environmental protection actions should be implemented immediately or that they will soon be necessary. Figure 28.3 highlights that employees (55 out of 69 respondents, 79.7%) prefer immediate environmental protection actions over students (210 out of 296 respondents, 70.9%). However, a greater number of students than employees believe that

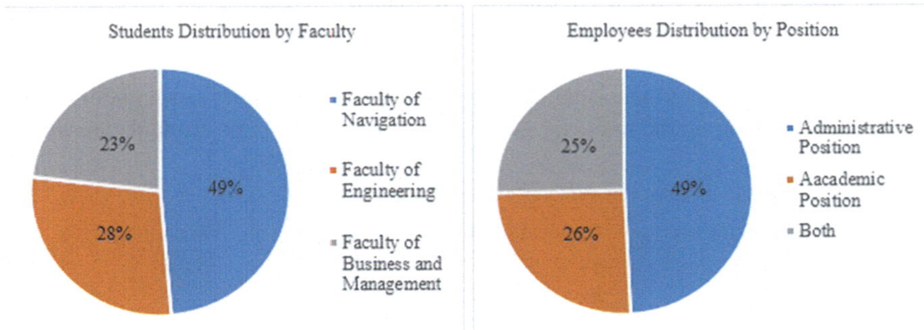

Fig. 28.1 Distribution by faculty and position

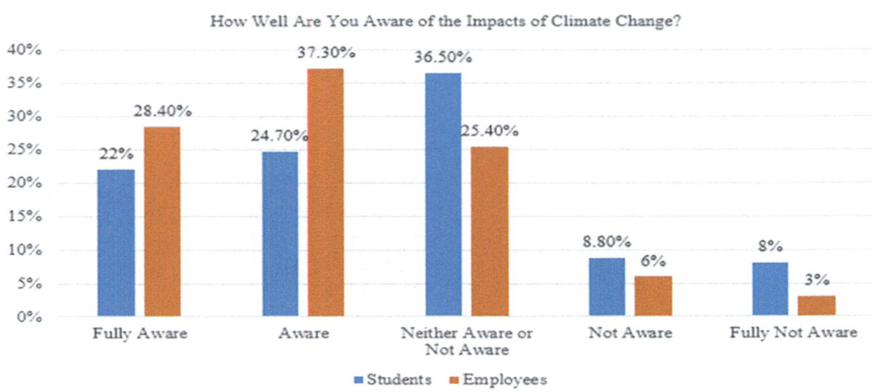

Fig. 28.2 How well are you aware of the impacts of climate change?

these actions will soon be necessary, 24.7% and 13% respectively. According to Fig. 28.3, a larger proportion of employees than students believe that this problem has no impact on their generation (4.3% and 2.4%, respectively).

According to the responses to the question "How environmentally friendly are you?", the majority of 68 employees say they are very (48.5%) or relatively (35.3%) eco-friendly. The majority of students consider themselves eco-friendly. According to 45.1% of 295 respondents, they are very eco-friendly, and 29.5% believe that they are more eco-friendly than non-eco-friendly. Students have a greater rate of non-eco-friendliness. Two percent are completely unfriendly to the environment, while 5.1% are less friendly (Fig. 28.4).

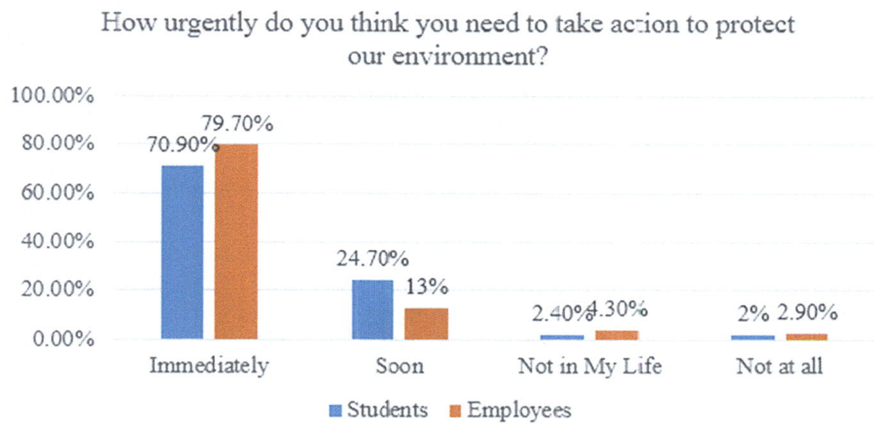

Fig. 28.3 How urgently do you think you need to take action to protect our environment?

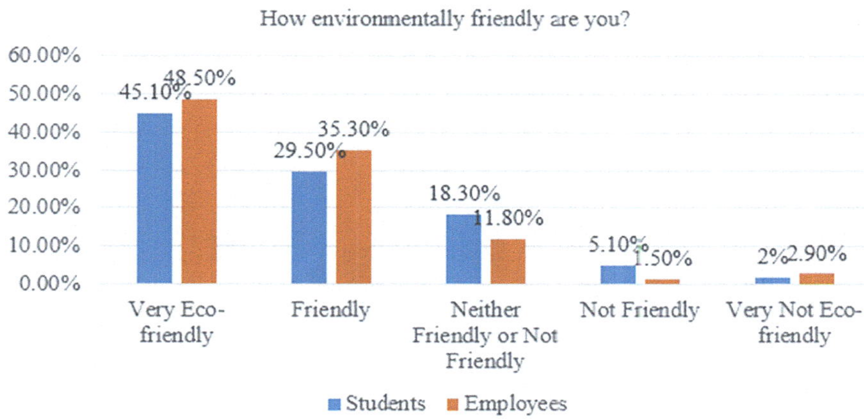

Fig. 28.4 How environmentally friendly are you?

According to the answers given to the question "How often do you encounter environmental protection campaigns (within the country)?", the majority of employees (43.5%) responded almost never, while most students responded once a month (40.7%). It should be highlighted that both daily and weekly environmental protection campaigns are recognized by a greater proportion of students than employees (Fig. 28.5).

As shown in Fig. 28.6, students are more likely to be informed about environmental protection campaigns than employees. The majority of employees, 59.4%, learn about environmental campaigns from television programs, and for students, Facebook is the best way to receive information. Students obtain information from TikTok, Instagram, TV shows, and local projects almost equally.

According to Table 28.4, both students and employees have heard of the UN Sustainable Development Goals. Six out of every ten employees or students are aware of the United Nations Sustainable Development Goals. Since there is a considerable relationship between the level of understanding of climate change and attitudes toward environmental protection, if appropriate information is provided,

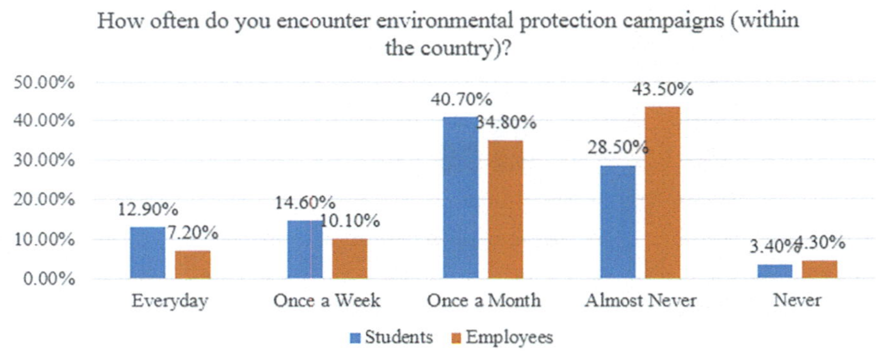

Fig. 28.5 How often do you encounter environmental protection campaigns (within the country)?

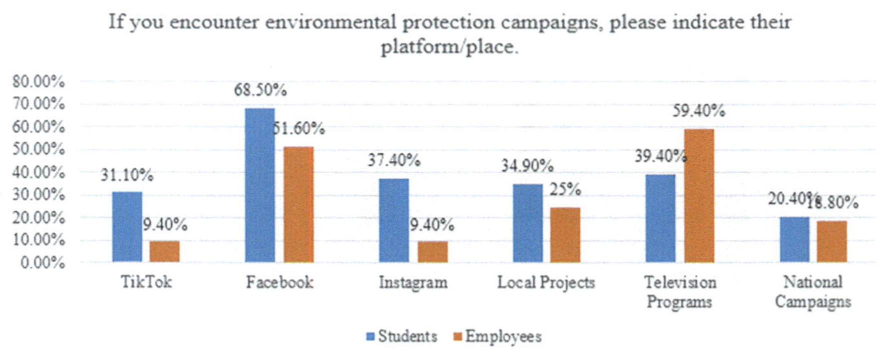

Fig. 28.6 If you encounter environmental protection campaigns, please indicate their platforms/place

students and employees can acquire good attitudes toward environmental conservation.

As shown in Table 28.5, the rates of participation in environmental activities among students and employees are almost equal. One hundred and eighty-eight (63.5%) of the 296 students, and 42 (60.9%) of the 69 employees participated in the environmental protection campaigns at least once. Environmental events in which students and employees participated or were organizers predominate campaigns to clean sea and river shorelines, roadways, and the environment in general, participation in environmental clubs, waste management, tree planting campaigns, etc.

Out of 68 BSMA employees, 24 are highly informed about green transportation, and 18 are informed (Fig. 28.7). Out of 286 students, 24.1% are highly informed about green transportation, and 22.4% have good knowledge of it. 22.1% of employees and 29.3% of students are not aware of green transport, which is a rather high and unfavorable result.

The survey revealed that 42% of employees drive to the academy, while 34.8% walk. Only 15.9% of employees use public transportation The primary mode of transportation for students (53.3%) is public transportation. 24.4% of students walk, while 17.4% of students drive. The percentage of students and employees that

Table 28.4 Do you know what the current UN SDGs are?

Answers	Students	Percentage (%)	Employees	Percentage (%)
Yes	172	58.5	42	60.9
No	122	41.5	27	39.1
Total	294	100	70	100

Table 28.5 Have you ever participated in an environmental protection campaign?

Answers	Students	Percentage (%)	Employees	Percentage (%)
Yes	188	63.5	42	60.9
No	108	36.5	27	39.1
Total	296	100	70	100

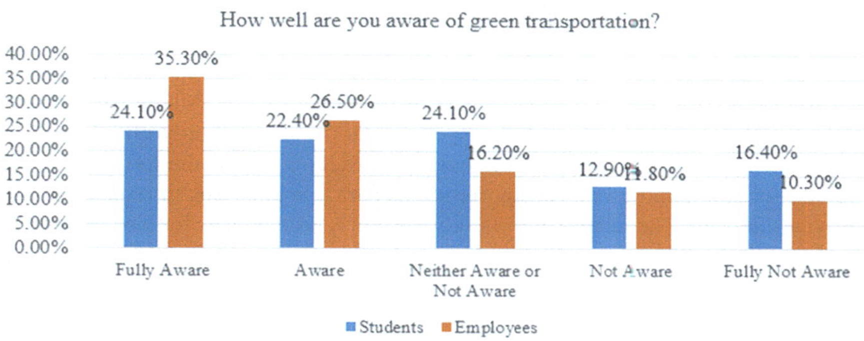

Fig. 28.7 How well are you aware of green transportation?

commute by bicycle, electric scooter, or scooter is quite low (2.4% and 2.9%, respectively).

28.4 Conclusion

According to the findings of the study, 91% of employees and 72% of students are aware of climate change issues. Almost half of the students are well aware of the environmental impact of climate change. The level of awareness is higher among the employees. Also, employees (95%) are more interested in environmental issues than students, but in both cases, the majority believe that actions that protect the environment should be implemented immediately or that essential measures should be taken soon. Six out of every ten employees or students have heard of the United Nations Sustainable Development Goals. Indicators of their participation in environmental initiatives are nearly similar.

Students hear more about environmental campaigns than employees. To improve environmental awareness, it is required to increase the frequency of such campaigns or more efficiently manage the channels that provide information about them to the public, as social media is the key way of obtaining information for students and television shows for employees. Furthermore, nearly seven out of ten employees and six students at BSMA are interested in learning more about environmental protection. As a result, the academy should take the lead in ensuring that students and employees are well-informed on climate change.

According to the survey, 75% of students and 83.6% of employees believe themselves to be very eco-friendly or environmentally friendly. However, 22.1% of employees and 29.3% of students have little or no knowledge of green transport. To get to the BSMA, 42% of employees and 17.4% of students use the most environmentally unfriendly mode of transportation, the car. Fifty-eight percent of employees use public transportation to come to the academy almost never or never. To mitigate the effects of climate change, measures to minimize the use of non-eco-friendly transportation in academics must be implemented.

The results revealed that both sub-levels are prepared to participate in the necessary actions to resolve the issues. Based on research analyses, BSMA can lead the way in providing accurate climate change information and taking action to support employees and students in the protection of the environment. Among the actions are: implementation of an energy-saving policy to improve energy efficiency and consumption habits; establishing internal infrastructure within the territory of the academy for small and green transportation as part of campaigns to reduce the use of private cars; etc. The results of the study will serve as the basis for developing the Batumi State Maritime Academy's environmental impact mitigation plan.

References

IMO. (2021). *Fourth IMO GHG study 2020: Full report*. IMO.

IPCC. (2022). Climate change 2022: Impacts, adaptation and vulnerablity. In H.-O. Pörtner, D. C. Roberts, M. Tignor, E. S. Poloczanska, K. Mintenbeck, A. Alegría, M. Craig, S. Langsdorf, S. Löschke, V. Möller, A. Okem, & B. Rama (Eds), *Contribution of Working Group II to the Sixth Assessment Report of the Intergovernmental Panel on Climate Change*. Cambridge University Press. https://doi.org/10.1017/9781009325844

Keleş, O. (2017). Investigation of pre-service science teachers' attitudes towards sustainable environmental education. *Higher Education Studies, 7*(3), 171. https://doi.org/10.5539/hes.v7n3p171

Marine Environment Protection Committee. (2018). *Initial IMO strategy on reduction of GHG emissions from ships* (Vol. 304(72)). MEPC.

MEPA. (2021). *Georgia's 2030 climate change strategy (mitigation)*. Government of Georgia.

Paris Agreement. (2015, December 12). *Paris Agreement to the United Nations Framework Convention on Climate Change*. T.I.A.S. No. 16-1104.

UNFCCC. (2022). *United Nations Climate Change Annual Report 2021*. United Nations Climate Change.

Chapter 29
Advancing Sustainable Aviation: Integration of Emerging Energy Technologies in Aviation Infrastructure

Omar Alharasees, Utku Kale, and Daniel Rohacs

29.1 Introduction

The aviation industry, a linchpin of global connectivity, stands at the forefront of a transformative era driven by an imperative for environmental stewardship. In 2019, the collective emissions from airports globally reached a staggering 920 million tonnes of CO_2, accentuating the sector's substantial carbon footprint (Graver et al., 2020). This environmental challenge has spurred a paradigm shift, exemplified by the pioneering research and innovations from organizations such as the International Air Transport Association (IATA) (Ploetner et al., 2017). As airports worldwide grapple with the pressing need for ecologically responsible practices, the integration of cutting-edge renewable energy technologies within airport infrastructures becomes paramount (Dimitriou & Karagkouni, 2022). Industrial initiatives, exemplified by the strategic investments and collaborations fostered by organizations like the Clean Sky 2 Joint Undertaking, underscore the industry's commitment to a greener future (Tanaka et al., 2023).

In the pursuit of sustainable aviation, the exploration of novel energy sources is fundamental. Advances in lignocellulosic biomass solutions find resonance in industrial endeavours such as the efforts led by the Advanced Biofuels Association (Wang et al., 2019). This sectoral initiative emphasizes the scalability and efficacy of biomass-derived fuels, a pertinent consideration for aviation's transition away from conventional fossil fuels (Rao et al., 2020). Simultaneously, industrial collaborations, such as those facilitated by the hydrogen council, propel the integration of

O. Alharasees (✉) · U. Kale · D. Rohacs
Department of Aeronautics, and Naval Architecture, Faculty of Transportation Engineering and Vehicle Engineering, Budapest University of Technology and Economics, Budapest, Hungary
e-mail: oalharasees@edu.bme.hu; kale.utku@kjk.bme.hu; rohacs.daniel@kjk.bme.hu

M. Z. Sogut, M. Koray (eds.), *Energy Rationality and Management for Decarbonization*, Green Energy and Technology,
https://doi.org/10.1007/978-3-031-68497-5_29

hydrogen fuel cells into ground support equipment across the aviation landscape (Kar et al., 2022). These strategic advancements underscore the global applicability of renewable energy solutions within the intricate web of aviation operations. The sectoral momentum is further echoed by organizations like the International Renewable Energy Agency (IRENA) (Asmelash et al., 2020), which actively shapes the discourse on the integration of renewable energy in aviation.

The landscape of sustainable aviation expands further with innovations at the intersection of technology and biology. Pioneering endeavours led by research institutions and organizations like the European Commission's Joint Research Centre delve into artificial photosynthesis, providing potential solutions to the industry's carbon-neutral aspirations (Joint Research Centre (European Commission), 2022). The transformative synergy between innovative technology and sustainable aviation practices is echoed in collaborations like the European Aviation Safety Agency's initiatives in fostering sustainable aviation technologies (Sanfourche & Fischer, 2022). Amidst these advancements, the imperative for efficient energy storage is underscored by industrial milestones achieved by companies specializing in advanced flywheel technology, exemplified by organizations like Beacon Power (Li & Palazzolo, 2022). This not only ensures grid stability and reliability but also serves as a critical milestone in resilient and sustainable airport operations, echoing the ethos of industrial players committed to advancing the sustainable aviation agenda.

29.1.1 Background

The aviation sector, crucial for global connectivity, confronts rising environmental concerns from greenhouse gas emissions. Amsterdam Airport Schiphol's adoption of concentrated solar technology on terminal roofs (Sreenath et al., 2020), reducing energy consumption by 15%, aligns with IATA's goal for a 50% cut in net aviation carbon emissions by 2050 (Zhou, 2022), showcasing industry-wide commitment to sustainability.

Amidst the aviation sector's environmental concerns, lignocellulosic biomass has gained prominence. Oslo Airport's innovative biomass boiler, meeting 80% of heating demands, showcases scalable biomass solutions, reducing reliance on traditional fuels (Baxter et al., 2020). In the Asia-Pacific, Brisbane Airport pioneers hydrogen fuel cells in ground support equipment (Ng et al., 2021), a global exemplar of renewable energy integration in aviation operations.

Optimizing geothermal potential curtails carbon footprints in airport infrastructures. Reykjavik's Keflavik International Airport in Iceland utilizes Earth's heat for heating and cooling, reducing reliance on traditional energy (Sætórsdóttir & Hall, 2019). This geothermal district heating system serves as a model for other geographically suitable airports worldwide. Artificial photosynthesis gains traction in aviation, with Swiss-based Climeworks pioneering atmospheric carbon dioxide capture for synthetic fuel production (Okonkwo et al., 2023). This exemplifies a transformative synergy between innovative technology and sustainable aviation practices, presenting a potential solution for the industry's carbon-neutral aspirations.

Vortex bladeless wind energy revolutionizes wind power utilization (Francis et al., 2021). The oscillating design, distinct from traditional rotation, minimizes noise and visual impact, presenting the potential for seamless integration in airport landscapes (Asre et al., 2022). This installation's success mirrors airports' increasing focus on sustainable energy and harmonious coexistence with communities. Additionally, energy storage gains prominence for reliable power supply. Amsterdam Airport Schiphol's use of water-based storage emphasizes the viability (Zhou, 2022), enhancing resilience, flexibility, and mitigating energy demand fluctuations.

Efficient renewable energy storage is crucial in sustainable aviation, demonstrated by Singapore Changi Airport's advanced thermal systems (Kim et al., 2012), enhancing efficiency and reducing peak demand. Active ventilation and natural shielding, exemplifying multifaceted strategies for optimized energy consumption (Carratt et al., 2020). These airports serve as sustainability models, emphasizing technology, infrastructure, and operational convergence to reduce the industry's environmental impact.

Helsinki Airport strategically deploys heat pump systems, leveraging seasonal temperature variations (Rasku & Kiviluoma, 2018). Extracting heat for cooling in warmer months and vice versa for heating optimizes energy efficiency, showcasing climate-specific renewable energy utilization. Aligning with the Airports Council International's (ACI) Airport Carbon Accreditation program (Comendador et al., 2019).

In sustainable aviation, Incheon International Airport in South Korea innovatively integrates natural and active shielding methods. Employing landscaping and architectural design strategically minimizes exposure to adverse weather (Baek et al., 2016), reducing the need for excessive heating or cooling. The installation of active shielding technologies, such as wind barriers and energy-efficient facades, exemplifies a holistic energy conservation approach. Incheon International Airport's success underscores merging architectural ingenuity with renewable energy applications for enhanced sustainability.

In aviation's renewable energy integration, energy storage is pivotal. Airports adoptions of mechanical storage systems with advanced flywheel technology is still a novel step towards sustainability (Rimpel et al., 2021), ensuring grid stability. Changi Airport in Singapore prioritizes water in sustainable aviation (Kim et al., 2012). Implementing water storage methods and advanced heat exchange systems ensures efficient temperature regulation and fosters closed-loop water utilization. Changi's commitment extends to water conservation, with natural ventilation systems leveraging water features for cooling. This holistic integration showcases multifaceted strategies airports employ to harmonize environmental stewardship with operational excellence.

29.1.2 Aims and Objectives

This chapter meticulously examines the integration of emerging energy technologies in aviation and transportation, aiming to address the industry's environmental challenges. By scrutinizing emerging technologies such as lignocellulosic biomass, concentrated solar, enhanced geothermal, artificial photosynthesis, vortex bladeless wind energy generation, heat pump, hydrogen, and various energy storage methods, the chapter seeks to outline their applicability, limitations, and potential impact. Providing real-life examples, the research explores the scalability and seamless integration of these technologies within airport landscapes renewable emerging energy integration in the aviation industry, exemplified by real-world instances, goes beyond theoretical paradigms. Airports are shaping a future where environmental stewardship intertwines with operational efficiency. The exploration of these examples and subsequent patent analysis establishes a foundation for understanding the intricate interplay between technological innovations, operational exigencies, and the relentless pursuit of sustainable aviation. The analysis encompasses the suitability of each technology in aviation, emphasizing challenges. A patent review from 2010 to 2022 provides additional depth. The ultimate goal is to offer informed recommendations for policymakers, airport authorities, and stakeholders, fostering a sustainable trajectory for the aviation industry.

29.2 Aviation's Emerging Energy Frontier: Technologies and Storage Innovations

29.2.1 New Sources, Methods, and Emerging and Future Technologies

29.2.1.1 Lignocellulosic Biomass/Cellulosic Ethanol

Lignocellulosic biomass, consisting of cellulose and hemicellulose derived from non-food sources, represents a prospective avenue for sustainable aviation (Voß et al., 2023). The application of lignocellulosic biomass in aviation involves its conversion into cellulosic ethanol, a renewable alternative sourced from non-edible biomass. This holds particular relevance for the aviation sector, providing an avenue to develop biofuels that mitigate reliance on conventional fossil fuels. Realizing applications in aviation infrastructure, the integration of biomass-derived fuels seeks to address both energy security and environmental stewardship. Current initiatives, exemplified by projects such as INEOS Bio's cellulosic ethanol plant (Chandel et al., 2018), underscore ongoing efforts to bring these technologies to fruition.

29.2.1.2 Concentrated Solar

Concentrated solar power, relying on solar collectors and concentrating mirrors to focus sunlight, emerges as a technology with diverse applications in the aviation and broader transportation sectors (Goyal et al., 2022). The concentrated solar technology demonstrates the potential for addressing energy demands within airport infrastructures, offering an alternative to conventional grid-based electricity. Future implementations envision expanded solar integration within airport facilities, with examples such as the Crescent Dunes Solar Energy Project indicating the viability of large-scale solar installations (Boretti et al., 2020).

29.2.1.3 Enhanced Geothermal

The applicability of enhanced geothermal systems in aviation extends beyond mere energy generation, influencing broader transportation infrastructure (Sowiżdżał et al., 2022). Through the extraction of steam from underground hot rocks, enhanced geothermal systems offer continuous and reliable power for various airport operations. The adoption of geothermal systems aligns with the aviation industry's sustainability commitment, with potential applications extending to both ground and air-based electric vehicles, fostering a cohesive and sustainable ecosystem.

29.2.1.4 Artificial Photosynthesis

Artificial photosynthesis, emulating natural processes to convert sunlight into synthetic fuels, holds substantial promise for revolutionizing the aviation sector's fuel landscape (Li et al., 2022). As an alternative to traditional hydrocarbon-based aviation fuels, artificial photosynthesis provides a pathway for carbon-neutral air travel. Potential applications in aviation involve the development of synthetic fuels derived from carbon dioxide, aligning with the industry's commitment to carbon neutrality. Companies like Joule Unlimited exemplify pioneering efforts in realizing the potential of artificial photosynthesis for sustainable aviation.

29.2.1.5 Vortex Bladeless Wind Energy Generation

Vortex bladeless wind energy generation, characterized by its oscillating turbine design, not only minimizes environmental impact but also offers strategic applicability within airport landscapes (Turkoglu & Toloui, 2020). The technology's low noise and visual footprint make it conducive to on-site energy generation without compromising the aesthetics of airport environments. Future implementations in aviation envision expanded utilization, potentially powering auxiliary systems and contributing to the broader electrification of airport ground operations.

29.2.1.6 Heat Pump

Heat pump systems, instrumental in efficient heating and cooling for airport buildings, demonstrate applicability in both infrastructural development and aviation operations. Leveraging seasonal temperature variations, as observed in Helsinki Airport's heat pump deployment (ur Rehman et al., 2020), these systems optimize energy efficiency. Future implementations involve the widespread integration of heat pumps in aviation facilities, providing a climate-specific model for renewable energy utilization. The strategic deployment of heat pump systems aligns with the broader goal of achieving carbon neutrality within the aviation sector, as mandated by international initiatives.

29.2.1.7 Hydrogen

Hydrogen, as a clean energy carrier, holds multifaceted applicability in the aviation, infrastructure, and transportation sectors (Yusaf et al., 2023). Its potential as a fuel source for hydrogen-powered aircraft and ground vehicles positions it as a cornerstone in the aviation industry's transition towards decarbonization. The molecular hydrogen gas (H_2) serves as both an energy carrier and a fuel, with Airbus exploring conceptual designs for hydrogen-powered aircraft (Stauch & Müller, 2022). The ongoing development of hydrogen infrastructure within airports signifies a pivotal step towards establishing a hydrogen-based aviation ecosystem, providing a sustainable alternative to traditional aviation fuels.

29.3 Energy Storage (S)

29.3.1 Mechanical Storage

Mechanical energy storage systems, such as advanced flywheel technology, play a crucial role in ensuring grid stability and reliability within airport infrastructures (Rimpel et al., 2021). The applicability of mechanical storage in aviation extends beyond mere energy balancing, contributing to the mitigation of challenges associated with the intermittent nature of renewable energy sources. Future implementations envision the seamless integration of mechanical storage in aviation facilities, enhancing overall energy resilience and supporting the transition toward sustainable and resilient airport operations.

29.3.2 Thermal Storage

Thermal storage systems, exemplified by Singapore Changi Airport's advanced implementations, contribute to reduced peak energy demand and enhanced overall energy efficiency (Rasku & Kiviluoma, 2018). The future implementation of advanced thermal storage technologies in aviation infrastructure holds the potential to optimize energy consumption and address the sector's specific requirements. Thermal storage emerges as a critical component in achieving sustainable aviation, ensuring a consistent and reliable power supply for diverse operational needs within airports.

29.3.3 Electrical Storage

Electrical storage, encompassing various battery technologies, serves as a portable and flexible power source with broad applications within aviation and transportation (Vontzos & Bargiotas, 2023). Aviation integrates electrical storage for emergency power and electric ground vehicles. Ongoing advancements, like solid-state batteries, drive emissions-free airport operations and the shift to electric aircraft, crucial for achieving carbon-neutral air travel.

29.3.4 Water Storage

Water storage methods, involving reservoirs for potential energy storage, find relevance in both hydroelectric power generation and sustainable aviation practices (Su & Karney, 2014). The future implementation of water storage in aviation envisions its utilization in conjunction with hydropower systems to meet energy demands within airport infrastructures. The strategic integration of water storage exemplifies a multifaceted approach, ensuring efficient temperature regulation, fostering closed-loop water utilization, and contributing to overall sustainability in aviation.

29.3.5 Heat Exchange

Heat exchange, a fundamental process in thermal management systems, plays a pivotal role in energy storage and utilization within aviation (Liu et al., 2021). Future implementations involve the refinement of heat exchange technologies to enhance efficiency in capturing and redistributing thermal energy. Applied in various aviation systems, from climate control in terminals to innovative thermal storage solutions, heat exchange continues to be at the forefront of sustainable airport practices. Ongoing research and advancements aim to optimize heat exchange processes, ensuring energy conservation and operational efficiency.

29.3.6 Natural Ventilation

Natural ventilation, harnessing ambient airflow for cooling and maintaining indoor air quality, holds substantial promise in aviation infrastructures (Rodríguez-Sanz et al., 2020). The future application of natural ventilation in aviation facilities aims to reduce energy consumption, enhance passenger comfort, and contribute to overall sustainability. Strategic architectural designs, incorporating natural ventilation systems, will become integral in airport expansions and new constructions, aligning with global initiatives for energy-efficient and environmentally conscious aviation practices.

29.3.7 Active Ventilation

Active ventilation, employing mechanical systems for controlled airflow, stands as a critical component in optimizing air quality and climate control within aviation environments (Tang et al., 2023). Future implementations of active ventilation technologies will focus on energy-efficient systems to strike a balance between operational functionality and sustainability. The strategic deployment of advanced ventilation systems in aviation terminals and critical operational areas ensures adherence to stringent environmental standards while meeting the industry's growing energy demands.

29.3.8 Natural Shielding

Natural shielding, integrating landscape and architectural features to protect against adverse weather conditions, anticipates increased application in aviation infrastructure. The strategic use of natural shielding in future airport designs will aim to minimize the need for excessive heating or cooling, enhancing energy conservation. Incorporating natural elements, such as windbreaks and greenery, will become synonymous with sustainable airport planning, promoting harmonious coexistence with the environment.

29.3.9 Active Shielding

Active shielding, characterized by technological interventions like wind barriers and energy-efficient facades, envisions widespread adoption within aviation facilities. Future implementations will prioritize advanced materials and innovative designs to enhance the effectiveness of active shielding methods. The integration of active shielding technologies will not only contribute to energy conservation but also reinforce the aviation industry's commitment to reducing ecological footprints. As airports evolve, active shielding will emerge as a key element in the holistic pursuit of sustainable and resilient aviation operations.

In summary, the exploration of these advanced energy storage methods provides a comprehensive understanding of their applications, future implementations, and pivotal roles within the aviation and transportation sectors. As the industry strives for sustainability, the seamless integration of these technologies becomes paramount for achieving a greener and more environmentally conscious aviation landscape.

29.4 Patent Analysis

This section conducts a comprehensive analysis of the integration of emerging energy technologies in aviation and transportation across eight leading countries globally: China (CN), the USA (US), Japan (JP), Korea (KR), Australia (AU), Spain (ES), Canada (CA), and Russia (RU). Examining patents from 2000 to 2022, the systematic search and analysis of patents within the specified timeframe were carried out using the (Espacenet—patent search, 2023) Patent database.

The study delves into prevalent trends, technological advancements, and emerging innovations tailored for airports and interconnected transportation modes. The insights aim to inform policymakers, airport authorities, scholars, and stakeholders, fostering informed decisions for sustainable energy adoption in the aviation industry.

In the landscape of emerging energy technologies, the study emphasizes lignocellulosic biomass/cellulosic ethanol, concentrated solar, enhanced geothermal, artificial photosynthesis, vortex bladeless wind energy generation, heat pump, and hydrogen. Figure 29.1 depicts a comparative analysis across leading countries: China, the USA, Japan, Korea, Australia, Canada, and Russia, illustrating their engagement with these energy types. The United States and China emerged as pivotal players, with China displaying pronounced interest in hydrogen and concentrated solar energy. Notably, the United States exhibits a particular focus on lignocellulosic biomass/cellulosic ethanol and artificial photosynthesis. This strategic divergence highlights the nuanced preferences of major global players in advancing sustainable energy within the aviation and transportation sectors.

In the realm of energy storage, encompassing mechanical, thermal, and electrical methods, alongside water, heat exchange, natural ventilation, active ventilation,

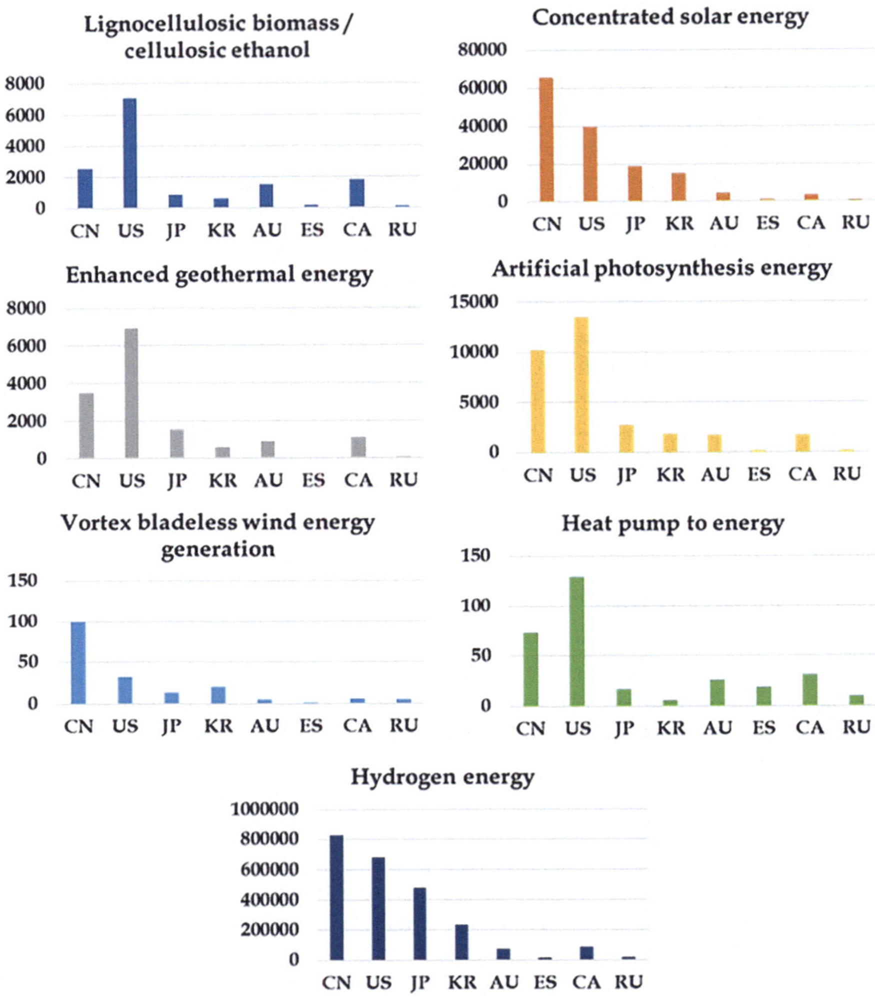

Fig. 29.1 Emerging energy technologies for the main eight countries

natural shielding, and active shielding, China emerges as a frontrunner in various facets. Figure 29.2 provides a comparative overview across leading countries, emphasizing China's prominent role in most energy storage types.

Notably, the United States takes the lead in active ventilation, showcasing a specific focus on this method. Japan excels in natural shielding, and both the United States and Japan lead in active shielding. China also expresses a keen interest in active shielding, showcasing a holistic approach to diverse energy storage strategies. This nuanced analysis highlights major global players' distinct contributions

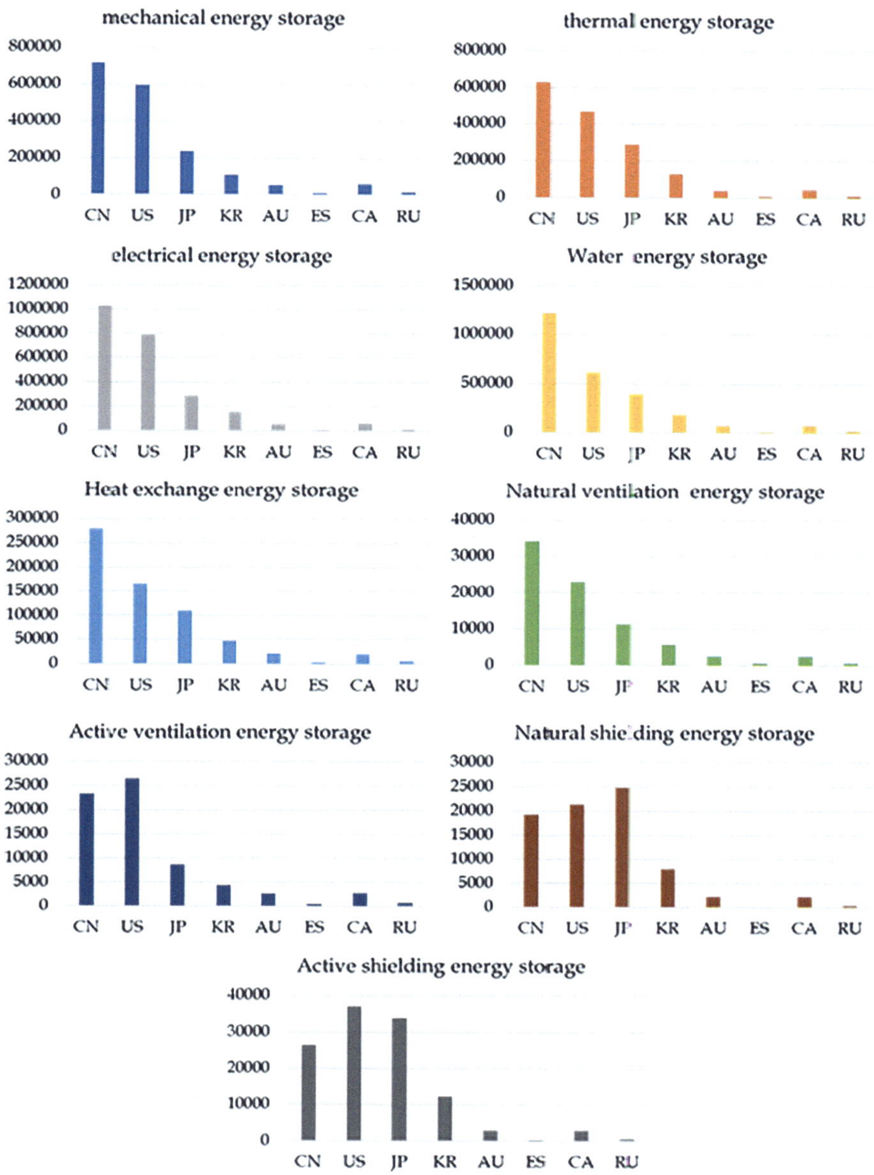

Fig. 29.2 Energy storge types for the main eight countries

and strategic orientations in advancing sustainable energy storage within aviation and transportation contexts.

Figure 29.3 provides a temporal review of global engagement with various energy types in aviation and transportation from 2010 to 2022. Notably, energy

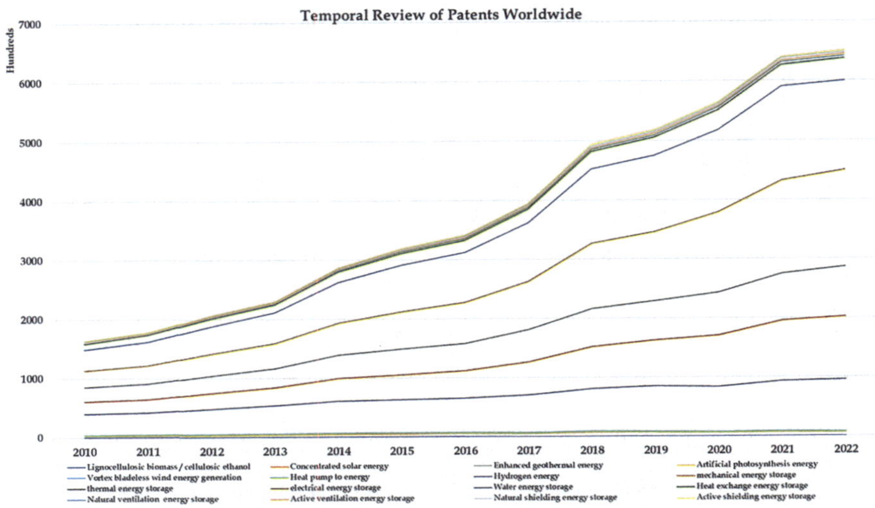

Fig. 29.3 Temporal comparison for the energy types between 2010 and 2022 worldwide

storage technologies, particularly active shielding, natural shielding, water storage, and mechanical storage systems, have garnered heightened interest over the years. The data reflects a significant surge in the exploration and adoption of these storage methods. Conversely, some technologies, such as lignocellulosic biomass/cellulosic ethanol and hydrogen energy, received relatively less attention in the earlier years but have experienced a noteworthy uptick in interest in recent times. This temporal analysis underscores the evolving dynamics and shifting priorities in the pursuit of sustainable energy solutions within the aviation and transportation sectors.

In summary, the patent analysis unveils the evolving landscape of global energy innovation in aviation and transportation. Over the years, a notable emphasis on energy storage technologies, especially in active shielding, natural shielding, water storage, and mechanical storage systems, has become apparent. The increased interest in novel solutions like lignocellulosic biomass/cellulosic ethanol and hydrogen energy reflects a dynamic shift towards more diverse and sustainable approaches. This analysis informs strategic decision-making, paving the way for the integration of cutting-edge sustainable energy practices in the aviation industry.

29.5 Conclusion

This comprehensive exploration of emerging energy technologies in aviation and transportation underscores pivotal findings crucial for steering the industry toward sustainable practices. Across the examined energy types, a nuanced global landscape emerges, with the United States and China playing central roles, each emphasizing distinct priorities. China's pronounced interest in hydrogen and concentrated solar energy aligns with its commitment to a diversified energy portfolio. In contrast, the United States strategically focuses on lignocellulosic biomass/cellulosic ethanol and artificial photosynthesis, reflecting a unique trajectory.

The analysis of energy storage technologies highlights China's frontrunner position, notably in active shielding, while the United States leads in active ventilation. Japan excels in natural shielding, emphasizing the multifaceted nature of energy storage strategies globally. This nuanced perspective contributes valuable insights for policymakers, airport authorities, and stakeholders, guiding decisions toward sustainable energy adoption.

Temporal trends from 2010 to 2022 reveal heightened interest in energy storage methods, particularly active shielding, natural shielding, water storage, and mechanical storage systems. This dynamic shift in focus reflects a growing commitment to diverse and resilient energy solutions. Noteworthy is the recent upswing in attention toward lignocellulosic biomass/cellulosic ethanol and hydrogen energy, signalling a progressive shift in priorities.

In conclusion, the patent analysis illuminates the evolving dynamics of global energy innovation in aviation and transportation. The increased emphasis on energy storage technologies, coupled with a growing interest in novel solutions, marks a pivotal moment for the industry. These insights pave the way for informed decision-making, fostering the integration of cutting-edge sustainable energy practices and shaping a greener future for aviation.

Acknowledgement "Project No. 2022-2.1.1-NL-2022-00012 has been implemented with the support provided by the Ministry of Culture and Innovation of Hungary from the National Research, Development and Innovation Fund, financed under the National Laboratories funding scheme."

References

Asmelash, E., et al. (2020). Role of IRENA for global transition to 100% renewable energy. *Lecture Notes in Energy, 74*, 51–71. https://doi.org/10.1007/978-3-030-40738-4_2/TABLES/1

Asre, C. M., Kurkute, V. K., & Kanu, N. J. (2022). Power generation with the application of vortex wind turbine. *Materials Today: Proceedings, 56*, 2428–2436. https://doi.org/10.1016/J.MATPR.2021.08.228

Baek, S., Kim, H., & Chang, H. J. (2016). Optimal hybrid renewable airport power system: Empirical study on Incheon International Airport, South Korea. *Sustainability, 8*(6), 562. https://doi.org/10.3390/SU8060562

Baxter, G., Srisaeng, P., & Wild, G. (2020). The use of aviation biofuels as an airport environmental sustainability measure: The case of Oslo Gardermoen Airport. *MAD—Magazine of Aviation Development, 8*(1), 6–17. https://doi.org/10.14311/MAD.2020.01.01

Boretti, A., Nayfeh, J., & Al-Kouz, W. (2020). Validation of SAM modeling of concentrated solar power plants. *Energies, 13*(8), 1949. https://doi.org/10.3390/EN13081949

Carratt, A., Kokogiannakis, G., & Daly, D. (2020). A critical review of methods for the performance evaluation of passive thermal retrofits in residential buildings. *Journal of Cleaner Production, 263*, 121408. https://doi.org/10.1016/J.JCLEPRO.2020.121408

Chandel, A. K., et al. (2018). The path forward for lignocellulose biorefineries: Bottlenecks, solutions, and perspective on commercialization. *Bioresource Technology, 264*, 370–381. https://doi.org/10.1016/J.BIORTECH.2018.06.004

Comendador, V. F. G., Valdés, R. M. A., & Lisker, B. (2019). A holistic approach to the environmental certification of green airports. *Sustainability, 11*(15), 4043. https://doi.org/10.3390/SU11154043

Dimitriou, D., & Karagkouni, A. (2022). Assortment of airports' sustainability strategy: A comprehensiveness analysis framework. *Sustainability, 14*(7), 4217. https://doi.org/10.3390/SU14074217

Espacenet—patent search. (2023). Retrieved June 13, 2023, from https://worldwide.espacenet.com/patent/

Francis, S., Umesh, V., & Shivakumar, S. (2021). Design and analysis of vortex bladeless wind turbine. *Materials Today: Proceedings, 47*, 5584–5588. https://doi.org/10.1016/J.MATPR.2021.03.469

Goyal, N., Aggarwal, A., & Kumar, A. (2022). Concentrated solar power plants: A critical review of regional dynamics and operational parameters. *Energy Research & Social Science, 83*, 102331. https://doi.org/10.1016/J.ERSS.2021.102331

Graver, B., Rutherford, D., & Zheng, S. (2020, October). CO_2 emissions from commercial aviation: 2013, 2018, and 2019. *International Council on Clean Transportation*, 36. Retrieved from https://theicct.org/publications/co2-emissions-commercial-aviation-2020

Joint Research Centre (European Commission). (2022). *Renewable fuels of non-biological origin in the European Union*. European Commission. https://doi.org/10.2760/76717

Kar, S. K., Harichandan, S., & Roy, B. (2022). Bibliometric analysis of the research on hydrogen economy: An analysis of current findings and roadmap ahead. *International Journal of Hydrogen Energy, 47*(20), 10803–10824. https://doi.org/10.1016/J.IJHYDENE.2022.01.137

Kim, Y. D., et al. (2012). Thermal analysis and performance optimization of a solar hot water plant with economic evaluation. *Solar Energy, 86*(5), 1378–1395. https://doi.org/10.1016/J.SOLENER.2012.01.030

Li, X., & Palazzolo, A. (2022). A review of flywheel energy storage systems: State of the art and opportunities. *Journal of Energy Storage, 46*, 103576. https://doi.org/10.1016/J.EST.2021.103576

Li, C., et al. (2022). Light-driven synthetic biology: Progress in research and industrialization of cyanobacterial cell factory. *Life, 12*(10), 1537. https://doi.org/10.3390/LIFE12101537

Liu, X., et al. (2021). Energy saving potential for space heating in Chinese airport terminals: The impact of air infiltration. *Energy, 215*, 119175. https://doi.org/10.1016/J.ENERGY.2020.119175

Ng, K. S., Farooq, D., & Yang, A. (2021). Global biorenewable development strategies for sustainable aviation fuel production. *Renewable and Sustainable Energy Reviews, 150*, 111502. https://doi.org/10.1016/J.RSER.2021.111502

Okonkwo, E. C., et al. (2023). Developing integrated direct air capture and bioenergy with carbon capture and storage systems: Progress towards 2 °C and 1.5 °C climate goals. *Energy Conversion and Management, 296*, 117687. https://doi.org/10.1016/J.ENCONMAN.2023.117687

Ploetner, K. O., et al. (2017). Technological and operational scenarios on aircraft fleet-level towards ATAG and IATA 2050 emission targets. In *17th AIAA Aviation Technology, Integration, and Operations Conference, 2017*. https://doi.org/10.2514/6.2017-3771

Rao, A. G., Yin, F., & Werij, H. G. C. (2020). Energy transition in aviation: The role of cryogenic fuels. *Aerospace, 7*(12), 181. https://doi.org/10.3390/AEROSPACE7120181

Rasku, T., & Kiviluoma, J. (2018). A comparison of widespread flexible residential electric heating and energy efficiency in a future Nordic power system. *Energies, 12*(1), 5. https://doi.org/10.3390/EN12010005

Rimpel, A., et al. (2021). Mechanical energy storage. In *Thermal, mechanical, and hybrid chemical energy storage systems* (pp. 139–247). Elsevier. https://doi.org/10.1016/B978-0-12-819892-6.00004-6

Rodríguez-Sanz, Á., et al. (2020). Airports and environmental sustainability: A comprehensive review. *Environmental Research Letters, 15*(10), 103007. https://doi.org/10.1088/1748-9326/ABB42A
Sætórsdóttir, A. D., & Hall, C. M. (2019). Contested development paths and rural communities: Sustainable energy or sustainable tourism in Iceland? *Sustainability, 11*(13), 3642. https://doi.org/10.3390/SU11133642
Sanfourche, J. P., & Fischer, M. (2022). Interview with Luc Tytgat, EASA Director, Strategy and Safety Management Directorate. *CEAS Aeronautical Journal, 13*(4), 809–812. https://doi.org/10.1007/S13272-022-00618-X/METRICS
Sowiżdżał, A., Starczewska, M., & Papiernik, B. (2022). Future technology mix—Enhanced geothermal system (EGS) and carbon capture, utilization, and storage (CCUS)—An overview of selected projects as an example for future investments in Poland. *Energies, 15*(10), 3505. https://doi.org/10.3390/EN15103505
Sreenath, S., Sudhakar, K., & Yusop, A. F. (2020). Airport-based photovoltaic applications. *Progress in Photovoltaics: Research and Applications, 28*(8), 833–853. https://doi.org/10.1002/PIP.3265
Stauch, A., & Müller, A. (2022). Technology assessment for sustainable aviation. *Management for Professionals, F359*, 23–50. https://doi.org/10.1007/978-3-030-90895-9_2/TABLES/3
Su, P. A., & Karney, B. (2014). Micro hydroelectric energy recovery in municipal water systems: A case study for Vancouver. *Urban Water Journal, 12*(8), 678–690. https://doi.org/10.1080/1573062X.2014.923919
Tanaka, K., et al. (2023). Environmental-impact assessments at airport level of Clean Sky 2 technologies. *Journal of Physics: Conference Series, 2526*(1), 012012. https://doi.org/10.1088/1742-6596/2526/1/012012
Tang, H., et al. (2023). Unlocking ventilation flexibility of large airport terminals through an optimal CO_2-based demand-controlled ventilation strategy. *Building and Environment, 244*, 110808. https://doi.org/10.1016/J.BUILDENV.2023.110808
Turkoglu, K., & Toloui, M. (2020). Novel approach towards energy utilization: Re-purposing jet turbine wake flow as a mean of renewable energy source for airports. *AIAA Scitech 2020 Forum*, 1 Part F. https://doi.org/10.2514/6.2020-2271
ur Rehman, H., et al. (2020). EU Emission targets of 2050: Costs and CO_2 emissions comparison of three different solar and heat pump-based community-level district heating systems in Nordic conditions. *Energies, 13*(16), 4167. https://doi.org/10.3390/EN13164167
Vontzos, G., & Bargiotas, D. (2023). A regional civilian airport model at Remote Island for smart grid simulation. *Lecture Notes in Intelligent Transportation and Infrastructure, F1378*, 183–192. https://doi.org/10.1007/978-3-031-23721-8_15/FIGURES/5
Voß, S., Bube, S., & Kaltschmitt, M. (2023). Aviation fuel production pathways from lignocellulosic biomass via alcohol intermediates—A technical analysis. *Fuel Communications, 17*, 100093. https://doi.org/10.1016/J.JFUECO.2023.100093
Wang, M., et al. (2019). Biomass-derived aviation fuels: Challenges and perspective. *Progress in Energy and Combustion Science, 74*, 31–49. https://doi.org/10.1016/J.PECS.2019.04.004
Yusaf, T., et al. (2023). Sustainable hydrogen energy in aviation—A narrative review. *International Journal of Hydrogen Energy*. https://doi.org/10.1016/J.IJHYDENE.2023.02.086
Zhou, Y. (2022). Low-carbon transition in smart city with sustainable airport energy ecosystems and hydrogen-based renewable-grid-storage-flexibility. *Energy Reviews, 1*(1), 100001. https://doi.org/10.1016/J.ENREV.2022.100001

Index

A

Aircraft, 163, 357, 359, 365, 393–399, 512, 513

Airport, 393, 399, 507–515, 519

Artificial intelligence (AI), 107–121, 163, 469

Artificial neural network (ANN), 84–103, 469

Atmosphere, 27, 46, 178, 188, 204, 257, 258, 361, 375, 410, 491

B

Batumi State Maritime Academy (BSMA), 491–494, 497–504

Behavior, 3–13, 16, 109, 115, 127, 132, 203, 211, 216, 254, 359, 427, 433, 437, 438, 450, 498

Biofuels, 174, 183, 184, 288, 289, 291, 292, 296, 298–300, 302–305, 510

BIST 50, 125–139

Building energy performance, 417

Building environment, 274–285

C

Carbon dioxide (CO_2), 77, 78, 111, 172, 178, 196, 201, 229, 311, 366, 373, 375, 378, 383, 491–493, 508, 511

Carbon management, 207–221, 463

CI engine, 85–86, 288–305

Circular chain, 207–221

Circular economy, 208, 210, 211

Climate change, 2, 77, 78, 80, 108, 112, 118, 161, 178–180, 187, 189, 208, 274, 338, 366, 373, 374, 378, 379, 410, 427, 429, 431, 438, 445, 446, 467, 491, 492, 497–504

Cold ironing, 77–82

Combustion, 84–103, 160, 162, 167, 171, 178, 190, 288–291, 297–300, 302–304, 340, 347, 351, 378, 387, 390, 462

Compressed air energy storage (CAES), 356, 366, 358, 369

Consumption, 2, 50, 84, 109, 125, 170, 183, 211, 251, 274, 314, 340, 357, 377, 398, 415, 428, 446, 469, 493, 504

Content analysis, 134

COVID-19, 179–184, 186–191, 445, 446

Cyclic variations, 96–100, 103, 304

D

Decarbonization, 2, 3, 23–72, 121, 159–174, 196–204, 207–221, 374, 378, 379, 436, 437, 512

Decision-making, 6, 71, 109, 117–121, 163, 173, 215, 221, 318, 410, 468, 469, 471, 485, 518, 519

Desalination, 148

Diethyl ether (DE), 288–305

E

Economic, 2, 24, 84, 125, 163, 179, 207, 228, 251, 275, 338, 357, 373, 401, 429, 445, 467

M. Z. Sogut, M. Koray (eds.), *Energy Rationality and Management for Decarbonization*, Green Energy and Technology,
https://doi.org/10.1007/978-3-031-68497-5

Efficiency, 4, 24, 81, 85, 107, 132, 144, 162, 184, 197, 212, 227, 249, 274, 288, 314, 338, 356, 378, 383, 394, 414, 427, 449, 468, 493, 509
Emerging energy innovations, 507–519
Emissions, 9, 24, 78, 88, 111, 126, 160, 178, 196, 207, 225, 274, 288, 311, 356, 373, 382, 393, 415, 428, 450, 467, 491, 497, 507
Energy, 1, 23, 78, 84, 107, 125, 144, 162, 178, 198, 207, 225, 249, 274, 288, 311, 338, 357, 379, 383, 393, 401, 414, 427, 445, 467, 492, 507
Energy and exergy analysis, 274, 275
Energy Charter Treaty (ECT), 401–404, 406, 408–411
Energy consumption, 4, 12, 52, 59, 84, 88, 108, 110–115, 120, 121, 125, 126, 128, 131–133, 137, 138, 187–190, 215, 225, 230, 249–251, 253, 254, 259, 260, 274, 285, 311, 312, 314, 315, 317, 319, 320, 322, 329, 331, 338, 339, 342, 347, 383, 395, 397–401, 414–416, 418, 419, 428–433, 435, 436, 438, 446–448, 450, 451, 454–460, 468–472, 474, 476–478, 482–485, 492, 493, 508, 509, 513, 514
Energy demand, 2, 30, 38, 39, 49, 55, 56, 111, 178–181, 184, 274, 314, 338, 339, 342, 346–350, 417, 434, 446, 448, 449, 455, 467, 485, 509, 511, 513, 514
Energy efficiency (EE), 1–16, 42, 51, 52, 111, 115, 117, 118, 126, 127, 131–135, 138, 145, 155, 160–167, 169, 171–174, 188, 190, 191, 204, 219, 221, 225–230, 274, 275, 277, 304, 311–335, 341, 367, 377, 399, 410, 413–422, 427–439, 445–463, 467–472, 474–478, 480, 483–486, 494, 504, 509, 512, 513
Energy management, 3, 4, 8, 9, 11–13, 107–121, 125–139, 161, 163, 171–174, 201, 328, 338, 347, 379, 414, 445–447, 449–457, 461, 463, 468–478, 483–485, 494
Energy rationalization, 374, 378, 379
Energy saving, 13, 24, 34–35, 108, 111, 127, 132, 230, 317, 330, 331, 338, 367, 395, 418, 420, 427, 436, 449, 455, 468, 469, 471–474, 476, 494, 504
Energy technologies, 184, 187, 188, 410, 421, 507–519
Energy trends, 185–187
Environment, 1, 24, 78, 85, 107, 126, 161, 180, 201, 208, 230, 249, 274, 288, 313, 356, 375, 394, 420, 427, 476, 492, 497, 511
Environmental attitudes, 497–504
Environmental awareness, 2, 131, 498, 504
Exergy, 19–72, 145, 147–150, 152, 153, 156, 157, 214, 215, 219–221, 228, 230, 239–243, 245, 273–285, 287–305, 332, 334, 357
Exergy analysis, 35, 145, 230, 284
Exergy destructions, 24–29, 31–33, 35, 39, 45, 46, 48, 55, 60–62, 66, 67, 147, 149, 150, 152, 157, 214, 215, 220, 279, 284, 285, 294, 298, 299, 332, 333
Exhaust, 228, 229, 231–233, 236, 237, 239, 240, 245, 252, 253, 288–305, 340, 347, 374, 375, 378, 379, 391

F

Fair and equitable treatment, 402, 403, 407, 409, 410
Foreign investment disputes, 401, 402
Fossil fuel, 26, 28, 30, 41, 42, 50, 61, 84, 112, 128, 161, 162, 178–183, 186, 187, 189, 196, 209, 211, 221, 251, 289, 327, 342, 357, 358, 360, 361, 366, 379, 402, 411, 413, 421, 428, 429, 434, 438, 445, 446, 448, 467, 469, 507, 510
Free-cooling, 311–335
Freshwater, 146, 148, 150, 155, 157, 315
Fuel burn, 95
Fuel consumption, 85, 88, 90, 162, 163, 166, 168–172, 196, 201, 209, 212, 216, 218, 219, 228, 229, 289, 342, 357, 359, 367, 377, 378, 383, 429, 433, 446

G

Geothermal, 44, 45, 47, 133, 144–157, 184, 190, 226, 274, 416, 419, 508, 510, 511, 515
Global warming, 2, 24–31, 35, 37–41, 78, 81, 82, 157, 164, 178, 179, 186, 196, 198, 204, 208, 251, 361, 373, 410, 413, 414, 450, 467, 491, 492, 494
Green buildings, 52–53, 421
Greenhouse gas (GHG) emissions, 24, 27, 38, 126, 161, 170, 173, 179, 180, 189, 196, 197, 359, 383, 404, 413, 422, 427–433, 437–439, 447, 449, 469–471, 491, 492, 494, 497, 508
Green transport, 499, 503, 504

Ground source heat pump (GSHP), 58, 274–278, 281, 283–285

H
Heat demand management, 341–343, 347
HVAC system, 35, 53, 249–254, 271, 319, 320, 340, 418, 419
Hydroprocessed waste cooking oil, 84–103

I
Indirect expropriation, 402–405, 409, 410
International Maritime Organization (IMO), 160, 161, 163, 167, 171, 173, 174, 196, 202, 204, 229, 375, 377, 383, 492, 497
ISO 50001, 9, 126, 127, 132, 135, 446, 447, 450, 453, 463, 471–474

M
Management, 2, 53, 108, 125, 172, 179, 202, 208, 311, 338, 362, 374, 391, 403, 419, 428, 447, 470, 499, 513
Marine genset, 229, 232, 233, 236, 237, 245
Marine systems, 200–202, 204
Maritime, 77, 78, 81, 159–162, 164, 169, 171, 173, 174, 202, 209, 211, 212, 225, 228, 229, 357, 361, 365–369, 373–375, 379, 383, 492, 494
MARPOL, 171, 229, 376, 377, 382, 383

N
Nearly-avoidable carbon dioxide emissions, 24, 50, 59, 62

O
Organic Rankine cycle (ORC), 44, 145, 225–245

P
Paris Agreement, 25, 27–30, 39, 42, 46, 78, 185, 186, 208, 209, 404, 410, 411, 428, 467, 491, 492, 497
Passive houses, 51, 415
Patent landscape analysis, 510, 515–519
Peltier effect, 260
Performance ratio (PR), 146, 154–157
Photovoltaic applications, 196–204
Photovoltaic technologies, 197, 198, 200, 202, 204
Pine oil (PO), 288–305
Pneumatic propulsion, 355–369
Pollution, 1, 2, 4, 78–80, 86, 103, 179, 201, 208, 209, 215, 220, 251, 288, 328, 333, 335, 342, 356, 366, 373–376, 378, 379, 427–429, 432, 436, 438
Ports, 77–82, 198, 379, 383
Pre-conditioned air systems, 393–399
Process cooling, 311–335
Propulsion, 160–164, 167–169, 171, 173, 174, 355–364, 366, 367, 369, 378, 384
PV panel, 27, 28, 42, 59, 64–66, 198–200, 259, 262, 263, 267, 269–271

R
Radiant heating and cooling systems, 274–285
Rational exergy management model (REMM), 31–33, 35, 40, 52, 67, 274
Recuperator, 230, 237–240, 242–245
Reducing friction, 164
Renewable diesel, 87
Renewable energy, 24, 42–50, 52, 59, 108, 112, 118, 126, 131–135, 162, 169, 180, 183–185, 187–191, 196–198, 202, 204, 253, 274, 275, 405–410, 413–419, 421, 428, 437, 447, 450, 463, 467, 507–509, 512
Renewable energy systems and equipment, 24, 42–50, 52, 196, 198, 417, 419
Responsibility, 1–16, 24, 26, 28, 31, 33, 35–41, 45, 46, 49–51, 53, 55–57, 62, 63, 65, 71, 112, 114, 125, 160, 164, 409, 419, 434, 453, 454, 457, 498
Road transportation, 212, 427–439

S
Second law of thermodynamics, 23–31, 50, 327, 332
Settlements, 339, 402, 411
Ship, 77–81, 160–169, 171–174, 196, 200–202, 204, 225, 228, 229, 232, 233, 236, 237, 245, 358, 361, 366, 367, 374–379, 381–384, 391, 492
Ship design, 159–174
Shipping, 160–164, 167, 169, 173, 174, 196, 229, 359, 366, 373–379, 382, 383, 497
Ship stability, 160, 163
Slow steaming, 381–391

Solar energy, 42, 47, 184, 201, 226, 251–253, 271, 275, 407, 515, 519
Sulzer, 381–384, 387
Sustainability, 2–4, 16, 23–31, 34, 37, 107, 112, 114, 120, 121, 126–128, 131–132, 137–139, 157, 160–164, 171–173, 179, 190, 191, 207–221, 229, 296, 297, 305, 311–327, 331, 332, 335, 357, 363, 420, 421, 433–435, 437, 438, 445–451, 457, 463, 468–470, 472, 484, 498, 508, 509, 511, 513–515
Sustainable aviation infrastructure, 507–519
Sustainable buildings, 29, 41, 52
Systematic literature review, 107–121

T

Thermodynamic analysis, 145, 148, 151, 216, 219, 221, 278–284
Thermo-economic analysis, 145, 343–345
Thermoelectric devices (TEC), 253–255, 259–261, 271
Transition, 1–16, 87, 112, 133, 161, 174, 178–191, 209, 237, 347, 357, 366, 377–379, 393–399, 401–411, 416, 417, 419, 445, 447–457, 471, 507, 512
Transport sector, 430–432, 435, 436, 438, 439
Trigeneration, 338–351
Trim change, 163
Turbocharger, 383–388, 390, 391
Türkiye, 201

V

Vehicles, 1, 85, 86, 109, 129, 163, 183, 188, 190, 211, 212, 229, 355–369, 413, 428–431, 433, 434, 436–438, 511–513
Ventilation, 59, 111, 250, 367, 398, 418, 462, 509, 514–516, 519

W

Waste heat, 25, 32, 41, 44, 45, 48, 50–51, 53, 54, 59, 69, 70, 171, 225–245, 339, 345, 347, 348, 351
Watermelon seed oil (WSO), 288–292, 296, 300, 304

Z

Zero energy buildings (ZEBs), 51, 59, 415–422

The manufacturer's authorised representative in the EU is Springer
Nature Customer Service Centre GmbH, Europaplatz 3, 69115 Heidelberg,
Germany. If you have any concerns regarding our products, please
contact ProductSafety@springernature.com

Printed and bound by CPI Group (UK) Ltd, Croydon, CR0 4YY

30/09/2026

02225592-0020